Community Engagement in the Online Space

Michelle Dennis
Adler University, USA

James Halbert
Adler University, USA

A volume in the Advances in Social Networking and Online Communities (ASNOC) Book Series

Published in the United States of America by
IGI Global
Information Science Reference (an imprint of IGI Global)
701 E. Chocolate Avenue
Hershey PA, USA 17033
Tel: 717-533-8845
Fax: 717-533-8661
E-mail: cust@igi-global.com
Web site: http://www.igi-global.com

Library of Congress Cataloging-in-Publication Data

Names: Dennis, Michelle, DATE- editor. | Halbert, James, DATE- editor.
Title: Community engagement in the online space / Michelle Dennis, and James Halbert, editors.
Description: Hershey, PA : Information Science Reference, [2023] | Includes bibliographical references and index. | Summary: "This book evaluates key issues and practices pertaining to community engagement in remote settings including an analysis of various community engagement efforts within online universities, campuses and programs"-- Provided by publisher.
Identifiers: LCCN 2022027229 (print) | LCCN 2022027230 (ebook) | ISBN 9781668451908 (hardcover) | ISBN 9781668451946 (paperback) | ISBN 9781668451915 (ebook)
Subjects: LCSH: Online social networks. | Information technology--Social aspects. | Communities. | Organizational sociology.
Classification: LCC HM742 .C6464 2023 (print) | LCC HM742 (ebook) | DDC 302.30285--dc23/eng/20220707
LC record available at https://lccn.loc.gov/2022027229
LC ebook record available at https://lccn.loc.gov/2022027230

This book is published in the IGI Global book series Advances in Social Networking and Online Communities (ASNOC) (ISSN: 2328-1405; eISSN: 2328-1413)

British Cataloguing in Publication Data
A Cataloguing in Publication record for this book is available from the British Library.

For electronic access to this publication, please contact: eresources@igi-global.com.

Advances in Social Networking and Online Communities (ASNOC) Book Series

Hakikur Rahman
Ansted University Sustainability Research Institute, Malaysia

ISSN:2328-1405
EISSN:2328-1413

Mission

The advancements of internet technologies and the creation of various social networks provide a new channel of knowledge development processes that's dependent on social networking and online communities. This emerging concept of social innovation is comprised of ideas and strategies designed to improve society.

The **Advances in Social Networking and Online Communities** book series serves as a forum for scholars and practitioners to present comprehensive research on the social, cultural, organizational, and human issues related to the use of virtual communities and social networking. This series will provide an analytical approach to the holistic and newly emerging concepts of online knowledge communities and social networks.

Coverage

- Best Practices for Mobile Computing
- Broadband Infrastructure and the New Wireless Network Solutions
- Introduction to Mobile Computing
- Generation of Municipal Services in Multi-Channel Environments
- Networks and Knowledge Communication in R&D Environments
- Measuring and Evaluating Knowledge Assets
- Distributed Knowledge Management Business Cases and Experiences
- Knowledge Management System Architectures, Infrastructure, and Middleware
- Communication and Management of Knowledge in R&D Networks
- Citizens' E-participation in Local Decision-Making Processes

IGI Global is currently accepting manuscripts for publication within this series. To submit a proposal for a volume in this series, please contact our Acquisition Editors at Acquisitions@igi-global.com or visit: http://www.igi-global.com/publish/.

The Advances in Social Networking and Online Communities (ASNOC) Book Series (ISSN 2328-1405) is published by IGI Global, 701 E. Chocolate Avenue, Hershey, PA 17033-1240, USA, www.igi-global.com. This series is composed of titles available for purchase individually; each title is edited to be contextually exclusive from any other title within the series. For pricing and ordering information please visit http://www.igi-global.com/book-series/advances-social-networking-online-communities/37168. Postmaster: Send all address changes to above address.

Titles in this Series

For a list of additional titles in this series, please visit: www.igi-global.com/book-series/advances-social-networking-online-communities/37168

Handbook of Research on Bullying in Media and Beyond
Gülşah Sarı (Aksaray University, Turkey)
Information Science Reference • © 2023 • 603pp • H/C (ISBN: 9781668454268) • US $295.00

Handbook of Research on Technologies and Systems for E-Collaboration During Global Crises
Jingyuan Zhao (University of Toronto, Canada) and V. Vinoth Kumar (Jain University, India)
Information Science Reference • © 2022 • 461pp • H/C (ISBN: 9781799896401) • US $295.00

Information Manipulation and Its Impact Across All Industries
Maryam Ebrahimi (Independent Researcher, Germany)
Information Science Reference • © 2022 • 234pp • H/C (ISBN: 9781799882350) • US $215.00

E-Collaboration Technologies and Strategies for Competitive Advantage Amid Challenging Times
Jingyuan Zhao (University of Toronto, Canada) and Joseph Richards (California State University, Sacramento, USA)
Information Science Reference • © 2021 • 346pp • H/C (ISBN: 9781799877646) • US $215.00

Analyzing Global Social Media Consumption
Patrick Kanyi Wamuyu (United States International University – Africa, Kenya)
Information Science Reference • © 2021 • 358pp • H/C (ISBN: 9781799847182) • US $215.00

Global Perspectives on Social Media Communications, Trade Unionism, and Transnational Advocacy
Floribert Patrick C. Endong (University of Calabar, Nigeria)
Information Science Reference • © 2020 • 300pp • H/C (ISBN: 9781799831389) • US $195.00

Electronic Hive Minds on Social Media Emerging Research and Opportunities
Shalin Hai-Jew (Kansas State University, USA)
Information Science Reference • © 2019 • 358pp • H/C (ISBN: 9781522593690) • US $205.00

Hidden Link Prediction in Stochastic Social Networks
Babita Pandey (Lovely Professional University, India) and Aditya Khamparia (Lovely Professional University, India)
Information Science Reference • © 2019 • 281pp • H/C (ISBN: 9781522590965) • US $195.00

701 East Chocolate Avenue, Hershey, PA 17033, USA
Tel: 717-533-8845 x100 • Fax: 717-533-8661
E-Mail: cust@igi-global.com • www.igi-global.com

Table of Contents

Preface xiv

Section 1
Community Engagement: Expansion Through the Online Venue

Chapter 1
What Is Community Engagement 1
La Toya L. Patterson, Chicago State University, USA

Chapter 2
Ethics in Online Community Engagement Among Marginalized Rural Groups 17
Tshimangadzo Selina Mudau, University of KwaZulu-Natal, South Africa

Chapter 3
Analyzing the Hong Kong Philharmonic Orchestra's Facebook Community Engagement With the Honeycomb Model 31
Suming Deng, The University of Hong Kong, Hong Kong
Dickson K. W. Chiu, The University of Hong Kong, Hong Kong

Chapter 4
Application of the AIDA Model in Social Media Promotion and Community Engagement for Small Cultural Organizations: A Case Study of the Choi Chang Sau Qin Society 48
Xinyu Jiang, The University of Hong Kong, Hong Kong
Dickson K. W. Chiu, The University of Hong Kong, Hong Kong
Cheuk Ting Chan, Independent Researcher, Hong Kong

Chapter 5
Co-Constructing Belongingness: Strategies for Creating Community and Shared Purpose Online – The Social Construction of Community and Meaning 71
Lilya Shienko, Adler University, USA
Barton David Buechner, Adler University, USA

Section 2
Community Engagement in Online Higher Education

Chapter 6
Remote Community Engagement in Higher Education 91
Paul M. Huckett, Johns Hopkins University, USA
Nathan Graham, Johns Hopkins University, USA
Heather Stewart, Johns Hopkins University, USA

Chapter 7
Teachers' Unique Knowledge to Effectively Integrate Digital Technologies Into Teaching and Learning: Community Engagement to Build in the Online Space 127
Abueng R. Molotsi, University of South Africa, South Africa
Leila Goosen, University of South Africa, South Africa

Chapter 8
An Extendable Functioning Waiting Room Solution for Online Education Platforms: The Beginning of an Online Efficiency Era 149
Kaan Kırlı, Yeditepe University, Turkey
Cagla Ozen, Yeditepe University, Turkey

Chapter 9
Creating a Community in the Course Room Through Discussion Posts 178
Tricia Mazurowski, Adler University, USA

Chapter 10
Thirty Days in 2020 and How They Future-Proofed a University 194
Michael Graham, National Louis University, USA
Bettyjo Bouchey, National Louis University, USA

Chapter 11
From Loneliness to Belonging Post Pandemic: How to Create an Engaged Online Community and Beyond 209
Andrea D. Carter, Adler University, USA
Lilya Shienko, Adler University, USA

Section 3
Applying Principles of Remote Community Engagement

Chapter 12
Undergraduate Perspectives on Community-Engaged Service During COVID-19: Exploring the Differences Between In-Person and Remote Tutoring Experiences 234
Craig Allen Talmage, Hobart & William Smith Colleges, USA
Kathleen Flowers, Hobart & William Smith Colleges, USA
Peter Budmen, Hobart & William Smith Colleges, USA
Alexander Cottrell, Hobart & William Smith Colleges, USA
Jonathan Garcia, Hobart & William Smith Colleges, USA
Jasmine Webb-Pellegrin, Hobart & William Smith Colleges, USA

Chapter 13
Remote Engagement Through Cohort Mentors 254
Donna DiMatteo-Gibson, Adler University, USA

Chapter 14
A Case Study in Meeting the Pandemic Moment Through NGO-University Collaboration: Creating a Global Health Virtual Practicum 269
Paul Shetler Fast, Mennonite Central Committee, USA
Brianne F. Brenneman, Goshen College, USA
Wade George Snowdon, Mennonite Central Committee, USA

Chapter 15
Belonging, Performance, and Engagement: What Tomorrow's Leaders Need to Know 285
Andrea D. Carter, Adler University, USA

Compilation of References 307

About the Contributors 356

Index 362

Detailed Table of Contents

Preface xiv

Section 1
Community Engagement: Expansion Through the Online Venue

This section defines community engagement in the online space, providing examples to illustrate the ways in which community engagement has expanded through the use of technology and the application of creativity.

Chapter 1
What Is Community Engagement 1
La Toya L. Patterson, Chicago State University, USA

This chapter provides an overview of community engagement within higher education institutions and describes the process of applying for and receiving the Carnegie Foundation for the Advancement of Teaching Community Engagement Classification. The author discusses the reasons for seeking the classification, reclassification, and clarification of the language used in the application process are discussed. The chapter also explores the importance of institutionalizing community engagement, the benefits that are received by the institution, faculty and students, and the challenges to institutionalizing community engagement. The author highlighted steps for institutionalizing community engagement within higher education institutions and pitfalls to avoid when trying to institutionalize community engagement within higher education institutions.

Chapter 2
Ethics in Online Community Engagement Among Marginalized Rural Groups 17
Tshimangadzo Selina Mudau, University of KwaZulu-Natal, South Africa

This chapter presents the virtual community engagement within rural areas. The methodological implications of this chapter are drawn from empirical research conducted before and during the emergence of the COVID-19 pandemic in the rural areas of Makhado in South Africa. The chapter describes how virtual community engagement impacts communication issues such as social norms, culture, and respect for community leadership structures. Critical discourse analysis was employed to expose the problems of power, position, and dominance during virtual community engagement as a data generation method. Findings are based on technological and other contextual socio-cultural factors in rural communities. A mixture of both virtual and telephonic engagement is the most acceptable option to manage and minimize social injustice among marginalized groups. The chapter closes with recommendations on promoting the voices and rights of the marginalized or vulnerable groups when conducting virtual community engagement.

Chapter 3
Analyzing the Hong Kong Philharmonic Orchestra's Facebook Community Engagement With the Honeycomb Model 31
Suming Deng, The University of Hong Kong, Hong Kong
Dickson K. W. Chiu, The University of Hong Kong, Hong Kong

Online social media has gradually become a trend. Many companies, brands, and organizations attach great importance to social media for community engagement, aiming for marketing, reputation, etc. This study analyzes the effectiveness of the Hong Kong Philharmonic's (HKPhil) Facebook social media with the seven aspects of the honeycomb model. Through quantitative and qualitative analysis of the problems encountered, the authors found that the video posts on Facebook are more active than posts without videos, and the comments are generally positive. HKPhil's engagement on Facebook is good, with about 1.9% average participation rate. This research help understand whether Facebook engagement is effective and can reflect the audience's expectations, suggesting improvement strategies of social media to attract and engage more audiences and discover potential users.

Chapter 4
Application of the AIDA Model in Social Media Promotion and Community Engagement for Small Cultural Organizations: A Case Study of the Choi Chang Sau Qin Society 48
Xinyu Jiang, The University of Hong Kong, Hong Kong
Dickson K. W. Chiu, The University of Hong Kong, Hong Kong
Cheuk Ting Chan, Independent Researcher, Hong Kong

This study explores the AIDA model (attention, interest, desire, action) for social media promotion and community engagement for small cultural organizations. The internal situation and external environment were first analyzed with the SWOT analysis augmented with PEST analysis. Then, the authors show how the AIDA model can be used in social media marketing to improve public awareness, engagement, and thus participation in the organization's activities. As the global economy is getting linked to the internet and social media, utilizing the AIDA model for small cultural organizations contributes to effective information dissemination and increases interactions in the targeted community. The rise of social media has also triggered smaller organizations to consider how to survive under dynamic changes and fulfill their mission through better community engagement.

Chapter 5
Co-Constructing Belongingness: Strategies for Creating Community and Shared Purpose Online – The Social Construction of Community and Meaning 71
Lilya Shienko, Adler University, USA
Barton David Buechner, Adler University, USA

Online environments have become a fact of life for education and business life, but participants do not always experience virtual space as a place where they truly belong. This case example illustrates some ways that social construction concepts, particularly the "communication perspective" of the coordinated management of meaning (CMM) theory, were applied to the development and enactment of an inclusive collaborative online learning program. This online interactive space was envisioned as a "virtual home" and support group for fellowship participants for the development and presentation of their individual projects. Perspectives of program participants and organizers are both represented in this case example to help shed light on the way that the group space evolved over the course of the fellowship and lessons learned from the process. The use of concepts from CMM theory is underscored to reveal the dynamics of communication theory as a source of adding life and dimension to an online collaborative space, contributing to a sense of belongingness.

Section 2
Community Engagement in Online Higher Education

This section focuses on various strategies for and examples of community engagement through online higher education. Included is a consideration of effective engagement in the online classroom, the use of technology tools to create connections, and emerging practices to support community engagement within the post-secondary student population.

Chapter 6
Remote Community Engagement in Higher Education 91
Paul M. Huckett, Johns Hopkins University, USA
Nathan Graham, Johns Hopkins University, USA
Heather Stewart, Johns Hopkins University, USA

This chapter examines strategies implemented to increase a sense of community for faculty and students in an online graduate engineering program. Facilitating community sites of interaction and shared knowledge creation—elements of the community of practice (CoP) framework—comprised the most valued additions for members of the learning community. With these improved sites of interaction, faculty and students benefited from participation in the learning community with their online peers and contributed to a community of practice in their degree program. Early data and outcomes suggest that higher education administrators can implement specific strategies to increase learners' and teachers' sense of community, facilitating engagement with the school, academic programs, and peers despite being geographically dispersed.

Chapter 7
Teachers' Unique Knowledge to Effectively Integrate Digital Technologies Into Teaching and Learning: Community Engagement to Build in the Online Space 127
Abueng R. Molotsi, University of South Africa, South Africa
Leila Goosen, University of South Africa, South Africa

In order to provide readers with an overview of, and summarize, the content, the purpose of this chapter is stated as to evaluate teachers' unique knowledge, which they need to effectively integrate digital technologies into teaching and learning. This is built on teachers' use of the technological, pedagogical, and content knowledge (TPACK) model to improve their delivery of lessons. It is important to note that the research reported on in this chapter is positioned against the background of community engagement in the online space.

Chapter 8
An Extendable Functioning Waiting Room Solution for Online Education Platforms: The Beginning of an Online Efficiency Era 149
Kaan Kırlı, Yeditepe University, Turkey
Cagla Ozen, Yeditepe University, Turkey

When COVID-19 spread from China, the majority of people witnessed that the form of education changed. This situation has increased the need for online education platforms. Although videoconference platforms such as Zoom and Google Meet became popular to meet the requirement, these platforms have some drawbacks. To illustrate, the majority of students are not active and cannot take a proactive role during the lectures. Furthermore, they stay idle and get bored on the waiting rooms of these platforms. To fill this gap, an extendable functioning waiting room was created by using ASP.NET framework via Visual Studio platform. Proposed waiting room present a solution for enhancing user experience (UX) by increasing user engagement on the online education platform. This chapter would be interesting for not only educators who use online learning platforms but also videoconference platform users and developers.

Chapter 9
Creating a Community in the Course Room Through Discussion Posts 178
Tricia Mazurowski, Adler University, USA

Online education has become increasingly popular due to the convenience of learning in an online environment. Online learning allows adult learners to continue their studies remotely. Discussion boards are used in an online learning environment in lieu of lectures from an instructor. Creating a community in an online learning environment through the use of discussion boards is important to engage adult learners. Discussion boards allow students to engage with and learn from their peers and from the instructor. This chapter will explore how to create a community in the course room through discussion posts to increase student engagement in an online learning environment. Online educational programs often experience a higher attrition rate since students may feel isolated from their peers and their instructors. Increasing engagement in the course room through discussion posts is important to ensure that adult learners are engaged in the course room. Interaction with peers and the instructor is instrumental in increasing engagement through online learning.

Chapter 10
Thirty Days in 2020 and How They Future-Proofed a University 194
Michael Graham, National Louis University, USA
Bettyjo Bouchey, National Louis University, USA

The COVID-19 pandemic was a catalyst for change in myriad ways. While nearly all the societal changes were negative, the crisis also brought about positive adaptations in response to this crisis. National Louis University's (NLU's) initial response to the crisis, beginning in February 2020 and through subsequent decisive actions, kept the university vibrant throughout the pandemic. Perhaps more importantly, these actions created a foundation for long-term institutional success. The steps included technological enhancements and intentional activities focused on ensuring that institutional personnel would remain connected to each other and to the students that NLU serves. Many of these changes are now a permanent part of the institution as the world moves past the COVID-19 era. Utilizing NLU as an exemplar, recommendations will be given designed to help college and university leadership evolve in this new hybrid reality. Ultimately, the institutions that can effectively adapt to this hybrid existence will be the ones most capable of continued success in an uncertain future.

Chapter 11
From Loneliness to Belonging Post Pandemic: How to Create an Engaged Online Community and Beyond 209
Andrea D. Carter, Adler University, USA
Lilya Shienko, Adler University, USA

The effects of isolation, loneliness, and ostracism on social engagement and online communities has been acknowledged throughout the pandemic by researchers. This chapter contextualizes and normalizes the behaviours associated with isolation and loneliness and draws attention to belonging methodology. Belonging indicators and practices are revealed and described so as to break the patterns many are struggling with. Finally, the chapter provides context for engagement, collaboration, and creating community post pandemic.

Section 3
Applying Principles of Remote Community Engagement

This section provides examples and guidance for the effective application of best practices for leadership with relevance to remote community engagement.

Chapter 12
Undergraduate Perspectives on Community-Engaged Service During COVID-19: Exploring the Differences Between In-Person and Remote Tutoring Experiences 234
Craig Allen Talmage, Hobart & William Smith Colleges, USA
Kathleen Flowers, Hobart & William Smith Colleges, USA
Peter Budmen, Hobart & William Smith Colleges, USA
Alexander Cottrell, Hobart & William Smith Colleges, USA
Jonathan Garcia, Hobart & William Smith Colleges, USA
Jasmine Webb-Pellegrin, Hobart & William Smith Colleges, USA

This chapter chronicles a rapid pivot in community engagement from in-person tutoring to remote (also called virtual) tutoring and a gradual shift back to in-person tutoring during the COVID-19 pandemic. The chapter provides reflections from both staff members in a community engagement and service-learning office and college students who served as tutors of local children and youth during the COVID-19 pandemic. Readers are encouraged to reflect upon two main areas of interest in this chapter. First, dynamics between parents and tutors altered during COVID-19 as children and youth received virtual tutoring from home. Second, the importance of training and professional development for tutors regarding how to use different technologies and engagement strategies for virtual tutoring is imperative. Lessons learned from the Tutor Corps program are shared via reflections from both staff and tutors in hopes that others will share their experiences in rapidly innovating their community engagement work during COVID-19.

Chapter 13
Remote Engagement Through Cohort Mentors 254
Donna DiMatteo-Gibson, Adler University, USA

Working virtually can be a lonely experience. Ways to engage employees is so important to uncover especially with so many organizations working virtually. One strategy that the researcher utilized to cultivate engagement was through the use of cohort mentors. This chapter will explore research on mentorship and engagement. The discussion will also present the research that has been done at an online university that embraced engagement through the use of cohort mentors. Doctoral students can benefit extensively from individually administered mentorship programs. This mentorship can be particularly beneficial for doctoral students completing their degrees in the online format. A doctoral cohort mentor model was implemented at the online campus of a mid-sized university in 2019. The model was expanded to include formal advising. Lessons learned and future directions will be discussed.

Chapter 14
A Case Study in Meeting the Pandemic Moment Through NGO-University Collaboration:
Creating a Global Health Virtual Practicum 269
Paul Shetler Fast, Mennonite Central Committee, USA
Brianne F. Brenneman, Goshen College, USA
Wade George Snowdon, Mennonite Central Committee, USA

During the COVID-19 pandemic, international exchange and service programs scrambled to adapt while students were newly engaged on topics of global health. Meeting the moment, Goshen College, a liberal arts college in northern Indiana renowned for its study abroad and community-engaged learning programs, and Mennonite Central Committee, a faith-based international relief, development, peace non-profit organization working around the world, came together to pilot a new partnership approach to web-based community-engaged learning. This chapter explores the learnings from this creative and successful pilot of a global public health virtual practicum, demonstrating the potential for mutually beneficial collaboration between higher education community-engaged learning programs and humanitarian organizations to meet the unique needs of diverse students and community-based organizations, while lowering entry costs and reducing barriers of engagement for students.

Chapter 15
Belonging, Performance, and Engagement: What Tomorrow's Leaders Need to Know 285
Andrea D. Carter, Adler University, USA

The effects of leadership on performance and engagement have long been evaluated and measured. Communities, organizations, and educational institutions seek to empower and support behaviors that lead to results, but the context of how is fundamentally different post-pandemic. Looking forward, organizational and institutional success will depend on the leader's ability to create a belonging environment. With followership no longer driven by rational rewards, leaders must learn how to connect with members' emotional needs to motivate and inspire action. This chapter examines the importance of adopting new leadership behaviors, the cost of membership loneliness and exclusion, and the impact of trust, accountability, and empathy. The chapter concludes with indicators and tactics to create belonging environments that drive engagement and high performance. In the absence of consideration of the concepts discussed in this chapter, mediocrity will rule, and a twenty percent engagement rate has the potential to become the global standard impacting performance worldwide (Gallup, 2022). These executed concepts and tactics will support community members to behave differently, empowering each other to rebound and thrive using tools that meet emotional needs and empower positive change.

Compilation of References 307

About the Contributors 356

Index 362

Preface

Due to technological advances and changes in perspective, the online space has risen in popularity as a relevant venue for anything from business meetings to higher education to holiday parties. Community engagement in the online space offers multiple benefits, key among which is the ability to reach large groups of individuals efficiently, thereby harnessing the power of a multitude of diverse voices. From an historical perspective, communities have been defined in terms of geographic locale, shared interests or characteristics, and shared needs. For instance, a neighborhood can be thought of as a community. So too, individuals passionate about the same political issue, survivors of a particular disease, and individuals enrolled in the same college may all be defined as communities. In cases where geographic locale is not the defining feature of the community, it has still impacted vital factors pertaining to the community, such as the perceived closeness of connections, frequency of interactions, and the potential to make an impact. Although relationships and interactions characterize communities, they do not - in and of themselves- characterize engagement.

Engagement can be defined as internally motivated participation - with gusto. In other words, one does not need to be engaged to participate, but those who are engaged will participate in such an active manner that they are inherently more likely to make an impact on the communities of which they are a part. For instance, online communities of practice where engagement is observed are associated with significantly more learning (e.g., Ray, Kim, & Morris, 2014). Community engagement may be defined as working together within a group connected by shared interests, characteristics, needs, or placement to bring about positive changes that meet the goals of the community and its members (Thompson et al., 2022). Examples of community engagement include organizing a rally to support a political candidate, serving a holiday meal to individuals facing food insecurity who reside in one's neighborhood, working with a community agency to provide reduced-cost therapy services, volunteering to tutor college students, or organizing a student group on a college campus.

BENEFITS OF COMMUNITY ENGAGEMENT

In the United States, community engagement came to the forefront in large part as a response to unjust treatment, which was and continues to be, leveled at individuals based on the groups of which they are a member. The civil rights movement of the 1960's and 1970's saw many communities come together in support of equality, and many unfair policies and laws were changed due to the community engagement efforts that brought together individuals in support of community improvement. There are many positive outcomes of community engagement, key among which is the increased support that may be provided to underserved populations, which often follows such endeavors.

Supporting Underserved Populations

One example of supporting underserved populations that was brought about through community engagement can be seen in health equity initiatives. Communities have come together to rally against unfair resource allocation, engaging in letter-writing campaigns, organizing protests, writing articles, and mobilizing community members. In response to this community-based advocacy, some funding has been generated to support community health centers, thereby providing avenues of support for underserved populations. Admittedly, community health centers still lack the funding to support underserved populations fully, and the majority of support still comes from altruistic professionals who work tirelessly and without adequate compensation to ensure that community members receive health services. Despite this, the fact that community mental health centers even exist is a testament to the power of community engagement; communities came together and demanded support for individuals whose health needs were unmet, and change began.

Community engagement impacts health equity in many ways (Braithwaite, Akintobi, Blumenthal, & Langley, 2020). For instance, Pratt & de Vries (2018) found that community engagement, in the form of global health research, served to advance health equity. Thompson et al. (2022) found that community engagement improved access to quality healthcare among marginalized individuals. Coronado, Chio-Lauri, Dela Cruz, & Roman (2021) found that community engagement served to moderate health disparities among Filipino Americans.

A second example that illustrates the power of community engagement pertains to educational equity in the K-12 space. Due to the inequitable allocation of resources, many children lack the basic materials necessary to learn, such as textbooks, paper and pencils, and a safe environment in which to interact with peers and teachers. Efforts at community engagement in support of children have led to some improvements. Despite this work, inequity exists in the United States and many other countries. There is more work to be done in this regard, but community engagement has and will continue to make a difference, increasing awareness and demanding justice for children. Further, equitable data inclusion can potentially increase equity among underserved populations. In addition to supporting underserved populations, community engagement can improve well-being through its impact on social interest.

Social Interest

Social interest is an Adlerian concept, which may be considered an intuitive sense of empathy connecting all humans (Ansbacher, 1968; King & Shelley, 2008). Social interest may develop throughout one's life and leads to behaviors such as aiding, cooperating, and volunteering. The experience of social interest contributes to many positive outcomes, such as mental well-being, emotional growth, life satisfaction, and happiness. In other words, those who are able to engage with communities for the good of the community tend to reap rewards they may never have expected to experience.

For instance, technology acceptance combined with a community feeling, a concept that aligns with social feeling, has been linked to learner satisfaction among distance education students (Ilgaz & Aşkar, 2013). Further, a high level of parental community feeling has been shown to impact later achievement among adolescents (Froiland & Worrell, 2017). In sum, community engagement is an impactful way to make nurture positive change, support groups and individuals, and promote individual well-being. Technological advances allow for the expansion of communities, leading to near-endless possibilities for connection, advocacy, and community improvement.

FACTORS CONTRIBUTING TO REMOTE COMMUNITY ENGAGEMENT

Technology supports rich dialogue, fortified with diverse viewpoints, contributing to more effective collaborations. For instance, online community engagement has been demonstrated to be effective in leadership development among teenage students (Pierce, Blanton, & Gould, 2018). McInroy, McCloskey, Craig, & Eaton (2019) found that community engagement in the online space was preferred by marginalized youth. Further, online community engagement has been shown to impact health indicators among populations of diabetic individuals (Litchman, Edelman, & Donaldson, 2018).

Over the past several years, multiple global phenomena have led to the need for community engagement, such as the COVID-19 pandemic and the racial justice movement. The former forced all communities to employ creative solutions to support community engagement in the online space.

The global pandemic, COVID-19, shut down communities across the world, making it necessary to engage while confined (Cattapan, Acker-Verney, Dobrowolsky, Findlay, & Mandrona, 2020). One consequence of the global pandemic, COVID-19, emergency remote teaching, differs drastically from online education, the latter of which is a practice that allows significant opportunities for student engagement (Bozkurt & Sharma, 2020; Dennis, 2020; Knudson, 2020). Emergency remote instruction is characterized by temporary changes in the mode of educational delivery that is directly precipitated by crisis conditions.

When delivering emergency remote instruction, the intention is always to return to the previous delivery format when the crisis conditions shift. As such, the primary focus of this type of instruction tends to be continued delivery of content – not engagement. Methods of emergency remote instruction may include lecture delivery via Zoom, for example. By comparison, online education represents the intentional preparation of content for delivery via remote means. Design is a key feature of online instruction. Through intentional design, instructional objectives may be carefully aligned to course resources and assessments to ensure that student learning outcomes may be assessed and achieved. Further, the design process includes a consideration of opportunities for interaction. Students in online courses interact with content, one another, and the instructor. When designed deliberately, these interactions facilitate engagement in these three areas, thereby contributing to the learning experience in positive ways. In addition to student learning, engagement in online courses has many potential outcomes.

First, in terms of the student experience, engaging courses tend to lead to a positive student experience. Students who experience engagement in online courses are more likely to perceive value, gain mentorship, develop meaningful relationships, and remain connected to the institutions they attend (Dennis, 2020), thereby contributing to persistence and graduation rates. Engagement in online courses also impacts the faculty experience. Engagement with students contributes to satisfaction among online adjunct faculty. Further, engagement in online courses has been shown to impact faculty retention (Dennis, 2020), which consequently benefits institutions by saving them valuable resources that would be needed to recruit and train new faculty.

Online community engagement may be sustained effectively through organization and collaboration (Dodd, 2017; (Mohmmed, Khidhir, Nazeer, & Vijayan, 2020), the application of best practices for online community development (France-Harris, Burton, & Mooney, 2019; Kuem, Khansa, & Kim, 2020), and the establishment of trust (Kang, Shin, & Gong, 2016). In addition to the global pandemic, online movements have flourished over the past several years due in part to the ease with which group connections may be sustained through virtual means.

It is uncertain whether community engagement work can be truly equitable (Telles, 2019), but it has certainly expanded its reach in recent years, thereby contributing to robust online movements fueled, in part, by the goal of justice, as defined by the group (Weisman, 2021). Technology makes it easier for groups to engage across large spaces. For instance, in the wake of the last election, communications shared via social media were numerous and frequent. These social media-fueled connections indeed represent community engagement for people on both sides of the aisle. In the aforementioned example, the argument could be made that a strong vision of a value for equity was the defining feature of the community, although other characteristics may also apply. The community came together in solidarity, rallied for their chosen party, and engaged in multiple activities aimed at securing the core community goal of their favored candidate.

This, in turn, led to many diverse voices expressing similar sentiments, which consequently reinforced the core objective of the community engagement surrounding the event. Two much smaller-scale recent instances of remote community engagement relevant to racial justice include online education in correctional facilities (Dennis & Halbert, 2022) and online social justice practicum (Dennis, Gordon, DiMatteo-Gibson, & Halbert, 2022).

Online Education in Correctional Facilities

Individuals in correctional facilities represent an historically underserved population, to the detriment of society. This group of individuals has faced erroneous restrictions in their access to literature, for instance. Further, access to higher education at the bachelor's degree level has typically not been offered to those who are incarcerated. Racial disparities in conviction and sentencing have led to racial disparities among incarcerated populations. As such, access to higher education for incarcerated individuals represents an opportunity to level the playing field for members of racial groups who have historically been marginalized.

A number of challenges are associated with offering higher education programming to incarcerated individuals, many of which center around resources (Brosens, Croux, & De Donder, 2019). As many correctional facilities are located in rural areas, it is challenging for universities to harness the necessary resources to transport faculty to the facilities. Further, safety-related restrictions at the facilities may not always support the full delivery of programming. Despite these challenges, some universities do support programming within correctional facilities.

As outlined above, the global pandemic, COVID-19 disrupted higher education, forcing programs to switch to remote delivery rapidly. This was a significant challenge for programs being delivered in correctional facilities, as technology within the facilities often did not allow for the continuation of learning (Armstrong, 2020). Online education is a viable option, and some fully online programs have been implemented within correctional facilities through careful planning and collaborative remote community engagement. For instance, an online pilot course was delivered in a correctional facility during the heart of the global pandemic, highlighting best practices for online course delivery in correctional facilities (Dennis & Halbert, 2022). After the pilot, adjustments were made to allow full engagement among students and online faculty. These improvements may be seen as a form of community engagement, as they involved representatives of the university and the correctional facility coming together as a community of educators who value education for incarcerated individuals. Through brainstorming to remove barriers, the formation of extended communities through the incorporation of other governmental agencies, and regular interaction to evaluate and move forward shared goals, a successful online educational experience was formed.

Virtual Practicum

Another example of remote community engagement is remote practica in online programs. Practicum, especially in the area of social justice, has been utilized in clinical programming, where nursing, counseling, or social work students have been provided with training in self-awareness and presented with formal sites where they may learn first-hand about the concepts they are exploring. Practicum may also be delivered remotely.

A significant body of research demonstrates the ability of learners to fully engage in online communities of practice (Abedini, Abedin, & Zowghi, 2021; Alexandra & Fahmi-Choirisa, 2021). Online practicum has been successfully employed to train teachers (Alghamdi, 2022; Archer-Kuhn, Judge-Stasiak, Letkemann, Hewson, & Ayala, 2022; (Vasiliki & Psoni, 2021), counselors, social workers (Baciu & Trancă, 2021) and engineers (Benitz, & Yang, 2020). Dennis, Gordon, DiMatteo-Gibson, & Halbert (2022) analyzed the launch of an online practicum in eight non-clinical online programs to raise self-awareness, encourage reflection, and provide a remote venue for service to communities.

TARGET AUDIENCE

This book will be relevant to a wide variety of individuals, as the authors present content that may serve as a guide for all who hope to bring about positive change within communities by expanding the concept of community across physical space. Leaders and change-makers from all disciplines may find the examples and lessons learned here helpful as they organize groups to improve the human condition. Further, administrators will find the content presented regarding connection and engagement in online programming to support continued evaluation and improvement. Students will likely find this content relevant. As such, faculty may find this book or the chapters contained therein suitable for use in undergraduate and graduate-level courses on a wide variety of subjects.

CHAPTER CONTRIBUTIONS

Chapter contributions are organized around the following themes:

- Expanding community engagement through the online venue
- Community engagement in online higher education
- Applying principles of remote community engagement

The first section defines community engagement in the online space, providing examples to illustrate the ways in which community engagement has expanded through the use of technology and the application of creativity.

Chapter 1 provides an overview of community engagement, presenting content on guidelines for receiving the Carnegie Foundation for the Advancement of Teaching Community Engagement Classification. This chapter summarizes the benefits of community engagement to students, faculty, and the institution, closing with an analysis of challenges and considerations for institutionalizing community engagement within higher education.

Chapter 2 focuses on community engagement in rural areas, highlighting opportunities for virtual engagement through an analysis of research conducted in rural areas of Makhado in South Africa before and during the COVID-19 pandemic. The findings of the investigations presented indicate that a combination of virtual and telephonic engagement may effectively promote social justice among marginalized groups in this area.

Chapter 3 establishes the importance of online social media within a wide variety of settings by presenting the results of a study that focused on the impacts of the Hong Kong Philharmonic's Facebook-based media on engagement, utilizing the Honeycomb Model. Findings indicate that video posts are more effective than posts with no video attached, producing higher levels of engagement. Implications are discussed in the context of user engagement theories.

Chapter 4 presents research on the Attention, Interest, Desire, and Action (AIDA) Model and its ability to promote community engagement within small cultural organizations. Evaluations of the promotion of public awareness, engagement with content, and participation in activities are evaluated. Included is an analysis of factors impacting the survival of small mission-driven organizations.

Chapter 5 presents research on how online communities support a shared sense of belongingness by analyzing the communication processes' dynamics. The social experience is explored in the context of relevant theory, and guidelines for online community development are presented.

The next section focuses on various strategies and examples of community engagement through online higher education. Included is a consideration of effective engagement in the online classroom, using technology tools to create connections, and emerging practices to support community engagement within the post-secondary student population.

Chapter 6 presents a strategic approach to increasing the perceived community in an online graduate engineering program at Johns Hopkins University. Data are presented demonstrating positive outcomes of the approach, indicating that administrators may improve the sense of community for faculty and learners alike through the facilitation of engagement with peers, programs, and the school as a whole irrespective of geographic locale.

Chapter 7 establishes best practices for supporting teachers serving the Dinaledi cluster of Bojanala district, North-West province, South Africa. Included is an analysis of crucial community engagement initiatives within online programs. The chapter closes with evaluating considerations for optimizing best practices in remote settings.

Chapter 8 presents research on implementing an extendable functioning waiting room, which was instituted to nurture engagement following the start of the COVID-19 pandemic. Included is an analysis of how technology may foster engagement among learners and faculty alike.

Chapter 9 establishes strategies for the targeted use of online discussion boards to build community. Best practices are presented in the context of theories of adult learning. Included is a consideration of the community's role in reducing isolation.

Chapter 10 presents lessons learned through the positive adaptation to the COVID-19 pandemic at National Louis University. Included is a consideration of critical steps for the enhancement of technology and the implementation of intentional activities. Recommendations for the evolution of colleges and universities are presented.

Chapter 11 explores the impacts of isolation and loneliness within communities, presenting research on how belongingness may be facilitated through social engagement in online communities. Included is an analysis of key indicators of belonging and practices which may support its continual nurturance.

The final section provides examples and guidance for the effective application of best practices for leadership with relevance to remote community engagement.

Chapter 12 presents research on remote tutoring. Included is an evaluation of parent-tutor dynamics during in-person and virtual tutoring. Lessons learned from the Tutor Corps program, including strategies for effective engagement, are shared.

Chapter 13 establishes best practices for the mentorship of cohorts by presenting an analysis of a virtual cohort mentor model implemented through the Online Campus of Adler University. Faculty, administrator, mentor, and student impacts are discussed in the context of relevant theory. The role of virtual advising as a complement to the mentor model is explored.

Chapter 14 presents best practices and lessons learned pertaining to remote community engagement during the COVID-19 pandemic by presenting findings of a successful pilot of a Global Public Health Virtual Practicum. Included are considerations for the effective collaboration between universities and community agencies to support student learning.

Chapter 15 establishes guidelines for future leaders, presenting research on belonging, performance, and engagement in the context of relevant theory. The chapter closes with an evaluation of community needs as they pertain to social justice and growth.

Through the contributions contained herein, the editors aim to present a cohesive and growth-centered argument for adopting virtual means to achieve community engagement. Technology now supports full engagement without the need for travel. Technology allows all to connect and to nurture each other – setting a place for all at the table – without walls and without boundaries. Through virtual means, diverse voices may be centered, and communities may grow in magnitude, thereby supporting full engagement and making real change possible.

Michelle Dennis
Adler University, USA

James D. Halbert
Adler University, USA

REFERENCES

Abedini, A., Abedin, B., & Zowghi, D. (2021). Adult learning in online communities of practice: A systematic review. *British Journal of Educational Technology*, *52*(4), 1663–1694. doi:10.1111/bjet.13120

Alexandra, Y., & Fahmi Choirisa, S. (2021). Understanding college 'students' e-loyalty to online practicum courses in hospitality programs during COVID-19. *Journal Of Learning Development in Higher Education*, *21*(21). Advance online publication. doi:10.47408/jldhe.vi21.627

Alghamdi, J. (2022). Equipping Student Teachers with Remote Teaching Competencies Through an Online Practicum: A Case Study. In M. S. Khine (Ed.), *Handbook of research on teacher education* (pp. 187–206). Springer. doi:10.1007/978-981-19-2400-2_12

Ansbacher, H. L. (1968). The concept of social interest. *Journal of Individual Psychology*, *24*(2), 131. PMID:5724576

Archer-Kuhn, B., Judge-Stasiak, A., Letkemann, L., Hewson, J., & Ayala, J. (2022). The self-directed practicum: An innovative response to COVID-19 and a crisis in field education. In R. Baikady, S. M. Sajid, V. Nadesan, & M. R. Islam (Eds.), *The Routledge handbook of field work education in social work* (pp. 531–540). doi:10.4324/9781032164946-40

Armstrong, M. (2020). Moving classes online is hard—Especially in prisons: Higher education in prison programs get creative to keep classes going. *Slate.* https://slate.com/technology/2020/05/remote-learning-prisons.amp

Baciu, E. L., & Trancă, L. M. (2021). Re-framing challenges as opportunities: Moving a social work practicum program in an online format and making it work. *Social Work Research*, *1*, 179–191.

Benitz, M. A., & Yang, L. (2020). Adapting a community engagement project in engineering and education to remote learning in the era of COVID-19. *Advances in Engineering Education*, *8*(4), 1–8.

Bozkurt, A., & Sharma, R. C. (2020). Emergency remote teaching in a time of global crisis due to Coronavirus pandemic. *Asian Journal of Distance Education*, *15*(1), 1–4.

Braithwaite, R. L., Akintobi, T. H., Blumenthal, D. S., & Langley, W. M. (2020). *The Morehouse Model: How one school of medicine revolutionized community engagement and health equity*. JHU Press. doi:10.1353/book.75006

Brosens, D., Croux, F., & De Donder, L. (2019). Barriers to prisoner participation in educational courses: Insights from a remand prison in Belgium. *International Review of Education*, *65*(5), 735–754. doi:10.100711159-018-9727-9

Cattapan, A., Acker-Verney, J. M., Dobrowolsky, A., Findlay, T., & Mandrona, A. (2020, October 1). Community engagement in a time of confinement. *Canadian Public Policy*, *46*(S3), S287–S299. doi:10.3138/cpp.2020-064

Coronado, G., Chio-Lauri, J., Dela Cruz, R., & Roman, Y. M. (2021). Health disparities of cardiometabolic disorders among Filipino Americans: Implications for health equity and community-based genetic research. *Journal of Racial and Ethnic Health Disparities*, 1–8. PMID:34837163

Dennis, M. (2021). Best practices for emergency remote education. In A. Bozkurt (Ed.), *Handbook of research on emerging pedagogies for the future of education: Trauma-informed, care, and pandemic pedagogy* (pp. 82–11). IGI-Global. doi:10.4018/978-1-7998-7275-7.ch005

Dennis, M., Gordon, E., DiMatteo-Gibson, D., & Halbert, J. D. (2022). Social justice practicum in nonclinical online programs: Engagement strategies and lessons learned. *Journal of Leadership, Accountability and Ethics*, *19*(3), 147–156.

Dennis, M., & Halbert, J. D. (2022). Effective online course delivery in correctional settings: A pilot. *Journal of Higher Education Theory and Practice*, *22*(8), 91–99.

Dodd, A. (2017). Finding the community in sustainable online community engagement: Not-for-profit organization websites, service-learning and research. *Gateways: International Journal of Community Research & Engagement*, *10*, 185–203. doi:10.5130/ijcre.v10i0.5278

France-Harris, A., Burton, C., & Mooney, M. (2019). Putting theory into practice: Incorporating a community engagement model into online pre-professional courses in legal studies and human resources management. *Online Learning*, *23*(2), 21–39. doi:10.24059/olj.v23i2.1448

Froiland, J. M., & Worrell, F. C. (2017). Parental autonomy support, community feeling and student expectations as contributors to later achievement among adolescents. *Educational Psychology*, *37*(3), 261–271. doi:10.1080/01443410.2016.1214687

Ilgaz, H., & Aşkar, P. (2013). The contribution of technology acceptance and community feeling to learner satisfaction in distance education. *Procedia: Social and Behavioral Sciences*, *106*, 2671–2680. doi:10.1016/j.sbspro.2013.12.308

Kang, M., Shin, D., & Gong, T. (2016). The role of personalization, engagement, and trust in online communities. *Information Technology & People*, *29*(3), 580–596. doi:10.1108/ITP-01-2015-0023

King, R. A., & Shelley, C. A. (2008). Community feeling and social interest: Adlerian parallels, synergy and differences with the field of community psychology. *Journal of Community & Applied Social Psychology*, *18*(2), 96–107. doi:10.1002/casp.962

Knudson, D. (2020). A tale of two instructional experiences: Student engagement in active learning and emergency remote learning of biomechanics. *Sports Biomechanics*, 1–11. doi:10.1080/14763141.2020.1810306 PMID:32924795

Kuem, J., Khansa, L., & Kim, S. S. (2020). Prominence and engagement: Different mechanisms regulating continuance and contribution in online communities. *Journal of Management Information Systems*, *37*(1), 162–190. doi:10.1080/07421222.2019.1705510

Litchman, M. L., Edelman, L. S., & Donaldson, G. W. (2018). Effect of diabetes online community engagement on health indicators: Cross-sectional study. *JMIR Diabetes*, *3*(2), e8603. doi:10.2196/diabetes.8603 PMID:30291079

McInroy, L. B., McCloskey, R. J., Craig, S. L., & Eaton, A. D. (2019). LGBTQ+ youths' community engagement and resource seeking online versus offline. *Journal of Technology in Human Services*, *37*(4), 315–333. doi:10.1080/15228835.2019.1617823

Mohmmed, A. O., Khidhir, B. A., Nazeer, A., & Vijayan, V. J. (2020). Emergency remote teaching during Coronavirus pandemic: The current trend and future directive at Middle East College Oman. *Innovative Infrastructure Solutions*, *5*(3), 1–11. doi:10.100741062-020-00326-7

Pierce, S., Blanton, J., & Gould, D. (2018). An online program for high school student-athlete leadership development: Community engagement, collaboration, and course creation. *Case Studies in Sport and Exercise Psychology, 2*(1), 23-29.

Pratt, B., & de Vries, J. (2018). Community engagement in global health research that advances health equity. *Bioethics*, *32*(7), 454–463. doi:10.1111/bioe.12465 PMID:30035349

Ray, S., Kim, S. S., & Morris, J. G. (2014). The central role of engagement in online communities. *Information Systems Research*, *25*(3), 528–546. doi:10.1287/isre.2014.0525

Telles, A. B. (2019). Community engagement vs. racial equity: Can community engagement work be racially equitable? *Metropolitan Universities*, *30*(2), 95–108. doi:10.18060/22787

Thompson, H. M., Clement, A. M., Ortiz, R., Preston, T. M., Wells Quantrell, A. L., Enfield, M., King, A. J., Klosinski, L., Reback, K. J., Hamilton, A., & Milburn, N. (2022). Community engagement to improve access to healthcare: A comparative case study to advance implementation science for transgender health equity. *International Journal for Equity in Health*, *21*(1), 1–14. doi:10.118612939-022-01702-8 PMID:35907962

Vasiliki, B., & Psoni, P. (2021). Online teaching practicum during COVID-19: The case of a teacher education program in Greece. *Journal of Applied Research in Higher Education*. Advance online publication. doi:10.1108/JARHE-07-2020-0223

Weisman, M. (2021). Remote community engagement in the time of COVID-19, a surging racial Justice movement, wildfires, and an election year. *Higher Learning Research Communications, 11,* 6.

Section 1

Community Engagement: Expansion Through the Online Venue

This section defines community engagement in the online space, providing examples to illustrate the ways in which community engagement has expanded through the use of technology and the application of creativity.

Chapter 1
What Is Community Engagement

La Toya L. Patterson
Chicago State University, USA

ABSTRACT

This chapter provides an overview of community engagement within higher education institutions and describes the process of applying for and receiving the Carnegie Foundation for the Advancement of Teaching Community Engagement Classification. The author discusses the reasons for seeking the classification, reclassification, and clarification of the language used in the application process are discussed. The chapter also explores the importance of institutionalizing community engagement, the benefits that are received by the institution, faculty and students, and the challenges to institutionalizing community engagement. The author highlighted steps for institutionalizing community engagement within higher education institutions and pitfalls to avoid when trying to institutionalize community engagement within higher education institutions.

INTRODUCTION

The practice of community engagement is not a new phenomenon in academia. There has been a desire by higher education institutions to gain recognition for their work with community engagement. Since its inception in 2006, several higher education institutions have sought the community engagement classification and made strides to achieve the classification (Mehta & Ahmed, 2018). A total of 317 colleges and universities were awarded the classification, which showed that these colleges and universities have institutionalized community engagement in their identities, culture, and commitments (Driscoll, 2014). In higher education institutions, community engagement was narrowly defined as community outreach and partnership where they are necessary and essential players and actors. Community engagement had deep roots within higher education institutions. These roots extended to the philosophical orientation geared toward community engagement work and practical application (Falk & Olwell, 2019). There were several definitions for the term community engagement. However, the term *community engagement* was selected to capture the broadest idea of collaboration between higher education and the community

DOI: 10.4018/978-1-6684-5190-8.ch001

as well as promote inclusivity (Driscoll, 2009). "Community engagement is the collaboration between institutions of higher education and their larger communities (local, regional/state, national, global) for the mutually beneficial exchange of knowledge and resources in a context of partnership and reciprocity" (Community on Public Purpose in Higher Education, 2022).

The community engagement classification allowed for higher education institutions to claim an identity that revolved around community engagement through a classification that is "the best practices that have been identified nationally" (Saltmarsh & Johnson, 2020, p. 108). It needs to be noted that the term "community engagement" was not an umbrella term trying to capture any activities that are correlated with civic education, experiential education, or campus engagement with local, regional, national, or global communities. Also, community engagement was not intended to be institutional commitments ranging from investments to procurements, employment, economic development, and outreach (Saltmarsh & Johnson, 2018). The development of community engagement was designed to help address elements of the higher education institution's mission and distinctiveness, which are not represented by the national data on colleges and universities (Driscoll, 2014). The primary aim or concern of community engagement was not a self-serving one, it concerned community improvement (Falk & Olwell, 2019).

Seeking Community Engagement Classification

There were distinct reasons why higher education institutions sought the Carnegie Community Engagement Classification. One of the most prevalent reasons was to have a structured process of institutional self-assessment and self-study. The classification was often sought after because it provided the opportunity for higher education institutions to legitimize community engagement work that was not otherwise publicized or recognized (Manok, 2018). The motivation to seek the classification was attributed to an attitude shift in higher education, which had moved beyond an interest in the financial aspect of engagement to a more social role in higher education, dominance of an audit culture where there was a development of accountability tools to monitor the institution's performance and exhibiting institutional values to the stakeholders, and market-based incentives where the institution could separate themselves from competitors and demonstrate their superior level of performance (Saltmarsh & Johnson, 2020). Completing the application process helped develop an individual identity around community engagement. It did not matter if the institution was a research university, community college, or a liberal arts institution, they could be community engaged, which created a distinctive identity for itself (Manok, 2018).

COMMUNITY ENGAGEMENT AS A CLASSIFICATION

The community engagement classification was the framework that guided increasingly more significant and intentional practices to engagement action. Additionally, it was required that institutions measured, described, and evaluated engagement across the institutions (Gelmon, et al., 2018). The classification of community engagement was an elective classification, which meant that higher education institutions were not mandated to participate; it was strictly voluntary. The elective classifications were developed to be complementary to the basic classifications. This provided higher education institutions with the opportunity to have an institutional identity associated with community engagement (Saltmarsh & Johnson, 2020). The community engagement classification was developed in part due to the growing community engagement that occurred in American higher education institutions, who were seeking legitimacy through recognition by higher education power brokers. The classification of community engagement was created for the following reasons: 1) to recognize the diversity of institutions and their approaches to community engagement; 2) to help institutions to engage in a process of inquiry, reflection, and self-assessment; and 3) honor institutions' achievements but also promote ongoing development of their programs (Saltmarsh & Johnson, 2020). Evidence of campus practice, structures, and policies designed to deepen community engagement are required for the community engagement application (Welch & Saltmarsh, 2013).

Framework of Documentation

The classification of community engagement was developed in 2005 through the Foundation leadership of President Lee Schulman and under the direction of Amy Driscoll (Saltmarsh & Jonson, 2020). The process of documentation was intensive and required collaboration between many higher education institutions and community participants (Driscoll, 2009). The Carnegie framework required higher education institutions to respond to two major sets of questions, which required documentation of engagement with the community, Foundational Indicators and Curricular Engagement (Zuiches et al., 2008). *Foundational Indicators* were divided into two sections- Institutional Commitment and Institutional Identity, and Culture (Saltmarsh & Johnson, 2018). Foundational indicators were required to demonstrate how they met institutional identity culture; this was done by demonstrating examples of mission, marketing, celebrations and rewards, and leadership (Driscoll, 2014). An example of institutional identity and culture would be the higher education institution indicating community engagement was a priority in their mission; this was demonstrated by providing direct quotes from the institution's mission statement. Institutional commitment required that institutions provide evidence pertaining to the institution's budget, infrastructure, strategic planning, and faculty development that supported community engagement (Driscoll, 2008).

For curricular engagement, higher education institutions were required to provide documentation of the following activities: service learning, internships, student leadership and research, and faculty scholarship (Driscoll, 2014). The framework described curricular engagement as "the teaching, learning and scholarship that engages faculty, students, and community in mutually beneficial and respectful collaboration" (Saltmarsh & Johnson, 2018, p. 6). Institutions were to demonstrate curricular engagement by addressing community needs, deepen students' civic and academic learning, enhancing the community's well-being and enrichment of scholarship for the institution (Driscoll, 2008). Lastly, institutions also had to demonstrate *Outreach and Partnerships.* Questions pertaining to outreach highlighted programs and institutional resources that were provided for the community, though this was not considered engagement; however, it was complementary to engagement activities. Partnership questions were geared toward mutuality and reciprocity in relationships. Institutions were asked0 to demonstrate this evidence by completing "the partnership grid," which captured the sense of the institution's extensive partnerships that showed reciprocity and mutual benefit (Saltmarsh & Johnson, 2018). Since higher education institutions had various philosophies, approaches, and contexts, the documentation framework was created to accommodate these variations (Driscoll, 2008).

Community Engagement Application: Documentation Process

For this elective classification, it required data collection and documentation that involved the pertinent areas of an institution's mission, identity, and commitments and involved the efforts by the invested participating institutions (Commission on Public Purpose in Higher Education, 2022). The application process enforced by the Carnegie Foundation, required that documentation be extensive and substantive, highlighted qualities, activities, as well as institutional provisions that showed the institution's approach to community engagement (Zuiches et al., 2008). The process also asked applicants to provide self-reported evidence geared toward the individual institution and its communities, which was heavily descriptive (Saltmarsh & Johnson, 2020). The purpose of the classification was to capture the wide range of practices that fulfill the purposes of community engagement across diverse types of institutions (Noel & Earwicker, 2015).

When higher education institutions were putting together the application for the classification, evidence was gathered and reflected on, understanding the institution's areas of strengths and weaknesses regarding institutional engagement helped with improving the practices and advancement of community engagement on campus (Saltmarsh & Johnson, 2020). With each application, the data and narratives showed there was room for improvement and expansion in relation to institutional capacity and the practice related to data collection and evaluation (Gelmon et al., 2018). However, the application process for the community engagement classification served another purpose. The application aided in bringing the dissimilar parts of the institution's campus together; meaning there was a unified agenda, which served as a catalyst for change, fostered alignment within the institution for community-based teaching, learning, and scholarship. Additionally, seeking out the classification allowed for identifying practices that can be shared across the institution (Saltmarsh & Johnson, 2020). Campuses that sought out the community engagement classification had to take a full inventory of their efforts in engagement so that they could address the questions that were posed by the Carnegie Foundation (Noel & Earwicker, 2015). Lastly, higher education institutions applied for the classification because it provided legitimacy for their community engagement work that otherwise would not have received public recognition and visibility (Saltmarsh & Johnson, 2020).

The community engagement classification required assessment practices, and these practices had to meet a broad range of areas, which included community's perception of institutional engagement, recorded and tracked institutional wide engagement data, the impact of community engagement on students, faculty, community, and institution, identified and assessed student learning outcomes in curricular engagement; lastly, provided feedback for partnership (Saltmarsh, 2017). For the assessment practices, the earlier applications had limited examples of tracking mechanisms to record data such as "number of students, number of hours of service, number and kinds of community sites, kinds of engagement, and number of faculty" (Gelmon et al., 2018, para. 3). The assessment portion of the application helped higher education institutions to have understanding and approaches that would assist with achieving the assessment goals (Saltmarsh & Johnson, 2020). There were more than two dozen assessments that had been published to measure community engagement in higher education institutions. These assessments tools were substantial in the areas of complexity, scope, process, structure, and focus. Applicants continued to struggle with assessments questions regarding the actual impact of community engagement. Though higher education institutions were asked to provide assessment practices and an example of the impact of community engagement on students, community, faculty, and institution, the examples were not authentic. The institutions tended to report statistics of growth in the number of students or faculty (Gelmon et al., 2018).

Higher education institutions ran into challenges when they assessed community engagement. Ward et al. (2013) reported that for the assessment of the 2006 applications, Amy Driscoll observed three areas that were challenging for the institutions. One of the challenges revolved around community involvement. The weakness involved the assessment of the community's needs for and perceptions of higher education institution's engagement as well as formed roles for the community to create the institution's plan for community. Another weakness was the ambiguity of how institutions achieved reciprocity within the community. The assessment of community engagement in general was a challenge and specific for categories of engagement. Lastly, faculty engagement also proved to be a challenge. There was a lack of support for faculty that engaged in community engagement, faculty development support and faculty recruitment and hiring practices. Higher education institutions grappled with another challenge concerning assessment, which data systems would be useful for tracking and measuring and assessing community engagement across all dimensions. The community engagement classification required ongoing and systematic data collection; therefore, various assessment tools had been developed for said purposes. There should have been collaboration between internal and external partners to choose a system that replicated their goals and agenda but also be customizable for local purposes (Gelmon et al., 2018).

Another reviewed area on the community engagement classification application is infrastructure. Welch and Saltmarsh (2013) stated that substantial infrastructure was a part of community engagement office, center or division and a vital component of highly engaged campus (Janke & Domagal-Goldman, 2017). Since the early 1990s infrastructure had been the focus for campuses and continued to be a critical focus area. When the classification referred to "coordinating infrastructure," it is defined as a place where facilitation of engagement occurred on the campus. Coordinating infrastructure was also important for establishing a culture of assessment and accountability involving engagement work (Saltmarsh & Johnson, 2020). Infrastructure was considered an important asset of change because it reinforced an identity in a structural and meaningful way (Dierberger et al., 2019). Having a campus infrastructure that supported both service learning and community engagement, required understanding and working within the context of how the higher education institution's mission affected the policy and actions, how individuals responded to change, and how management decisions were made and disseminated (Pigza & Troppe, 2003).

Incentives and rewards regarding faculty work was another area that is reviewed on the application. Faculty members felt that their higher education institution discredited their community engagement work (Vuong et al., 2017). Faculty motivation for community engagement had a rich and large literature on motivation and behavior of the faculty, academic careers, academic culture, and faculty reward systems. When faculty members were being reviewed for merit salary increases or tenure, their community engagement work was viewed as being less rigorous than what was considered traditional forms of work conducted by faculty (Vuong et al., 2017). When there had been discussion of faculty rewards for community engagement, there was difficulty creating a campus culture of community engagement due to not having clear incentives for faculty to rank their roles, which consisted of research, teaching, and service in promotion and tenure criteria. When there were no defined incentives then there were disincentives (Saltmarsh, 2017). The community engagement classification was looking for evidence that provided clear policies that recognized community engagement that occurred within teaching and learning, and in research and creative activity. Additionally, the Carnegie Foundation also pushed campuses to revise their current promotion and tenure policies, in hopes of initiating study, conversation, and reflection on promoting and rewarding the scholarship of engagement (Saltmarsh & Johnson, 2020). Moore and Mendez (2014) reported that higher education institutions revised their tenure and promotion policies to reflect the definition of scholarship and shine light on engaged scholarship on par with traditional examples of research.

Community engagement offered untapped possibilities for alliance with other campus priorities and initiatives that would help achieve greater impact. An example of this was a first-year program that combined community engagement as well as learning communities where community engagement was integrated into the design. It also incorporated diversity initiatives that clearly tied active and collaborative community-based teaching and learning to the success of the underrepresented students (Saltmarsh, 2017). When higher education institutions became more intentional with explicitly connecting community engagement to their strategic priorities, better community engagement was institutionalized and transformed the campus culture (Saltmarsh & Johnson, 2020). Through the Carnegie Community Engagement Classification application process, it was revealed that there were specific areas that campuses needed to pay closer attention to fully realize engagement. There were areas that proved to be challenging depending on the culture and context of the campus. As campuses advanced their engagement goals or agendas, they considered the importance of developing an environment that valued engagement for the new generation of scholars, a new generation of faculty members-that are a representation of what had been a historically underrepresented population- who had also been engaged scholars as undergraduate and graduate students and expected the campus to allow them to thrive in teaching, research, and service (Saltmarsh, 2017).

The application examined the partnership between higher education institutions and the community. When there was shared information about campus-community partnerships, it led to a discussion of shared goals, recognition of the connected agenda, and the development of common language that was used for service-learning and community engagement (Pigza & Troppe, 2003). Partnerships entailed a high level of intentional practices that were geared toward reciprocity and mutuality. Higher education institutions had to have an ongoing commitment to initiating and collaborating, two-way partnerships, strategies for systematic communication, mutually beneficial partnerships, and authentic collaboration (Saltmarsh, 2017).

The documentation process for the Carnegie Community Engagement Classification provided higher education institutions with a blueprint to create long-term institutionalization of community engagement, aligning campus programs, units, structures, and policies. Additionally, the documentation process was a tool for helping to improve higher education institution's central purposes: creating and disseminating knowledge through research, teaching, and learning through undergraduate education, and fulfilling public purpose. The application process was not just about obtaining the classification but more so providing higher education institutions with the ongoing opportunity to examine, assess, document, and reflect on their community engagement practices (Saltmarsh & Johnson, 2020).

INSTITUTIONALIZATION OF COMMUNITY ENGAGEMENT IN HIGHER EDUCATION

Beere et al., (2011) reported that the term *community* had been defined in a geographical sense such as a town, state, or county. Community can even be denoted as individuals' identity or status; and by online connections such as Facebook community or distant learners. Defining the term community and determining the focus of community was pertinent because it aided in the institutional planning process. Institutionalization of service learning was multifaceted and defined the work and goals of the stakeholders (Bringle & Hatcher, 2000). Institutionalization typically occurred when there was a critical and sustained feature of the institution as opposed to being viewed as peripheral work to the institution. Repositioning community engagement from the margins and making it the focus and institutionalized role in higher education institutions, required there to be commitment and critical reflection regarding values and purposes of the institution. When there was institutionalization, "regular dissemination of campus-community partnership will lead to conversations of shared goals, recognition of connected agendas, and development of a common language germane to service-learning and community engagement" (McRae, 2015, p. 117). Also, institutionalized communication assisted higher education institutions to recognize cross-campus support and to create ways to continue campus cultural change (Pigza & Troppe, 2003).

Institutionalization of service learning from a faculty perspective was located within the course and curriculum development, activities involving faculty development, recognition, and rewards recognition, understanding and support for service learning by board faculty, and scholarship on service learning. For students, institutionalization of service learning was observed through service and service-learning scholarships, classes that promoted service-learning, student culture, fourth credit options, as well as transcripts that documented cocurricular. Institutionalization from a community relationship standpoint, involved agency resources that were paired with those of the institution to develop reciprocal, enduring and diverse partnerships that reciprocated community interests and academic goals (Bringle & Hatcher, 2000).

Higher education institutional leaders accepted community engagement as a strategy for connecting the classroom, campus, and community. However, administrators were at a stalemate between their desire to cultivate a culture of learning and service, and the realities of their resources and the infrastructure of the institution (Hutson et al., n.d.). Progress had been made when it came to the infusion of engagement in curriculum through community-based learning, service-learning, and outreach and partnership. Nevertheless, there continued to be difficulty surrounding full institutionalization of community engagement into the institution. Institutions that were low on institutionalization of service learning, but were strong in other areas of civic engagement, ran the risk that the community activities would be separated from the institutions' core activities, were conducted by a few individuals, and had minimal effect on work, activities, and culture of the faculty, staff, and students (Bringle et al., n.d). Community engagement had been shown to contribute to increased retention rates in higher education institutions (Hutson et al., n.d.). According to Eckel et al. (1998), community engagement was transformational to be institutionalized (Cunningham & Smith, 2020).

Eckel et al. (1998) completed a national study that examined institutional change, specifically changes that were "transformational." Demonstrating transformational change included change that a) changed the culture of the institution by changing underlying assumptions and institutional behaviors, processes, and products; b) deep and pervasive, impacting the whole institution; c) intentional; and d) happened over time (Manok, 2018). Cunningham and Smith (2020) reported that Holland (1997) created the Holland matrix for evaluating institution's commitment to community engagement Additionally, Furco, Weerts, Burton, and Kent's (2009) assessment rubric was one of the most used rubrics in higher education institutions. The rubric consisted of the following areas: 1) mission and philosophy; 2) faculty support; 3) student support; 4) community partnership; and 5) institutional support. However, not all the dimensions listed had to be fully operationalized for a campus to institutionalize community engagement (Hutson et al., 2020). Other matrices were developed as well to be used in the institutionalization process. Pigza and Troppe (2003) itemized nine indicators and methods for supporting engagement in higher education institutions. In 2009 Holland created benchmarks that aided in the process of institutionalized efforts (Dostilio & Welch, 2019).

It was essential that higher education institutions authentically engaged with the community, had shared beneficial partnership as they evaluated the progress demonstrated in the first wave of community-engaged institutions that were classified by Carnegie Foundation for the Advancement of Teaching (Cunningham & Smith, 2020). The application process for the classification was often used by higher education institutions to provide a blueprint for institutionalizing community engagement (Dostilio & Welch, 2019).

How To Institutionalize Community Engagement in Higher Education

To create institutionalize community engagement, it was imperative that it was a part of fabric of the institution and inserted into the culture and priorities of the institution. If not, there was a possibility for community engagement to lose momentum and be disregarded upon changes in administration (Cunningham & Smith, 2020). Engagement had to be included in the mission statement, strategic planning, funding, and other incentives for faculty members, and acknowledged community-based research and service learning as valid approaches (McRae, 2015). It had been argued that the mission statement of the institution had to include "rationale, direction, motivation, and commitment for the institution to involve itself in community engaged work" (Cunningham & Smith, 2020, p. 55). The literature regarding community engagement recognized various perspectives about how it should have been organized and supported. To get community engagement to thrive within higher education institutions, it had to align with or fit with the institution's core activities.

The institutionalization of community engagement required commitment from faculty, students, and community partners, which helped to maintain relationships and that were furthered developed (McRae, 2015). There were several factors that played a role in the institutionalization of community engagement in higher engagement. One of the factors was the support of administration. The support of administration was recognized or evident through infrastructure and financial resources. Through infrastructure and financial resources, it sent the message to the faculty, students, and community that the institutionalization of engagement was taken seriously and encouraged. Providing a centralized location in the institution where the coordination of community engagement could occur was important. Not only was it important, but it also demonstrated that the work of engagement is an institutional effort, not just a movement or a particular interest of a department or individual (Cunningham & Smith, 2020).

According to Farner (2019), the acceptance of institutionalizing within higher education institutions involved boundary expansion. *Boundary expansion* was when there was a legitimized difference between the organization and the innovation and there was agreement to take on those differences (Sandmann & Weerts, 2008). To facilitate institutionalization there needed to be change agents, they were the individuals that would negotiate relationships, power, and information. One of the change agents that was needed were boundary spanners. *Boundary spanners* were the change agents that were positioned to address any adaptive challenges; they were the ones that negotiated the wants and needs of parties that were involved with disseminating knowledge as well as creating knowledge. The framework created by Sandmann and Weerts (2010) identified four roles of boundary spanners. The roles included community-based analytical people, technical experts, engagement champions, and internal engagement advocates. The boundary spanners activities that occurred on the individual level were pertinent to the operationalizing of institutionalization process. However, on the organizational level, comprehending how the separate institutional factors aligned to impact the process of institutionalization was critical (Farner, 2019).

Process for Institutionalizing Community Engagement

Ward et al. (2013) created a five-step model or approach for institutionalizing community engagement:

1. Engaged knowledge generation that is applied, problem centered, entrepreneurial, and networked
2. Clear understanding of the characteristics and values of community-engaged scholarship
3. An integrated role for faculty and includes teaching, research, and service

4. Knowledge generation that occurs in diverse ways and with multiple stakeholders
5. Aligning community engagement with institutional identity, mission, place, work, and reward policies

There had to be a change with the culture of the institution for institutionalizing to occur. The change would allow for individuals of the institution and community partners to observe engagement as an integrative approach to teaching, service, and research (McRae, 2015). Campuses addressed the necessary changes needed to perform the collaborative turn for "teaching and learning, research, scholarship, and creative activity, and outreach and service" (Saltmarsh, 2017, p.11). The academic and administrative teams' plan for developing truly institutionalized community engagement had ties to the priorities of the institution. Therefore, the plan had to show a connection to the mission statements, goals, and strategic priorities as well as other institutional practices that included outcomes for student learning. Also, having the community engagement plans connected to the higher education institution's priorities helped guarantee relevance as well as get administrators and faculty to buy-in; this helped them to perceive the plan as important and be a catalyst to develop and implement those priorities (Cunningham & Smith, 2020).

BENEFITS OF INSTITUTIONALIZING COMMUNITY ENGAGEMENT IN HIGHER EDUCATION INSTITUTIONS AND ONLINE SPACES

An abundance of benefits came from community engagement on an individual level, such as personal satisfaction to resume-building and networking, learning skills, and making friends (Falk & Olwell, 2019). Community engagement addressed several community issues, which included poverty, homelessness, mental health, education, and health and wellness. Community engagement had a social justice perspective; therefore, the human well-being and the wellness of the community are intertwined with issues of "equity, access, power, and privilege" (p. 64). It was believed that community engagement supported student development in a variety of ways. Service-learning had an impact on a student's personal and interpersonal growth, application of course content, civic identity, and critical thinking skills (Falk & Olwell, 2019). From an institutional perspective, community engagement was beneficial because when there was evidence that the higher education institution was fulfilling their commitment to the community as well as their mission, it led to increased alumni, community relations, and improved revenue streams. Additionally, higher education institutes also benefitted through public relations, marketing, and branding (Falk & Olwell, 2019). If there was a positive image of the institution it led to more contracts with public and private services and more donors were willing to invest in the institution. Additionally, when the institution contributed to the community, it was more likely that the business and community leaders would have been advocates for the institution. The institutionalizing of community engagement had benefits for both public colleges and universities and urban universities. For public colleges and universities, institutionalization of community engagement created legislative support and increased appropriations. For urban universities, there was a commitment to maintain local quality of life; this involved partnering with communities to remedy problems such as drug rings, high crimes, gangs, and abandonment of buildings (Beere et al., 2011). Community engagement was more so about partnership and alliance than charity. Through community engagement, there were opportunities for creating and promoting partnerships, which created an experience of learning and growth for all participants that were involved whether they be researchers, faculty members, students, nonprofit leaders, or community members (Falk & Olwell, 2019).

The higher education institution served the needs of the community in a variety of ways; this was usually done by sharing facilities or providing programs for the community. There was a likelihood for institutions to share their libraries, wellness centers, athletics, museums, art galleries, classrooms, and meeting rooms. Programs were offered by colleges and universities, which were either on or off campus. Some of the programs that were offered could be summer camps for kids, academic camps, health clinics, as well as mental health facilities. Cultural activities were offered such as musical programs, theatrical productions, or forums for debates on controversial topics (Beere et al., 2011).

Institutionalized community engagement had an academic and student impact. Academically, service-learning or community engagement improved students' learning outcomes. Research showed that service participation caused an increase of academic performance (i.e., GPA, critical thinking skills, and writing skills), values, leadership as well as self-efficacy (Reilly & Langley-Turnbaugh, 2021). Students received substantial benefits from being a part of service learning and volunteering. The learning that was observed from the service-learning students illustrated that the learning was deeper and not just students regurgitating back facts about a topic or subject. Other benefits included an appreciation of ethical issues that impacted the world of practice, better understanding and sense of self, increased self-confidence, career readiness, and networking connections (Beere et al., 2011). Even when students graduated from the institution there were still benefits that were gained from service-learning or community engagement. Students had an increased understanding of social justice issues that were taking place in the world, personal insight, and long-term cognitive development from participating in service-learning. Research highlighted those students that participated in service-learning were politically engaged, took accountability of issues, became better citizens that were more tolerant (Reilly & Langley-Turnbaugh, 2011). Also, Falk and Olwell (2019) highlighted that community engagement is mutually beneficial from an educational and human development standpoint because it encouraged individual growth and transformation for all that were involved.

Engagement with the community also produces teaching benefits. Community engagement played a role in what faculty members chose to emphasize within the classroom; this was important because it helped faculty to remain current and kept course content up to date. Additionally, it provided a connection between the courses they were teaching and what was going on in the world; powerful examples were used within the classroom and enriched faculty's teaching (Beere et al., 2011). There was also a development of stronger relationships with faculty members among students. This was particularly true of minority students because community engagement caused a decrease in cultural dissonance and a sense of belonging in the institution (Reilly & Langley-Turnbaugh, 2021). Regarding research, community engagement provided faculty members with new ideas for research studies and allowed testing of new theories and access to research sites and research data that may have not been available. There was a level of satisfaction reached through engagement. Faculty members took on a new challenge, utilized their expertise to make a difference in the world, and tested their theories. Lastly, faculty members decreased intellectual isolation by acquiring new colleagues from their institution as well as other institutions, and the community at large (Beere et al., 2011).

When both faculty and students engaged in the community, they were among the strongest advocates for creating expansion of public engagement. Students appreciated an experience that let them apply the concepts that they learned in the classroom to the real world, specifically when students felt that it made them more competitive in the job market. When community leaders and organizations of the community had positive experiences with local higher education institutions, they encouraged more community engagement work. Due to communities having an array of issues, when colleges and universities assisted with addressing those issues, the communities pushed for more connections (Beere et al., 2011).

Implementing community engagement in online courses or spaces was an effective way to provide opportunities for engagement, collaboration, and interaction, which provided a feeling of connectedness for students to their communities (Simunich, 2017). It has been shown that community engagement in online spaces can benefit perceived learning, satisfaction with courses, and student engagement and persistence. When students had a sense of connectedness and psychological closeness they were better prepared to be engaged in online spaces and had higher order thinking and knowledge (Dolan, Kain, Reilly, & Bansal, 2017). There is a symbiotic relationship between e-service learning and online instruction. Online courses provided students the opportunity to engage in service-learning when they were not able to or chose to not attend traditional face-to-face courses. Lastly, e-service learning was organized in a way where real-world challenges and scenarios were implemented, so that students could master skills and work collaboratively to find solutions online. Service learning as a pedagogy positively influenced retention and engagement and increased understanding of academic content, civic/community engagement application, and connection to personal growth (Germain, 2019).

CHALLENGES OF INSTITUTIONALIZING COMMUNITY ENGAGEMENT IN HIGHER EDUCATION INSTITUTIONS

There were some challenges when discussing the institutionalizing of community engagement. There was a disconnect between senior management that were concerned with priorities and policies whereas faculty members were concerned with creating activities and research that supported students and community partners interests (McRae, 2015). Institutionalizing community engagement was an example of challenging change within higher education institutions. Campuses were described as "brimming with change initiatives," disconnected, and missed opportunities for collaboration on campus (Falk & Olwell, 2019, p. 65). The current engagement movement redirected higher education institutions to its original purpose, which was to address social issues and build community to promote a just and democratic society. However, there was the argument that the movement had caused isolation in the efforts of a single faculty member with the academic departments. Working on institutionalization demonstrated that there was a shift from the pedagogical framework, which was primarily concentrated on teaching, learning, and scholarship that involved faculty and students to a more political context collaborating with administrators and other leaders of the institution and in the community. Despite the effectiveness of institutionalization, there was still fear that institutionalizing created a loss of work's soulful nature (Dostilio & Welch, 2019).

There were broader challenges that were presented by community engagement, which did not involve just enhancing the curriculum but also clarifying the mission, creating broad-based institutional support for community engagement, better community input into higher education institutions, reconceptualizing scholarships and its assessment, recognition of faculty roles and rewards, and assessing institutional successes (Bringle & Hatcher, 2000). The institutionalization challenges were significant, which included limited resources, increased teaching responsibilities, unresponsive reward structures, hesitance to change the culture, and other commitments by support services and being disconnected from one another. There are several pitfalls that should be avoided when trying to institutionalize community engagement within a higher education institution:

1. Make sure to connect community engagement with the core elements of the higher education institution. The relevancy of community engagement needs to be seen by people from all parts of the institution.
2. There needs to be open discussion about the process, seek out feedback from the leadership team at all levels, and act on the feedback. Stay within the institutions culture of consultation, faculty governance, etc.
3. Have students to be a part of the planning. There should not be an assumption that faculty's understanding and needs will be the same as the students.
4. Build upon the strengths of the institution.
5. Identify win-win situations for the institution. Observe how the needs of the community, student, faculty, and institution can be met within the same project and framework.

Despite there being pitfalls in the process of institutionalizing community engagement, it was still a worthwhile process that was beneficial for all higher education institutions (Letven et al., 2001).

SOLUTIONS AND RECOMMENDATIONS

To address the challenges that were brought on by the process of the institutionalization of community engagement, higher education institutions were challenged to develop a model that was inclusive of all stakeholders, faculty members, and community. Usually, community engagement happened in a vacuum among these constituents rather than incorporating everyone and being cooperative. Also, higher education leaders advanced community engagement through the establishment of Campus Compact coalition, whose mission was to advance the public purpose of higher education institutions by "deepening their ability to improve community life and to educate students for civic and social responsibility" (Hutson et al., n.d., p. 12). Since administration needed to be a part of the process, it was important that the administrative units be involved in the efforts as well. Having administration involved ensures that the offices across the institution's campus were supportive of having staff members engaged in service activities within the community (Cunningham & Smith, 2015).

FUTURE RESEARCH DIRECTIONS

There had been continual evolvement by higher education institutions in the United States to meet the needs and demands of the time. Additionally, higher education institutions had to address the ever-growing societal challenges. An emerging trend for community engagement involved the impact of COVID-19 pandemic. During the COVID-19 crisis, universities around the globe started assembling their knowledge and resources to respond to the societal needs that were created by the pandemic. University leaders along with policymakers had to consider the long-lasting effects of COVID-19 on universities' external communities, local and regional (Farnell, 2020). Another area of focus for the future is social entrepreneurship as the new service-learning. Social entrepreneurship had the ability to address social problems on a larger scale that were experienced across nonprofit, business, or government sectors. There was much for social entrepreneurship to gain by allying with service learning.

Successful social entrepreneurs closely aligned with some of the characteristics of service learning such as beliefs in the capacity of all individuals contribute to social and economic development, seeing problems as opportunities, having the passion and determination to develop innovative solutions to social issues (Jacoby, 2015). Noel and Earwicker (2015) suggested that future research should look into the following areas: longitudinal study that observes the long-lasting effects of applying for or receiving the Carnegie community engagement classification, comparison of successful and unsuccessful classification application while identifying any strategies that enabled success or gaps that led to unsuccessful submission, and to study the quality of community engagement at Carnegie classified institutions, that used measurement tools. Lastly, utilizing an asset-based framework for service learning provides the students as well as partners with the opportunity to look for strengths, resources, and capabilities. To assist students with seeing themselves as civic leaders that had the capabilities to address the dauting social issue of our time. To assist with this effort, there had to be an intentional effort to create an asset-based service-learning experience that engaged everyone involved (Baur et al., 2015).

The ongoing challenge for higher education institutions was making community engagement central to the institutions. There had to be an effort on behalf of the institution to adopt methodologies from the margins to the forefront (Fitzgerald et al., 2012). This chapter discusses the importance of community engagement and provides the steps that institutions can use to institutionalize community engagement and make it the focal point of the institution. The institutionalization of community engagement must be included in the fabric of the institution, meaning that community engagement must be embedded into the mission statement, teaching, research, and service; must be reciprocal and mutually beneficial; and welcome the civil democracy process and values (Bringle & Hatcher, 2011).

CONCLUSION

The rise of community engagement presents universities and colleges with an assortment of possibilities to be vital in the role as key institutions in social development and transformation (Shuib & Azizan, 2017). Engagement, when considered as an innovative concept in higher education, has many levels of compatibility and profitability even if the institutions are similar in type (Sandmann & Weerts, 2008). Due to the complexity of societal issues and the effects that it has on the community and higher education institutions, the traditional way of approaching community engagement would not suffice (Fitzgerald et al., 2016). Addressing adaptive challenges required the institution to have change agents that were willing to negotiate the wants and needs of those involved while spreading knowledge. Community engagement was imperative to higher education institutions and continued work needed to occur to make community engagement connected to the classroom, campus, and the surrounding community. Not all higher education institutions had the values and norms that were compatible for the institutionalization of engagement (Sandmann & Weerts, 2008). Commitment by faculty, students and community partners or stakeholders is what helps with the institutionalization of community engagement within higher education institutions. To continue to address societal issues, higher education institutions must engage private and public stakeholders and partners to achieve their goals and resolve community issues.

REFERENCES

Baur, T., Kniffin, L. E., & Priest, K. L. (2015). The future of service-learning and community engagement: Asset-based approaches and student learning in first-year courses. *Michigan Journal of Community Service Learning*, 89–92.

Beere, C., Votruba, J. C., & Wells, G. W. (2011). *Becoming an engaged campus: A practical guide for institutionalizing public engagement*. John Wiley & Sons, Inc.

Bringle, R. G., & Hatcher, J. A. (2000). Campus-community partnerships: The terms of engagement. *The Journal of Higher Education*, *71*(3), 273–290. doi:10.2307/2649291

Bringle, R. G., & Hatcher, J. A. (2011). Student engagement trends over time. In H. E. Fitzgerald, C. Burack, & S. D. Seifer (Eds.), Handbook of engaged scholarships: Contemporary landscapes, future directions: Vol. 2. Community-campus partnerships (pp. 411-430). Michigan State University Press.

Byrne, J. V. (2016). Outreach, engagement, and the changing culture of the university. *Journal of Higher Education Outreach & Engagement*, *20*(1), 53–58.

Commission on Public Purpose in Higher Education. (2022). public-purpose.org

Cunningham, H. R., & Smith, P. C. (2020). Community engagement plans: A tool for institutionalizing community engagement. *Journal of Higher Education Outreach & Engagement*, *24*(2), 53–68.

Dierberger, J., Everett, O., Kehrberg, R. S., & Greene, J. (2019). University, school district, and service-learning community partnerships that work. In J. A. Allen & R. Reiter-Palmon (Eds.), *The Cambridge handbook of organizational community engagement and outreach* (pp. 244–260). Cambridge University Press. doi:10.1017/9781108277693.014

Dolan, J., Kain, K., Reilly, J., & Bansal, G. (2017). How do you build community and foster engagement in online courses? *New Directions for Teaching and Learning*, *151*(151), 45–60. doi:10.1002/tl.20248

Dostilio, L. D., & Perry, L. G. (2017). An explanation of community engagement professionals as professionals and leaders. In L. D. Dostilio (Ed.), *The community engagement professional in higher education: A competency model for an emerging field*. Campus Compact.

Dostilio, L. D., & Welch, M. (2019). *The community engagement professional's guidebook: A companion to the community engagement professional in higher education*. Campus Compact.

Driscoll, A. (2008). Carnegie's community-engagement classification: Intentions and insights. *Change. The Magazine of Higher Learning*, *40*(1), 38–41. doi:10.3200/CHNG.40.1.38-41

Driscoll, A. (2009). Carnegie new community engagement classification: Affirming higher education's role in community. In L. R. Sandmann, C. H. Thornton, & A. J. Jaeger (Eds.), *Institutionalizing community engagement in higher education: The first wave of Carnegie classified institutions* (pp. 5–12). Wiley Periodicals, Inc.

Driscoll, A. (2014). Analysis of the Carnegie classification of community engagement: Patterns and impact on institutions. In D. G. Terkla & L. S. O'Leary (Eds.), *Assessing civic engagement* (pp. 3–16). Wiley Periodicals, Inc. doi:10.1002/ir.20072

Falk, A., & Olwell, R. (2019). Institutionalizing community engagement in higher education: The community engagement institute. In J. W. Moravec (Ed.), *Emerging education futures: Experiences and visions from the field* (pp. 59–82). Education Futures.

Farner, K. (2019). Institutionalizing community engagement in higher education: A case study of processes toward engagement. *Journal of Higher Education Outreach & Engagement, 23*(2), 147–152.

Fitzgerald, H. E., Bruns, K., Sonka, S. T., Furco, A., & Swanson, L. (2012). The centrality of engagement in higher education. *Journal of Higher Education Outreach & Engagement, 16*(3), 7–27.

Gelmon, S. B., Holland, B. A., & Spring, A. (2018). *Assessing service-learning and civic engagement: Principles and techniques*. Campus Compact.

Germain, M. (2019). *Integrating service-learning and consulting in distance education*. Emerald Publishing Limited. doi:10.1108/9781787694095

Giles, D. E. (2008). Understanding an emerging field of scholarship: Toward a research agenda for engaged, public scholarship. *Journal of Higher Education Outreach & Engagement, 12*(2), 97–106.

Hernandez, K., & Pasquesi, K. (2017). Critical perspectives and commitments deserving attention from community engagement professionals. In L. D. Dostilio (Ed.), *The community engagement professional in higher education: A competency model for an emerging field* (pp. 56–78). Compact Campus.

Hutson, N., York, T., Kim, D., Fiester, H., & Workman, J. L. (n.d.). Institutionalizing community engagement: A quantitative approach to identifying patterns of engagement based on institutional characteristics. *Journal of Community Engagement and Higher Education, 11*(2), 3-15.

Jacoby, B. (2015). *Service-learning essentials: Questions, answers, and lesson learned*. John Wiley & Sons, Inc.

Janke, E. M., & Domagal-Goldman, J. M. (2017). Institutional characteristics and students civic outcomes. In J. A. Hatcher, R. G. Bringle, & T. W. Hahn (Eds.), *Research on student civic outcomes in service learning: Conceptual frameworks and methods*. Stylus Publishing, Inc.

Letven, E., Ostheimer, J., & Statham, A. (2001). *Institutionalizing university-community engagement*. Retrieved February 22, 2022, from https://journals.iupui.edu/index.php/muj/article/download/19907/19601/27754

Manok, G. (2018). Key lessons and guiding questions. In J. Saltmarsh & M. B. Johnson (Eds.), *The elective Carnegie community engagement classification: Constructing a successful application for first-time and re-classification applicants* (pp. 50–54). Campus Compact.

McRae, H. (2015). Situation Engagement in Canadian Higher Education. In O. Delano-Oriaran, M. W. Penick-Parks, & S. Fondie (Eds.), *The SAGE sourcebook of service-learning and civic engagement* (pp. 401–406). SAGE Publications, Inc. doi:10.4135/9781483346625.n68

Mehta, S. S., & Ahmed, I. (2018). Planning academic community engagement courses. In H. K. Evans (Ed.), *Community engagement best practices across the disciplines: Applying course content to community needs* (pp. 1–16). Rowman& Littlefield.

Moore, T. L., & Mendez, J. P. (2014). Civic engagement and organizational learning strategies for student success. *New Directions for Higher Education, 165*(165), 31–40. doi:10.1002/he.20081

Noel, J., & Earwicker, D. P. (2015). Documenting community engagement practices and outcomes: Insights from recipients of the 2010 Carnegie community engagement classification. *Journal of Higher Education Outreach & Engagement, 19*(3), 33–62.

Pigza, J., & Troppe, M. (2003). Developing an infrastructure for service-learning and community engagement. In *B. Jacoby & Associates, Building partnerships for service- learning* (pp. 106–130). Jossey-Bass.

Reilly, S., & Langley-Turnbaugh, S. (2021). *The intersection of high-impact practices: What's next for higher education?* Lexington Books.

Sandmann, L. R., Furco, A., & Adams, K. R. (2019). Building the field of higher education engagement. In L. R. Sandmann & D. O. Jones (Eds.), *Building the field of higher education engagement: Foundational ideas and future directions*. Stylus Publishing.

Sandmann, L. R., & Weerts, D. J. (2008). Reshaping institutional boundaries to accommodate an engagement agenda. *Innovative Higher Education, 33*(3), 181–196. doi:10.100710755-008-9077-9

Saltmarsh, J. (2017). A collaborative turn: Trends and directions in community engagement. In J. Sachs & L. Clarke (Eds.), *Learning through community engagement: Vision and practice in higher education* (pp. 3–16). Springer. doi:10.1007/978-981-10-0999-0_1

Saltmarsh, J., & Johnson, M. B. (2018). An introduction to the elective Carnegie community engagement classification. In J. Saltmarsh & M. B. Johnson (Eds.), *The elective Carnegie community engagement classification* (pp. 1–18). Campus Compact.

Saltmarsh, J., & Johnson, M. (2020). Campus classification, identity, and change: The elective Carnegie classification for community engagement. *Journal of Higher Education Outreach & Engagement, 24*(3), 105–114.

Shuib, M., & Azizan, S. N. (2017). University-community engagement via m-learning. In M. Shuib & K. Y. Lie (Eds.), *The role of the university with a focus on university-community engagement*. EPUB.

Simunich, B. (2017). Service learning in online courses. In C. Crosby & F. Brockmeier (Eds.), *Community engagement program implementation and teacher preparation for 21st century education.* IGI Global. doi:10.4018/978-1-5225-0871-7.ch009

Ward, E., Buglione, S., Giles, D. E., & Saltmarsh, J. (2013). The Carnegie classification for community engagement. In P. Benneworth (Ed.), *University engagement with socially excluded communities* (pp. 285–308). Springer. doi:10.1007/978-94-007-4875-0_15

Welch, M., & Saltmarsh, J. (2013). Best practices and infrastructures for campus centers of community engagement. In A. Hoy & M. Johnson (Eds.), *Deepening community engagement in higher education: Forging new pathways*. Palgrave MacMillian. doi:10.1057/9781137315984_14

Vuong, T., Hoyt, L., Newcomb-Rowe, A., & Carrier, C. (2017). Faculty perspectives rewards and incentives for community-engaged work. *International Journal of Community Research and Engagement, 10*, 249–264.

Zuiches, J. J., Cowling, E., Clark, J., Clayton, P., Helm, K., Henry, B., ... Warren, A. (2008). Attaining Carnegie's community engagement classification. *Change, 40*(1), 42–45. doi:10.3200/CHNG.40.1.42-45

Chapter 2
Ethics in Online Community Engagement Among Marginalized Rural Groups

Tshimangadzo Selina Mudau
https://orcid.org/0000-0002-6826-8316
University of KwaZulu-Natal, South Africa

ABSTRACT

This chapter presents the virtual community engagement within rural areas. The methodological implications of this chapter are drawn from empirical research conducted before and during the emergence of the COVID-19 pandemic in the rural areas of Makhado in South Africa. The chapter describes how virtual community engagement impacts communication issues such as social norms, culture, and respect for community leadership structures. Critical discourse analysis was employed to expose the problems of power, position, and dominance during virtual community engagement as a data generation method. Findings are based on technological and other contextual socio-cultural factors in rural communities. A mixture of both virtual and telephonic engagement is the most acceptable option to manage and minimize social injustice among marginalized groups. The chapter closes with recommendations on promoting the voices and rights of the marginalized or vulnerable groups when conducting virtual community engagement.

INTRODUCTION

This chapter aims to discuss the application of the objectives and issues of ethics in community engagement in rural areas. The methodological implications of this chapter are drawn from empirical research conducted among teenage mothers in the rural areas of Makhado in South Africa before and during the emergence of the COVID-19 pandemic. This chapter will discuss how virtual community engagement impacts the local social norms and values of communicating with the rural local leaders. Although local authorities are not the central unit of analysis, they influence studies since they are the gatekeepers who provide consent to conduct research studies. The emergence of the COVID-19 pandemic required

DOI: 10.4018/978-1-6684-5190-8.ch002

reconsidering how life, in general, was to be approached, including research (Bueddefeld et al., 2021). Similarly, adherence to the non-pharmaceutical measures of preventing the spread of the virus was followed to save lives. The chapter is inspired by the empirical findings of a study conducted before the COVID-19 lockdown in South African rural communities to advance possible ways of conducting virtual community engagement studies. Furthermore, the chapter describes how the application of virtual/online community engagement impacts communication issues of social norms, culture, rights, and respect of community leadership structures, and such issues affect the full participation of the marginalized groups. Data generation and analysis will be guided by van Dyk's (2008) critical discourse analysis. Critical discourse analysis is employed to expose power, position, and dominance issues during data generation and virtual community engagement among the marginalized group in rural areas (van Dyk, 2008). According to Fairclough (1989), language is embedded in social practices made of relations connected to culture, production, social consciousness, and identities that create social discourses and promote and legitimize some above the other. Those in socially legitimized positions use power to exclude, marginalize and oppress those considered weak and less qualified for community benefits. The chapter closes by providing recommendations relating to areas of further studies, promoting the voices and rights of the marginalized or vulnerable groups when conducting virtual community engagement.

BACKGROUND

The definition of community engagement is highly contested. Scholars have struggled to locate one common definition of the concept of community engagement (Nkoana & Dichaba, 2016). Some of the reasons include the complexities related to the concept of community and engagement, the context and application of the activities in community engagement, and scholarly preferences in the choice and use of the concepts (Nkoana & Dichaba, 2016; Mtawa et al., 2016). Bidandi et al. (2022) defined community engagement as activities performed by universities to support or uplift communities from their needs. In this study, community engagement is a community-based process and practice characterized by a reciprocal, collaborative commitment to solving a common problem and sharing and exchanging knowledge. Higher education institutions are tasked with the three-pronged responsibilities of teaching, learning, research, and community engagement in their daily operations (Tshishonga, 2020). In most universities, community engagement as the third leg of operation has been trailing behind, particularly on the engagement of communities (Mtawa et al., 2016; Bidandi et al., 2022).

The principles of community engagement require that researchers ensure the redress of the skewed relationship of master-servant with the communities (Nkoana & Dichaba, 2015; Tshishonga, 2020). The redress can be attained by ensuring mutual learning, respect, and understanding throughout the process. The iterative process of community engagement includes planning, identifying, and prioritizing points of action together, reflections, and critical evaluation of implemented action for further refinement as suitable to the context (Mtawa et al., 2016). In essence, what separates community engagement research from other research designs is its ability to promote the voices of the collaborated communities from start to finish. Communities experience the liberty and the democracy to choose or identify the problem, decide when and how the process will unfold, provide community resources, and catalyze agency and ownership of the solutions in an open public platform (Black et al., 2018). So, the emergence of virtual community engagement threatens such freedom and democracy from the community members.

African countries are lagging in embracing the Fourth Industrial Revolution (Corrigan, 2020). Scholars have recorded challenges contributing to such slow progress in technology among most if not all African countries. Corrigan (2020) reported that in 2018, 51.4% of the population were internet users, with the highest number of users in Europe (80.1%) and the Americas (74.6%), followed by the former Soviet Union states at 69.9, the Arab States at Asia (49.5%), and the Pacific (46.2%). On the contrary, only 26.3% of Africans are using internet resources. The challenges faced by Africa include human and material factors such as lack of skills and capacity to use and purchase gadgets, absence, inconsistent electricity supply, and poor and no internet coverage (Hove & Grobbelaar, 2020). The majority of African states, whether rural or urban, lack information technology infrastructure devices and skills which impact communication across all aspects of life (Mpungose, 2020; Mishra et al., 2020). A plethora of studies have shared various infrastructural developmental challenges, skills gaps, illiteracy, and socioeconomic inequalities faced by those living in rural areas (Shifa et al., 2021; Mishra et al., 2020). In the event of virtual research studies, those issues of sidelining and marginalization are perpetuated. In most instances, most of those marginalized are children, youth (especially girls and teenage mothers), and women in rural areas. Since universities were to ensure that all operations continued even during the COVID-19 lockdown, research studies across various fields were also expected to continue (Adebisi et al., 2021). In the process, ethical review boards had to ensure the protection of citizens during research studies by allowing studies to be conducted virtually, whether telephonically or on all forms of media, including Teams and Zoom meetings.

The global emergence of the coronavirus, popularly known as the COVID-19 pandemic, in December 2019 shook the world due to high infection and death rates (Sun et al., 2020; Cristofoletti & Pinheiro, 2022). Most governments and political heads of structures devised strategies to contain the virus's spread, including limiting and avoiding physical meetings, closing borders, and most operations in business, education, and travel, among others. In the education sector, schools and institutions of higher learning were closed, and academic calendars and other activities had to be realigned to stop the infection's spread. Institutions of higher education such as universities are expected to play triple roles, including teaching and learning, research, and community engagement (Mugabi, 2015; Adebisi et al., 2021).

Community engagement entails collaboration with communities, individuals, and organizations to address or solve a mutually identified problem. According to the tenets of community engagement, the known structures of powers should seek to redress political positions during engagement such that the voices of the marginalized, the powerless, the voiceless, and those usually spoken for are promoted and protected to ensure social justice and transformation (Musesengwa & Chimbari, 2017). In the process of engagement, the skewed relationship between the researcher and the researched is brought to equity in the collaboration process.

Marginalization is a condition or circumstance in one's life which prohibits full participation or involvement in social, political, economic, and related developmental activities that render the vulnerable to exclusions and inequalities (Wilson et al., 2018). Those marginalized are placed at the furthest point of life in various aspects. In most instances, the minority group within the wider community uses formal or informal privileges to exercise socially legitimated positions of power to exclude or deny the majority some community rights (Women United Nations, 2020). For instance, in the case of geographical distributions, it is acknowledged that the majority of the population is in the rural areas (Statistics South Africa [Stats SA], 2021; Valentine et al., 2021). However, these communities are usually placed inferior to those in the urban areas. A rural community is defined as a group of people sharing common values, standards, and norms in place located outside modern development and characterized by poor access to services and poor infrastructural development, such as electricity, water, and means of communication (Pillay & Luckan, 2019). Within the communities, boundaries such as economic position, education, race, ethnicity, age, sociopolitical, and masculinity are used consciously or unconsciously to create barriers to full community participation (Baah et al., 2019). For example, among Africans, women are less involved in the major decisions of their families and the general community even if they form the majority (Chauke, 2015). In the case of economic opportunities in sub-Saharan Africa, 31.3% of women are unemployed compared to 27.2% of males, even though women are the majority in the country and are more qualified than men (Stats SA, 2019). For instance, a study by van Rensburg (2021) confirmed that South African women are still silenced in academic boardrooms and denied promotional opportunities while men receive favours, even though women are more educated than men.

The Global Gender Gap Report (2021) measures gender gaps based on four sub index components: economic participation and opportunity, educational attainment, health and survival, and political empowerment. The global report revealed that most countries are doing well in educational attainment at 95%, health and survival at 96%, economic participation and opportunity at 58%, and political empowerment at 22%. According to the report, sub-Saharan Africa and the Middle East have educational gender gaps higher than 10%. In the case of political positions, the space is highly gendered in rural and urban areas. Women's representation in traditional local councils is still lagging (Chauke, 2015). Generally, various social-cultural issues pose gender divides, preventing some groups from participating in community activities. For example, a study by Aramunya and Cheben (2016) in Kenya revealed that some women could not join the community project because culturally married women are not allowed to mix with other men except their husbands.

In this chapter, van Dyk's critical discourse analysis (CDA) will be employed to expose issues of power, position, and dominance during data generation during virtual community engagement. CDA is concerned with exposing social discourses and the use of power, texts perpetuating inequality, dominance, and its related abuse (van Dyk, 2008). Those in the lower position of power and dominance are usually less assertive in expressing their views or perceptions when engaging with those regarded as in superior positions (Wodak, 2001). According to Fairclough (1989), society creates social hegemonic practices that are subtly accepted and become embedded in everyday social life so that those in socially legitimated positions continue to impose and abuse those regarded as subordinate. In using CDA, the researcher seeks to deconstruct social practices to co-create a contextual transformation that encourages full participation of all. For example, in Japan, Sybing (2022) used CDA to deconstruct authoritarian teacher-learner interaction to balance social and classroom communication, resulting in active participation. The social hegemonic practices include text and language, which may perpetuate dominance and power and inhibit the full participation of the marginalized during community engagement. Through CDA, the research plays a reflective role in assessing social discourse practices that impose social injustices such as dominance, marginalization, exclusions, and power abuse. In South Africa, Mahlomaholo (2009) used CDA among academic students to deconstruct their personal, educational, social and cultural views leading to transformative self-awareness to respect others and be accommodative of views of others. Visible and invisible acts of dominance and power are exposed and critically analyzed to redress the injustice suffered by the minor or marginalized group.

COMMUNITY ENTRY AND ETHICAL-RELATED ISSUES

Community engagement is an iterative process where the co-researchers generate and analyze data concurrently (Denzin & Lincoln, 2018). This chapter is based on empirical findings where participatory action research (PAR) was conducted guided by the community engagement approach. PAR is cyclic and popularly known for its messy, back-and-forth actions which involve plan-act-reflect-evaluate-replan (Kemmis et al., 2014). Community engagement and PAR entail narration in reporting the findings. Throughout the steps, the researcher must ensure data quality and employ all the necessary research tools in planning, implementation, reflection, and evaluations. The methodology in the chapter is presented from community entry which is the planning, the implications of virtual meetings on community engagement, and then the research trustworthiness.

In the case of the project in this chapter, researchers gained access to the community through the village chief who represents the authority structure in rural areas. However, the entry was done telephonically through the representative leader who reports to the chief. The difference during the COVID-19 lockdown was that community entry was done telephonically, which was not permitted before. The community leaders expressed willingness to have the project in the community but emphasized safety measures. This is because community leaders are essential gatekeepers with authority to permit the conduct of the study. Such leaders are important because they carry the influence to access community members who are potential participants and other community facilities. Although community leaders are regarded as traditional and located in rural areas, they are at a certain level of modernity and are receptive to using technological advancement as alternative means of communication addressing social issues. According to Black et al. (2018), respecting community members includes allowing them to choose the virtual method they are most comfortable with. Community leaders expressed the concern that they could not manage sophisticated platforms such as Zoom or Teams but minimal use of the telephonic interviews. The choice came with the warning that, as researchers, we should consider the socioeconomic status of the community members and that no one is intentionally excluded based on a lack of resources, but safer physical meetings are the best option. As researchers, we explored effective and efficient ways to conduct the study within the communication infrastructural context of the community. Telephonic communications included arranging meetings and feedback between the lead researcher and the liaising community member.

When studies are conducted through physical collaborations, some local norms or rites are followed in the presence of the chiefs. These include paying a small fee that serves as a token of respect. During the virtual community engagement, such token becomes "temporarily suspended" and is only paid whenever possible. As researchers, we learned after our first telephonic engagement that we have to arrange for the delivery of the token at the most convenient time. This showed that the local leaders were flexible and accommodative based on the conditions of the COVID-19 lockdown measures. This occurred because consultation and debriefing sessions wherein the importance of study and saving lives through social distancing were mutually understood and promoted. Additionally, community members and researchers must decide and clarify from the inception stage of the study how communication costs will be managed. Adhikari (2020) indicates that basic needs such as airtime and internet data should be identified and provided as compensation to maximize the project outcomes.

All research studies are guided by ethical principles aimed at protecting and advocating for the rights of the participants. In the case of community engagement, researchers are expected to redress positions and power (Machimana et al., 2021). Like any research study, rural leaders as gatekeepers are given respect starting from community entry and all issues are within their power or authority (Adhikari et al., 2020). Observing the ethical principle of research is a continuous process from the start of the proposal drafting to the last part of disseminating findings or results (Mudau, 2019; Machimana et al., 2021). In the case of community engagement, issues of social justice and observing and respecting cultural norms and values are the primary guiding principles that determine the future of the study

Implications of the COVID-19 Lockdown on Community Engagement

Generally, there is evidence of significant advancement and development in research approaches, methodologies, and perspectives across disciplines. Drawing from the eight moments of qualitative research, it is acknowledged that there has been an evolution of research across the decade (Denzin & Lincoln, 2018; Harvey, 1989). Several studies have proven a close relationship between social and general life development and transformation and research findings because one impacts and influences the other (O'Driscoll, 2018; Mudau, 2019; Ramusetheli, 2019; Adebisi et al., 2021). The COVID-19 lockdown brought the resurgence of the postmodern moment, the second moment of qualitative research, where researchers questioned ways of conducting research. In the case of community engagement during the COVID-19 lockdown periods, both the communities and stakeholders revisited their traditional collaboration methods. The social norms were critically analyzed and deconstructed to construct safer ways to conduct collaborative research in rural areas.

Paying homage was no longer done the conventional traditional way where females are expected to cover their head and kneel (Mudau, 2019), but the tone, pace during the telephonic conservations, and giving the chief more chance to speak were the determining factors. Conducting community engagement research studies during the COVID-19 lockdown robs the general community of the beauty of experiencing the dynamics of sharing and exchanging knowledge between the participants (locals) and the researchers. This is because, during collaboration meetings, researchers must ensure adherence to the principle of beneficence and minimize risks. For example, gatherings were at a certain point, limited to fifty people in one indoor venue or hundred or more in the open space, depending on the lockdown level (Jee, 2020). Large gatherings that attract as many as possible were then replaced by a few participants, mainly accessed through the preference of those who recruited them or complied with the inclusion criteria. More so, the general community cannot participate in community research findings dissemination which serves multiple purposes. Such purposes include an awareness of local strength, appreciation of social assets in skills, knowledge, materials, and capacity building that leads to maximum transformation.

Researchers should ensure that community engagement studies adhere to the bioethics principles of respect for persons, beneficence, and justice (Musesengwa & Chimbari, 2017). Amid the COVID-19 pandemic, extra safety measures had to be employed to protect the health of the participants and the general community members (Corrigan, 2020; Jee, 2020; Adebisi et al., 2021; Edwards, 2022). Social distancing limited the possibility of all members being part of the study. The opportunity to give feedback in the general community assembly becomes replaced by a few members directly involved in the research and selected community leaders. In the case of the COVID-19 pandemic, factors that minimize full community involvement and participation impact the dynamics of knowledge sharing and creation and social cohesion during the feedback sessions in the general community assembly. In particular, the involvement of the elderly in the research was highly impacted because extra precautions had to be employed to minimize the spread of the virus since they are the high-risk groups. Research collaborations are continued with the confidence that those available represent the interests of the community (Musesengwa & Chimbari, 2017).

DATA GENERATION AND ANALYSIS PROCESS

This chapter draws insights from the empirical findings of the study community engagement project conducted in 2017-2018 during a PhD project (Mudau, 2019) and a follow-up project in 2021 (in progress). The first project was conducted face-to-face, allowing the researcher to observe and address hegemonic social practices that sought to perpetuate social injustice, silence the marginalized voices, sideline their thought, and crush their rights. Such social discourses included refusing teenage mothers to be part of the research team, using verbal pressure through shouting, and constant staring as a way of silencing the speaker based on privileged power. This is because, based on the Vhavenda social norms, young people cannot express their views or thoughts in the presence of adults or elders (Ramusetheli, 2019). These hegemonic social practices legitimize some positions and powers, thus privileging one above the other (van Dijk, 2008). So, in conducting the virtual community engagement, the researcher had to reflect on the socially constructed texts and language (verbal and non-verbal) that may prevent full participation and expression of thoughts. As Wodak (2001) indicated, those oppressed and silenced may not know or have the ability to resist such practices.

Community engagement is an iterative process where the co-researchers generate and analyze data concurrently (Denzin & Lincoln, 2018). Research among marginalized groups entails or demands that the researcher critically engages in data generation methods that will promote inclusion, democracy, and social justice. Some reflective points of reference may include:

- What ICT tools are most available among the community members?
- What ICT knowledge do they possess?
- How best can data be generated to ensure and promote inclusivity and social justice?
- How can the voices and rights of marginalized groups such as children, youth, and women be represented?
- How can the researcher redress epistemic injustice?
- How can local social norms and values related to traditional chiefs and leaders be observed in the virtual platforms?
- Which infrastructure is available in the community, and how can the identified lack be compensated?

The researcher's position, especially among the vulnerable group, is always higher than the engaged groups because they are regarded as a knower and well-resourced concerning the researched group (Verma & Singh, 2015). According to Gaventa and Cornwall (2008), power is inherent in every human encounter, and those with knowledge are socially regarded as superior to others. Therefore, it is important that in employing CDA during data generation and analysis, the researcher should critically reflect and engage in critical dialogue to raise awareness and address elements of oppression, abuse, marginalization, inequality, and social injustice. The researcher also becomes the neutral sounding board to absorb and reflect on participants' identified elements of power. According to Tenorio, 2011, the analysis of discourses and texts during the data generation process is done to pin-out significant texts as warning signs of discomfort and inequality.

In virtual community engagement, there are elements of participation bias based on the fact that those without gadgets depend on the haves during virtual interviews. In the case of a community engagement with a demographically diverse group composed of young adults, the voices of youth and the vulnerable group will be silent or softer. Softer voices will always miss the notice of the researcher, whose role is to listen, negotiate, facilitate and promote the rights of all participants (Verma & Singh, 2015). As Karamunya and Cheben (2016) alluded, socio-cultural issues impact participant interaction, especially when there is a mixture of children, youth, and adults. In employing critical discourse analysis, the researcher becomes an active listener and an active data collection and analysis tool to respond and fill the missing gap.

Knowledge is power (Gaventa & Cornwall, 2008), Universities as knowledge creators and distributors are socially regarded as powerhouses and those associated with such structures through employment or affiliation are automatically positioned higher than others. The researcher carries a double sword of the engaged research scholar seeking to redress skewed power relations and wishes to give the community participants their leading role (role exchange) and also a negotiator who tactfully advocates for softer voices, especially of a young one who may be exposed to marginalization on the other side of the call.

So, in face-to-face data generation meetings, the researcher can observe the social discourses hidden in text and other social structures and manage them through critical dialogue.

- Limited access to ICT gadgets and knowledge impact virtual community engagement. This implies that the researcher must be considerate to enhance inclusion by using a blended approach, thus both virtual and physical meetings according to the context.
- The use of innovative methods to easy access communities does not remove the important structures such as the Chiefs, social norms and values, culture, and traditions of the local communities.
- Rural community leaders are adaptive to innovation and change. The leaders demonstrated the willingness to halt the spread of the virus by allowing virtual meetings and emphasis on regulation adherence.
- In community engagement, the researcher plays an advocacy role in promoting those who are regarded as inferior and voiceless.
- Trustworthiness in emancipatory approaches is not an end but a means thereof. This is because the researcher ensures critical reflection and dialogue to question and practices, seek contextual meaning for communicative mutual understanding

ENHANCING SOCIAL JUSTICE AND BENEFICENCE TO THE MARGINALIZED GROUP

Researchers are not only to ensure that studies adhere to the ethical principles which are outlined in the research proposals submitted for clearance to the institutional review boards (Keshtgar, Hania & Sharif, 2022) but also conscious of their instrumental role needed to promote humanity. Researchers must constantly enhance social justice to ensure the maximum benefit for participants during the study. Through continued reflections, the researcher should consider the human dynamics and contextual factors that may impact the engagement between the communities and the marginalized individuals or groups (Schreiber & Tomm-Bonde, 2015). Bearing the empirical findings of the study this researcher conducted in 2017-2018 (Mudau, 2019), wherein social injustices and marginalization were observed from the community entry, and throughout the study, the researcher found it challenging to conduct virtual community engagement. This is because of the socio-cultural hegemonic discourses embodied in text and talk from those regarded as "positionally legitimate" (adults and community leaders) to impose upon the marginalized teenage mothers. There are social discourses such as verbal and eye suctions which are locally understood and legitimized, and this is where the researcher will observe and play the advocacy role to enhance critical dialogue and raise awareness that will lead to the deconstruction of such texts and co-creation of socially owned inclusive and shared social spaces. Then in conducting virtual community engagement, the blended platform must be used.

RESEARCH TRUSTWORTHINESS

Qualitative research employs various measures to attain research quality. In CDA, data trustworthiness is ensured through accessibility and completeness (Mullet, 2018). Research is accessible when it is readable by its audience since the participants use their language within their locality. Completeness is when data collection produces no new information. In this study, continual researcher reflections minimized research bias by ensuring that preconceived knowledge and cultural factors do not influence data generations and analysis. As a researcher sharing the same culture with the participants, this researcher had to consistently reflect on the methodological and theoretical principles of research when participants were reluctant to take charge of the meeting when encouraged to do so. This was to enhance catalytic authenticity by redressing the skewed positions of legitimate institutions and actions (Mullet, 2018). It was important to promote the principles of social justice and enhance the capacity development of teenage mothers by promoting their voices and inclusion when leaders thought of excluding them. Kemmis et al. (2014) asserted that in action research, quality is attained when the study changes the lives of the participants.

As Mahlomaholo and Nkaone (2002) alluded, social research cannot be conducted as an end to itself, but it should promote social usefulness. In case, this study employed action research through community engagement to ensure contextual transformation, freedom and social justice when the research process, despite the COVID-19 restrictions, community norms relating to addressing local leaders, preventing the spread of the disease, and quality of findings are attained. Furthermore, data dependability was ensured through member-checking and critical dialogue to collaboratively understand the meaning of words, text, and observed body movements such as silence over the phone, and when young girls would look down during interactions in the meetings. To ensure freedom of self-expression in this study, adults and teenagers were integrated when readiness to do so was established. This researcher had to play both the facilitation and the observant role by constantly scanning through traces of dominance and silencing the voices of teenage mothers to position them at the centre of their space. According to Kemmis et al. (2014), this is the practice architecture when the researcher observes how different participants occupy spaces such as the language used, taking turns, and whose ideas are sanctioned or promoted. For example, in Mudau (2019) teenage mothers found it mostly easy to express their pain in the absence of parents. To enhance quality, the researcher played a mediator role to ensure a common communicative space for all participants by assessing readiness to integrate both groups after the necessary preparations and consultations of both groups.

RECOMMENDATIONS

Based on the presentations above, virtual community engagement is possible when researchers continuously reflect on community engagement principles. In short, researchers should ensure that:

- Continuous communication with those identified as the community representatives and liaising agents.
- Virtual meetings should continue to promote the voices and rights of marginalized or vulnerable groups.
- The risk of beneficence entails not only preventing the risk of infection during the pandemic but also ensuring prevention of limit to individual expression of experiences and the possibility of developing locally-owned solutions, capacity building, and transformation.
- There should be measures to ensure social justice in the representation of all classes, especially youth, women, and children. This entails that their thought, voices, and rights are well-represented.
- Communication resources such as airtime or data bundles and lack of ICT skills are not hindrances to full participation and inclusion of the marginalized groups, thus perpetuating epistemic injustice.
- Researchers should ensure continuous reflection and balance of safety so that research findings are trustworthy and certified credible by the participants as co-researchers. In this case, an environmental safety assessment should guide perfect timing for a blended approach so that physical meetings are strategically arranged.
- The interactive process with the community leaders during the engagement guides the researcher on the best possible option to conduct the study where the risk of beneficence is reflected upon. In this process, the researcher takes the lead to ensure the study is morally and ethically conducted as expected by the review boards.

CONCLUSION

This chapter presented that virtual community engagement was necessary and will still be necessary even during the pandemic as long as all collaborators uphold ethical measures. Whether conducted virtually or physically, community engagement remains a capacity-building tool that promotes and appreciates transformative civic education. The most important factor is that researchers should always reflect on the objectives and principles of community engagement for successful transformative community partnerships. Additionally, when collaborating with marginalized communities or groups through virtual community engagement, the researcher must always consider the contextual factors that need to be alleviated and corrected so that exclusion is not perpetuated. As O'Driscoll (2018) recommended, a one-size-fits-all approach cannot be acceptable in community engagement as various factors impact the study process. Critical emancipatory practices are recommended so that social justice is promoted.

REFERENCES

Adebisi, Y. A., Rabe, A., & Lucero-Prisno Iii, D. E. (2021). Risk communication and community engagement strategies for COVID-19 in 13 African countries. *Health Promotion Perspectives, 11*(2), 137–147. doi:10.34172/hpp.2021.18 PMID:34195037

Adhikari, B., Pell, C., & Cheah, P. Y. (2020). Community engagement and ethical global health research. *Global Bioethics, 31*(1), 1–12. doi:10.1080/11287462.2019.1703504 PMID:32002019

Baah, F. O., Teitelman, A. M., & Riegel, B. (2019). Marginalization: Conceptualizing patient vulnerabilities in the framework of social determinants of health— An integrative review. *Nursing Inquiry, 26*(1), e12268. doi:10.1111/nin.12268 PMID:30488635

Bidandi, F., Anthony, A. N., & Mukong, C. (2022). Collaboration and partnerships between South African higher education institutions and stakeholders: A case study of a post-apartheid university. *Discover Education, 1*(1), 1–14. doi:10.100744217-022-00001-2 PMID:35795019

Black, G. F., Davies, A., Iskander, D., & Chambers, M. (2018). Reflections on the ethics of participatory visual methods to engage communities in global health research. *Global Bioethics, 29*(1), 22–38. doi:10.1080/11287462.2017.1415722 PMID:29434532

Bueddefeld, J., Murphy, M., Ostrem, J., & Halpenny, E. (2021). Methodological bricolage and COVID-19: An illustration from innovative, novel, and adaptive environmental behavior change research. *Journal of Mixed Methods Research, 15*(3), 437–461. doi:10.1177/15586898211019496

Chauke, M. T. (2015). The role of women in traditional leadership with special reference to the Valoyi tribe. *Studies of Tribes and Tribals, 13*(1), 34–39. doi:10.1080/0972639X.2015.11886709

Corrigan, T. (2020). *Africa's ICT Infrastructure: Its present and prospects*. Academic Press.

Cristofoletti, E. C., & Pinheiro, R. (2022). Taking stock: The impacts of the COVID-19 pandemic on university-community engagement. *Industry and Higher Education*. doi:10.1177/09504222221119927

Denzin, N. K., & Lincoln, Y. S. (2018). *The SAGE handbook of qualitative research* (5th ed.). SAGE Publications.

Edwards, S. J., Silaigwana, B., Asogun, D., Mugwagwa, J., Ntoumi, F., Ansumana, R., Bardosh, K., & Ambe, J. (2022). An ethics of anthropology-informed community engagement with COVID-19 clinical trials in Africa. *Developing World Bioethics*, dewb.12367. doi:10.1111/dewb.12367 PMID:35944158

Fairclough, N. (1989). *Language and Power*. Longman.

Gaventa, J., & Cornwall, A. (2008). Power and knowledge. The Sage handbook of action research: Participative inquiry and practice, 2, 172-189.

Harvey, D. (1989). *The condition of postmodernity: An enquiry into the origins of cultural change*. Academic Press.

Hove, P., & Grobbelaar, S. S. (2020). Innovation for inclusive development: Mapping and auditing the use of ICTs in the South African primary education system. *South African Journal of Industrial Engineering, 31*(1), 47–64. doi:10.7166/31-1-2119

Jansen van Rensburg, S.K. (2021). Doing gender well: Women's perceptions on gender equality and career progression in the South African security industry. *SA Journal of Industrial Psychology/SA Tydskrif vir Bedryfsielkunde, 47*(0), a1815. doi:10.4102/sajip.v47i0.181

Jee, Y. (2020). WHO international health regulations emergency committee for the COVID-19 outbreak. *Epidemiology and Health, 42*, 42. doi:10.4178/epih.e2020013 PMID:32192278

Karamunya, J., & Cheben, P. (2016). Socio-cultural factors influencing community participation in community projects among the residents Inpokot South Sub-County, Kenya. *American Based Research Journal, 5*(11).

Kemmis, S., McTaggart, R., & Nixon, R. (2014). *The action research planner: Doing critical participatory action research*. Springer Science. doi:10.1007/978-981-4560-67-2

Keshtgar, A., Hania, M., & Sharif, M. O. (2022). Consent and parental responsibility: The past, the present and the future. *British Dental Journal, 232*(2), 115–119. doi:10.103841415-022-3877-7 PMID:35091615

Machimana, E. G., Sefotho, M. M., Ebersöhn, L., & Shultz, L. (2021). Higher education uses community engagement-partnership as a research space to build knowledge. *Educational Research for Policy and Practice, 20*(1), 45–62. doi:10.100710671-020-09266-6

Mahlomaholo, S. (2009). Critical emancipatory research and academic identity. *Africa Education Review, 6*(2), 224–237. doi:10.1080/18146620903274555

Mahlomaholo, S., & Nkoane, M. (2002). The case for emancipatory qualitative research on assessment of quality. *Education as Change, 6*(1), 69–84.

Mpungose, C. B. (2020). Emergent transition from face-to-face to online learning in a South African university in the context of the coronavirus pandemic. *Humanities and Social Sciences Communications, 7*(1), 1–9. doi:10.105741599-020-00603-x

Mtawa, N. N., Fongwa, S. N., & Wangenge-Ouma, G. (2016). The scholarship of university-community engagement: Interrogating Boyer's model. *International Journal of Educational Development, 49*, 126–133. doi:10.1016/j.ijedudev.2016.01.007

Mudau, T. S. (2019). *Enhancing self-regulation among teenage mothers: A university-community engagement approach* [Unpublished PhD thesis]. University of the Free State.

Mullet, D. R. (2018). A general critical discourse analysis framework for educational research. *Journal of Advanced Academics, 29*(2), 116–142. doi:10.1177/1932202X18758260

Musesengwa, R., & Chimbari, M. J. (2017). Experiences of community members and researchers on community engagement in an Ecohealth project in South Africa and Zimbabwe. *BMC Medical Ethics, 18*(1), 1–15. doi:10.118612910-017-0236-3 PMID:29237440

Nkoana, E. M., & Dichaba, M. M. (2016). Are we heading in the right direction? Towards excellence in educational practices. *South Africa International Conference on Education 2016, 15*(7), 213-227.

O'Driscoll, D. (2018). *Transformation of marginalised through inclusion*. University of Manchester.

Pillay, N., & Luckan, Y. (2019). The rural school as a place for sustainable community development. *Sustainable Urbanisation through Research, Innovation and Partnerships*, 342-353.

Ramusetheli, M. D. (2019). *The relevance of nyambedzano as an effective process for promoting morality among the youth* [Unpublished PhD thesis]. University of Venda.

Republic of South Africa. (2021). *Department of Statistics South Africa. Marginalized Groups Indicator Report, 2019*. Statistics South Africa.

Schreiber, R., & Tomm-Bonde, L. (2015). Ubuntu and constructivist grounded theory: An African methodology package. *Journal of Research in Nursing, 20*(8), 655–664. doi:10.1177/1744987115619207

Statistics South Africa. (2019). *Quarterly Labour Force Survey*. Retrieved from https:// www.statssa.gov.za/publications/P0211/P02114thQuarter2019.pdf

Sun, L., Tang, Y., & Zuo, W. (2020). Coronavirus pushes education online. *Nature Materials, 19*(6), 687–687. doi:10.103841563-020-0678-8 PMID:32341513

Sybing, R. (2022). Dialogic validation: a discourse analysis for conceptual development within dialogic classroom interaction. *Classroom Discourse*, 1-17.

Tenorio, E. H. (2011). Critical discourse analysis: An overview. *Nordic Journal of English Studies, 10*(1), 183–210. doi:10.35360/njes.247

Tshishonga, N. S. (2020). Forging University social responsibility through community engagement in higher education. In S. Chhabra & M. Kumar (Eds.), *Civic engagement frameworks and strategic leadership practices for organization development* (pp. 96–115). IGI Global. doi:10.4018/978-1-7998-2372-8.ch005

Valentine, A., Gemin, B., Vashaw, L., Watson, J., Harrington, C., & LeBlanc, E. (2021). Digital learning in rural K–12 settings: A survey of challenges and progress in the United States. *Research Anthology on Developing Effective Online Learning Courses*, 1987-2019.

van Dyk, T. A. (2008). *Discourse and context*. Cambridge University Press.

Verma, R. B. S., & Singh, A. P. (2015). The abstract book of 3rd Indian social work congress. In *Community engagement, social responsibility and social work profession*. Rapid Book Service.

Wilson, D., Heaslip, V., & Jackson, D. (2018). Improving equity and cultural responsiveness with marginalized communities: Understanding competing worldviews. *Journal of Clinical Nursing, 27*(19-20), 3810–3819. doi:10.1111/jocn.14546 PMID:29869819

Wodak, R. (2001). The discourse-historical approach. In R. Wodak & M. Meyer (Eds.), *Methods of critical discourse analysis* (pp. 63–94). SAGE Publications.

Women, U. N. (2020). Commission on the status of women. *Fiftieth Session, 27*.

KEY TERMS AND DEFINTIONS

Community Engagement: Is a community-based process and practice characterized by a reciprocal, collaborative commitment to solving a common problem and sharing and exchanging knowledge.

Marginalized Group: Is a group of people experiencing a condition or circumstance in life which prohibits their full participation or involvement in social, political, economic, and related developmental activities, leading to exclusions and inequalities.

Rural Community: Is a group of people sharing common values, standards, and norms in place located outside modern development and characterized by poor access to services, and poor infrastructural development, such as electricity, water, and means of communication.

Rural Leader: Is a recognized traditional or civic leader within the rural community.

Virtual Community Engagement: Is a community engagement process that is undertaken through any social media platform, including the use of telephonic calls.

Chapter 3
Analyzing the Hong Kong Philharmonic Orchestra's Facebook Community Engagement With the Honeycomb Model

Suming Deng
The University of Hong Kong, Hong Kong

Dickson K. W. Chiu
https://orcid.org/0000-0002-7926-9568
The University of Hong Kong, Hong Kong

ABSTRACT

Online social media has gradually become a trend. Many companies, brands, and organizations attach great importance to social media for community engagement, aiming for marketing, reputation, etc. This study analyzes the effectiveness of the Hong Kong Philharmonic's (HKPhil) Facebook social media with the seven aspects of the honeycomb model. Through quantitative and qualitative analysis of the problems encountered, the authors found that the video posts on Facebook are more active than posts without videos, and the comments are generally positive. HKPhil's engagement on Facebook is good, with about 1.9% average participation rate. This research help understand whether Facebook engagement is effective and can reflect the audience's expectations, suggesting improvement strategies of social media to attract and engage more audiences and discover potential users.

INTRODUCTION

With the continuous development of social media and mobile technologies, user information behavior has changed because of the Internet (Yu et al., 2021; Wang et al., 2016; Lam et al., 2019; Lau et al., 2020; Fong et al., 2020; Dong et al., 2021). Brands and organizations need to establish good social networks

DOI: 10.4018/978-1-6684-5190-8.ch003

and connect with users, as social media can help promote brands, turn potential users into formal users, and engage them (Dhir & Torsheim, 2016; Swani, Milne, Brown, Assaf, & Donthu, 2017; Wang et al., 2022; Jiang et al., 2022).

According to the official website of the Hong Kong Philharmonic Orchestra (HKPhil), it is one of the most influential symphony orchestras in Asia, attracting more than 200,000 music lovers a year, with musicians from all over the world (Wang et al., 2022). Its performances have also been well received worldwide and won the Gramophone Orchestra of the Year Award. HKPhil has opened a school of education, which has attracted many students and teachers to participate. HKPhil holds concerts in schools, hospitals, and other public communities every year to outreach the public to directly serve users and bring musical happiness to the audience (Yu, Chiu, & Chan, 2022).

The predecessor of the HKPhil was the Sino-British Symphony Orchestra, which was renamed HKPhil in 1957 and officially became a professional orchestra in 1974 (Yu, Chiu & Chan, 2022). The main sponsors of HKPhil are the Swire Group and the Jockey Club Foundation. Since 2006, Swire has helped the Hong Kong Orchestra promote excellent art, popularize Chinese and foreign music, and allow the people to enhance the aesthetics of music and enhance Hong Kong's reputation worldwide. The Hong Kong Economic and Trade Office also supports the Hong Kong Orchestra. With funding from the government, patrons, and foundations, HKPhil launches new popular programs, music education courses, and community programs every year. In 2015, HKPhil completed a grand European tour covering London, Zurich, Eindhoven, Birmingham, Berlin, and Amsterdam. In 2017, HKPhil also toured Asia and Oceania, performing in Seoul, Daban, Singapore, Melbourne, Sydney, and other places. In 1978, HKPhil recorded its first record with Philips Records. In 1997, the orchestra was selected for the album "Heaven, Earth and Earth: Symphony 1997" to celebrate the return of Hong Kong. HKPhil closely cooperated with the Hong Kong Opera House, art festivals, dance troupes, and other institutions. The educational activities of HKPhil are diverse and open. Excellent curriculum design and teachers attract more than 40,000 students and teachers to participate each year. The Hong Kong Philharmonic Academy collaborates with HKPhil and the Hong Kong Academy of Music.

Social media tools are essential for cultural organizations to improve their reputation and attract new audiences (HKPhil, 2021; Zhang et al., 2022). HKPhil uses seven social media tools: Facebook, Twitter, Bilibili, YouTube, Instagram, WeChat, and Weibo. Facebook has its largest number of fans, with 55,000 subscribers as of December 2021, which aligns with the fact that Facebook has the largest social media user base in Hong Kong (Chan et al., 2020; Lam et al., 2022). The second is Instagram (10,000), followed by YouTube (9,680), Twitter (2,273), and Weibo (4,976). The Bilibili video site has only 1,364 fans and subscriptions. Among the social media tools, HKPhil has relatively more fans on international social media, such as Facebook, Twitter, and Instagram.

As affected by COIVD-19, social media users have increased sharply, causing 40% of Internet users to spend more time using social media (GlobalWeblndex, 2020; Huang et al., 2021; 2022). For non-government organizations (NGOs), it is both a challenge and an opportunity. COVID-19 has made it impossible for international orchestras like HKPhil to hold worldwide concerts online (Yu, Chiu, & Chan, 2022). Many performance venues are locked down and do not allow crowd gatherings, making it difficult for HKPhil to perform concert performances (Wang et al., 2022; Yu, Lam, & Chiu, 2022). This study mainly collected some recent social media data from HKPhil's Facebook pages in November and December 2021 and uses the seven components of the honeycomb model (Kietzmann, Hermkens, McCarthy, and Silvestre, 2011) to analyze its recent effectiveness in the key aspects of community engagement: sharing, presence, relationships, reputation, identity, groups, conversation. Regarding these seven aspects, this study points out the problems existing in HKPhil's Facebook and suggests strategies and solutions for overcoming them. Based on the honeycomb model, this study raises the following questions.

RQ1, What are the content characteristics shared by HKPhil on Facebook?

RQ2. How does HKPhil engage its user community on Facebook and meet its expectations?

RQ3. What needs to be improved about running HKPhil's Facebook?

LITERATURE REVIEW

Overview of Social Media

Social media is a tool platform. It allows users to share information, including images, videos, links, websites, and so on (Tess, 2013). Social media at this stage include blogs, forums, vlogs, Facebook, Instagram, Twitter, Wechat, and so on (Meshi, Tamir, & Heekeren, 2015). Social media communication can create hotspots and topics for people. The technical support of the social network is Web 2.0 (Patrut & Patrut, 2013), and social media is developed based on the support of the masses and technology. According to the analysis of the July 2020 Global Digital Insights Report released by 'We Are Social' and 'Hootsuite' (Kemp, 2020), there are about 4 billion social media users worldwide, accounting for about 50% of the global population, which is an increase compared to 2019 nearly ten percent (Global Statshot, 2020). The number of social media users is snowballing. From 2019 to 2020, an average of more than 1 million people signed up on social media every day. Since 2019, the number of new social media users in China has exceeded 376 million. Ninety-nine percent of users use social media through mobile phones, but the styles and purposes of people using social platforms are diverse, such as daily communication, business information, learning, and entertainment (Dong et al., 2021; Lam et al., 2019).

Social Media Data

Social media data refers to information from social networks. Such information has rich quantitative data (Lam et al., 2019), such as the number of likes, the number of favorites, the number of comments, the number of web page clicks, the number of new fans, the viewing time, etc. Further, it contains qualitative data, such as comment sharing and profile information. Social media can improve interpersonal and mass communication (Dong et al., 2021; Au et al., 2021). Thus, marketers can use these social media data to attract customers and make good social media strategies by analyzing these data, as there are many social media data from famous Facebook, Twitter, WeChat, Weibo, Instagram, etc. (Lam et al., 2019; Chan et al., 2020). For example, marketers can focus on service and product scopes by identifying users of specific tags, greatly saving costs and improving communication efficiency (Cheng et al., 2020; Ni et al., 2021). Marketers can also accurately advertise on social media based on location, gender, and age to attract target groups. Further, social media algorithms can push relevant content published by organizations to some users based on tags and usage behaviors (Au et al., 2021).

COVID-19 Increases the Number of Users Using Social Media

Studies have shown that COVID-19 has caused 40% of Internet users to spend more time using social media (Huang et al., 2021; 2022; Ho et al., 2022). In the first quarter of 2022, the average daily time of social media users is about two hours (Chaffey, 2022). With the spread of COVID-19, the use of social media by Internet users worldwide has become longer. Social media allows people to stay connected in a state of separation. In the epidemic context, social media has connected products, services, and users worldwide, creating many possibilities in the market. During COVID-19, many companies cannot allow employees to return to work offline to reduce the risk of infection. Therefore, workers need to work online, which requires social media to ensure effective work communication, so the time of using social media increases. Second, the closure of public facilities has prevented people from enjoying entertainment in public environments during the epidemic (Yu, Lam, & Chiu, 2022). For example, public brick-and-mortar stores such as cinemas, playgrounds, cafes, and restaurants have been closed due to the epidemic, and people can only convert offline entertainment activities to online. Therefore, using social media can provide people's needs for online entertainment activities, and users have increased their use of social media during the epidemic (Wang et al., 2022; Yu, Chiu, & Chan, 2022). For example, online social media-based entertainment activities such as live broadcasts, short video sharing, and delivery of goods have become frequent during the epidemic.

Comparison of Traditional Media and Social Media

Traditional media is based on professionally generated content (PGC) and regularly releases information to the public (Bruhn, 2012), including newspapers, television, radio, newspapers, magazines, etc. In contrast, the production model of social media is based on user-generated content (UGC). With the gradual increase in the user number of social networking sites, content production for social media marketing is gradually shifting from UGC to PGC (Mangold & Faulds, 2009). Key differences between social media and traditional media include the following.

1. Different professional requirements: As the professional threshold for entering traditional media is relatively high, but the threshold for social media is relatively low, social media is more inclined to interface design (Ho, Chan, & Chiu, 2022; Au, Ho, & Chiu, 2021).
2. Timeliness is not the same: The production time of traditional media is usually longer than social media because social media content tends to be shorter and general expectations (Han, 2020).
3. Once published in traditional media, it is difficult to edit and correct, but social media can be edited in time. Further, conventional media editing involves higher human resources, material resources, and time costs (Han, 2020).
4. Traditional media are monopolized by governments and big media companies (Rajendran, 2014), but all users can easily post content on social media (Au et al., 2021).

Social Media and Community Engagement

Social media participation can help brands measure the effectiveness of social media activities (Wang et al., 2022). It shows the customer's reaction to the brand to a certain extent. If the customer engagement on social media is not high, the customer is not interested in the brand. If social media involvement is high, this situation can help increase brand image and reputation (EIU, 2007; Wang et al., 2022). Social media participation also affects potential customers as high participation allows more potential customers to join the brand community discussion while transforming potential customers into formal customers. For example, if a potential user sees a lot of comments on a post made by a brand, the potential user will become interested and enthusiastic and stay on the brand page for a longer time.

Many indicators can reflect customer engagement (Wirtz et al., 2010). For example, the Bilibili video website has many likes, favorites, forwards, coins, comments, and other data, reflecting participation. Facebook's engagement can be measured through likes, reactions, shares, and clicks on signs. Facebook's participation rate refers to the number of participating post actions (likes, sharing, etc.) divided by the number of brand followers. Different social media calculate the degree of engagement based on their individual characteristics. Mari Smith, a social media expert, said that the average social media participation rate on Facebook is only 2%, while SEO authority Michael Leander (2014) noted that more than 1% of Facebook content should be considered good. Statista's Facebook engagement data shows that the average follower engagement rate in the fourth quarter of 2020 is only 0.19%. According to Scrunch, a reasonable Twitter engagement rate is considered 0.02% to 0.09% (Mee, 2019; Garcia-Rivera, 2022). The average participation rate in 2020 is 0.045%. Instagram is a platform with a relatively high participation rate. A reasonable Instagram engagement rate is between 1% and 5%. Instagram engagement depends on the industry, the size or type of audience, and the content style. HootSuite said that the average Instagram engagement rate in 2020 was 4.59%. A study by Statista in July 2020 found that the average participation rate of TikTok Internet celebrities in the United States is close to 18%. The average participation rate of ordinary users on TikTok is between 3% and 9%.

METHODOLOGY

This research includes quantitative and qualitative data from HKPhil's official Facebook account in November and December 2021 to reflect the current local lockdown situation (Yu, Lam, & Chiu, 2022; Meng et al., 2022). There are 30 qualitative data. Attributes are the number of likes, shares, comments, and whether the post is in the form of a video (0 means no, 1 means yes). The qualitative data is the content of the comments below the post. These public data are used for this study, as shown in Table 1 and Figure 1.

Figure 1. Qualitative data

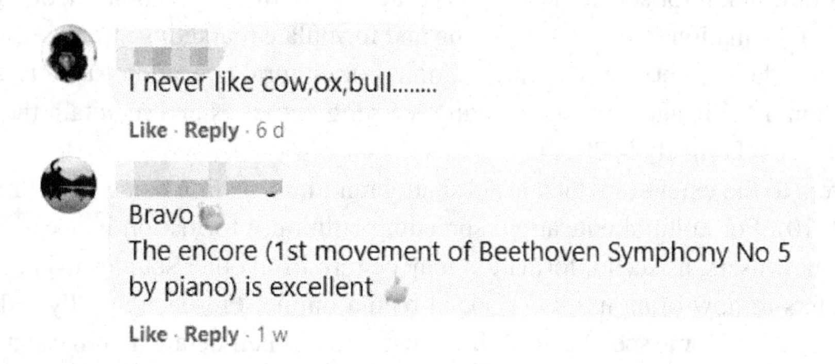

Table 1. Table of quantitative data

Post date	likes	comments	shares	video
18-Dec	197	3	12	0
17-Dec	41	1	3	1
17-Dec	38	1	13	1
21-Dec	18	0	0	0
14-Dec	7	0	0	0
12-Dec	488	12	62	1
11-Dec	301	3	19	0
8-Dec	270	21	35	1
8-Dec	12	0	0	0
7-Dec	179	3	4	0
6-Dec	98	0	5	0
30-Nov	207	6	103	0
27-Nov	223	5	22	0
25-Nov	88	1	12	1
24-Nov	185	2	13	1
24-Nov	160	0	29	0
18-Nov	66	6	22	1
17-Nov	112	2	10	0
16-Nov	193	9	15	0
14-Nov	13000	42	36	1
13-Nov	12000	436	370	1
13-Nov	396	3	52	1
12-Nov	578	11	15	1
12-Nov	238	1	51	0
10-Nov	66	0	8	1
8-Nov	375	9	65	0
6-Nov	552	16	33	0
5-Nov	12	0	0	0
5-Nov	9	0	0	0
3-Nov	106	2	5	0

As the honeycomb model (Kietzmann et al., 2011) analyzes seven major components of social media to evaluate the effectiveness of social media engagement, this research analyzes the following seven aspects of Facebook:

- Identity refers to the degree to which users or organizations can display or disclose their information (Kietzmann et al., 2011), such as location, anniversaries, and contact information. As identity is the core information for social media, it is at the core of the honeycomb model. Brands can use this personal information to target advertising and formulate marketing strategies, while users can search for their desired services and information. For cultural education, such user identities create social capital to engage learners in future learning activities and maintain the community of cultural practices (Fong et al., 2020).
- Sharing refers to the extent to which users share brand information using social media (Hoffman & Fodor, 2010). For cultural education, spreading cultural information is necessary to draw the attention of new users, as such information may be rare from other sources (Mak et al., 2022).
- Presence refers to how often users use social media online. Presence usually reflects whether a user is online or not. Some social media allows users to set their detailed online status. (Kietzmann

et al., 2011). For existing users, sharing useful cultural and learning information help maintain their identity and presence through useful or interesting content and conversations (Fong et al., 2020; Cheng et al., 2020; Lam et al., 2022).

- Relationships refer to the degree to which users are connected in social media activities, which can be cultivated through interaction and guidance. (Kietzmann et al., 2011). Good relationships are essential to maintain the community of practice and engage learners (Lei et al., 2021; Lam et al., 2022).
- Groups refer to the degree to which social media allows users to form a group or community. Groups allow like-minded users to gather for comment posting on brand topics and start discussions around brand content. (Kietzmann et al., 2011). For cultural learning, such groups refer to communities of practice and small interest groups for special related topics, which have shown to be effective aids for learning (Lei et al., 2021).
- Reputation refers to the background and reputation of social media users. Brands hope to keep high-quality users in their use of social media. Users who are often active under brand posts are favored, e.g., high-quality user comments, likes, and reposts. They may also create topics independently, discover some brand hotspots on their own, and then promote the brand's influence. (Kietzmann et al., 2011). For cultural learning, learners are usually keen on watching or learning from reputable grandmasters and performing organizations (Lei et al., 2021; Mak et al., 2022; Lo et al., 2019).
- Conversation refers to the extent to which users communicate and discuss with each other on social media. Conversations can be system messages, instant chats, videos, or group chats. For cultural learning, such communication helps the user community establish a platform to share their opinions and ideas, which is essential for the community of practice and solves learners' problems through mutual help (Lei et al., 2021). Brand organizers and educators can also keep members of the group community active and understand their needs through replying and posting, creating discussion topics, etc. (Kietzmann et al., 2011).

RESULTS AND ANALYSIS

Identity

HKPhil revealed its identity information on Facebook with an authentic profile for this public figure, providing contact information, such as telephone, email, and website. The category is performing arts. For page transparency, HKPhil's first post was published on April 6, 2009, as seen on Facebook, with multiple social media managers. Advertisements from others are not allowed on this page to better protect HKPhil's identity. On the about page of Facebook, users can link to the official website page to learn more about HKPhil. Facebook shows how many followers it has and which people it has followed. For example, renowned followers include Lang Lang, BBC music magazine, and New York Symphony Orchestra. This strong identity can engage learners in future learning activities and maintain the community of cultural practices

Relationships

HKPhil has 55k followers and has followed 43 users. For privacy reasons, Facebook does not allow users to see all followers' identity information. However, all 43 individuals followed by HKPhil can be

Figure 2. Times of likes, comments, and shares in the monitored period

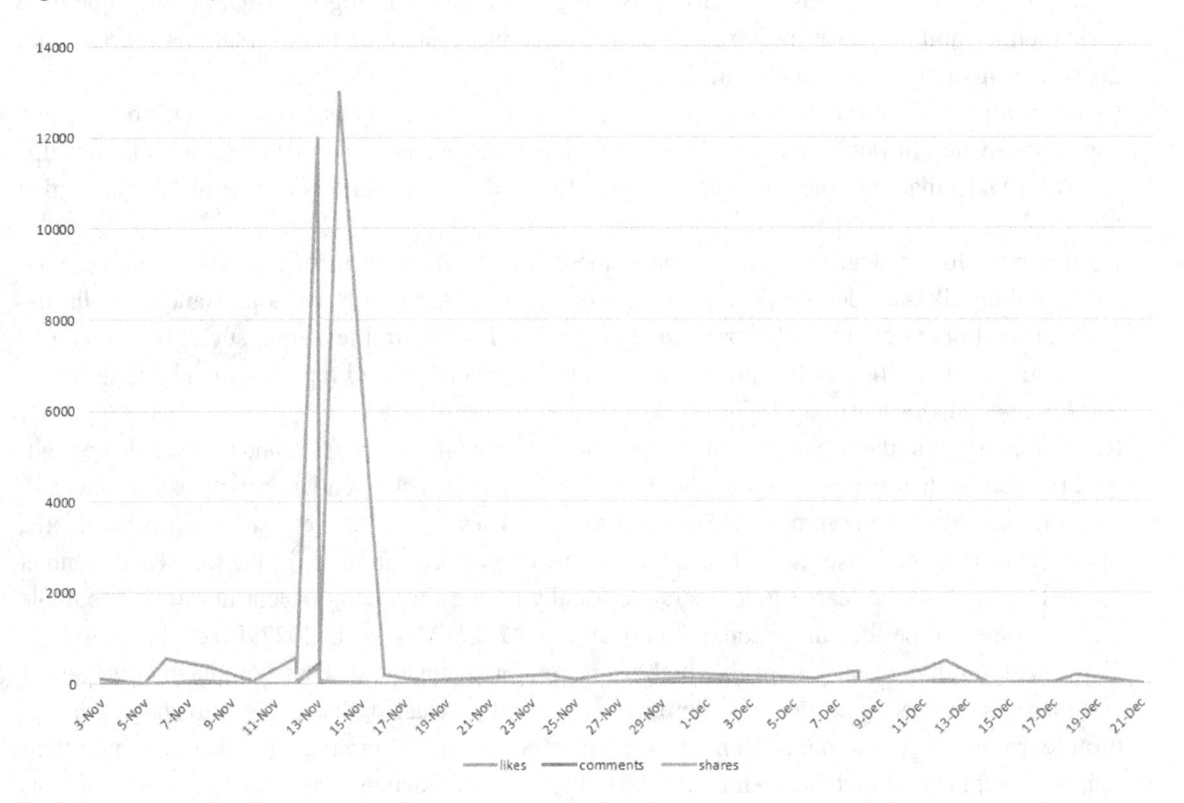

accessed by clicking on them, among which 36 are music-related. Among these users, only two HKPhil followers also follow HKPhil forming a mutual-follow relationship, Lang Lang and Lio Kuokman. They are musicians who have worked with HKPhil and thus have a strong relationship.

Sharing

HKPhil shared 30 posts during the two months of the monitoring period, which are more frequent than other social media platforms. These posts were music-related, including pictures, videos, text, and links. Users also shared the comments under the post. Almost all comments were positive, reflecting these posts satisfied and engaged members. Because of Christmas in December, there were some musical performances related to Christmas, including concert information, interviews with performers, recordings of concerts and performances, music awards, and related advertisements. A total of 1014 posts have been shared (see Figure 2). Among them, advertisement posts are the least, and videos of concert performances are the most. The average participation rate of fans of HKPhil is 1.9%, using the metrics of the total number of likes, comments, and shares in the two months divided by the number of followers. Michael Leander (2014) noted that Facebook's participation rate of over 1% is good, reflecting the fans of HKPhil enjoyed good participation in November and December. Regarding sharing, the Facebook algorithm spreads the content well (Wong et al., 2022).

Reputation

Because of the privacy protection of followers on Facebook, the research cannot investigate the information background of the followers of HKPhil. The reputation of the Hong Kong Philharmonic can be reflected by its number of fans on Facebook (55k) and overwhelmingly positive comments. However, there is still a significant gap in the number of fans compared with famous musicians such as pianist Lang Lang (580k).

Conversation

There are three ways of communication on Facebook: private messages, comments, and group discussions. HKPhil sometimes communicates with other users by leaving comments under their posts.

Figure 3. Screenshot of comments of 'Hong Kong Music, Starry Night, Symphony'

During the monitoring period, the audience generally hoped that the performance of HKPhil could be further improved, though they were satisfied with the performance and concert (see Figure 3). Most comments included positive words such as beautiful, bravo, thank, good, amazing, and so on. Figure 3 shows a screenshot of the comments from the post of 'Hong Kong Music, Starry Night, Symphony' on November 13, 2021.

Groups

Searching for groups of HKPhil yielded no result, indicating that HKPhil might not desire to use this mechanism to connect people having similar interests, themes, relationships, and identities. HKPhil should build more brand-related groups for more intimate music-related discussions and feedback and let fans interact with the content through group discussions online. There is also a possibility that hidden social communities were not discovered.

Presence

The Facebook of HKPhil states the online contact time: Monday to Friday, 9:00 am to 5:45 pm (12:30 noon to 1:30 pm is lunch break), which may not be adequate as users usually deal with entertainment outside the office hours. According to the presence of the post, HKPhil was currently preparing for a Christmas show during the period monitored.

SOLUTIONS AND SUGGESTIONS

Enhancing User Groups

Currently, there is no open group related to HKPhil on Facebook, resulting in a performance or concert being insufficiently discussed. The discussions related to the posts are quite scarce. In addition to responding to comments below posts, HKPhil can establish community groups with different themes and tags, such as a Baroque style, classical music style, animation style, pop style, and so on. Comments and discussions on posts are basically around the theme so that music teachers, learners, volunteers, and other fans can form a community of practice with HKPhil (Lei et al., 2021; Mak et al., 2022). Community groups can gather users with similar interests to make discussions more relevant, interesting, and engaging (Chan et al., 2020; Tse et al., 2022), especially for learners and teachers (Lei et al., 2021). HKPhil can place more precise performance advertisements for different groups to increase the exposure of the performance to different fan segments, and activities like master classes.

Posting More Music Content

HKPhil did not provide enough performance content, which is also not quite timely. Among 30 posts in two months, only 11 videos were posted, and four videos were of the same event about the Hong Kong Music Starry Night Symphony. Because the feature of Facebook is to arouse the interest of viewers or other users through pictures and videos, if there are too many plain-text advertisements and information, users may not want to browse or share (Cheng et al., 2020; Chan et al., 2020)

Thus, HKPhil can post a series of preview videos or short pieces on Facebook after every performance. If performances include live broadcasts, related promotions can be posted on Facebook for advertisement and later for archiving and educational purposes (Sun et al., 2022; Mak et al., 2022). If it involves privacy or copyright restrictions of the performance content, HKPhil can extract a short piece and place it on Facebook with the musicians' consent. Such content is especially useful for teachers as demonstration pieces to students (Lei et al., 2021). On HKPhil's Facebook page, data showed that if a post had a video, the number of comments, likes, and reposts increased. The total number of likes for video-based posts was about seven times those without videos, comments about nine times, and sharing about twice. Therefore, HKPhil should post more videos to increase brand exposure and engage users.

Increasing Responsiveness

HKPhil's response speed to user comments is not timely. If there are too many Facebook private messages, HKPhil can customize the private message auto-reply function or keyword responses to impress users. For example, custom replies can be accompanied by some holiday blessings near Christmas and other festivals to increase the users' cozy feelings. Responses based on specific keywords in the content of the private messages can help answer frequently asked questions. For example, if a user asks about the performance time of a particular musician, official keyword responses can provide answers immediately based on the musician's name, performance name, and other indexes.

Facebook's messages cannot rely solely on smart replies or artificial intelligence because the content of automatic responses is relatively rigid, and manual customer service also needs to receive customers. Although smart answers can deal with some users' problems, it is best to stipulate that every time a user sends a message, customer service staff should be reminded to solve issues that smart replies cannot solve promptly. HKPhil should hire multiple customer services and recruit volunteers and interns with quicker responses instead of just robot functions (Sun et al., 2022; Lin et al., 2022; Tsang & Chiu, 2022). As educating the newer generation can create social capital for sustainable development, HKPhil should particularly arrange for staff or volunteers to respond to questions related to music teaching and learning (Yu, Chiu, & Chan, 2022).

Following Other Musicians

HKPhil is following too few fans, and these posts are infrequent. The administrators of HKPhil can check out the posts from other musician accounts on Facebook, who may post more professional and interesting comments. If other musicians or organizations can interact with HKPhil, they can exchange and share posts and comments to form a larger community, bridging different audience groups and enriching the content. The audience, especially music teachers and learners, will be more excited to see this kind of linkage and exchange and may propose collaborations for more interesting future programs. Different audiences have diverse tastes and preferences, and they may like multiple musicians or performances. If preferences are similar, HKPhil can attract new fans and learners from other musicians and vice versa, resulting in a win-win situation (Yue, Chiu, and Chan, 2022).

Fostering More Discussions

Although an average participation rate of 1% on Facebook can be considered good fan participation, only 1.9% is not competitive and may reflect a relatively low engagement. HKPhil may prepare a small budget for fans to offer gifts, souvenirs, and tickets to award some high-quality comments. Some prizes can be granted to attract more participation. However, creating hot discussion topics, such as hot spots for concerts, performances, and musicians, is still essential to foster attention and discussions. Posting diverse and entertaining content such as musicians' lives, cultural lifestyles, environment protection, and funny music knowledge can make fans and learners feel closer and more freely express their thoughts, thus engaging them in the community (Cheng et al., 2020; Chung et al., 2020; He et al., 2022).

CONCLUSION

This study analyzed how HKPhil uses Facebook through the honeycomb model and discussed problems and solutions. Under COVID-19, online social media further developed. At this challenging time of COVID, HKPhil should use social media better and more, such as Facebook, which Hongkongers and music lovers worldwide favor. Regarding the seven aspects of the honeycomb model, this study has found that most Facebook comments were positive, and followers' participation rate was still satisfactory. However, few musicians or music organizations have followed HKPhil, there were few discussions and no discussion groups, and HKPhil's replies were relatively slow. There were relatively few video posts that could effectively engage their audience. Thus, HKPhil should interact more with followers, fellow musicians, and organizations, post more video posts, and create more interesting topics to increase fan activities and engagement. If HKPhil can use the recommendations of this study on Facebook, HKPhil may better understand the expectations and preferences of their fan community and engage them in their future development direction. Further, engaging teachers and learners through measures suggested in this study is crucial for the community's sustainable development.

As for limitations, the amount of data is small, though it better reflects the current COVID-19 situation and festive event promotion and interactions. We plan to expand the coverage, but the difficulty remains with HKPhil's relatively low general usage. Next, we would analyze and compare other orchestras' social media pages. On the other hand, we are interested in studying the social media of libraries, museums, and archives through similar methodologies for their promotion and user engagement.

REFERENCES

Au, C. H., Ho, K. K. W., & Chiu, D. K. W. (2021). The role of online misinformation and fake news in ideological polarization: Barriers, catalysts, and implication. *Information Systems Frontiers*. Advance online publication. doi:10.100710796-021-10133-9

Bruhn, M., Schoenmueller, V., & Schäfer, D. B. (2012). Are social media replacing traditional media in terms of brand equity creation? *Management Research Review*, *35*(9), 770–790. doi:10.1108/01409171211255948

Chaffey, D. (2022). *Global social media statistics research summary 2022*. https://www.smartinsights.com/social-media-marketing/social-media-strategy/new-global-social-media-research/

Chan, M. M. W., & Chiu, D. K. W. (2022). Alert Driven Customer Relationship Management in Online Travel Agencies: Event-Condition-Actions rules and Key Performance Indicators. In A. Naim & S. Kautish (Eds.), *Building a Brand Image Through Electronic Customer Relationship Management* (pp. 286–303). IGI Global. doi:10.4018/978-1-6684-5386-5.ch012

Chan, T. T. W., Lam, A. H. C., & Chiu, D. K. W. (2020). From Facebook to Instagram: Exploring user engagement in an academic library. *Journal of Academic Librarianship*, *46*(6), 102229. doi:10.1016/j.acalib.2020.102229 PMID:34173399

Chan, T. T. W., Lam, A. H. C., & Chiu, D. K. W. (2020). From Facebook to Instagram: Exploring user engagement in an academic library. *Journal of Academic Librarianship*, *46*(6), 102229. doi:10.1016/j.acalib.2020.102229 PMID:34173399

Cheng, W. W. H., Lam, E. T. H., & Chiu, D. K. W. (2020). Social media as a platform in academic library marketing: A comparative study. *Journal of Academic Librarianship*, *46*(5), 102188. doi:10.1016/j.acalib.2020.102188

Chung, C., Chiu, D. K. W., Ho, K. K. W., & Au, C. H. (2020). Applying social media to environmental education: Is it more impactful than traditional media? *Information Discovery and Delivery*, *48*(4), 255–266. doi:10.1108/IDD-04-2020-0047

Dhir, A., & Torsheim, T. (2016). Age and gender differences in photo tagging gratifications. *Computers in Human Behavior*, *63*(October), 630–638. doi:10.1016/j.chb.2016.05.044

Dong, G., Chiu, D. K. W., Huang, P.-S., Lung, M. M., Ho, K. K. W., & Geng, Y. (2021). Relationships between Research Supervisors and Students from Coursework-based Master's Degrees: Information Usage under Social Media. *Information Discovery and Delivery*, *49*(4), 319–327. doi:10.1108/IDD-08-2020-0100

Dong, G., Chiu, D. K. W., Huang, P.-S., Lung, M. M., Ho, K. K. W., & Geng, Y. (2021). Relationships between Research Supervisors and Students from Coursework-based Master's Degrees: Information Usage under Social Media. *Information Discovery and Delivery*, *49*(4), 319–327. doi:10.1108/IDD-08-2020-0100

Economist Intelligence Unit (EIU). (2007a). *Beyond loyalty: Meeting the challenge of customer engagement, part 1*. EIU.

Economist Intelligence Unit (EIU). (2007b). *Beyond loyalty: meeting the challenge of customer engagement, part 2*. EIU.

Economist Intelligence Unit (EIU). (2007c). *The engaged constituent: meeting the challenge of engagement in the public sector, part 1*. EIU.

Economist Intelligence Unit (EIU). (2007d). *The engaged constituent: meeting the challenge of engagement in the public sector, part 2*. EIU.

Fong, K. C. H., Au, C. H., Lam, E. T. H., & Chiu, D. K. W. (2020). Social network services for academic libraries: A study based on social capital and social proof. *Journal of Academic Librarianship*, *46*(1), 102091. doi:10.1016/j.acalib.2019.102091

Gao, W., Lam, K. M., Chiu, D. K. W., & Ho, K. K. W. (2020). A Big data Analysis of the Factors Influencing Movie Box Office in China. In Z. Sun (Ed.), *Handbook of Research on Intelligent Analytics with Multi-Industry Applications* (pp. 232–249). IGI Global.

Garcia-Rivera, D., Matamoros-Rojas, S., Pezoa-Fuentes, C., Veas-González, I., & Vidal-Silva, C. (2022). Engagement on Twitter, a Closer Look from the Consumer Electronics Industry. *Journal of Theoretical and Applied Electronic Commerce Research*, *17*(2), 558–570. doi:10.3390/jtaer17020029

Kemp, S. (2020). *Digital 2020: July Global Statshot*. Available at: https://datareportal.com / reports / digital-2020

Han, R., & Xu, J. (2020). A Comparative Study of the Role of Interpersonal Communication, Traditional Media and Social Media in Pro-Environmental Behavior: A China-Based Study. *International Journal of Environmental Research and Public Health*, *17*(6), 1883. doi:10.3390/ijerph17061883 PMID:32183217

He, Z., Chiu, D. K. W., & Ho, K. K. W. (2022). Weibo Analysis on Chinese Cultural Knowledge for Gaming. In Z. Sun (Ed.), *Handbook of Research on Foundations and Applications of Intelligent Business Analytics* (pp. 320–349). doi:10.4018/978-1-7998-9016-4.ch015

Ho, K. K. W., Chan, J. Y., & Chiu, D. K. W. (2022). Fake News and Misinformation During the Pandemic: What We Know, and What We Don't Know. *IT Professional*, *24*(2), 19–24. doi:10.1109/MITP.2022.3142814

Hoffman, D., & Fodor, M. (2010). Can you measure the ROI of your social media marketing? *MIT Sloan Management Review*, *52*(1), 55–61.

Huang, P. S., Paulino, Y., So, S., Chiu, D. K. W., & Ho, K. K. W. (2021). Special Issue Editorial - COVID-19 Pandemic and Health Informatics (Part 1). *Library Hi Tech*, *39*(3), 693–695. doi:10.1108/LHT-09-2021-324

Huang, P. S., Paulino, Y., So, S., Chiu, D. K. W., & Ho, K. K. W. (2022). Special Issue Editorial - COVID-19 Pandemic and Health Informatics (Part 1). *Library Hi Tech*, *40*(2), 281–285. doi:10.1108/LHT-04-2022-447

Jiang, X., Chiu, D. K. W., & Chan, C. T. (2022). Application of the AIDA model in social media promotion and community engagement for small cultural organizations: A case study of the Choi Chang Sau Qin Society. In M. Dennis & J. Halbert (Eds.), *Community Engagement in the Online Space*. IGI Global.

Kietzmann, J. H., Hermkens, K., McCarthy, I., & Silvestre, B. (2011). Social media? Get serious! Understanding the functional building blocks of social media. *Business Horizons*, *54*(3), 241–251. doi:10.1016/j.bushor.2011.01.005

Kotler, P., Rackham, N., & Krishnaswamy, S. (2006). Ending the war between sales and marketing. *Harvard Business Review*, *84*(7/8), 68. PMID:16846190

Lam, A. H. C., Chiu, D. K. W., & Ho, K. K. W. (2022). *Instagram for student learning and library promotions? A quantitative study using the 5E Instructional Model.* Aslib Journal of Information Management.

Lam, E. T. H., Au, C. H., & Chiu, D. K. W. (2019). Analyzing the use of Facebook among university libraries in Hong Kong. *Journal of Academic Librarianship*, *45*(3), 175–183. doi:10.1016/j.acalib.2019.02.007

Lau, K. S. N., Lo, P., Chiu, D. K. W., Ho, K. K. W., Jiang, T., Zhou, Q., Percy, P., & Allard, B. (2020). Library, Learning, and Recreational Experiences Turned Mobile: A Comparative Study between LIS and non-LIS students. *Journal of Academic Librarianship*, *46*(2), 102103. doi:10.1016/j.acalib.2019.102103

Leander, M. (2014). *What Is a Good Facebook Engagement Rate on a Facebook? Here Is a Benchmark for You.* Michael Leander Company. Available at https://www.michaelleander.me/blog/facebook-engagement-rate-benchmark

Lei, S. Y., Chiu, D. K. W., Lung, M. M., & Chan, C. T. (2021). Exploring the Aids of Social Media for Musical Instrument Education. *International Journal of Music Education*, *39*(2), 187–201. doi:10.1177/0255761420986217

Lin, C.-H., Chiu, D. K. W., & Lam, K. T. (2022). Hong Kong Academic Librarians'. *Attitudes Towards Robotic Process Automation*. Advance online publication. doi:10.1108/LHT-03-2022-0141

Lo, P., Chan, H. H. Y., Tang, A. W. M., Chiu, D. K. W., Cho, A., Ho, K. K. W., See-To, E., He, J., Kenderdine, S., & Shaw, J. (2019). Visualising and Revitalising Traditional Chinese Martial Arts – Visitors' Engagement and Learning Experience at the 300 Years of Hakka KungFu. *Library Hi Tech*, *37*(2), 273–292. doi:10.1108/LHT-05-2018-0071

Leung, T. N., Hui, Y. M., Luk, C. K. L., Chiu, D. K. W., & Kevin, K. K. W. (2022). An Empirical Study on the Aid of Facebook for Japanese Learning. *Library Hi Tech*.

Mak, M. Y. C., Poon, A. Y. M., & Chiu, D. K. W. (2022). Using Social Media as Learning Aids and Preservation: Chinese Martial Arts in Hong Kong. In S. Papadakis & A. Kapaniaris (Eds.), *The Digital Folklore of Cyberculture and Digital Humanities* (pp. 171–185). IGI Global. doi:10.4018/978-1-6684-4461-0.ch010

Mangold, W. G., & Faulds, D. J. (2009). Social media: The new hybrid element of the promotion mix. *Business Horizons*, *52*(4), 357–365. doi:10.1016/j.bushor.2009.03.002

Mee, G. (2019). *What is a good engagement rate on Twitter?* https://www.scrunch.com/blog/what-is-a-good-engagement-rate-on-twitter

Meng, Y., Chu, M.Y., & Chiu, D.K.W. (2022). The impact of COVID-19 on museums in the digital era: Practices and Challenges in Hong Kong. *Library Hi Tech*. doi:10.1108/LHT-05-2022-0273

Meshi, D., Tamir, D. I., & Heekeren, H. R. (2015). The emerging neuroscience of social media. *Trends in Cognitive Sciences*, *19*(12), 771–782. doi:10.1016/j.tics.2015.09.004 PMID:26578288

Ni, J., Chiu, D. K. W., & Ho, K. K. W. (2022). Exploring Information Search Behavior among Self-Drive Tourists. *Information Discovery and Delivery*, *50*(3), 285–296. doi:10.1108/IDD-05-2020-0054

Patrut, M., & Patrut, B. (2013). *Social media in higher education: teaching in Web 2.0*. Information Science Reference. doi:10.4018/978-1-4666-2970-7

Rajendran, L., & Thesinghraja, P. (2014). The impact of new media on traditional media. *Middle East Journal of Scientific Research, 22*(4), 609–616.

Sun, X., Chiu, D. K. W., & Chan, C. T. (2022). Recent Digitalization Development of Buddhist Libraries: A Comparative Case Study. In S. Papadakis & A. Kapaniaris (Eds.), *The Digital Folklore of Cyberculture and Digital Humanities* (pp. 251–266). IGI Global. doi:10.4018/978-1-6684-4461-0.ch014

Swani, K., Milne, G. R., Brown, B. P., Assaf, A. G., & Donthu, N. (2017). What messages to post? Evaluating the popularity of social media communications in business versus consumer markets. *Industrial Marketing Management, 62*(April), 77–87. doi:10.1016/j.indmarman.2016.07.006

Tess, P. A. (2013). The role of social media in higher education classes (real and virtual) – A literature review. *Computers in Human Behavior, 29*(5), A60–A68. doi:10.1016/j.chb.2012.12.032

Tsang, A. L. Y., & Chiu, D. K. W. (2022). Effectiveness of Virtual Reference Services in Academic Libraries: A Qualitative Study based on the 5E Learning Model. *Journal of Academic Librarianship, 48*(4), 102533. doi:10.1016/j.acalib.2022.102533

Tse, H. L., Chiu, D. K., & Lam, A. H. (2022). From Reading Promotion to Digital Literacy: An Analysis of Digitalizing Mobile Library Services With the 5E Instructional Model. In A. Almeida & S. Esteves (Eds.), *Modern Reading Practices and Collaboration Between Schools, Family, and Community* (pp. 239–256). IGI Global. doi:10.4018/978-1-7998-9750-7.ch011

Wang, J., Deng, S., Chiu, D. K. W., & Chan, C. T. (2022) Social Network Customer Relationship Management for Orchestras: A Case Study on Hong Kong Philharmonic Orchestra. In Social Customer Relationship Management (Social-CRM) in the Era of Web 4.0. IGI Global. doi:10.4018/978-1-7998-9553-4.ch012

Wang, P., Chiu, D. K. W., Ho, K. K., & Lo, P. (2016). Why read it on your mobile device? Change in reading habit of electronic magazines for university students. *Journal of Academic Librarianship, 42*(6), 664–669. doi:10.1016/j.acalib.2016.08.007

Wirtz, B. W., Schilke, O., & Ullrich, S. (2010). Strategic development of business models: Implications of the Web 2.0 for creating value on the internet. *Long Range Planning, 43*(2-3), 272–290. doi:10.1016/j.lrp.2010.01.005

Wong, J., Chiu, D. K. W., Leung, T. N., & Kevin, K. K. W. (in press). Exploring the Associations of Addiction as a Motive for Using Facebook with Social Capital Perceptions. *Online Information Review*.

Yang, Z., Zhou, Q., Chiu, D. K. W., & Wang, Y. (2022). Exploring the factors influencing continuance intention to use academic social network sites. *Online Information Review, 46*(7), 1225–1241. Advance online publication. doi:10.1108/OIR-01-2021-0015

Yu, H. Y., Tsoi, Y. Y., Rhim, A. H. R., Chiu, D. K., & Lung, M. M. W. (2021). (in press). Changes in habits of electronic news usage on mobile devices in university students: A comparative survey. *Library Hi Tech*. Advance online publication. doi:10.1108/LHT-03-2021-0085

Yu, H. H. K., Chiu, D. K. W., & Chan, C. T. (2022). Resilience of symphony orchestras to challenges in the COVID-19 era: Analyzing the Hong Kong Philharmonic Orchestra with Porter's five force model. In W. Aloulou (Ed.), *Handbook of Research on Entrepreneurship and Organizational Resilience During Unprecedented Times*. IGI Global. doi:10.4018/978-1-6684-4605-8.ch026

Yu, P. Y., Lam, E. T. H., & Chiu, D. K. W. (2022). (in press). Operation management of academic libraries in Hong Kong under COVID-19. *Library Hi Tech*. Advance online publication. doi:10.1108/LHT-10-2021-0342

Zhang, Q., Huang, B., & Chiu, D.K.W., & Ho, K. W. (2015). Learning Japanese through social network sites: A case study of Chinese learners' perceptions. *Micronesian Educators*, *21*, 55–71.

Zhang, Y., Lo, P., So, S., & Chiu, D. K. W. (2020). Relating Library User Education to Business Students' Information Needs and Learning Practices: A Comparative Study. *RSR. Reference Services Review*, *48*(4), 537–558. doi:10.1108/RSR-12-2019-0084

KEY TERMS AND DEFINITIONS

COVID-19 (Coronavirus Disease 2019): A contagious disease caused by a virus called the severe acute respiratory syndrome coronavirus 2 (SARS-CoV-2). The first known case was identified in Wuhan, China, in December 2019, and the disease spread worldwide, leading to the COVID-19 pandemic.

Customer Relationship Management: A combination of practices, strategies, and technologies for organizations to manage and analyze customer interactions and data throughout the customer lifecycle, aiming to improve customer service relationships and assist in customer retention and increase sales revenue.

Digitalization: A process of moving the libraries to provide services digitally, using digital technologies to alter the mode of library services and enhance the quality of services to users in the case of libraries.

Honeycomb Model: A model that includes important forces behind the social media ecology within which all social media marketers, users, and platforms cooperate with the following seven aspects, sharing, presence, relationships, identity, conversations, reputation, and groups.

Hong Kong Philharmonic Orchestra (HKPhil): The most funded Hong Kong orchestra by the Hong Kong Government was established in 1947. HKPhil was the first orchestra from Asia to win the prestigious *Gramophone Orchestra of the Year Award* in 2019.

Social Media: Interactive network technologies and digital channels that facilitate the creating and sharing of information, ideas, interests, images, videos, links, and other information.

Chapter 4

Application of the AIDA Model in Social Media Promotion and Community Engagement for Small Cultural Organizations:

A Case Study of the Choi Chang Sau Qin Society

Xinyu Jiang
The University of Hong Kong, Hong Kong

Dickson K. W. Chiu
https://orcid.org/0000-0002-7926-9568
The University of Hong Kong, Hong Kong

Cheuk Ting Chan
Independent Researcher, Hong Kong

ABSTRACT

This study explores the AIDA model (attention, interest, desire, action) for social media promotion and community engagement for small cultural organizations. The internal situation and external environment were first analyzed with the SWOT analysis augmented with PEST analysis. Then, the authors show how the AIDA model can be used in social media marketing to improve public awareness, engagement, and thus participation in the organization's activities. As the global economy is getting linked to the internet and social media, utilizing the AIDA model for small cultural organizations contributes to effective information dissemination and increases interactions in the targeted community. The rise of social media has also triggered smaller organizations to consider how to survive under dynamic changes and fulfill their mission through better community engagement.

DOI: 10.4018/978-1-6684-5190-8.ch004

INTRODUCTION

The field of public relations and marketing communication has been substantially changed with the success of online social networks, further enhanced by mobile technologies (Yu et al., 2021; Wang et al., 2016; Lam et al., 2019; Lau et al., 2020; Fong et al., 2020; Dong et al., 2021). As information is more readily available and accessible to the public, the interactive and ubiquitous digital world connects organizations and their customers more engagingly (Chan et al., 2020; Lam et al., 2022). As more and more people from various lifestyles spend time on social media and media consumption is gaining popularity, scholars and practitioners worldwide are trying to leverage consumers' current online behaviors and find more appropriate marketing strategies (Wang et al., 2022; Deng et al., 2022). Globally, the number of social media users is expected to reach nearly 3.43 billion by 2023 (Statisca, 2020), while that number in China is estimated to be close to 800 million (Statista, 2019). Organizations are increasing their social media marketing budget and deciding on digital marketing channels to engage their user community (Wang et al., 2022; Deng et al., 2022; Lam et al., 2022). Among these channels, social listening has shown to be an effective strategy on the most popular online social platforms for marketers, such as Facebook, Instagram, and Twitter (HubSpot, 2020). Furthermore, many organizational successes come from increased media exposure, better website visits, and engaging a widened range of potential users through social media marketing (Yu, Chiu, & Chan, 2022).

On the other hand, COVID-19 has struck a double blow to social and economic development, resulting in extremely tight public budgets in the coming years (Yu, Chiu, & Chan, 2022; Leung et al., 2022; Huang et al., 2021; 2022). Thus, public agencies, non-government organizations (NGOs), and smaller organizations should reconsider how to promote their services better and fulfill their missions with a modest budget. However, traditional marketing channels are costly in operations and maintenance. Under this context, smaller cultural organizations that may not have broad audiences or provide living-essential services need to leverage low-cost social media solutions to gain support and participation from their target communities (Wang et al., 2022).

Notably, cultural heritage protection is considered a way of regenerating social ties and strengthening public feelings of belonging and identity (Mak et al., 2022; Sun et al., 2022; Jiang et al.; 2019; Lo, Cheuk, et al., 2021). In this regard, cultural organizations have long been engaged in promoting national identity to locals, especially with traditional Chinese culture in Hong Kong (Lo et al., 2019). They serve as preservers, promotors, and educators to bridge the gaps in cultural perceptions and enrich people's lives by revitalizing ancestral knowledge and engaging with traditional culture (Lo et al., 2019; Mak et al., 2022). They naturally constitute intangible assets of society, and this study attempts to help these small cultural organizations spread, preserve, and educate traditional Chinese culture by engaging more potential audiences (Wang et al., 2022; He et al., 2022).

Over recent years, museums, libraries, and other cultural organizations have realized the benefits and necessity of digital marketing and support technology involvement (Cho et al., 2017; Jiang et al., 2019). They have created an enabling, participatory, and engaging environment favored by the public. The information and promotion of such environments can be spread through various social media initiatives to engage target users, provide education, and enhance outreach, marketing, crisis communication, and feedback (Chung et al., 2020; Lam et al., 2022). However, many obstacles exist, especially in turbulent times like the COVID-19 pandemic, such as limited availability of public resources, economic slowdown, complexities of intercultural interactions, and rapid technology changes (Huang et al., 2021; 2022; Yu, Chiu, & Chan, 2022). Deploying effective social media strategies represents a fundamental challenge and opportunity for small cultural organizations (Deng et al., 2022; Wang et al., 2022).

Although some scholars have studied and explored promotion in the cultural industry (Polk, 2018; Deng et al., 2022), scant studies focus on small cultural organizations, especially using the AIDA model on social media for organizations in East Asia. Therefore, this chapter aims to apply a systematic approach to social media marketing management for a small cultural institution to analyze the selected case comprehensively and present suggestions on applying the AIDA marketing communication model to improve its marketing and user engagement process on social media.

CASE BACKGROUND

This paper analyzes the Choi Chang Sau Qin Making Society (蔡昌壽斲琴學會), founded in 2011 in Hong Kong, which aims at protecting and disseminating the Qin (or Guqin, Chinese 7-string plucked instrument) crafting art passed from Master Xu. The society's predecessor is the Qin-carving study group, established in July 1993 at a small company "Choi Fook Kee Instruments." Embedded in the spirit of engagement and dedication, the organization features a community-based organization (Wexler, 2005). The study group and courses have evolved and become a part of the Choi Chang Sau Qin Making Society's services. The importance of the society is that Hong Kong is the last place where Qin culture is well preserved after the Cultural Revolution in mainland China, while the organization's owner, Master Choi, is the world's leading Qin craftsman. This cultural organization helps grow legions of Qin lovers and enables supporters to learn the Qin-making art and culture systematically. Master Choi has taught the next generation a more sophisticated understanding of the traditional instrument and has taken steps toward the community's traditional cultural and artistic life in the strokes of cutting, drilling, engraving, and burnishing.

There are two main types of activities in this organization. The primary one is essential training and coaching services, which are unique because only this organization provides them. Other activities usually complement the primary ones and are called supportive activities, namely promotional campaigns, alignment, operations management, and fundraising activities (Cheung et al., 2021). The organization posts its promotional videos and related documentaries on YouTube (see: https://www.youtube.com/channel/UCnAHf8j8r1ZyVmnlGL9LcqQ) and has developed a website (https://www.ccs.org.hk/) introducing relevant information on Qin history, event reports, promotional video clips, and documentary links. Yet, the website has only simple interactive features, such as fundraising and the latest news reports.

As an organization missioned with cultural preservation and education, it needs to exploit marketing communications and promotions to engage members and implement activities (Tse et al., 2022). To ensure long-term, sustained growth, it should take the advice from Oakes, Dennis, and Oakes (2013) to pay more attention to social media forums and metaphysical branding promotions to arouse customers' interest in the art and engage them through related inner impetus for self-improvement. More importantly, through social media promotion, the organization can develop, engage, and educate target audiences (mainly local communities and young people) (Butler, 2000; Fong et al., 2020; Mak et al., 2022; Wang et al., 2022; Lam et al., 2022).

LITERATURE REVIEW

This literature review aims to investigate and extend the best appropriate theories and practices of marketing for other cultural institutions. This study applies the AIDA marketing communication model and its practical significance to develop a sustainable marketing scheme instead of stop-gap social welfare support activities for the Choi Chang Sau Qin Society.

Social Media Marketing

Social media (SM) is a broad concept including social networking sites (SNS, e.g., Plus, Twitter, Facebook), user-generated content (UGC, e.g., Tik-Tok, YouTube), virtual world (e.g., Second Life), virtual games community, microblog (i.e., Plurk, Twitter, Pinterest, Flickr) and blogs (Stepaniuk, 2017). It refers to an interactive application of Web-based 2.0 to encourage more users to use and share links to the website across social media and networking sites and facilitate information exchange and creation (Kaplan & Haenlein, 2010). While the term social media, believed to be first used by Chris Sharpley, was coined in the same year, Web 2.0 became a mainstream concept. Compared to "Web 1.0," which emphasizes the storage and transmission of vast amounts of information in cyberspace, "Web 2.0" supposedly renders the whole process interactive and focuses on the ability of people to work collaboratively. It has endowed online interpersonal communication, exhibited enormous vitality and influence, and gradually shown a vital marketing function, especially for the younger generation (Lam et al., 2022; Deng et al., 2022). As a communication tool, social media not only owns traditional integrated marketing communication (IMC) functions to establish the relationship between organizations and their consumers but also provides an opportunity for consumers to exchange information, views, and perspectives with one another, which can be regarded as the extension of word of mouth (Bickhoff et al., 2014, p. 105).

AIDA Model

In cognitive marketing, the AIDA model explains the quality and pathways of valid information. This concept was first proposed by the sales pioneer Elias. St. Elmo Lewis, in the late 1800s (Barry, 1987), according to three successful advertisement principles: to draw attention to consumers, arouse their interest, and convince them to believe (Lewis, 1909). Until late 1904, the system concept was conceptualized and theorized, while the acronym AIDA stands for Attention, Interest, Desire, and Action (Dukesmith, 1904).

For years, the AIDA model has been widely applied in marketing and advertising (Lee & Hoffman, 2015) as a crucial element of the Promotional part of the 4Ps Marketing Mix model (Product, Price, Place, and Promotionk, Jobber & Ellis-Chadwick, 2012). More recently, marketers have employed it as a communication technique (Lamb et al., 2012; Lee & Hoffman, 2015) to drive consumers to purchase or potential consumers to consider (McElhone, 2020). It improves user experience and raises retention rates.

Building websites for nonprofit organizations in the early 21st century was not promising due to the lack of advanced expertise and staff (e.g., Kent, Taylor, & White, 2003; Saxton, Guo, & Brown, 2007). In contrast, the emergence of social media (agile and cooperative means, inclusive processes) upended the early belief of the Internet, which cannot be used in marketing strategies for public sectors (Macias, Hilyard, & Freimuth, 2009). Then, Facebook, Twitter, and the like have broken the monopoly of blogs. Bortree and Seltze (2009) and Greenberg and MacAulay (2009) researched social media applied to public sectors and indicated two main functions of public sectors using social media: sharing information and establishing dialogue. Nevertheless, organizations had no idea how to operationalize social media for consumer communication. Lovejoy and Saxton (2012) addressed this question by analyzing the utilization of Twitter in 100 nonprofit organizations, the first study to classify social media messages by organizations. The proposed message model provided guidance and incentives for public organizations to communicate on social media.

Temporal (2015) suggested that the public sector has entered the world of branding for differentiation and attraction since strong brands differentiate and attract people to them rather than having to chase after them. His book further underscored the importance of social media in elaborating the brand strategy for every public sector. Compared to traditional ways of communication (i.e., advertising, direct sales) for public sectors with their audience, two-way exchange, revealed in this book, would be a trend in future communication between the sector and the audience. This book also discusses the necessity of integrating social media with traditional ways of communication for public sectors and then expounds on ten basic metrics measuring results:

- Breadth (community scale and its growth)
- Depth (conversion and viewing)
- Direct Engagement (engagement volume and responsiveness)
- Loyalty (return community)
- Customer Experience (sentiment, indicators, survey feedback)
- Campaigns
- Strategic outcomes

Further, Men and Muralitharan (2017) conducted experiments in China and the US and proposed a social peer communication model, stressing that organizations should monitor peer communication on social media. They gave examples to illustrate marketing techniques for future practices, such as launching online campaigns to gather like-minded fans (Men & Tsai, 2013; Deng et al., 2022), using big-data analytics to deliver targeted advertising and better understand the audience with their preference (Deng et al., 2022; He et al., 2022; Gao et al., 2021), gamification to motivate and engage customers (Men & Tsai, 2014).

Recently, Leijerholt, Biedenbach, and Hultén (2019) have provided insights into the current trends in branding development in the public sector and how these affect both social relationships and decision-making. They clearly distinguished between market and brand direction in public sectors while reviewing relevant literature from external and internal branding perspectives. From the internal perspective, they indicated the internal branding efforts that target employees matter while effective communication among employees needs further study. Their external analysis has recognized the critical role of external stakeholders and communities in providing critical information and positive branding impact. Leijerholt et al. (2019) also pointed out the importance of studying different marketing strategies specific to stakeholders with different levels of social status and different requirements. Particular attention is necessary for digital marketing communication efforts exploited in public sectors when digital challenges and difficulties facing the community grow ever more complex and intricate, such as ineffective communication strategies (Cheng et al., 2020) and social media misinformation (Au et al., 2021).

Recent literature also focuses on modern technologies integrated into the sustainable development strategies of libraries, museums, and art galleries (Ho & Chiu, 2022; Jiang et al., 2019). As a new-type productive form, social media has also brought much significant and farsighted innovation and replenished both the theories and practices of marketing. Plenty of literature has suggested networking as a critical factor in those marketing plans (Fong et al., 2020; Lam et al., 2019; Lam et al., 2022).

Exploratory empirical research related to art institutions (three orchestras) in Germany identified the constraints and opportunities within viral marketing on social media and showed their different priorities, options, visions, and approaches (Hausmann, 2012). They presented three proposals based on three social media. For the blog used as a public venue for the orchestra, more fresh and catchy content was suggested. Microblogs were recommended as a mixture of instant messages and status notifications. As for NRW-forum Dusseldorf on Twitter, the marketing result was less than satisfactory, despite a sheer amount of reading. However, the authors spelled out a few specific policies and amendments, building on existing channels, but no serious consideration to innovative proposals such as more marketing channels and detailed implementation strategies for orchestras.

In addition, Albuquerque (2018) systematically analyzed museums in Portugal and the UK and identified the effect of strategies on audience participation and support, highlighting the marketing orientation of museums and other public sectors, particularly more public-focused activities without obstacles. A fundamental principle was that cultural institutions should adopt different strategies aligned with their management priorities. Accordingly, social media would inevitably gain more momentum in the marketing strategies of public sectors, though it might prosper at a different rate in different places.

As cultural diversity, characterized by the cross-cultural features from both the East and the West, contributes to the unique identity of Hong Kong, Hongkongers should synergize the essence of Eastern and Western culture and seek a new direction amidst clashes and contradictions (Wang et al., 2022; Jiang et al., 2019; Lo, Cheuk, et al., 2021). On the other hand, governments, public organizations, and NGOs are embracing the progressive integration of business concepts into their communication, promotions, and services (Chan et al., 2022; Cheung et al., 2021; Tse et al., 2022; Wang et al., 2022). Thus, Choi Chang Sau Qin Making Society should choose the right social media strategies based on Hong Kong's socio-cultural background and the organization's mission and management priorities. This study proposes several theories and marketing models to aid in developing their marketing strategies. The AIDA model was one of the earliest attempts introduced on an experimental basis for the social media marketing plan for cultural tourism and other cultural industries as a classic marketing concept to explore the impact of the various media and cultures on tourist destinations (Lin & Huang, 2006; Hudson, Wang & Gil, 2011). For instance, with the effective deployment of the AIDA model, Lin & Huang (2006) successfully classified network information and studied user behaviors. Further, Wang & Gil (2011) explored how films trigger the desire to travel using AIDA.

Li and Yu (2013) developed an AIDA-based integration model for a bank at a university, achieving notable success in practice. This innovative model was grounded in research and has reached a broad audience. The advertising and activities from the bank were developed according to the AIDA model and succeeded in engaging and attracting students, suggesting that the marketing concept yielded reputable results in the student market.

Polk (2018) applied the AIDA model to the education areas, especially for the informational and transactional classes, to develop students' motivation for studies and build interactive relationships with others. Unique teaching materials and technical aids were necessary to attract students (Lei et al., 2021). Authenticity and skepticism also were mentioned in the process of the AIDA model. However, this application only targeted adult learners without overlooking the youth and children, who are also the target groups of cultural communication.

Literature reviews show that other cultural institutions, especially NGOs and nonprofit organizations, have hardly applied the AIDA model in their operations. However, if these organizations want wider recognition to bring more benefits, they have to accumulate followers and cultivate them to accomplish great missions together, and social media can be a powerful tool to attract more users, especially young people, to be involved in the activities and programs. Therefore, applying the AIDA model in NGOs and cultural sectors can be rewarding.

The literature review has demonstrated the overwhelming advantages of social media marketing in public sectors. While AIDA implies that *attention* leads to *interest*, which leads to *desire,* and finally, *action*, it does not fully consider the context. Therefore, this paper embeds PEST (political, economic, socio-cultural, and technological) in the SWOT (strength, weakness, opportunities, and threat) analysis of the overall situation and analyzes the need for sustainable development of the organization in a multicultural context in Hong Kong. This approach fills the gap in environmental factors and expands the frontiers in marketing NGO cultural institutions, especially for engaging communities for education. This study combines the AIDA model with social media marketing strategies to explore the provision of content for engaging and educating target audiences, which is innovative.

ANALYSIS OF CULTURAL INSTITUTIONS IN HONG KONG

Due to the complexity and multilayered nature of market phenomena and the importance of cultural institutions in disseminating information, making the right decisions in the strategy and tactics requires a sophisticated understanding of external and internal factors. Accordingly, this chapter introduces the PEST-embedded SWOT analysis to provide an analytical framework for the marketing development strategies of cultural institutions.

As an indispensable diagnostic tool for the modern strategic management of sustainable development, the SWOT analysis has yielded promising results in multiple fields. For example, Mohezar et al. (2017) employed a specific SWOT analysis on medical tourism in Malaysia with a corresponding development strategy. Herrera-Franco et al. (2020) conducted an optimization proposal on the reputation of higher education after a clear picture of the strengths and weaknesses was obtained from the SWOT analysis. Moreover, the SWOT analysis was applied to develop marketing activities for an academic library (Smith, 2011), where both external and internal conditions restricted its development.

Considering an organization's external conditions are incredibly dynamic and uncertain amid the pandemic (Huang et al., 2021; 2022), the PEST further helps analyze the macro environment during the SWOT analysis. With such a combination, Blery et al. (2010) reviewed the effectiveness of the marketing practices initiated by WWF (Worldwide Fund for Nature). Similarly, Aslan et al. (2012) developed a new organizational structure plan according to corporate policies to meet the world trade requirements of the textile industry. Anatoly and Aleksey (2016) presented a SWOT matrix (embed with PEST) at public catering companies while creating a new assessment methodology for the quality of services.

By utilizing the SWOT analysis embedded with PEST, the paper analyzes both external and internal environments in which the organization operates, as shown in Figure 1. It developed an outreach/marketing scheme based on the AIDA model to reach even broader audiences for cultural promotion and education.

Figure 1. SWOT Analysis Embedded with the PEST

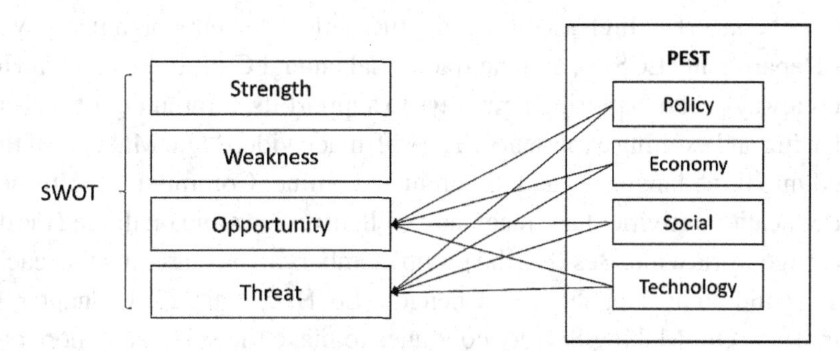

Analysis of Strengths (S)

This section focuses on reviewing the internal marketing-related factors instead of analyzing all the favorable internal elements of Choi Chang Sau Qin Making Society.

Keen interest in Hong Kong has laid the foundation for trusting relationships with the audience. The heritage brand of "Choi Fook Kee," the only Qin maker in Hong Kong, is famous for its high quality and true artistry. Stepping into the instrument factory, people can see the plaques of honor from around the world covering a wall, noting its reputation and image. The owner of "Choi Fook Kee," Mr. CHOI Cheung-Sau, is the only apprentice and sole transmitter of Xu Wenjing's qin-carving craftsmanship. Besides, the Choi Chang Sau Qin Making Society was entitled Protected Unites of China's Intangible Cultural Heritage at National Level with Master Choi, the last professional Qin maker globally. Moreover, Master Choi was honored as a Representative Inheritor of the art of Zhuóqín at the National Level in 2008, drawing further attention to the ICH in Hong Kong. A good reputation, a socially responsible corporate image, and many honorable titles have shaped the cornerstones of public trust.

Interactive instruction in the Qin-carving research classes tends to ensure greater participation. It is undeniable that numerous cultural institutions undertake education liabilities, exploit marketing communication, and promote interpretation and popularization (Wang et al., 2022). The first qin-carving study group was established at "Choi Fook Kee Instruments" in 1993 to inherit the carving art. Small class sizes ensure the quality of teaching, while audience engagement can be enhanced by the cross-fertilization of minds and interactive learning patterns (Patsiaouras, Veneti & Green, 2018).

Unique Qin instruments are custom-made for personalized service. The Qin is one of the "Four Arts" (the other three are chess, calligraphy, and painting). Qin has been a Chinese plucked string instrument for over 3,000 years, and Qin music has been a minority interest but elegant art shared by some cultural scholars and collectors (Lo, Hsu, et al., 2021, Chapter 9). There is a low demand for Qin instruments, with fewer than 1,000 expert players worldwide. However, master Choi has clung to Qin making and carving (or Zhuoqin) and its education even if his store is the only one in Hong Kong. Because of being purely handmade, each Qin instrument is unique and valuable. The distinctive nature of cultural goods, services, and activities as vehicles of identity has made Choi's store the one and only.

By improving the technical service level and carrying out various market promotions, Mr. Choi has stably raised the brand's public profile. Festivals, exhibitions, concerts, and many other proactive outreach initiatives are conducted regularly, including a photography and instrument exhibition held in the Hong Kong Heritage Museum entitled "The Legend of Silk and Wood: A Hong Kong Qin Story," the Hong Kong Central Library talk and symposium entitled "Subject Talk on Hong Kong Memory," lecture tours and local concert in higher education, other cultural events organized by the Leisure and Cultural Services Department (LCSD), and the traditional annual Culture Festival in Hong Kong.

The Society has always been exploring new ways to support its community of consumers and make learning knowledge fun and exciting experience. The self-made video "Qin Makers" of the 22nd Century has been translated into three languages and disseminated online. Community multi-media centers and culturally specific education activities have received excellent reviews and facilitated the dissemination of Qin-making knowledge worldwide. Besides, the group members are researching the teaching methods of Qin music for young children in Hangzhou and Chengdu (Lo, Hsu, et al., 2021, Chapter 9). Additionally, to appeal to children, the Qin Making Society continues to make the self-made more children-friendly.

Analysis of Weakness (W)

Human resources shortage. The organization, staffed by volunteers, has been carrying out its mandate by facilitating Qin art inheritance and cultural transmission, but it lacks early childhood education staff and communication projects for teenagers. Besides, although volunteers are from different backgrounds and with different experiences, they are all middle-aged or older but highly intellectual. This inevitably causes difficulty building a bridge with people living in other segments of society or younger age groups (Mak et al., 2022). A public cultural organization requires enthusiastic volunteers, reasonable financial allocations, and diverse staff composition to better adapt to emerging market needs (Cheung et al., 2021).

Lack of access to marketing channels and distinctive marketing strategies. The Society has already embraced information technologies in management, promotion, education, and marketing, such as uploading materials to video-sharing websites like YouTube and posting photos on Facebook (Mak et al., 2022). Nevertheless, it does not fully understand the new digital landscape. Different social media channels appeal to different visual audiences. For example, some people prefer to receive information via an image on Instagram, while others may prefer video content on YouTube or other forms of content on Twitter (Chan et al., 2020). However, the social media presence of the Society was limited to a few platforms. Besides, in the Society's current approach, users are alienated as they do not have effective interaction with resource providers. For instance, there are no message fields or other interactive features on its website, and the Society has turned off comments on YouTube.

Inadequate space and limited resources. Hong Kong has limited space, small homes, and very high real estates price and rent. However, since the Qin is a relatively sizeable wood-based instrument, adequate storage and playing space are necessary. Proper facilities, plus an exhibition area large enough for visitors and customers, also need space. Currently, the place can only hold a few collections and a classroom designed for no more than ten persons at once, which hinders preserving, transmitting, and inheriting the culture.

Analysis of Opportunities(O)

Policy

The Convention for Safeguarding the Intangible Cultural Heritage and the ICH Funding Scheme. UNESCO places a high priority on safeguarding cultural heritage and indigenous knowledge of traditional cultures. The Convention for Safeguarding of the Intangible Cultural Heritage was developed by the United Nations Educational Scientific and Cultural Organization (UNESCO) on October 17, 2003, emphasizing the transmission and enhancement of leading culture through formal and informal education. Since the implementation of the Convention in 2006, a specific unit has been established for better operations. The relevant practice team has directly participated in developing Hong Kong's policies while implementing the convention. Its mission is to cultivate public awareness and responsibility for intangible cultural heritage through a series of Hong Kong educational activities similar to Chinese martial arts (Mak et al., 2022).

Later, this unit evolved into an office (ICHO) in May 2015, where the concept of cultural promotion was first proposed with the mission of the ICHO. Notably, to facilitate community involvement, the ICHO has developed the Funding Scheme in Hong Kong and guarantees supervision of this community activity through its resources and cultural services, reflecting Hong Kong's concern for intangible cultural protection and commitment to accelerating the development of living heritage.

Hong Kong remains steadfast in its commitment to being an international cultural metropolis. Hong Kong's arts education policy aims to nurture top talent with "generic skills" of creativity. As we know, arts and music are recognized as a source of creativity and innovation, and policymakers make arts and music education into the official curricula of primary and secondary schools. Traditional Chinese music then gets the opportunity to be visible to the young generation in Hong Kong. Meanwhile, the government has made efforts to build itself as an international cultural metropolis and came up with a policy recommendation report designed to propel the process of traditional culture protection in 2003. The cultural policy aims to foster public participation in the cultural process and increase intangible cultural heritage exposure (Lo, Hsu, et al., 2021).

Economy

Rapid development of electronic commerce. The advent of e-commerce in Hong Kong has heralded a monumental shift in the retail industry, bringing new opportunities and challenges to traditional retailers. According to available figures, retailers, especially many small businesses, have struggled to overcome the economic woes originating from the recession. Many traditional industries have been transforming and developing new business models, and e-commerce has become a new way to grow revenue. At the same time, COVID-19 opens a window of e-commerce opportunities due to gathering restrictions and lockdowns (Yu, Lam, & Chiu, 2022), resulting in a surge in online users and changes in commodity consumption patterns that transform promotion and communication.

Society

Cultural heritage consciousness among locals. Conscious of the importance of regional intangible culture, some local communities have engaged actively in culture preservation, appreciation, and learning (Lo, Hsu, et al., 2021; Mak et al., 2022). One prominent example of community action is the most famous annual Jiao festival, which came only third on the list of intangible cultural heritage (ICH) in China, prompted by the Cheung Chau Rural Community (on a small island in Hong Kong). There are other instances of community participation, such as the Chiu Chow Community, which has invested efforts in the nomination of its Yu Lan Ghost Festival. Moreover, the Hau clan in Sheung Shui has promoted the traditional Hung Shing Festival. More and more communities continue to be sensitized by the ICH Office and its implementation of the Convention on the positive effects of ICH through rallies, community outreach, public education, and awareness-raising activities.

Technology

High degree of Internet penetration. Based on available data, as of January 2020, Hong Kong had an estimated Internet user base of 6.79 million, and the penetration rate of the Internet accounted for 91%. The average daily use of each person's Internet reached six hours and 16 minutes, indicating that the technologies like Wi-Fi, 4G, and 5G have made the Internet a ubiquitous part of locals' daily lives (Wang et al., 2021; Yip et al., 2021). The popular websites are Google for searching, YouTube with streaming videos, Facebook for social connection needs, and Wikipedia for knowledge reference (Lau et al., 2020).

More active social media users. According to Kemp (2020), since 2019, there has been a four percent increase in active social media users, with that number in Hong Kong being 5.8 million. The penetration rate is over 78%, 1.5 times the global average (49%). Every day, the time is almost equally divided between television (2h 33m) and social media (2h). The commonly used social media are Facebook, YouTube, WhatsApp, Instagram, and WeChat. The usage of the younger generation and college students is even higher (Dong et al., 2021; Lam et al., 2022).

Mature and safe smart payment system technology. Instead of face-to-face dealings by cash or card transactions, smart payment has more features (i.e., secured transactions, invoice processing, vendor management, and information exchange), contributing to more flexible and immediate transaction models while increasing trade security. Many online merchants and users have grown consistently, especially during the pandemic. For instance, AlipayHK has over 62,000 active small and medium businesses using the Alipay app in Hong Kong, with the active rate rising by a sharp 50 percent from last year (Kshetri, 2020). Meanwhile, the total value of the consumer (B2C) e-commerce market in Hong Kong increased 8% in 2019 compared to 2018, reaching a total market size of $14.00 billion. Currently, AlipayHK, WeChat Pay, and PayPal are the most commonly used payment tools.

Development and application of e-learning technologies. In a reaction to the "knowledge economy society," online and mobile education has become a favorite of both schools and institutions, especially for higher education and cultural institutions, because of its cost-effectiveness and time and spatial flexibility, especially due to the current COVID-19 lockdown (Tse et al., 2022; Lam et al., 2022; Yao et al., 2022).

Analysis of Threats (T)

Modern commercialization and civilization bring threats. In contrast, NGO or nonprofit organizations, which are approachable to citizens, are making efforts to make traditional culture longstanding and attractive (Lo et al., 2019; Mak et al., 2022). However, global intangible cultural heritages have been challenged and threatened by the invasion of western cultures, and citizens are losing interest in traditional Chinese culture (Mak et al., 2022; Sun et al., 2022). Besides, the environment of traditional culture has dramatically changed and is quickly impacted by present social values, imposing a challenge to preserve and inherit Qin making and carving art. Therefore, avoiding excessive utilitarian business exploitation and deeply discovering inner cultural content is necessary to guarantee the Qin Making Society's sustainable and stable development. The attractiveness of traditional folk arts can be radiant to the public (Lo, Hsu, et al., 2021, Chapter 9).

Economic

Recession and economic restructuring. As with Singapore, Malaysia, and Thailand, Hong Kong also sees its small exported-led economy shrinkage. In 2019, due to large protests and frequent insurgencies during the rising Sino-US trade tension 2019, Hong Kong was shaken by the worst economic crisis in 10 years. Worse still, given the COVID-19 outbreak in Hong Kong and the likelihood of the virus's mutation, it is projected that Hong Kong's economy will continue to be sluggish in 2021 due to a massive lockdown (Yu, Lam, & Chiu, 2022). The grave economic crisis has made it difficult for cultural centers and musical performances to operate effectively (Yu, Chiu, & Chan, 2022). As they are losing cultural profits, public interest groups have to slash expenditures due to limited budgets, which indicates that organizations have to review the institutional and systemic frameworks to function better with a tight budget (Yu, Chiu, & Chan, 2022).

Social

Social imbalances in arts participation in Hong Kong. Cultural agencies and industries, mainly traditional cultural bodies in Hong Kong, are generally preferred by middle-aged people. On the other hand, despite the high art displayed in most Hong Kong art institutions, the younger generations seem to be more interested in Western and modern cultures. With technology continually improving, the younger generation is no longer content with standard ways of participation. They prefer activities combined with advanced technology and interactive patterns (Lei et al., 2021; Tse et al., 2022). However, there is a concern that younger audiences will not respond positively to the Qin Making Society's experience.

Table 1. SWOT MIX

<table>
<tr><th colspan="2">Strength</th><th colspan="2">Weakness</th></tr>
<tr><td colspan="2">• Fanaticism for art in Hong Kong has laid the foundation for a trustworthy relationship with the audience.
• The utilization of interactive methods of instruction in the Qin-carving Research Class contributes to more participation.
• Unique guqin instruments are customized.
• The technical service level has improved, and various market promotions have been taken, raising the public profile.</td><td colspan="2">• There is a shortage of human resources.
• Marketing channels are limited and monotonous.
• The store is crowded and in need of more resources.</td></tr>
<tr><th colspan="2">Opportunity</th><th colspan="2">Threat</th></tr>
<tr><td>Political</td><td>• The Convention for the Safeguarding of Intangible Cultural Heritage and the ICH Funding Scheme is being implemented.
• Hong Kong remains steadfast in its commitment to being an international cultural metropolis.</td><td rowspan="2">Economic</td><td rowspan="2">• Hong Kong is facing a recession and economic restructuring, especially hard hit by COVID.</td></tr>
<tr><td>Economic</td><td>• Electronic commerce enjoys rapid growth.</td></tr>
<tr><td>Social</td><td>• The cultural heritage consciousness of local people is waking up.</td><td rowspan="2">Social</td><td rowspan="2">• Imbalance number of people of different social statuses participating in arts.
• Chinese and Western cultures collide with each other.
• Hong Kong residents have a variety of leisure and recreational pursuits.</td></tr>
<tr><td>Technological</td><td>• Internet highly penetrates people's daily life.
• An increasing number of people are active on social media.
• Smart payment technology is mature and secure.
• E-learning technologies are well developed.</td></tr>
</table>

They are more likely to be immersed in online gaming, social media, and the immediate gratification of mobile videos (He et al., 2022). In such an environment, the authenticity and tradition of the Qin Making Society and the Qin would be challenged severely by the charms of the virtual world.

Choices and collisions between Chinese and Western cultures. More negative social attitudes and antagonisms caused by the immense economic disparity, marginalization of immigrant workers, ongoing fight for full democracy, etc., still exist in Hong Kong. The absence of Chinese cultural identity and the lack of direct access to traditional Chinese culture are problems for more than one generation of the Hong Kong masses. Even the few locals who have learned Chinese culture lack proper knowledge and understanding. These further contribute to a poor perception of Chinese culture among the youth, except for massive online games with Chinese cultural elements (He et al., 2022).

A variety of leisure and recreational pursuits. Hong Kong offers unlimited entertainment, leisure, and numerous activities for enriching civil life. Over the past hundred years, Hong Kong has witnessed an unprecedented population, a booming economy, and advanced technology, followed by outreach activities with its various amusement facilities, cultural venues, entertainment devices, and participatory approach. Mass entertainment requires less "ritualistic," more interactive, and more productive storytelling (Lo, Hsu, et al., 2021; Wang et al., 2022).

STRATEGIES FOR ESTABLISHING SUCCESSFUL QIN INHERITANCE AND ENGAGING CULTURAL PROMOTION AND EDUCATION IN HONG KONG

Marketing refers to the process of activities, including creating, communicating, delivering, and exchanging offerings, which has provided the consumers and the community with value (Lamb et al., 2012, p. 3). The goal is to use the appropriate marketing model to galvanize consumers' actions for purchase or engagement (Lamb et al., 2016, p. 6). The application of marketing strategies in the non-market sector, namely, social marketing, is becoming increasingly visible and strongly influences production and consumption patterns (Pashootanizadeh & Khalilian, 2018). Social marketing uses marketing philosophy, marketing techniques, and business intelligence tools to realize social values and cultural ideas (Andreasen, 2006). Under the globalized knowledge economy, marketing is not just attracting people with benefits (such as distributing coupons) but also sharing values, thus causing behavioral effects.

As information and communication technology advances, NGOs and cultural institutions must keep abreast of the times, adopt state-of-the-art technologies to make information open and instant, and launch new operations to remain competitive (Wang et al., 2022; Yu, Chiu, & Chan, 2022). In this regard, service marketing is one of the viable approaches. Institutions and organizations are increasingly finding ways to ensure effective producer-user communication while recognizing user needs and behaviors. They have used different models and strategies to improve messaging, branding, and opportunities for potential partnerships and global events (Pashootanizadeh & Khalilian, 2018). Notably, the AIDA model can be systematically applied to the cultural industry to spread and publicize to engage cultural learning. The model focuses on individual transactions and personal purchases and emphasizes that users accept the notion of products, services, and values. There are three critical parts of the marketing push in the digital age: creating and maintaining the presence of products and services, establishing relationships and social ties, and building mutual value-creating relationships with consumers (Rowley, 2002; Chan & Chiu, 2022). Therefore, AIDA is appropriate to the digital context, especially for the newer generation (Cheung et al., 2022).

Following the AIDA (attention, interest, desire, action) four-stage model, strategies are designed to reinforce brand recognition, which requires following the trend of constant innovation to draw the attention of others and appeal to others' desires and, finally, actions of learning culture (Hadiyati, 2016).

Attention

The prerequisite to drawing others' interest is to obtain their attention (Vakratsas & Ambler, 1999). The action phase typically involves an initial assessment of information channels and optimal allotment of information resources (Kahneman, 1973). People have recognized that compared to commercial advertising, embedded advertising in hit TV shows or films is quicker in promoting products (Vakratsas & Ambler, 1999). Besides, Alcañiz, García, and Blas (2009) emphasized the importance of pop culture and its positive impact on the audience's thoughts, beliefs, emotions, and even attitudes towards art goods or products. Therefore, when it comes to catching the audience's attention, in addition to traditional marketing methods, this study focuses on product placement and other smart marketing methods that mesh with pop culture (Wang et al., 2022). To gain the audience's attention, organizations need to know consumers' preferences on social media and thus select the most appropriate media channels to promote the brand. Digital 2020, an authoritative source of statistical data for making evidence-based decisions on resource allocation (Kemp, 2020), shows that young and middle-aged people are the largest social media groups, and the top three popular social media platforms are Facebook, YouTube, and WhatsApp. These platforms should make use of their advantages to attract target audience groups. For example, Choi Chang Sau Qin Making Society could cast "pearls" via Facebook and Twitter. Likewise, it can release a motion poster of activities, post short videos on YouTube, or send formal or informal messages through WhatsApp. After determining the audience and identifying the platform attributes, it is to consider some methods based on attention, such as:

- Strong headlines and graphics: attractive headlines with figures of speech like puns, hyperboles, and metaphors are useful.
- Artful use of colors like accent colors and colors in stark contrast would activate the incredible power of visual memory.
- Raising thought-provoking questions on topical issues can relate to the organization, and keeping an open mind can stimulate an interactive exchange of views.
- The audience would be more interested to see authentic and transparent information with statistics.

Summing up the analysis above, strategies in this phase are recommended as follows:

- Sign up for accounts on multiple social media to reach more people (Lam et al., 2019).
- Properly use tagging since other people with the same interest would easily find the organization by searching for the tags (Wang et al., 2022).
- Trigger and manage controversy in the products, services, or even cultural industry to stimulate talks on the organization. (Wang et al., 2022).
- Build understanding among potential consumers by using search engines for information and create cross-links with enough information to attract users. (Wang et al., 2022).
- Combine traditional promotion methods, such as roadshows with online photo-sharing and presentation and exchange fora with short videos posted on YouTube and Instagram (Lam et al., 2022).
- Create publicity and word-of-mouth marketing for the service (Chan et al., 2022).

Interest

Since interest is the prerequisite for motivation and desire (Strong, 1925), the second step is to attract considerable interest from users. This step involves using information media to approach audiences by putting on animation films or other content that shares the value of the products/services. When others would like to devote more time to know more information about the products and services, it would be a sign of success in this phase. In particular, it is difficult to engage learners or just audiences without keen interest, especially since Qin music and crafting are complex art (Mak et al., 2022). Here are some strategies according to the notion of interest, emphasizing the engagement of the audience and learners (Yu, Chiu, & Chan, 2022; Cheng et al., 2020; Lam et al., 2022; Wang et al., 2022).

- Provide concise information about the organization, including responsibility, services, and events.
- Possess a unique logo to be remembered or easily retrieved.
- Update regular information updates on social media platforms.
- Provide more online performances and recordings of Qin music, with adequate and easy-to-understand appreciation notes
- Provide easy-to-listen demonstrations and entry-level tutorials of Qin music playing to arouse more public interest
- Fostering social media discussion to engage beginners
- Anytime there is a social event related to the cultural industry, forward some content to raise public awareness promptly.
- Soft product placement. Begin a partnership with television series and hit films to establish the brand image.

Desire

The third step is essential to achieve real progress. This step requires marketers to successfully create users' needs and obtain users' confidence and trust in organizations. They also put plenty of temptation in consumers' way by recommending purchases. During this process, marketers usually need to provide examples that show consumers how to use products and others' feedback, with "before and after" techniques often conducted (Joseph, 2019). Moreover, the Scarcity Principle used since the 1960s (Cialdini, 2009) can be applied to better traction from the audience, among which the most common methods are Limited Selling and Hunger Marketing. This fundamental principle in economics states that an imbalance between supply and demand would fail to value the equation. Since resources are scarce, individuals and organizations are forced to make choices, thus creating economic events. In short, maintaining the scarcity of "diamonds" keeps consumers' desire and products' high prices and sets the mood and tone for potential consumers to some degree, making them feel compelled to purchase or engage. Thus, some methods are as follows (Cialdini, 2009):

- According to the scarcity principle, providing limited services and products, for instance, set criteria on admission requirements, periodic open internal materials, and welfare policy within a period. In particular, each Qin instrument handmade from individual wood is unique, scarce, and can be very expensive.

- Organize outreach events or initiatives in the community, city, region, or online to strengthen others' desires (Wang et al., 2022; Tse et al., 2022).
- Provide good pre-sale services for users to increase public acceptance and support.
- Manage the guestbook, oversee the comments forum, analyze customer feedback, and reply to messages timely (Lam et al., 2019; Wang et al., 2022).

Action

This is the last step in emphasizing the potential clients' motivation to take action (Lamb et al., 2012). During this phase, marketers often need to provide information such as a retail address, website address, and marketing policies to lead to purchase and participation. Marketers can also generate a sense of urgency through a price reduction. Meanwhile, mobile apps or materials often facilitate the action (Joseph, 2019; Chan et al., 2022). Some more strategies are as follows:

- Provide clear information, a replacement program, and details of past performance and reference information should be available (Wang et al., 2022).
- Freshers' party for new members.
- Clear process and appropriate techniques to address one-off shocks.

DISCUSSION AND CONCLUSION

Characterized by interaction and flexibility, social media have offered new ideas for public institutions, educators, and NGOs to establish good relationships with their audience, particularly with engagement (Sashi, 2012; Lam et al., 2022). Today, many organizations enjoy the convenience of interactive information on social media like Facebook, Instagram, and Twitter, which facilitates effective communication and develops personalized relationships with their customers (Lam et al., 2022; Wang et al., 2022; Deng et al., 2022; Chan & Chiu, 2022). Simultaneously, social media engagement involves two hierarchical levels of media consumption and proactive engagement positively related to information communication and satisfaction (Men & Tsai, 2013). Therefore, the AIDA model guides the provision of economically realistic means to strengthen impact and influence through social media.

Contemporary technology advancement has changed how users consume information and may distract focus by overloading information. How to dominate the competition and obtain steady traffic deserves careful consideration. This chapter aims to study this organization's marketing management and establish a strategic initiative to arouse a wider audience. Hence, the objective is to maximize the influence and impact through social media and accordingly yield some suggestions to contribute to cultural organizations for engaging audiences.

Cultural organizations should engage the audience in every step of the information consumption process, from the initial attention through arousing interest and obtaining a desire to the detailed negotiation, implementation, monitoring, and evaluation stage. Both traditional marketing tools and social media play a vital role in raising public awareness of preserving traditional culture and subsequently promoting appreciation and learning. Social media is undoubtedly an effective and affordable choice for public sectors to make this effort. Meanwhile, the AIDA model guides how the brand value can be raised through promotion activities with the integration of digital and traditional marketing approaches. Notably, public relations cannot be ignored, and a brand should maintain certain exposure to the country. Besides, organizations can enhance dissemination to a maximum level by following these recommendations: (i) developing skilled human resources and infrastructure and harnessing its limited public resources for cultural and industrial development; (ii) learning information and communication technology knowledge; (iii) taking steps to expand the current premises, including technical facilities and storage space; (iv) and broadening the channels to attract potential audiences with lofty ideals to come to site visits.

Although this case study covers one organization, the situations and problems are quite similar for other small NGOs. Also, many cultural and education organizations have similar problems, especially limited budgets, human resources, and lack of interest from the general public (Lo, Hsu, et al., 2022). For example, smaller special libraries and school libraries have similar user promotion and engagement problems. Even larger academic libraries may encounter problems in engaging their users on social media if the libraries' efforts are inadequate (Cheng et al., 2020; Lam et al., 2019). More importantly, this chapter demonstrates how such analysis can be applied to different education and cultural organization to customize their solutions. However, it is evident in many cases that forming a virtual community of practice helps engage audiences and learners and convert audiences to learners (Lei et al., 2021; Wang et al., 2022).

Yet, this chapter has some limitations, which require further studies and research. Notably, all analyses only cover the Qin Making Society in Hong Kong, with information selected from reports, websites, videos, and an interview with the manager. Besides, it discusses some main issues that need to be addressed to know the feasibility of the future marketing work of Choi Chang Sau Qin Making Society. We are interested in promoting other cultures and cultural organizations on social media (Wang et al., 2022; Deng et al., 2022; Yu, Chiu, & Chan, 2022; Mak et al., 2022). We are also interested in designing outreach and literacy programs for schools similar to Tse et al. (2022).

REFERENCES

Albuquerque, T. P. D. (2018). *Strategic marketing on museum audience attraction–a comparative study between Portugal and the UK* [Doctoral dissertation].

Alcañiz, E. B., García, I. S., & Blas, S. S. (2009). The functional-psychological continuum in the cognitive image of a destination: A confirmatory analysis. *Tourism Management*, *30*(5), 715–723. doi:10.1016/j.tourman.2008.10.020

Andreasen, A. R. (Ed.). (2006). *Social marketing in the 21st century*. Sage.

Aslan, I., Çınar, O., & Kumpikaitė, V. (2012). Creating strategies from tows matrix for strategic sustainable development of Kipaş Group. *Journal of Business Economics and Management*, *13*(1), 95–110. doi:10.3846/16111699.2011.620134

Au, C. H., Ho, K. K. W., & Chiu, D. K. W. (2022). The role of online misinformation and fake news in ideological polarization: Barriers, catalysts, and implication. *Information Systems Frontiers*, *24*(4), 1331–1354. Advance online publication. doi:10.100710796-021-10133-9

Barry, T. E. (1987). The development of the hierarchy of effects: An historical perspective. *Current Issues and Research in Advertising, 10*(1-2), 251-295.

Bickhoff, N., Hollensen, S., & Opresnik, M. (2014). *The quintessence of marketing: What you really need to know to manage your marketing activities*. Springer. doi:10.1007/978-3-642-45444-8

Blery, E. K., Katseli, E., & Tsara, N. (2010). Marketing for a nonprofit organization. *International Review on Public and Nonprofit Marketing*, *7*(1), 57–68. doi:10.100712208-010-0049-2

Bortree, D. S., & Seltzer, T. (2009). Dialogic strategies and outcomes: An analysis of environmental advocacy groups' Facebook profiles. *Public Relations Review*, *35*(3), 317–319. doi:10.1016/j.pubrev.2009.05.002

Butler, P. (2000). By popular demand: Marketing the arts. *Journal of Marketing Management*, *16*(4), 343–364. doi:10.1362/026725700784772871

Chan, M. M. W., & Chiu, D. K. W. (2022). Alert Driven Customer Relationship Management in Online Travel Agencies: Event-Condition-Actions rules and Key Performance Indicators. In A. Naim & S. Kautish (Eds.), *Building a Brand Image Through Electronic Customer Relationship Managementm*. IGI Global. doi:10.4018/978-1-6684-5386-5.ch012

Chan, T. T. W., Lam, A. H. C., & Chiu, D. K. W. (2020). From Facebook to Instagram: Exploring user engagement in an academic library. *Journal of Academic Librarianship*, *46*(6), 102229. doi:10.1016/j.acalib.2020.102229 PMID:34173399

Chan, V. H. Y., Ho, K. K. W., & Chiu, D. K. W. (2022). Mediating effects on the relationship between perceived service quality and public library app loyalty during the COVID-19 era. *Journal of Retailing and Consumer Services*, *67*, 102960. doi:10.1016/j.jretconser.2022.102960

Cheng, W. W. H., Lam, E. T. H., & Chiu, D. K. W. (2020). Social media as a platform in academic library marketing: A comparative study. *Journal of Academic Librarianship*, *46*(5), 102188. doi:10.1016/j.acalib.2020.102188

Cheung, T. Y., Ye, Z., & Chiu, D. K. W. (2021). Value chain analysis of information services for the visually impaired: A case study of contemporary technological solutions. *Library Hi Tech*, *39*(2), 625–642. doi:10.1108/LHT-08-2020-0185

Cheung, V. S. Y., Lo, J. C. Y., Chiu, D. K. W., & Ho, K. K. W. (2022). Predicting Facebook's influence on travel products marketing based on the AIDA model. *Information Discovery and Delivery*. Advance online publication. doi:10.1108/IDD-10-2021-0117

Chung, C., Chiu, D. K. W., Ho, K. K. W., & Au, C. H. (2020). Applying social media to environmental education: Is it more impactful than traditional media? *Information Discovery and Delivery*, *48*(4), 255–266. doi:10.1108/IDD-04-2020-0047

Cialdini, R. B. (2009). *Influence: Science and practice* (Vol. 4). Pearson Education.

Deng, S., & Chiu, D. K. W. (2022). Analyzing Hong Kong Philharmonic Orchestra's Facebook Community Engagement with the Honeycomb Model. In M. Dennis & J. Halbert (Eds.), *Community Engagement in the Online Space*. IGI Global.

Dong, G., Chiu, D. K. W., Huang, P.-S., Lung, M. M., Ho, K. K. W., & Geng, Y. (2021). Relationships between Research Supervisors and Students from Coursework-based Master's Degrees: Information Usage under Social Media. *Information Discovery and Delivery*, *49*(4), 319–327. doi:10.1108/IDD-08-2020-0100

Dukesmith, F. H. (1904). Three natural fields of salesmanship. *Salesmanship*, *2*(1), 14.

Fong, K. C. H., Au, C. H., Lam, E. T. H., & Chiu, D. K. W. (2020). Social network services for academic libraries: A study based on social capital and social proof. *Journal of Academic Librarianship*, *46*(1), 102091. doi:10.1016/j.acalib.2019.102091

Gao, W., Lam, K. M., Chiu, D. K. W., & Ho, K. K. W. (2021). A Big data Analysis of the Factors Influencing Movie Box Office in China. In Z. Sun (Ed.), *Handbook of Research on Intelligent Analytics with Multi-Industry Applications* (pp. 232–249). IGI Global. doi:10.4018/978-1-7998-4963-6.ch011

Goldberg, A., Hannan, M. T., & Kovács, B. (2016). What does it mean to span cultural boundaries? Variety and atypicality in cultural consumption. *American Sociological Review*, *81*(2), 215–241. doi:10.1177/0003122416632787

Greenberg, J., & MacAulay, M. (2009). NPO 2.0? Exploring the web presence of environmental nonprofit organizations in Canada. Global Media Journal: Canadian Edition, 2(1), 63-88.

Gross Domestic Product and its major components. (n.d.). *Hong Kong Economy*. https://www.hkeconomy.gov.hk/en/situation/development/index.htm

Hadiyati, E. (2016). Study of marketing mix and AIDA model to purchasing on line product in Indonesia. *British Journal of Marketing Studies*, *4*(7), 49–62.

Hausmann, A. (2012). Creating 'buzz': Opportunities and limitations of social media for arts institutions and their viral marketing. *International Journal of Nonprofit and Voluntary Sector Marketing*, *17*(3), 173–182. doi:10.1002/nvsm.1420

He, Z., Chiu, D. K. W., & Ho, K. K. W. (2022). Weibo Analysis on Chinese Cultural Knowledge for Gaming. In Z. Sun (Ed.), *Handbook of Research on Foundations and Applications of Intelligent Business Analytics* (pp. 320–349). IGI Global. doi:10.4018/978-1-7998-9016-4.ch015

Herrera-Franco, G., Carrión-Mero, P., Alvarado, N., Morante-Carballo, F., Maldonado, A., Caldevilla, P., Briones-Bitar, J., & Berrezueta, E. (2020). Geosites and Georesources to Foster Geotourism in Communities: Case Study of the Santa Elena Peninsula Geopark Project in Ecuador. *Sustainability (Basel, Switzerland)*, *12*(11), 4484. doi:10.3390u12114484

Huang, P. S., Paulino, Y., So, S., Chiu, D. K. W., & Ho, K. K. W. (2021). Special Issue Editorial - COVID-19 Pandemic and Health Informatics (Part 1). *Library Hi Tech*, *39*(3), 693–695. doi:10.1108/LHT-09-2021-324

Huang, P.-S., Paulino, Y. C., So, S., Chiu, D. K. W., & Ho, K. K. W. (2022). Guest editorial: COVID-19 Pandemic and Health Informatics Part 2. *Library Hi Tech*, *40*(2), 281–285. doi:10.1108/LHT-04-2022-447

Hudson, S., Wang, Y., & Gil, S. M. (2011). The influence of a film on destination image and the desire to travel: A cross-cultural comparison. *International Journal of Tourism Research*, *13*(2), 177–190.

Jiang, T., Lo, P., Cheuk, M. K., Chiu, D. K. W., Chu, M. Y., Zhang, X., Zhou, Q., Liu, Q., Tang, J., Zhang, X., Sun, X., Ye, Z., Yang, M., & Lam, S. K. (2019). 文化新語:兩岸四地傑出圖書館、檔案館及博物館傑出工作者訪談 [New Cultural Dialog: Interviews with Outstanding Librarians, Archivists, and Curators in Greater China]. Hong Kong: Systech publications.

Jobber, D., & Ellis-Chadwick, F. (2012). *Principles and practice of marketing*. McGraw-Hill Europe.

Joseph, C. (2019, February 5). *A.I.D.A model in marketing communication*. Small Business - Chron.com. https://smallbusiness.chron.com/aida-model-marketing-communication-10863.html

Kahneman, D. (1973). *Attention and effort* (Vol. 1063). Prentice-Hall.

Kaplan, A. M., & Haenlein, M. (2010). Users of the world, unite! The challenges and opportunities of Social Media. *Business Horizons*, *53*(1), 59–68. doi:10.1016/j.bushor.2009.09.003

Kemp, S. (2020). *Digital 2020: Hong Kong*. Available at: https://datareportal.com/reports/digital-2020-hong-kong

Kent, M. L., Taylor, M., & White, W. J. (2003). The relationship between Web site design and organizational responsiveness to stakeholders. *Public Relations Review*, *29*(1), 63–77. doi:10.1016/S0363-8111(02)00194-7

Kshetri, N. (2020). China's emergence as the global fintech capital and implications for southeast asia. *Asia Policy*, *27*(1), 61–81. doi:10.1353/asp.2020.0004

Lam, A. H. C., Chiu, D. K. W., & Ho, K. K. W. (2022). Instagram for student learning and library promotions? A quantitative study using the 5E Instructional Model. *Aslib Journal of Information Management*. Advance online publication. doi:10.1108/AJIM-12-2021-0389

Lam, E. T. H., Au, C. H., & Chiu, D. K. (2019). Analyzing the use of Facebook among university libraries in Hong Kong. *Journal of Academic Librarianship*, *45*(3), 175–183. doi:10.1016/j.acalib.2019.02.007

Lamb, C. W., Hair, J. F., & McDaniel, C. (2012). *Marketing* (12th ed.). Cengage Learning.

Lamb, C. W., Hair, J. F., & McDaniel, C. (2016). MKTG 10 (10th ed.). Cengage Learning.

Lau, K. S. N., Lo, P., Chiu, D. K. W., Ho, K. K. W., Jiang, T., Zhou, Q., Percy, P., & Allard, B. (2020). Library, Learning, and Recreational Experiences Turned Mobile: A Comparative Study between LIS and non-LIS students. *Journal of Academic Librarianship*, *46*(2), 102103. doi:10.1016/j.acalib.2019.102103

Lee, S. H., & Hoffman, K. D. (2015). Learning the ShamWow: Creating infomercials to teach the AIDA model. *Marketing Education Review*, *25*(1), 9–14. doi:10.1080/10528008.2015.999586

Lei, S. Y., Chiu, D. K. W., Lung, M. M., & Chan, C. T. (2021). Exploring the Aids of Social Media for Musical Instrument Education. *International Journal of Music Education*, *39*(2), 187–201. doi:10.1177/0255761420986217

Leijerholt, U., Biedenbach, G., & Hultén, P. (2019). Branding in the public sector: A systematic literature review and directions for future research. *Journal of Brand Management*, *26*(2), 126–140. doi:10.105741262-018-0116-2

Leung, T. N., Luk, C. K. L., Chiu, D. K. W., & Kevin, K. K. W. (2022). User perceptions, academic library usage, and social capital: A correlation analysis under COVID-19 after library renovation. *Library Hi Tech, 40*(2), 304–322. doi:10.1108/LHT-04-2021-0122

Lewis, E. S. E. (1909). The Duty and Privilege of Advertisng a Bank. *Bankers' Magazine, 78*(4), 710-711.

Li, J., & Yu, H. (2013). An Innovative Marketing Model Based on AIDA:-A Case from E-bank Campus-marketing by China Construction Bank. *I-Business*, *5*(03, 3B), 47–51. doi:10.4236/ib.2013.53B010

Lin, Y. S., & Huang, J. Y. (2006). Internet blogs as a tourism marketing medium: A case study. *Journal of Business Research*, *59*(10-11), 1201–1205. doi:10.1016/j.jbusres.2005.11.005 PMID:32287521

Lo, P., Chan, H. H. Y., Tang, A. W. M., Chiu, D. K. W., Cho, A., Ho, K. K. W., See-To, E., He, J., Kenderdine, S., & Shaw, J. (2019). Visualising and Revitalising Traditional Chinese Martial Arts – Visitors' Engagement and Learning Experience at the 300 Years of Hakka KungFu. *Library Hi Tech, 37*(2), 273–292. doi:10.1108/LHT-05-2018-0071

Lo, P., Cheuk, M. K., Ng, C. H., Lam, S. K., & Chiu, D. K. W. (2021). 文武之道【下冊】:以拳入哲 [The Tao of Arts and Warriorship: The Philosophy of Fists]. Systech Publications.

Lo, P., Hsu, W.-E., Wu, S. H. S., Travis, J., & Chiu, D. K. W. (2021). *Creating a Global Cultural City via Public Participation in the Arts: Conversations with Hong Kong's Leading Arts and Cultural Administrators*. Nova Science Publishers.

Lovejoy, K., & Saxton, G. D. (2012). Information, community, and action: How nonprofit organizations use social media. *Journal of Computer-Mediated Communication*, *17*(3), 337–353. doi:10.1111/j.1083-6101.2012.01576.x

Macias, W., Hilyard, K., & Freimuth, V. (2009). Blog functions as risk and crisis communication during Hurricane Katrina. *Journal of Computer-Mediated Communication, 15*(1), 1–31. doi:10.1111/j.1083-6101.2009.01490.x

Mak, M. Y. C., Poon, A. Y. M., & Chiu, D. K. W. (2022). Using Social Media as Learning Aids and Preservation: Chinese Martial Arts in Hong Kong. In S. Papadakis & A. Kapaniaris (Eds.), *The Digital Folklore of Cyberculture and Digital Humanities*. IGI Global. doi:10.4018/978-1-6684-4461-0.ch010

McElhone, R. (2020, January 25). *Using the AIDA model to get 'buy in'*. B Online Learning. https://bonlinelearning.com/using-the-aida-model-to-get-buy-in-elearning/

Men, L. R., & Muralidharan, S. (2017). Understanding social media peer communication and organization–public relationships: Evidence from China and the United States. *Journalism & Mass Communication Quarterly*, *94*(1), 81–101. doi:10.1177/1077699016674187

Men, L. R., & Tsai, W. H. S. (2013). Toward an integrated model of public engagement on corporate social networking sites: Antecedents, the process, and relational outcomes. *International Journal of Strategic Communication*, *7*(4), 257–273. doi:10.1080/1553118X.2013.822373

Men, L. R., & Tsai, W. H. S. (2014). Perceptual, attitudinal, and behavioral outcomes of organization–public engagement on corporate social networking sites. *Journal of Public Relations Research*, *26*(5), 417–435. doi:10.1080/1062726X.2014.951047

Mohezar, S., Moghavvemi, S., & Zailani, S. (2017). Malaysian Islamic medical tourism market: A SWOT analysis. *Journal of Islamic Marketing*, *8*(3), 444–460. doi:10.1108/JIMA-04-2015-0027

Oakes, S., Dennis, N., & Oakes, H. (2013). Web-based forums and metaphysical branding. *Journal of Marketing Management*, *29*(5-6), 607–624. doi:10.1080/0267257X.2013.774289

Pashootanizadeh, M., & Khalilian, S. (2018). Application of the AIDA model: Measuring the effectiveness of television programs in encouraging teenagers to use public libraries. *Information and Learning Science*, *119*(11), 635–651. doi:10.1108/ILS-04-2018-0028

Patsiaouras, G., Veneti, A., & Green, W. (2018). Marketing, art and voices of dissent: Promotional methods of protest art by the 2014 Hong Kong's Umbrella Movement. *Marketing Theory*, *18*(1), 75–100. doi:10.1177/1470593117724609

Polk, X. L. (2018). Marketing: The key to successful teaching and learning. *Journal of Marketing Development and Competitiveness*, *12*(2), 49–57.

Rowley, J. (2002). Information marketing in a digital world. *Library Hi Tech*, *20*(3), 352–358. doi:10.1108/07378830210444540

Sashi, C. M. (2012). Customer engagement, buyer-seller relationships, and social media. *Management Decision*, *50*(2), 253–272. doi:10.1108/00251741211203551

Saxton, G. D., Guo, S. C., & Brown, W. A. (2007). New dimensions of nonprofit responsiveness: The application and promise of Internet-based technologies. *Public Performance & Management Review*, *31*(2), 144–173. doi:10.2753/PMR1530-9576310201

Smith, D. A. (2011). Strategic Marketing of Library Resources and Services. *College & Undergraduate Libraries*, *18*(4), 333–349. doi:10.1080/10691316.2011.624937

Statista. (2019, December). *Statista TrendCompass 2020*. https://www.statista.com/study/69166/statista-trendcompass/

Statista. (2020). *Social media usage in the United Kingdom (UK)*. https://www.statista.com/study/21322/social-media-usage-in-t
he-united-kingdom-statista-dossier/

Stepaniuk, K. (2017). Blog content management in shaping pro recreational attitudes. *Journal of Business Economics and Management*, *18*(1), 146–162. doi:10.3846/16111699.2017.1280693

Strong, E. K. (1925). *The psychology of selling and advertising*. McGraw-Hill book Company, Incorporated.

Sun, X., Chiu, D. K. W., & Chan, C. T. (2022). Recent Digitalization Development of Buddhist Libraries: A Comparative Case Study. In S. Papadakis & A. Kapaniaris (Eds.), *The Digital Folklore of Cyberculture and Digital Humanities*. IGI Global. doi:10.4018/978-1-6684-4461-0.ch014

Temporal, P. (2015). *Branding for the public sector: Brand Communications Strategy*. John Wiley & Sons.

Tsang, A. L. Y., & Chiu, D. K. W. (2022). Effectiveness of Virtual Reference Services in Academic Libraries: A Qualitative Study based on the 5E Learning Model. *Journal of Academic Librarianship*, *48*(4), 012533. doi:10.1016/j.acalib.2022.102533

Tse, H. L., Chiu, D. K., & Lam, A. H. (2022). From Reading Promotion to Digital Literacy: An Analysis of Digitalizing Mobile Library Services With the 5E Instructional Model. In A. Almeida & S. Esteves (Eds.), *Modern Reading Practices and Collaboration Between Schools, Family, and Community* (pp. 239–256). IGI Global. doi:10.4018/978-1-7998-9750-7.ch011

Vakratsas, D., & Ambler, T. (1999). How advertising works: What do we really know? *Journal of Marketing*, *63*(1), 26–43. doi:10.1177/002224299906300103

Wang, P., Chiu, D. K. W., Ho, K. K., & Lo, P. (2016). Why read it on your mobile device? Change in reading habit of electronic magazines for university students. *Journal of Academic Librarianship*, *42*(6), 664–669. doi:10.1016/j.acalib.2016.08.007

Wang, W., Lam, E. T. H., Chiu, D. K. W., Lung, M. M., & Ho, K. K. W. (2021). Supporting Higher Education with Social Networks: Trust and Privacy vs. Perceived Effectiveness. *Online Information Review*, *45*(1), 207–219. doi:10.1108/OIR-02-2020-0042

Wexler, M. N. (2005). *Leadership in context: The four faces of capitalism*. Edward Elgar Publishing.

Yao, L., Lei, J., Chiu, D. K. W., & Xie, Z. (2022). Adult Learners' Perception of Online Language English Learning Platforms in China. In A. Garcés-Manzanera & M. E. C. García (Eds.), *New Approaches to the Investigation of Language Teaching and Literature*. IGI Global.

Yip, K. H. T., Chiu, D. K. W., Ho, K. K. W., & Lo, P. (2021). Adoption of Mobile Library Apps as Learning Tools in Higher Education: A Tale between Hong Kong and Japan. *Online Information Review*, *45*(2), 389–405. doi:10.1108/OIR-07-2020-0287

Yu, H. H. K., Chiu, D. K. W., & Chan, C. T. (2022). Resilience of symphony orchestras to challenges in the COVID-19 era: Analyzing the Hong Kong Philharmonic Orchestra with Porter's five force model. In W. Aloulou (Ed.), *Handbook of Research on Entrepreneurship and Organizational Resilience During Unprecedented Times*. IGI Global. doi:10.4018/978-1-6684-4605-8.ch026

Yu, H. Y., Tsoi, Y. Y., Rhim, A. H. R., Chiu, D. K., & Lung, M. M. W. (2021). Changes in habits of electronic news usage on mobile devices in university students: A comparative survey. *Library Hi Tech*. Advance online publication. doi:10.1108/LHT-03-2021-0085

Yu, P. Y., Lam, E. T. H., & Chiu, D. K. W. (2022). Operation management of academic libraries in Hong Kong under COVID-19. *Library Hi Tech*. Advance online publication. doi:10.1108/LHT-10-2021-0342

Chapter 5
Co-Constructing Belongingness:
Strategies for Creating Community and Shared Purpose Online - The Social Construction of Community and Meaning

Lilya Shienko
Adler University, USA

Barton David Buechner
Adler University, USA

ABSTRACT

Online environments have become a fact of life for education and business life, but participants do not always experience virtual space as a place where they truly belong. This case example illustrates some ways that social construction concepts, particularly the "communication perspective" of the coordinated management of meaning (CMM) theory, were applied to the development and enactment of an inclusive collaborative online learning program. This online interactive space was envisioned as a "virtual home" and support group for fellowship participants for the development and presentation of their individual projects. Perspectives of program participants and organizers are both represented in this case example to help shed light on the way that the group space evolved over the course of the fellowship and lessons learned from the process. The use of concepts from CMM theory is underscored to reveal the dynamics of communication theory as a source of adding life and dimension to an online collaborative space, contributing to a sense of belongingness.

DOI: 10.4018/978-1-6684-5190-8.ch005

INTRODUCTION

In this chapter, we discuss ways that online communities can be built, shaped, and enacted in ways that enhance a shared sense of belongingness by focusing attention on the dynamics of the communication process. As we spend more and more of our time online, it is increasingly evident just how much the quality of our social experience is shaped by the ways that we interact in virtual space. The online environment is particularly notable as a human construct, and therefore subject to the creative and constitutive input of participants to shape the quality of the experience. The notion of social reality as a human construct has been recognized in the human sciences for some time (Berger & Luckmann, 1967). The approaches to building community and belongingness online described in this chapter are informed by principles of social construction, and further draw on a body of communication theory referred to as the Coordinated Management of Meaning (CMM) (Pearce & Cronen, 1980). Unlike more traditional models of communication, CMM theory views communication as a dynamic and interactive process, not just the conveying of information from a sender to a receiver (Pearce, 2007). This approach, referred to as the "communication perspective" can be applied in helping groups to look at their processes of communication as the grounding for their experience of online social reality. By better understanding this process, and their role in it, members of a group can take ownership for creating a group space that is welcoming, inclusive, and creative. Communication with the group is then experienced as a constitutive process, in that the very act of communicating creates the meanings generated by the people involved.

The Virtual Social World

Technology and uses of new media have greatly expanded the complexity and scope of our social worlds. Today's information technology systems work at an exponential pace - we have emerging news, accumulated knowledge, and more at our fingertips and available online 24/7 to interact with whenever needed. We can anticipate that technology will continue to impact the field of education, business, and politics, and that forms of communication will continue to become more complex. There are no more limitations of distance, we are now free to connect with those many miles away, and discuss issues that transcend our local boundaries and social horizons. More than just reflecting the current events in our social worlds, these new forms of media can themselves become the message (Mcluhan, 1995), in the sense of adding to the capacity to amplify, conceal, or distort conflicts, affinities, and trends – whether positive or negative. Additionally, our expanded online social worlds let more diverse "others" into our lives, and along with them come the influences of other cultures, ideologies, and life experiences. This raises the possibility of needing more nuanced, or "cosmopolitan" forms of communication that allow for the accommodation of more richness, the toleration of ambiguity, and include more reflective space to promote deeper listening before responding (Penman, 2021). In this sense, the quality of our online experience can be shaped by the interaction of the characteristics of the space itself, and our ability to navigate this complexity. We may well ask ourselves how we can live in harmony within this expanded social world? Will we "belong" more simply by assimilating more information through media technology? Is it possible to mirror face-to-face interactions with others in the online environment, and will this evoke the same engagement and emotions? This presents some other practical questions: will individuals be able to bond effectively with each other at a distance? Will it be the same as for individuals who interact daily in person and share common experiences? To answer these and other emergent questions, it is helpful to look more deeply at the communicative processes of forming of relationships with diverse others, and making shared meaning of new information in conversation with them.

Creating Belongingness Online

As the world has moved even more deeply in recent years into the online environment as a primary modality for working, learning and collaborating, institutions and individuals have been forced to contend with ways to reclaim the sense of immediacy and intimacy that comes with physically gathering together. Many have found that they miss the accompanying rituals and norms that bind groups together at the interpersonal level, while at the same time appreciating new freedoms inherent in alternative forms of coordinating organizational activities from a distance. For the most part, the online collaborative spaces that have arisen in this era of forced experimentation have been defined more by the enabling technology than by the accompanying human qualities and practices. This can be attributed to these online environments being highly structured, pre-defined by technological experts and administrators, and driven by "legacy" ideas about required content, measurable objectives, and set agendas. In other words, in responding to the sudden disruption in our organizational practices, we have in many cases compressed (or suppressed) some aspects of our organizational social worlds in order to fit our activity into the virtual space, using the existing containers available at the time. The concept of belonging is becoming more and more important in a time when physical distance is an increasingly common fact of life for distributed learning and working groups, while at the same time, Diversity, Equity and Inclusion (DEI) are at the forefront of awareness in most organizations. Groups tend to operate on assumptions of acceptance of certain culturally-imbedded values and accepted practices, but this can be problematic when some within the group do not share the same cultural background, and run the risk of becoming marginalized.

The Inclusive Power of Co-Construction

The present chapter explores possibilities for online groups to co-create their own processes and identity while also accomplishing individual project objectives. This begins with a question: what if, instead of just asking participants to conform to established conventions of the online environment, we allow them to co-create some parts of it together? Taking this approach transforms the concept of online spaces from a vehicle for connection and repository of information into a truly interactive home for collaboration, in which an invitation to co-creation of the space itself is built into the agenda. This process of co-creation is an intentionally creative act, in which power relationships are de-emphasized and the focus is shifted away from accepting or relying on a traditional, pre-existing structure to the dynamic forces shaping the group's collective identity and sense of being together. The experience becomes less about adapting or fitting in, and more about making a contribution to creating social reality. By taking ownership of these formative aspects of the group's processes, members have the ability to bring more of their "whole selves" into the group in a more authentic and creative way, and are, in turn more likely to experience a sense of belonging.

LITERATURE REVIEW

The concepts and practices described in this chapter are derived from a robust body of human science literature concerning the social construction of reality. The underlying premise for this focus is that the online environment is increasingly defined by the interactions of participants and the ways that people are using the technology, and less impeded by constraints of the natural world. The literature of social construction concerns itself with the balance between the objective and subjective, and also takes into consideration contextual factors including culture, power relationships, and purpose (Gergen, 1999). Social constructionist concepts are also part of the literature of psychology, particularly Adlerian psychology which considers social feeling to be a requisite for mental health and well-being (Watts, Williamson, & Williamson, 2004).

Social Construction of the Lifeworld

Building on the phenomenological worldviews of Martin Heidegger, Hans Gadamer, and Alfred Schutz, Peter Berger and Thomas Luckmann expanded on the concept of the "lifeworld" as the ground for all human experience in everyday life as a process of the "sociology of knowledge" (Berger & Luckmann, 1966, p.3). In the social constructionist paradigm, human knowledge is created through a social process, which in turn shapes the way (social) reality is experienced (Gergen, 1999), including sense of self (identity) and personal beliefs. Social construction processes also operate at the systemic level, serving to create overarching constructs such as cultures, moral codes and collective identities. In a practical sense, these social construction processes are carried out through language and communication. As proposed by Gadamer, each individual operates from a "horizon of understanding" which can only be expanded through "dialogic relationship" with others, resulting in a "fusion of horizons" (Gergen, 1999, p. 144). Using this model, it is possible to envision dialogic processes which are capable of expanding, and thereby bridging, the scope and meaning of experiences among a diverse group.

Design Thinking

When we consider social worlds as constructs, one way to approach social construction would be to follow principles of engineering and design to make "better" or at least more intentional social worlds. One way to define the stages of design thinking is as follows (Dam, 2021):

Stage 1: Empathize—Research Your Users' Needs.
Stage 2: Define—State Your Users' Needs and Problems.
Stage 3: Ideate—Challenge Assumptions and Create Ideas.
Stage 4: Prototype—Start to Create Solutions.
Stage 5: Test—Try Your Solutions Out.

This design thinking structure promotes an orderly approach to creating a social system from a user centric point of view, guided by expert knowledge and controlled by the designer or a design team. The process includes preparatory steps of interviewing users and gaining their input, but the process itself is not necessarily participatory.

Disruptive Change

The phenomenon of "disruptive change" has gained relevance in many organizations, and learning how to make predictions and adapt has become something of an art - particularly in the time of the COVID-19 pandemic. Some implications of disruptive changes for organizations are that, if the disruption is seen as a threat, the tendency may be to over-react; conversely, if it is seen as more of an opportunity, the organization might underestimate what needs to be done and continue more or less as usual. (Gilbert & Bower, 2002). Another view of dealing with disruptive change is to see it as an opportunity for development and growth by clearing away forms and structures that no longer serve their intended purpose, sometimes referred to as positive disintegration (Tiller, n.d.). For purposes of this discussion, disruptive change can be seen as a form of releasing previous structures of the lifeworld, opening possibilities to co-create something new together.

The Co-Construction of Individual and Collective Identity

When considering the socialization of knowledge, it is also helpful to consider the way knowledge is imparted in education processes, and the balance between the individual and collective level of development. In most social systems, there is a balance to be struck between "individual freedom" and "group solidarity" (Belanger, 2015, 178). This in turn surfaces tensions between cognitivist and social constructionist approaches to learning, in which the former is seen to rely on imparting a particular way of thinking or behaving on learners, and the second is a more learner-directed approach to constructing identity and beliefs, drawing on both "internal and external resources" (Belanger, 2015, p. 47). The approach of co-construction balances these forces by assigning to the group the ability to direct the learning and socialization process, through dialogue among the group. This interpersonal approach, which is one of the core elements of Adlerian psychology, has an impact on overall well-being and mental health (Watts, Williamson & Williamson, 2004). It also effectively bridges between cognitive constructivist and social constructionist perspectives, shifting attention to intersubjectivity (Watts et al., 2004, p.9).

The Communication Perspective and CMM

If we consider that knowledge is created in social processes, it then becomes helpful to look at the specific forms and processes of communication that underlie the creative process. The Coordinated Management of Meaning (CMM) theory was developed in an effort to better understand and map out these processes, in a way that enables their practice (and outcomes) to become more intentional (Pearce & Cronen, 1980). This conceptual approach has been developed over a 45 year time span, and has been used in fields as disparate as family therapy and conflict management (Barge & Pearce, 2004). Over this time, CMM has progressively moved from being a mostly interpretive theory to become a "practical theory" (Cronen, 2001), offering a way of deconstructing problematic social interactions and mapping out approaches to imagine and create new alternatives. It has more recently been applied in Diversity, Equity, and Inclusion (DEI) work (Wasserman, 2014, Afuape, 2011) and in creating psychological resilience and addressing moral injury with military members and veterans (Buechner, Van Middendorp & Spann, 2018, MacKenzie, Steen and Buechner, 2018).

Here are some of the key concepts and models of CMM that will be used in this chapter. The source of these definitions is "Making Social Worlds (Pearce, 2007) unless otherwise indicated.

Communication as Constitutive

Communication can be seen as more than just the exchanging of information, but rather it is an act of creating social reality in relationships with others, referred to as the "communication perspective" (p. xiv). Within this paradigm of communication, meaning is made turn-by-turn in conversations with others. This includes choices on whether - and how- to be a part of the conversation. Recognizing this, we can then break down the critical moments in which the way we interact determines our future (collective) social reality. Communication and belonging are therefore inherently intertwined, as we cannot create a sense of belonging without having communication. For this purpose – taking the communication perspective is an invitation to see the processes of meaning making in communicating in a new light. When there is inability to coordinate these different ideas, that quality of belongingness is also diminished.

Critical Moments

The CMM concept of critical moments (or "bifurcation points") in our social interactions helps us to understand our role in shaping our social reality by "acting wisely" in episodes of communication, with the realization that "every conversation has an afterlife" (p. xi). In this sense, communication is generative, meaning that it is a process of doing and creating things, not just talking about them (p. xiii). Knowing that conversations are joint acts, are reciprocal, and have lasting consequences in shaping perceptions and setting up future interactions, shows the weight of our words for creating the social reality that remains after a particular episode of communication takes place. Recognizing and understanding the significance of the critical points in conversation helps us to take responsibility for how what one says next affects the unfolding pattern of interaction and potentially takes it in a different direction. Being aware of these points can lead one to control a situation that could end badly by changing the direction of the conversation to seek to understand the other person despite disagreement. Our conversations can have a better afterlife if we are self-aware of the critical moments and how the exchange shapes not only our identity – but our social worlds as well.

The "Relational Mind"

As we learn more about the communication dynamics behind the ways that we coordinate our actions and make meanings from them, our awareness increases of the way that we make shared meaning of events that we experience together, and how this serves to create a sense of "oneness" among groups (p.217). The "next level" of this turn from individual to collective-level thinking further reveals a shared quality of "relational mind" (p. 213), which evolves over time and can include elements of group identity and promote synergy among the group. This happens through a process of relationship-building that can be developed through the intentional process of story-telling, in which the group pays specific attention to "stories told" about "stories lived" and the intentionality of stories that are either told, or untold, heard or unheard, or in some cases, unknown (p.212). Thinking about the relational mind as an essential quality of a cohesive group can help us pay attention to the things (aspects of being together) that help members of a group feel like they truly belong. It may also help a group (including an online group) to become something more than just the sum of its individual parts.

Cosmopolitan Communication

Group cultures tend to vary with respect to the way they engage with those seen as outsiders. This can be seen as taking one of four approaches: monocultural, ethnocentric, modernistic, or cosmopolitan (Pearce, 1989, p. 168). At the monocultural level, everyone is seen as the same; within the ethnocentric view, others are acknowledged as different but inferior; in the modernistic view, others are tolerated and differences are minimized. Only at the level of Cosmopolitan Communication do we engage in coordinating with others in a way that fully respects the common humanity of the other, while also appreciating differences. In the cosmopolitan model of communication, interactions take the form of either "coherence, coordination, or Mystery" (Pearce, 1989, p. 169). Since we cannot always rely on coherence (culturally-based understanding) to base actions, more attention is generally paid to coordination, which allows for collaboration despite differences in understanding. Allowing space for emergence (or Mystery) is important, as it is the social space in which new shared meaning (co-constructed knowledge and meanings within the group) can take form.

The next section will describe the process of setting up and enacting a collaborative space for a fellowship program drawing on the key concepts of social construction and CMM theory described above. These concepts will then be re-visited, in the words of the participants, with emphasis on their importance to the way the group evolved and worked together by applying them.

CASE EXAMPLE: THE CMMI FELLOWS PROJECT

In the case study presented here, the organizers intuitively drew on ten years of experience working with previous fellows' cohorts, and had set out a general theme to be followed for individual fellows' proposals. They also added several new elements based on prior fellows' experience, and invited the new group to be a part of a "thought experiment" that was being tried by creating more open space for sharing, elements of conceptually-based group learning, and in invitation for the group to co-construct their final presentation together.

The group came together periodically over the fellows year in gatherings that started with open-ended discussions, progressed into a group learning space with individual project support, and culminated in an open webinar program presented by the group.

Background

The CMM Institute (CMMI) was established in 2010 to perpetuate scholarship and practice related to the Coordinated Management of Meaning (CMM) theory. The fellowship program was instituted in 2012. Each group of CMMI fellows has consisted of widely distributed scholar-practitioners, competitively selected, each working on individualized projects with some connection to CMM theory. In the early days of the fellows program, participants worked independently and came together at a conference to present their projects to the broader CMM community of scholarship and practice. Later, the practice of holding periodic preparatory meetings was added to create a sense of familiarity among the group and answer questions related to their use of the CMM theory. A peer-mentoring initiative was initiated to further increase support and connections ahead of the conference presentations. In response to the global COVID-19 pandemic in 2020, the face-to-face conference was replaced by an online meeting, and the online preparatory meetings took on a greater significance. For the the 2021-22 cohort, the call for proposals was specifically focused on using CMM principles to help bridge the fragmented social realities and identities made more visible by the pandemic, and a more formalized CMM-based course of study was added to the online meetings.

Group Process

The project structure was left open-ended as much as possible to help create a welcoming and accepting online environment for all of the fellows. Several key elements were progressively inserted into the process, such as peer feedback and recommendations for contacting others who were known to be pursuing similar interests. These things were done in a deliberately non-directive way, with the intention of keeping the locus of control within the group, and the group focused on supporting each others' project development and learning. Over the course of eight months, the fellows group met periodically (on a bi-weekly basis) at times that were proposed and agreed upon by the group. Meeting structure included individual "check-ins" from all participants, interspersed with short presentation on the theoretical concepts of the Coordinated Management of Meaning (CMM) and individual mentoring connections on applying these principles to fellows' individual projects. At each step in this process, the group members were invited to attend to the collaboration space, including meeting times and scheduling, and agenda. The group was also given the responsibility to develop a plan for the presentation of their projects to a broader community of practice within the communication field. The modalities for group collaboration were both synchronous and asynchronous, in the sense that meetings were held for those able to participate, and recorded for the benefit of those who could not. The *BaseCamp* collaborative platform was used to anchor the process, which provided for space to share resources and emerging thoughts, and for posting of responses to questions and activities. Constructive feedback was given each week by the group organizers on both content and process of the gatherings.

Shared Ownership

From the beginning of their time together, the group of fellows was given both the freedom and the responsibility for their collaboration together. This included a role in setting or modifying group processes, such as the times and dates to meet and the process of interacting among each other during the synchronous meetings. The intention was to cultivate a sense of belongingness by sharing personal stories behind individual projects, and giving the group some control over their group processes. While some wanted more connection time, everyone acknowledged that there was a need to balance quality and frequency of synchronous meetings with busy schedules of all participants, and subject-matter experts/ organizers. The group was also given an asynchronous space (on the "*BaseCamp*" platform) to share documents and comments on each other's projects as they developed.

Evolution of the Learning Space

The context, or online "container" for the collaborative work of this group of fellows was initially designed as a collaborative space, with intention to allow group members to employ models and concepts of CMM theory re-define and shape their process of convening and collaborating from a communication perspective. In the early meetings, discussion time was primarily allocated to sharing of personal stories and explaining the individual fellow's project concepts, inviting input from others in the group and the organizers. While members of the fellows group did not necessarily have a full grasp of CMM theory at the beginning of the project, they were invited to actively experiment with it along the way, as concepts were introduced. After the group had met for a few times and established their rhythm for meeting and interacting, content from "Making Social Worlds" was introduced through readings, more formalized postings, and discussion in the bi-weekly meetings. This included a rhythm and expectations for posting and receiving feedback on asynchronous postings, and sharing documents among the group. Several heuristic models drawn from the larger body of CMM theory were engaged as practical tools to help enlarge perspective and sharpen the "learning edge" of the group, and facilitate and deepen their process of coming together to support each other in learning. This took place on three levels; creating a shared understanding of their individual projects, grasping and using the CMM/social construction concepts, and understanding and appreciating each other on an interpersonal level.

Connecting With a Community of Practice

Fellows were invited to deepen their understanding of CMM theory through interactions with others, and encouraged to take an active role in the evolution of the theory through their own contributions and connections. As the circle of this conversation was expanded, concepts and practices of CMM were gradually introduced in response to conversational turns in the discussions. This began with a broad conceptual framework of the Coordinated Management of Meaning (CMM) theory, after which specific examples and conceptual models were brought into the conversation among the fellows, as connections emerged with their projects. This approach was designed to augment and support each individual fellow's project responsively through creating a supportive environment. This also included a peer mentoring process, through which the intention was that this would serve as an invitation to further collaboration, both within the group and within a broader community of practice. To this end, an informal peer mentoring process was initiated by the organizers, in which connections between the work of individual fellows and others were intentionally pointed out, and introductions made to known CMM scholars (including past-year fellows). Through participating in this expanding web of conversations, the fellows were progressively drawn to reflect on the implications of CMM theory for their individual projects, and invited to add their contributions to the continuing evolution of the theory.

Group Seminar Study: "Making Social Worlds

As a first-time initiative, the 2021-22 CMM fellows group process included an optional seminar on CMM theory, using the book *Making Social Worlds* (Pearce, 2007) as text. This seminar course was led by a past-year CMMI fellow, and was positioned not as a requirement but an "invitation" to join together in community to look at familiar things and see them in a new way by applying a different set of lenses from CMM theory in each session. Sharing among the group was done in both online posts in the *BaseCamp* online platform, and verbally in online *Zoom* conference call meetings. Participants were introduced to a set of conceptual tools or "heuristics" (p. 89) as tools to help them to look past the taken-for-granted meanings in conversational settings and group processes and call attention to these in a purposeful way. Within this context, each participant was invited to share their views on each chapter, and explore ways in which the theories may be applied in their projects. Through this process, particular chapters and concepts from this book stood out to students as particularly valuable; these will be briefly outlined in the next section on participant experience.

PARTICIPANT EXPERIENCE

This section examines the way that the students experienced their group process, including their forms of participation, and the practical engagement with key concepts of CMM theory through the realizations of the group. This CMMI fellows' group was diverse and inter-disciplinary in many ways, including participants from four countries, and schools ranging from psychology to intercultural relations. This level of diversity served to shape and expand the evolving identity of the group, as members connected their own cultures, lived experience and academic disciplines through the scaffolding of a unifying body of theory.

Application of CMM Concepts

The explanatory narrative and selected quotations shown here were extracted from postings in the asynchronous posting space of the CMMI fellows course. The comments shown are sampled from postings by various individual fellows.

Critical Moments

The concept that generated the most conversation in online activity and discussions was the concept of "critical moments" from the first Chapter of Making Social Worlds (Pearce, 2007). Critical moments are turning points or moments in conversations in which a comment or action determines the flow of the rest of the conversation or interaction. This can be in one conversation or a series over a period of time. It is these moments that determine how one understands the other, reacts to the other, and essentially sets the course as to whether one wants to continue interacting in the future. Relationships are built on critical moments in conversations - hence many of the participants found that they could relate the online conversations with other episodes of conversing in daily life with others. Awareness of the conceptual models of CMM aided them in visualizing how an enhanced understanding of the other can result in changes in life and relationships by acting into these turning points more consciously in conversations.

"The reason I personally valued the chapter (on critical moments) so much was due to its introduction of dialogic communication - whether in politics, organizations, or personal relations. Ordinary people without formal training can apply these tips to communicate more effectively... producing better results."

These concepts also were helpful as heuristics for diagnosing past cases of conversations gone wrong, where undesired outcomes were produced:

"The concept of critical moments helps us to better interpret real-life situations where barriers in communication cause a misunderstanding, and as a result, a sense of alienation or lack of belonging. This may stem from a conversation where individuals would see the problems only from their perspective – a common phenomenon and cause of conflict. In such circumstances, each individual misses the opportunity to change the course of action of the conflict, often with tragic results. This shows the weight that communication holds in our life."

Critical moments often involve opportunities to change the communication dynamic from a debate (where one side wins and the other loses) into a constructive dialogue, in which multiple realities can co-exist. The idea of taking a "Yes/And" approach to welcoming ideas instead of argumentation or evaluation enhanced awareness of overlapping concepts and connections between learning projects. This approach in turn opened space for more conversations sparked by curiosity of shared interests, and a thirst to learn new things together.

Relational Minds

The idea that received the most overall attention was the social construction concept of "relational minds" from the eighth Chapter of Making Social Worlds (Pearce, 2007). The concept that relationships to others form how we think, and how we form our social world in the first place, was new to many of the participants.

"Despite us having our own unique separate ways of thinking, we are a function of the environment to some extent, and (the relational minds concept) reinforces this idea as the quality of our relationships being equal to the quality of our lives."

Having an awareness of the qualities of relational minds reinforced the idea within the group of being mindful of the different worlds we come from when we converse with others and take another look at what relational minds really signify in judgments that are made or expressed in daily conversations.

"Reflecting on critical points (or moments) in (my) life, I also thought about the power of influence, for example how one chooses their career is highly influenced by communication with close ones, mentorship opportunities, and feedback on performance in particular areas (i.e., from relatives or instructors)."

The notion that relational minds are constructs (and can be co-constructed) also has implications for the sense of belongingness experienced by members of the group. Having a shared awareness of the influences that each member of the group brings with them from the social worlds that they have previously inhabited can create space for co-creating a relational mind among the group.

Asynchronous Posting Protocols

After some trial and error, the group came up with their own rhythm for sharing and commenting on each other's asynchronous posts. In the beginning, this was mostly used for "check-ins" by those who were not able to attend the synchronous meetings, after watching or listening to the recordings of the meetings. They also found some creative ways for balancing the tasks of building supportive relationships with the sharing of ideas, and giving and receiving feedback. Over time, this seemed to help the group with using the synchronous meeting time more for the development and sustaining of interpersonal relationships. The asynchronous features were then used more for the sharing and documenting of content related to concept development and project details. Participants generally found these asynchronous features to be a valuable addition to the bi-weekly meetings. They also engaged in reflection on the process of using them, making adjustments along the way.

"Posting online was convenient as I could write down all the thoughts I had while reading the chapter and outline important quotes to reflect upon later. The problem was that we did not engage with each other in real time each week and at times thoughts would get forgotten by the time we reached the bi-weekly meeting (at least for myself).... reviewing of the material at the beginning of meetings would be helpful – but this would come at the cost of having either shorter periods left for discussion, or longer meetings overall."

Overall, engaging both synchronous and asynchronous modalities for the group's processes added to the experience of belongingness among participants. However, there were some limitations to this, as will be addressed in the next section.

Addressing Shortcomings and Areas for Improvement

In addition to time management and time commitment issues, the group came to grips with the reality of the limitations of on-line collaborations.

"Of course, in real time, the conversations would have been more meaningful. I think a lot is left unsaid when we comment online despite having the ability to have more time to think and carefully choose what we want to say and look for more sources of reference to support our thoughts."

While the group became more efficient at using time together, there was still a sense of limitation, and this translated into something of a sense of urgency as well as a "felt sense" of paying attention specifically to the quality of belongingness among the group.

"If meetings were (more) frequent ...I think we would definitely build more rapport and insight into each other's perspective, goals, and character... having more dialogue between the group members would enhance (the) sense of belonging. I think even seeing the positive feedback from the individuals who did participate online built a stronger sense of belonging, as we felt acceptance, support, and confirmation that our thoughts were not just valid, but also impactful."

As having more frequent meetings was not a realistic expectation, redirecting attention to the quality of interactions (and the underlying implications for enhancing social connectedness) tended to push more of the information-sharing functions to the asynchronous space, leaving more synchronous time available for building relationships. The conceptual framing of CMM theory helped to reveal ways that other, more desirable realities can be created by more intentionally paying attention to context and the dynamics of interactions in shaping social outcomes.

"(sense of) belongingness comes through individual understanding of one another – to understand each individual's unique story or narrative that led to the project (they are working on), and why the individual chose particular CMM theories over others."

While acknowledging that there will never be enough time to accomplish everything, and the online environment comes with inherent limitations, the use of CMM principles for making the (online) collaborative space a more intentional social world has further potential to make the experience more positive, and also generative, creating new possibilities.

Impact of COVID-19 Pandemic

The theme for the CMMI fellows who participated in this project included the application of CMM principles to understanding the global impact of responses to the COVID 19 pandemic, and rebuilding social worlds afterwards. As of this writing, the post-pandemic social reality continues to emerge, often in unanticipated ways. Sorting out the enduring patterns in all of this will still take some time and reflection. The experience of the CMMI fellows in creating an online community during this time of disruption offers some of the thoughts and reflections about these implications, and how we might experiment with ways to "act into" an emerging future of online collaborative groups, from a communication perspective.

"I think COVID-19 in particular is a perfect example of a phenomenon that created changes in millions of people's lives around the world as people were stripped of the (social) comforts they were so used to (i.e., seeing co-workers every day and not having to conduct so many Zoom meetings) and having consistent in-person contact with their loved ones. Instead, we all suddenly found ourselves to be heavily dependent on technology (even more than before), living 24/7 in the virtual world."

The diversity and subjective nature of the individual responses to the pandemic are summed up in this comment:

"Some individuals felt empowered to express themselves online throughout the pandemic, whereas others felt their communication was cut short and a new sense of loneliness and panic as their in-person dialogue and online dialogue quality differed significantly."

This observation underscores the importance of creating an online space where everyone can feel at home with each other. Emphasizing shared interests and leaving space in agendas for "check-ins" where members can share personal stories and feelings is more than just a form of courtesy, it fulfills a necessary role in creating a sense of belongingness.

"Reading (in "Making Social Worlds") about how communication has affected the consequences of events such as critical events in history (i.e., responses from political leaders), has made me question the role communication plays without our social worlds and how our conversations impact our thinking, behavior, and decisions...online and in person."

At the beginning of the pandemic, it was common to hear people (mostly politicians and media commentators) say things like "we're all in this together." But as time went on, the inequities in how the pandemic affected people became more evident. Within the fellows group, having the conversational space to share and compare these personal impacts turned out to be an important part of creating and sustaining a more authentic form of shared community and grounding experience. Adding in some conceptual tools to analyze and discuss the role of both context and communication dynamics on the way that new social reality is being created also serves to alleviate some of the perceived helplessness in the face of such a large-scale, global disruption in our daily lives.

"I also realized the consequences of these communications - the toll on mental health, work productivity and the way we normalize communication and support today (is now being re-shaped) by our responses to the pandemic."

As we begin to adapt to what organizational and social life is becoming as a result all of the disruptive forces unleashed during the COVID-19 pandemic era, it is likely that we will find the opportunity to co-construct online collaborative space will become even more important.

"Online, it is simply easier to find space, to gather thoughts and convey them as one wants to without facing any threat, which real-time does not always permit."

The hopeful statement here is that, as more people have the experience of gathering in online spaces, we will continue to get better at creating more collaborative environments where participants can feel the comfort of human connection and a sense of belonging. These environments may be shaped by a growing understanding of the communicative forces in play, expanding our capacity to hear and learn from each other across pre-existing social divides and boundaries. As we learn to invite and pay attention to each other's stories, the online space may become more human and hospitable. This may also be something of an antidote to the "ZOOM Fatigue" phenomenon that many of us are feeling. Rather than just cutting back to fewer meetings, or making better use of time in the ones we have, paying attention to creating more authentic and fulfilling connections in our online spaces may hold potential for something better to emerge.

FINDINGS AND RECOMMENDATIONS

Findings

Throughout this project, the increasing awareness of the communication dynamics within the group led to the development of a generative capacity, or an expanded competence for engaging in constructive interpersonal dialogue, joint action, and cosmopolitan communication (Penman & Jensen, 2019). This in turn allowed them to create project designs with potential to engage authentic social change. Participants began the fellowship with a strong personal sense of wanting to create social change in specific areas, but over the course of the fellowship experience, came to embody a shared commitment to "make (better) social worlds" in a more general sense.

"Communication theory is so dynamic. It can be applied to so many fields! We rely more on communication theory than we realize..."

This included gaining an enthusiasm for the ways that other fellows in the group were using CMM theory. This includes the application of constructive conflict management strategies that are based on CMM principles.

"I wanted to utilize communication theory not only in my online working life and for my personal project touching on improving online communication post COVID-19.... but I also found that I now had a desire to study conflict management...."

Participants also envisioned and discussed ways of putting CMM concepts into practice outside of the online space for developing healthier interpersonal relationships in personal and professional life. Participants who would otherwise not engage in conversation with a psychotherapist in professional practice were able to gain an appreciation for how therapist-client communication is essential to promoting mental health.

"It was fascinating to read about the psychological/neurological/ behavioral/ and socio-emotional regulatory aspects that impact the individual, (and learn about) how coordinating with communication acts and meaning making can manage internal aspects (of identity and behavior) from perspective of a therapist."

Enhanced awareness of the social dynamics at work within a collaborative group leads members to become more conscious communicators among the group. This is especially valuable and necessary in online platforms, where individuals may experience less immediacy, less commonality, and therefore risk disconnection from their peers/audience.

"Many individuals... only see what they know rather than being open to possibilities of (a different way of seeing things). Seeking information that is novel, and having the sensitivity to "discern the critical moments" can lead to decisions that take into account the perspective of other individuals, too."

The use of a CMM-informed approach to meeting online allows each member of a group to take a step back outside of prescribed social roles, and engage in an ongoing process of reflecting and re-evaluate choices enacted in online meetings.

"The communication perspective provides a new way to think about belonging, to think about how we as a society form belonging in our communities and how we as individuals are shaped by our society and dialogue in the way we help others and ourselves belong in our social circles."

The shared experience of developing individual learning projects in a collaborative online setting was driven by creating open space for feedback and sharing ideas among an expanding circle as a supportive community of practice.

"The communication perspective provides a new way to think about belonging, to think about how we as a society form belonging in our communities and how we as individuals are shaped by our society and dialogue in the way we help others - and ourselves - (to feel that they truly) belong in our social circles."

The culminating experience of the final meeting (project presentations) was experienced as a celebration, and also setting up of (yet undefined) future connections to be continued. Participants found it enjoyable to consider how unique each project was, was and how the common framework of CMM theory helped to create cross-cutting connections between them.

"Idealistically, we think that belonging and bonding should have no bounds. Realistically, it is all dependent on the willingness of parties to make the effort to communicate. In this expanded and more complex and diverse environment, this will also require (a greater) capacity to keep an open mind, and seek to understand - not simply listen and plan a response."

Recommendations for Promoting Community Online

Here are some practical suggestions based on what was learned from the 2021-2022 CMMI fellows' process, and an invitation to further experimentation with applying the CMM/Social Construction concepts described in this chapter.

- As the group begins its online experience, include a brief information session on the various elements of learning that will be employed, and how any individual learning projects or objectives that are included might be coordinated. (In the CMMI fellowship, this included offers to connect participants with others in the community of practice who might have similar interest, and giving access to previous fellows' work posted on the Institute's website.)
- Build in or introduce contemporary topics as "conversation starters" to get people talking, preferably at the beginning of each meeting. This can be a part of the "check-in" process, where participants can include stories of their personal and emotional responses to topics of discussion as part of bringing their whole self into the group.
- Create space for presentations of learning projects in multiple formats, and opening the possibility for inviting further questions. (In the case of the CMM fellows' project, the organizers reinforced the notion of projects not having to be "finished" for the final presentation, seeing it instead as a report of "work in progress.")
- Use a minimalist approach to process - enough to guide interactions, but not imposing more structure than is necessary. (The CMMI fellowship organizers sometimes used the metaphor of a "jam session" in which musicians employ a minimal structure to allow them to interact as a group, but also create the space for individuals to take solos.)
- Help participants orient to the online environment, especially those who may not be as "tech savvy" as others. In educational settings, there may be some restrictions based on the use of highly structured Learning Management Systems (LMS) and some of these are more user-friendly and flexible than others. (The CMMI fellowship used an open-system platform (*BaseCamp*) which the company makes available to educators at no cost. This platform allows for the sharing of files, efficient distribution of announcements, and the posting of recordings of meetings for those not able to attend synchronously.)
- Experiment with new technology - to a point. Educators are realizing that younger students do not always find the use of point-to-point communication useful (such as email and LMS systems) and instead favor more interactive and multi-modal online environments and more immediate communication (i.e. text). While it's helpful to keep a known and tested platform in place as a point of reference, sometimes trying out suggestions from the group can be a valuable aspect of the "co-creation" ethic, and open up new possibilities for the group to work together.

CONCLUSION

Through the case example of a collaborative online fellowship process, the communication dynamics concepts of The Coordinated Management of Meaning (CMM) theory have shown to have a previously unrealized application in online learning.

The "short course" in CMM theory and "Making Social Worlds (Pearce, 2007) was an integral part of this experience. Coming to grips with the concepts of the "communication perspective" resulted in an expansion of the group's awareness of how their own relational identity was being formed, thus giving them a greater locus of control of their own process. For those interested in doing work with online groups using the concepts and model presented here, the descriptions in this chapter should be enough to get started. For a deeper examination of implications for analysis or development of change interventions for complex social problems, the authors refer you to the original text, and several others mentioned in the reference section.

Interactive communication technology (*Zoom* teleconferencing and *BaseCamp* collaborative sharing platform) were essential elements of this interactive learning and collaborating experience. However, it is important to remember that technology alone does not define the space. Communication technology can expand our thinking and exposure to other ideas from "outside" our social sphere, but it is still up to the participants to create an atmosphere in which it is possible to think more critically about our identity and beliefs; reflect on our role(s) in the social world, and act into critical moments in such a way as to make that world a better place. It is also important to understand the limits of technology - that overuse or saturation can decrease our attention spans, and dilute or diminish our motivation, or preclude us from reaching out in person to make connections. This makes it even more important to consider our sense of belongingness as being not just a side issue, but perhaps the critical lens and metric we can use to determine what social progress could look like.

It is particularly helpful to foreground the process of "storytelling" as a way to promote belonging in the online space. Sharing personal stories helps to create interpersonal connections among group members as well as illustrate abstract ideas, and make complex concepts easier to visualize. When our CMMI fellows and facilitators shared their stories in our online meetings, it brought everyone closer and increased the sense of understanding we had of each individual's interests, experiences, beliefs, and motivations. This deep understanding of each other was the first step towards creating a sense of acceptance and belonging.

Two of the essential ideas behind the approach to online collaboration modeled here are co-construction and belongingness. We consider them to be inextricably linked. The communication perspective of CMM provides some practical ideas of how a group's communication processes can be coordinated in ways that allow each member of the group to bring diverse perspectives to the conversation, be heard, and be a full participant in the group's collective identity. Belonging does not necessarily signify agreement. Many controversial topics can be discussed constructively, despite disagreement. It is then possible to learn about what inspires others, and find a way to accept the differences. In this way, a shared capacity for acceptance and mutual understanding through dialogue despite differences is created, which further expands the sense of belonging.

It is also important to consider that it is the sharing of the personal that creates and expands the level of social interest. This process invites the group members to make their individual experiences a part of the group consciousness and identity, and invites everyone to explore and uncover these connections together. We can only fully understand and appreciate the meaning of the deeper meanings of life when we can link our experiences and insights together with others.

ACKNOWLEDGMENT

The authors wish to recognize the CMMI fellowship organizers Beth Fisher-Yoshida, PhD, Columbia University; Ilene Wasserman, ICW Consulting; Sergej Van Middendorp, PhD, (Netherlands) and Barbara McKay, DPsych, (UK).

REFERENCES

Afuape, T. (2011). *Power, Resistance and Liberation in Therapy with Survivors of Trauma: To Have Our Hearts Broken*. Routledge.

Barge, K., & Pearce, W. B. (2004). A reconnaissance of CMM research. *Human Systems, 1*, 13-32. https://cmminstitute.org/wp-content/uploads/2020/12/Human-Systems-2004-A-reconnaissance-of-CMM-research-Barge_W.-Pearce.pdf

Belanger, P. (2015). *Self-construction and social transformation: lifelong, lifewide, and life-deep learning*. Hamburg, Germany: UNESCO Institute for Lifelong Learning.

Berger, P., & Luckmann, T. (1967). *The social construction of reality: A treatise in the sociology of knowledge*. Anchor Books.

Buechner, B., Van Middendorp, S., & Spann, R. (2018). Moral Injury on the Front Lines of Truth: Encounters with Liminal Experience and the Transformation of Meaning. *Journal of Schutzian Research, 10*, 51–84. doi:10.5840chutz2018104

Cronen, V. (2001). Practical Theory, practical art, and the pragmatic-systemic account of inquiry. *Communication Theory, 11*(1), 14–35. doi:10.1111/j.1468-2885.2001.tb00231.x

Dam, R. F. (2021). *Five stages in the design thinking process*. The Interaction Design Foundation. Retrieved June 2, 2022, from https://www.interaction-design.org/literature/article/5-stages-in-the-design-thinking-process

Gilbert, C., & Bower, J. (2002, May). Disruptive Change: When Trying Harder Is Part of the Problem. *Harvard Business Review*. PMID:12024762

Mcluhan, M. (1995). *Essential McLuhan*. Anansi.

Pearce, W. B. (2007). *Making social worlds: A communication perspective*. Blackwell.

Penman, R., & Jensen. (2019). *Making better social worlds: Inspirations from the Theory of the Coordinated Management of Meaning*. CMMI Press.

Penman, R. (2021). *A cosmopolitan sensibility: Compelling stories from a communication perspective*. CMM Institute Press.

Steen, S., Mackenzie, L., & Buechner, B. (2018). Incorporating Cosmopolitan Communication into Diverse Teaching and Training Contexts: Considerations from Our Work with Military Students and Veterans. *The Routledge Handbook of Communication Training*. https://www.routledge.com/The-Handbook-of-Communication-Training-A-Best-Practices-Framework-for/Wallace-Becker/p/book/9781138736528

Tiller, W. (n.d.). *The theory of positive disintegration by Kazimierz Dąbrowski*. Retrieved from http://www.positivedisintegration.com/

Wasserman, I. (2014). Strengthening interpersonal awareness and fostering relational eloquence. In B. Ferdman & B. Deane (Eds.), *Diversity at work: the practice of inclusion* (pp. 128–154). Jossey-Bass. https://cmminstitute.org/wp-content/uploads/2019/07/4-Wasserman-Strengthening-Interpersonal-Awareness-and-Relational-Eloquence-in-Ferdman-and-Deane-Diversity-at-Work-The-Practice-of-Inclusion.pdf

Watts, R., Williamson, D., & Williamson, J. (2004). *Adlerian Psychology: A Relational Constructivist Approach. Adlerian Yearbook. 2004*. Adlerian Society and Institute for Individual Psychology.

Section 2
Community Engagement in Online Higher Education

This section focuses on various strategies for and examples of community engagement through online higher education. Included is a consideration of effective engagement in the online classroom, the use of technology tools to create connections, and emerging practices to support community engagement within the post-secondary student population.

Chapter 6
Remote Community Engagement in Higher Education

Paul M. Huckett
Johns Hopkins University, USA

Nathan Graham
Johns Hopkins University, USA

Heather Stewart
Johns Hopkins University, USA

ABSTRACT

This chapter examines strategies implemented to increase a sense of community for faculty and students in an online graduate engineering program. Facilitating community sites of interaction and shared knowledge creation—elements of the community of practice (CoP) framework—comprised the most valued additions for members of the learning community. With these improved sites of interaction, faculty and students benefited from participation in the learning community with their online peers and contributed to a community of practice in their degree program. Early data and outcomes suggest that higher education administrators can implement specific strategies to increase learners' and teachers' sense of community, facilitating engagement with the school, academic programs, and peers despite being geographically dispersed.

INTRODUCTION

In this chapter, readers will learn about a strategic approach to increase a sense of community (SoC) in an online graduate engineering program at Johns Hopkins University with over 6,000 learners and 600 faculty. The approach detailed in this chapter was developed over two years in response to five years of annual survey data where students indicated that they did not feel a SoC. Improving a

DOI: 10.4018/978-1-6684-5190-8.ch006

SoC for students is associated with other areas of focus for the school, including increased retention and degree completion (Ehrenberg & Zhang, 2005; Jacoby, 2006; Jaeger & Eagan, 2011), greater commitment to the organization (Milliman et al., 2003), decreased work stress (Royal & Rossi, 1996), and increased collaboration, knowledge sharing, and communication (Andersen et al., 2013; Rovai, 2002). Early data and outcomes from the approach described in this chapter suggest that higher education administrators can implement specific strategies to increase a SoC for learners and teachers, facilitating engagement with the school, academic programs, and peers despite being geographically dispersed.

Two primary sites of student and faculty interaction and community development were implemented and observed over a three-year period from March 2019 to April 2022. The strategies implemented in this online graduate engineering program are based on the theoretical components of creating a community of practice (CoP) in digital environments outlined by Wenger et al. (2002). Three primary elements encourage the growth of a community: (a) domain, (b) community, and (c) practice (Sherer et al., 2003, p. 185).

- Domain: The shared interest or passion of a community. Membership in the domain requires a level of competence and commitment that distinguishes members from non-members.
- Community: The pursuit of shared interest as members seek to gain competence and recognition, including information sharing, engaging in conferences and events, and collaboration on a shared issue. This pursuit of a common interest enables members to learn from one another and increases their participation as they seek visibility and belonging within the community.
- Practice: The sustained interaction of the practitioners as they develop a set of resources: documentation, stories, ways of interacting, and tools.

For this chapter, a learning community can be viewed as a community of practitioners consisting of students and faculty who develop a SoC through sharing resources, knowledge, and goals. A SoC is used to explore feelings of connectedness and learning expectations within the CoP framework (see Theoretical Framework section).

A combination of end-of-year surveys and online content analysis was used to measure a SoC. The data used in this chapter comes from faculty and adult learners in an online graduate engineering division at Johns Hopkins University, which has offered online master's degrees through its part-time engineering division for over 20 years. At the time of this study in Spring 2022, there were 22 master's programs (20 can be fully completed online), 14 post-master's certificates, 11 graduate certificates, and 593 unique online courses taught by approximately 600 adjunct faculty with a student headcount of over 6,000, making it the second-largest part-time online engineering division in the United States.

STATEMENT OF THE PROBLEM

The challenge that higher education institutions (HEIs) offering online education face is creating a sense of community (SoC) for its faculty and students due to the lack of physical connectedness. Studies have found several benefits to the organization (the domain) and the learning experience by developing a SoC, including increased retention and degree completion (Ehrenberg & Zhang, 2005; Jacoby, 2006; Jaeger & Eagan, 2011), greater commitment to the organization (Milliman et al., 2003) and decreased work stress (Royal & Rossi, 1996), and increased collaboration, knowledge sharing, and communication (Andersen et al., 2013; Rovai, 2002). To fully understand this challenge, one must explore the evolution of the technologies and practices used in online education.

Advancements in educational technology in the 21st century created unique opportunities for HEIs to offer online courses and programs to students who no longer had to live in the same geographical area to learn in a physical classroom. While overall enrollment in higher education has declined over the past 20 years, online education has seen steady growth in the United States (Allen et al., 2016). Since 2000, online education course enrollments have increased annually (Seaman et al., 2018). For graduate students, the number has increased each year since 2012. However, while enrollments have grown in online programs due to flexibility and ease of access, student retention remains a concern, with attrition rates being 10-20% higher in online courses compared to onsite programs (Bart, 2012).

The convenience of online programs attracts students to apply and enroll. But once enrolled, they can become disconnected due to a lack of community interaction (Dziuban et al., 2013). To address attrition rates and improve learning outcomes, increasing students' SoC or feeling of connectedness in the program is a key component of engagement and an indicator for degree completion (Berry, 2017; deNoyelles et al., 2014). Universities have recently taken a more comprehensive approach to the student experience, moving beyond improving content delivery, technology, and student services to include opportunities for students in online courses to build a community (Allen & Seaman, 2014; Marinoni et al., 2020; Meyer, 2014).

Since faculty are the primary point of contact for most students as they navigate through their program, engaging and retaining faculty is an essential component of learning community development (Eagan & Jaeger, 2008). Across student populations and content delivery methods, regular interaction with faculty is vital to student engagement within the course and with the larger community (Andersen et al., 2013; Walker, 2016). This is particularly important for online programs, which are largely taught by part-time faculty. Studies have found that increases in part-time faculty have a negative impact on rates of student retention and degree completion (Ehrenberg & Zhang, 2005; Jacoby, 2006; Jaeger & Eagan, 2011). For faculty engaged in course delivery, SoC has been linked to an increased SoC at work (McGinty et al., 2008; Milliman et al., 2003) to greater commitment to the organization (Milliman et al., 2003) and decreased work stress (Royal & Rossi, 1996). Providing spaces for interactions between learners as well as faculty increases group collaboration, knowledge sharing, and communication (Andersen et al., 2013; Rovai, 2002). The opportunity for students to interact with faculty and peers by asking questions, sharing opinions, or disagreeing with varying points of view is vital to the learning experience (Garrison et al., 2001).

While there is a large body of research on SoC, there is no standard definition used across studies. Rovai (2002) specifies two core components of SoC—feelings of connectedness, and shared learning expectations—and reviews several common definitions of SoC. One such definition is from McMillan and Chavis (1996), who state that it is "a feeling that members have of belonging, a feeling that members matter to one another and to the group, that they have duties and obligations to each other and to the school, and that they possess shared expectations that members' educational needs will be met through their commitment to shared learning goals" (p. 9).

It is this summary of SoC that the authors feel is critical to developing a CoP for an online, graduate engineering program.

THEORETICAL FRAMEWORK

The following section will outline the main theoretical framework in this study—community of practice (CoP)—and several conceptual frameworks that support the development of a CoP, including social constructivism, Moore's three types of interaction, Bloom's Taxonomy of Educational Objectives, and adult learning theory, or andragogy. The CoP is examined in two main ways: (a) at the school/division level, and (b) at the course level.

Community of Practice

A CoP is a group of people with a shared interest, a craft, or a profession who develop a set of practices around their domain (Lave, 1991; Wenger, 1998). A CoP develops through the initiation and guidance of newcomers by existing members in a process that creates events, materials, and knowledge to further the practice (Lave, 1991). Online communities such as those involved in the development of learning experiences provide social organization and structure identity for those looking to enter or change status within the program (Wenger, 1998, p. 241). The social context that regulates community dynamics, as reflected in the practices of production of learning experiences, determines the symbolic and material resources, social and communication networks, and the existing power structures of the CoPs (Fairclough, 2001, p. 122).

The CoP and the objects or traces of the community members' experience are interpreted as products of a system of relationships between community members, activities, and the larger field. In the context of this chapter, the online learning experience operates as a central system for community development and participation, legitimating and conferring symbolic power. The learning community defines itself by its engagement with the system of online learning, which involves the realized form of the practices of the community through the two primary sites of interaction: the division and the course. Because participation in the learning community requires literacy ranging from the practices of the domain to the terminology of the craft, an elevated level of expertise and knowledge is required to fully join the CoP, and careful gatekeeping and interaction rituals are in place to ensure that the degree secures and maintains its ability to confer value and status (Collins, 2004, p. 164). As new students and faculty attempt to engage with the community, they are invited more broadly to the program through their participation in the system of online learning.

Practices in the production of learning experiences (e.g., collaborative creation of group presentations, office hour discussions, annotations, etc.) mirror the shifts in culture within the learning community. Structural changes within the institution result in increased engagement and participation in practices valued by the institution, and these shifting institutional demands force communities to realign existing knowledge to fit the new realities. Transformations in traces of the experiences of the community (knowledge products) reify the practices of the community (Lave, 1991). These practices (e.g., editing syllabi, attending office hours, and participating in discussions) and products (e.g., course materials, board meetings) of the learning community provide the body evidence for this chapter. Faculty and students are practitioners within the domain and are part of a CoP, where the practice is online learning. The practice of creating online learning experiences is the core platform and integral activity of the community. A CoP operates at three primary levels, each of which must be addressed to facilitate SoC within an online learning community: domain (e.g., graduate education), community (e.g., faculty, students, staff), and practice (e.g., producing learning experiences; Wenger & Snyder, 2000, p. 144).

Social Constructivism

John Dewey, a founder of constructivism, suggested that learners learn best by doing, and that learning should be grounded in individual and real-life situations. Dewey's work established that the learner must be interested in an issue, thus activating the learner's need and desire to resolve the issue (Duffy & Cunningham, 1996). In constructivist thought, the learner is an active participant in the environment and constructs meaning from their experiences in the world.

Lev Vygotsky posited that learners construct knowledge partly through a social process (Duffy & Cunningham, 1996). Social constructivism, also called socio-constructivism or historicism, is a branch of constructivism that focuses on the interactions of learners and instructors, the learning environment, and cultural contexts. This branch of constructivism places emphasis on social interaction as an integral component to learning. Vygotsky's concept of Zone of Proximal Development, or ZPD, is the foundation for an educational practice more popularly known as scaffolding. Scaffolding is a process of learning in which the individual increases levels of cognition through the help of a supporter, or someone with a higher level of knowledge. In the online environment, this person is often the instructor but also includes peer learners.

According to Berge (1995, 2008), the online instructor plays a "social role" by encouraging social interactions among learners to foster and build relationships. When applying social constructivist theory to the online learning environment, technology can provide a way to facilitate interaction so that learners interact and learn from each other and the instructor rather than in isolation. Interaction is a broad term that Moore's three types of interaction can help clarify.

Moore's Three Types of Interaction

Moore (1989) developed descriptions for three specific types of interaction that occur in a distance learning environment: (a) learner-content, (b) learner-instructor, and (c) learner-learner. Through the years, researchers have used his definitions to understand the overall concept of interaction and to "overcome the misunderstandings between educators who use different media" (Moore, 1989, p. 1).

First, in learner-content interaction, the content, or subject matter, is the foundation for learning. The instructor is usually considered the subject matter expert (SME) who delivers information about the content. In an online course, content is delivered through different mediums such as lectures, reading materials, videos, and interactions with the instructor and other learners, to name a few. Learning occurs when there are "changes in the learner's understanding, the learner's perspective, or the cognitive structures of the learner's mind" (Moore, 1989, p. 1) after interacting with the content.

Second, learner-instructor interaction generally involves the communication and facilitation of content by the instructor to the learner. The instructor ultimately determines if the learner understands the subject matter and can apply new information appropriately. In an online classroom, this type of interaction is helpful for many reasons. The instructor typically develops a curriculum to retain and encourage learner interest and motivation. The instructor may create presentations based on the developed curriculum or include other learning materials within the course. Learners access these materials and practice or apply their knowledge of the material. The evaluation process often occurs when instructors provide support and feedback to learners about their practice and application of the material, most often in the form of grades, which helps students better gauge mastery of the learning outcomes. Additional opportunities for this type of interaction include email correspondence, synchronous meetings, and discussion boards.

Learner-instructor interaction is particularly valuable to learners because they can interact with SMEs and draw on an instructor's professional and educational experiences.

Lastly, learner-learner interaction is at the heart of the social constructivist point of view included in the theoretical framework of this research paper. Learners interact with one another in a group setting (e.g., in a discussion board or small group project) or in a one-on-one situation (e.g., email correspondence or synchronous meeting space). This interaction occurs between the learners through group activities, course activities, or other activities and may or may not include instructor presence. These interactions can be synchronous or asynchronous in nature. However, facilitating learner-learner interactions depend on a variety of factors. Moore (1989) suggests that group interaction is desirable depending on the learner's "age, experience, and level of learner autonomy" (p. 3). He further notes that learner-learner interaction that encourages motivation may not be as important for adult or advanced learners (typically self-motivated learners) as for younger or novice learners. However, an activity that "acknowledges and encourages the development of their expertise but also tests it, and teaches important principles regarding the nature of knowledge and the role of the scholar as a maker of knowledge" (Moore, 1989, p. 3) is more appropriate for adult or advanced learners who bring their worldview and professional experiences into the learning environment. This suggests that adult learners benefit from a learning situation that accounts for their diverse needs.

Bloom's Taxonomy

In 1956, educational psychologist Dr. Benjamin Bloom, with his colleagues, created a classification of learning objectives in what is now known as Bloom's Taxonomy of Learning. The taxonomy, a hierarchical representation of cognitive levels (remembering, understanding, applying, analyzing, and evaluating), provides a "clear, concise visual representation" (Krathwohl, 2002) of alignment between what the learner is expected to do (i.e., educational objectives) and how the learner demonstrates mastery of the stated learning objectives (i.e., assessment and mastery). In education, the taxonomy has proved a fundamental resource for writing measurable student learning objectives (Mager, 1997; Marzano, 2009) In addition, it has provided a model for identifying learning outcome indicators and measuring results against those indicators (McNeil, 2011).

It is important to acknowledge that the students referenced in this study are adult learners who enter the learning environment with significant professional, educational, and life experiences that ultimately influence their learning and learner-learner interactions. Using Bloom's Taxonomy of Learning can help establish clear objectives for what the learners will achieve during the interactions in the course. The development of these learning objectives needs to align with the audience. This can be achieved using the adult learning theory as a foundational building block for the objectives and experience in the course.

Adult Learning Theory

The term *pedagogy* refers to the "art and science of teaching children" (Knowles, 1984, p. 52) and is often used to describe adult learning. Malcolm Knowles proposed andragogy as a theoretical model to differentiate between the learning needs of children and adults. Andragogy consists of the following six principles of adult learning.

1. The need to know: This principle refers to the importance of establishing a reason for learning. Knowles (1984) suggested that the facilitators of learning should make a case for the value of what is being presented, as adult learners will appreciate content relevant to their lives and the application of the material grounded in real-world experiences.
2. Learner's self-concept: Adult learners are responsible for their actions and behaviors. They are self-directed and may feel resentful in situations in which they are treated as dependents (Cranton, 1994; Knowles, 1984). For this reason, what works for younger learners is not always appropriate for adult learners.
3. Role of learner's experience: Adult learners enter a learning situation with the benefit of having a wide range of life experiences.
4. Readiness to learn: This principle assumes that learners must be in the right place at the right time for learning. Learners are ready to learn when the need to know something arises. This is why real-life applications are preferred for adult learners.
5. Orientation to learning: Adult learners are motivated when learning is centered around topics that will help them in their life situations because "adults are task-centered learners" (Knowles, 1984, p. 61).
6. Motivation: According to Knowles (1984), intrinsic motivation is more influential for adult learners than extrinsic motivation. Adult learners may be more motivated to succeed in a course to boost self-esteem or increase their quality of life and may not benefit from the kinds of external encouragement that a less mature audience might.

The program discussed in this study was designed for adult professionals already working in engineering who bring unique knowledge to the learning environment and learner-learner interactions. Understanding adult learning theory or andragogy is an important aspect of creating opportunities to develop a SoC for faculty and students within the two sites of interaction (the school and course (Yarbrough, 2018). The next section explores what the literature revealed about the impact of a SoC on the learning experience.

LITERATURE REVIEW

The following literature review provides context for developing a SoC within the two primary sites of interaction (school and course) discussed in this chapter. Increasing SoC among students engaged in online education requires addressing both the larger domain of the community (i.e., the school) and the localized site of interaction—the online course.

Online Course Interaction

McIsaac et al. (1999) argue "that interaction may well be the single-most-important activity in a well-designed distance education experience" (p. 2). Interaction within a course encourages learners to analyze alternative ways of thinking and acting. Through this participation, learners explore their own experiences within the context of the activity. The interactions with the instructor and fellow peers provide the opportunity for the social construction of knowledge within the learning community, which is vital in online programs where regular and substantive interactions lack physical interactions experienced in a face-to-face environment (Delahunty et al., 2014; Shea et al., 2006).

The lack of physical presence means that technology plays a crucial role in enabling interactions in an online course. Learning management systems (LMSs; e.g.., Blackboard, Moodle, and Canvas) offer tools such as blogs, wikis, course messages, discussion boards, and integrations with third-party collaborative chat platforms (e.g., Microsoft Teams and Slack), enabling students to flexibly engage their peers and instructors (i.e., asynchronously or synchronously).

In addition to technology that fosters interaction, organizations such as Quality Matters (QM) provide frameworks and best practices for how interactions in online courses should be designed. The QM Rubric Standards state, "activities for learner-learner interaction might include assigned collaborative activities such as group discussions" (QM, 2013). These "best practices" have been adopted and, in some cases, mandated as part of an online course design. For example, Oregon State University (2022, p. 4) requires that all courses include three forms of interaction:

- Student/content (discussion boards, readings, videos, and research projects)
- Student/instructor (discussion boards, response to assignments, and general discussion forum facilitated by the instructor)
- Student/student (discussion boards, team projects, peer-reviewed assignments, and blogs)

The Online Learning Consortium (OLC, 2022) identifies interaction as a "best practice" in their Five Pillars of Quality Online Education. In addition, the OLC Quality Scorecard includes the following in its teaching and learning criteria: Student-to-student and faculty-to-student interaction are essential characteristics and are encouraged and facilitated (OLC, 2014). The International Board of Standards for Training, Performance and Instruction provides online learning and instructor competencies that identify interaction as a critical component (Forshay et al., 1986).

As demonstrated in the literature, the inclusion of interactions is a critical element of an online learning experience. According to Murphy and Cifuentes (2001), the discussion board as an asynchronous interaction tool is the most used form of computer-mediated communication in education. This is due to its wide-ranging benefits, such as deriving feedback from a larger student population and more diverse participation, providing opportunities for deeper and more reflective student thought, and accommodating different student populations within a course or program (Jinhong & Gilson, 2014). Records on discussions and interactions are accessible within an LMS and thus easy to evaluate to provide an objective assessment method (Rourke et al., 1999).

Online Discussions

Online discussion boards have provided a platform for students to learn from one another, moving away from a teacher-centered approach toward one that is more student-centered (Kupczynski et al., 2012). This student-centered approach, as suggested by Davies and Graff (2005), improves learning and provides support, especially to those students who may need additional help from the instructor and classmates. Research also suggests that online discussions foster greater participation in certain learning situations. For example, nonnative English speakers were shown to participate in discussions more frequently because prolonged writing times enable them to think, write, edit, and post their responses (McIntosh et al., 2003).

Opportunities to interact in an online discussion board enable students to have a voice and to connect with all members of the course, empowering less responsive, or otherwise quiet students to overcome their reluctance to participate and to feel secure and part of a social community (Betty Cox & Becky

Cox, 2008; Swan & Shih, 2005). Online asynchronous discussions are less likely to be dominated by a single student which is a common occurrence in a face-to-face classroom discussion (Redmon & Burger, 2004). Kehrwald (2008) suggests that the social nature of discussions promotes supportive behavior and connectedness among students. Because of the increase in social presence, students are more likely to contribute.

Despite the perceived benefits on the use of discussion boards, the data on its impact on learning is varied. Several studies discovered that students who participated more frequently in an online discussion received more points, and ultimately, a higher grade in the course than those who spent little time interacting with their peers and the instructor(s) in the forum (Cheng et al., 2011; Kay, 2006; Masters & Oberprieler, 2004). On the contrary, in an undergraduate study conducted by Davies and Graff (2005), 122 first-year students were evaluated to see if their participation on the online discussion board had any correlation on their summative performance. Although students who performed poorly in the course were found to have interacted less frequently with the discussion board, greater interaction did not lead to significantly higher performance and better grades.

Swan (2002) suggests that the inclusion of a discussion forum in an online course does not automatically result in learning. Therefore, which factors impact the learning experience and learning itself, as measured by higher performance? Research suggests two critical components: instructor presence and quality of interaction.

Instructor Presence

Even if student-centered approaches appear to affect student learning and social connectedness, the literature revealed that instructor presence (or other staff such as graduate assistants) in the discussion board also makes a difference in student learning. For example, the number of postings made by a student can have little impact on grades, but students who direct postings to an instructor or teaching aid commonly achieve a higher grade than those who post more often to other students (Finegold & Cooke, 2006). This also benefits the instructor, as student interaction with an instructor in a discussion board has resulted in positive instructor evaluations (Du et al., 2011).

Instructor-led discussion prompts can positively impact deeper learning (Du et al., 2011). The way the instructor develops the discussion prompt is of critical importance (Magnuson, 2005; Williams et al., 2015). Furthermore, Williams et al. (2015) state that the "depth of thinking is more likely to occur when discussion prompts require students to put together an original project that challenges claims, synthesizes information from the group, and cocreates new understandings" (p. 61). Extended posts (i.e., posts that explore topics and solutions in depth) can lead to higher levels of divergent and convergent thinking versus surface-level posts (i.e., exchange of pleasantries); instructors can influence this by explicitly explaining what they expect from student responses (Williams et al., 2015). Because asynchronous discussions can be extended beyond a set time frame, deep and critical thinking can be achieved as long as the quality of the interaction is high.

Quality of Interaction

The quality of interaction, specifically the evidence of critical thinking, is crucial in online interactions (DeLoach & Greenlaw, 2007; Magnuson, 2005; Weltzer-Ward et al., 2009). Critical thinking can be defined as identifying a problem, exploring the problem, suggesting a solution, judging the solution, and implementing the solution (Dewey, 1998, as cited in Weltzer-Ward et al., 2009, p. 169). Critical thinking encourages experiential learning by bridging the gap between what is taught in academia and what is required to function effectively in the workplace and community (Lee, 2007). Achieving meaningful interactions that promote critical thinking requires more than having a discussion forum—it must include opportunities to engage in rich discussions, work collaboratively on group projects, and connect with the broader community. In the context of this study, the broader community is the engineering community.

Discussion Strategy

Weltzer-Ward et al. (2009) found that students tend to use opinions rather than reliable sources such as references, data, or theories as evidence to support their posts. Also common are low- and high-quantity responses containing a basic opinion and response to another student's post to meet a participation requirement (Kim et al., 2007). In this approach, students rarely respond to one another's opinions and often repeat others' points. This, coupled with the sometimes-chaotic structure of discussion boards (e.g., lack of turn-taking), often leads to interactions that lack coherence and depth (Brooks & Jeong, 2006; Herring, 1999). Other studies adduce that discussions in which factual information—rather than rich, meaningful topics—occurs frequently and is a prominent factor that influences the quality of the interaction (Darabi & Jin, 2013; DeLoach & Greenlaw, 2007; Ertmer et al., 2007).

Interactions involving conflicts of perspectives promote more critical thinking (Jeong, 2004), and ideally, discussion topics allow for a variety of opinions. Questions posed within these topics should be ambiguous and provide opportunities for students to scrutinize multiple viewpoints. In this way, students are required to respond with arguments that are multifaceted and have multiple solutions, which is at the heart of increasing critical thinking skills. Muilenburg and Berge (2002) posit four types of thinking that promote discussion: critical thinking, higher order thinking, distributive thinking, and constructive thinking. These types of thinking are both hierarchical and interrelated and "[t]he level of student thinking is directly proportional to the level of questions asked" (Muilenburg & Berge, 2002, p. 12).

Magnuson (2005) developed a discussion strategy focusing on including problem-based learning to promote experiential learning for students, as outlined below:

1. Determine the topic for the discussion board activity based on the content.
2. Determine the goals and objectives to be accomplished—the goals and objectives are tied to the content.
3. Decide what role the instructor will have and what role the learners will have.
4. Determine the methodology or how the learners can be engaged in the discussion board activity.
5. Sculpt the question to encourage higher-order critical thinking. Provide an experiential learning opportunity within the context of the activity. This question can be problem-based.
6. Consider how the learners might approach the problem/question posed on the discussion board.
7. Manage the discussion board.
8. Assess the learners' posts.
9. Reflect on the process.
10. Produce and provide any scaffolds that are required.

Research suggests that discussions enabling students to think critically result in a positive learning experience and impact learning depending on how the discussions are created and facilitated.

Group Collaborative Work

Courses that incorporate group collaborative work provide learners an opportunity to interact and contribute to their teams in a variety of ways to help craft solutions to challenges (Mentzer, 2014). In online learning, collaborative learning is connected to cognitively based instruction models, which stress student's active engagement and communication (De Miranda, 2004; Kelley, 2012; Kelley et al., 2020; Spector & Anderson, 2000). Accordingly, the need to incorporate collaboration work, particularly in engineering education, has risen to be one of the most important skills today—not only because of the need to engineer solutions, but also because of affective factors gained during collaborative work (Advance CTE, Association of State Supervisors of Math, Council of State Science Supervisors, & International Technology and Engineering Educators Association 2018; International Technology and Engineering Educators Association [ITEEA] 2020; Jones & Issroff, 2005). Some researchers have found that motivation as an affective element in student learning significantly contributes to collaborative learning (Jones & Issroff, 2005). Others have found that social interactions (Järvelä et al., 2008) and a strong SoC can motivate students within a group project (Reeves & Gomm, 2015).

Community-Engaged Work

Community-engaged learning is another pedagogical strategy that impacts the quality of interaction since it can enhance skills, deepen content knowledge, and increase the SoC (Hatcher & Bringle, 2010; Paquin, 2006). Community-engaged learning seeks to engage and accredit students within the curriculum for working in partnership with different organizations, most commonly to act on local societal challenges (Campus Engage, 2022). According to Goggins and Hajdukiewicz (2020), their community-engaged project enabled students to:

(i) develop the ability to identify, formulate and solve engineering problems in their field of study in a real world context; (ii) select and apply relevant methods from established engineering practice by critically using appropriate sources of information to pursue detailed investigations and research of technical issues in their field of study, (iii) recognise the importance of non-technical – societal, health and safety, environmental, and economic – constraints, and (iv) develop the ability to communicate effectively information, ideas, problems and solutions with engineering community and society at large. (p. 395)

Additional researchers have demonstrated that the ability to effectively interact and collaborate is critical in engineering education (Goggins & Hajdukiewicz, 2020; Han et al., 2022; Jacobs et al., 2021). However, being able to interact and collaborate within a team should be an educational outcome for HEIs. Based on the review of literature on the quality of interactions in online courses, the authors argue that incorporating more than discussion forums will lead to an increased SoC.

School-Wide Interaction

Research on the development of community in higher education highlights the importance of establishing spaces and activities that facilitate shared interactions between members (Wenger, 2011). The shared spaces and activities are formed around a central domain, which for this chapter is an online graduate engineering program. School-wide sites of interaction establish the space for collaboration and spaces to consecrate the output of the collaboration. Because participation in an online learning community requires developing expertise and knowledge, the spaces for school-wide interaction (e.g., community group meetups, web conferencing and chat platforms, conferences) are the primary infrastructure for community gatekeeping and interaction rituals, which confer status within the community (Collins, 2004, p. 176). Amin and Roberts (2008) note that within the community of practice (CoP) framework, there are "four types of collaborative work: craft or task-based work, professional practice, epistemic or high-creativity collaboration, and virtual collaboration" (p. 356). While the course is a site where these four types of collaborative work are formally expressed, informal communication and community building happens outside of the course at the school level. In particular, virtual collaboration, which is often initiated at the course level (e.g., group assignments, study groups), occurs on school-wide sites of interaction. Research on CoP and sense of community (SoC) has highlighted the role of the institution and decision makers in facilitating interaction at the larger domain level of the community.

The four main sites of student and faculty interaction were implemented at the school-wide level, and were developed to increase the opportunity for informal interaction and networking among members (community), and enhance the visibility of the larger community (domain), which provides an environment for the day-to-day practices of the community within courses. The four sites of interaction for students, which will be discussed in depth later in the chapter, include New Student Orientation, Student Community, Student Advisory Board, and Community Day. The purpose of each was to initiate members to the social organization of the online engineering program, and provide spaces for connection outside of (and around) their course work.

Similarly, the four sites of interaction for faculty include faculty course development cohorts, faculty community, Faculty Forward trainings, and two faculty meetings. These four sites of interaction mirror the student sites of interaction in that they include an orientation, community communication space, advisory group, and event. Throughout their career, and especially with part-time and online programs, faculty seek out professional development with peers (Sherer, Shea, & Kristensen, 2003, p. 184). Establishing sites to facilitate interaction between faculty can help better develop faculty and connect them to the practices of the community. Cox (2001) identified two of the most common categories of faculty development: cohort focused, and issue focused--where the cohort-focused groups address "teaching, development, and learning needs" and issue-focused groups address a school-wide teaching and learning issue (p. 73). With online education, faculty need not only development on effective instruction but also on how to integrate and utilize educational technology (Reilly et al, 2012, p. 100). Adding cohort-focused and issue-focused faculty development opportunities with sites for informal conversation and collaboration helps develop a set of shared practices for the faculty community.

METHODOLOGY

The central concept of this study is that establishing specific sites of interaction, and facilitating access to those sites, can increase sense of community (SoC) among students in an online graduate program. To understand the impact of the sites of interaction and facilitation strategies on students and faculty, it is necessary to examine the community engagement at the established sites and the products of their engagement (e.g., documents, discussion posts). To assess SoC at the two primary sites of interaction—the school and course—data from past student and faculty surveys (sampled between 2016 and 2021) and online content from instructional platforms will be the source for analysis. Engagement will be discussed using a microanalysis of online interactions (e.g., discussion content) and a macroanalysis of engagement trends (e.g., post activity, channel member totals) on the primary communication tool for the community (i.e., Microsoft Teams). The study includes two groups of practitioners: students and faculty. These groups were selected because they are the key actors within the domain of an online learning community.

To describe the change in SoC, the study will utilize a multiple method approach (Creswell, 1998). Content used for analysis in this study was selected based on the sites of interaction created by the authors of this chapter. These sites of interaction were strategically implemented using the community of practice (CoP) framework. A microanalysis of online data was conducted in order to examine the types of activity and practices of the community at the sites of interaction. Microanalysis (also sometimes referred to as close reading in other fields) of online content is a type of conversation analysis that examines community member interaction on a digital platform with the understanding that the interactions are embedded within the context of the communication tool (Giles et al, 2015). Microanalysis on the sites of interaction in this chapter, provides a method to understand the ways community members develop a shared SoC.

Research Design

The research design used in this chapter includes secondary data collected from five sources: annual graduating student survey, annual faculty meeting survey, faculty training and development sessions, annual student conference, and faculty and student activity on Microsoft Teams, which is the primary school-wide communication platform. The annual graduating student survey asks students questions about their experience throughout the process of getting a degree. The survey includes both open and closed-ended questions and is published by the program. This content was sampled over a five-year period for the graduating student survey between 2016 and 2021. The annual faculty meeting survey is administered at the end of the event and asks faculty who attended the Fall Faculty Meeting about their experience during the event, and it includes open and closed-ended questions. This survey data was sampled over a five-year period between 2016 and 2021. Faculty training and development data includes registration and attendance activity, topics of workshops, as well as dates. This data was sampled over a two-year period between 2021 and 2022. The annual student event, Connect to Campus (C2C), includes attendance and registration data, as well as a participant survey. This data was sampled throughout September, 2022. Finally, this chapter uses data from the school-wide communication platform, Microsoft Teams, and includes student and faculty activity on the platform.

The researchers then conducted a microanalysis of the online conversation and the survey open-ended responses, and a macroanalysis of the online activity and survey closed-ended responses. This approach was chosen to examine the engagement (activity data), and sentiment (survey data) of community members. By including both types of data, this chapter can better describe how students and faculty experience and respond to the implemented sites of interaction.

Discussion

The challenge that the online programs faced was a lack of cohesion around these community-building efforts for students and faculty at the school level. Over the last 5 academic years, a graduating student survey was sent to students with two questions about community to gauge the level of community felt by its remote students and their level of desire for community in their programs of study.

1. There is a sense of community among students.
2. A sense of community is important to me.

The average of responses over the last 5 academic years revealed that only 35% of students felt there was a SoC among students, and 55% a SoC in their remote programs of study was important to them.

For faculty, a survey was sent to those who attended a faculty meeting—a bi-annual faculty event. Overall, 84% of respondents stated that "Networking" was the most useful component of the faculty meetings.

Based on the survey response, which highlighted a lack of SoC for students and a desire for faculty to network, the authors collected the efforts made to build an online CoP. Prior to the period of this study, between 2001 and 2018, the division offered select methods for students and faculty to collaborate with their peers, including:

1. **Students**
    a. Town hall: An annual town hall led by members of the school's Dean's Office. All online students were invited to attend.
    b. Curriculum-based community: Some online programs, such as the Applied Biomedical Engineering program, required onsite attendance at various points in the curriculum.
    c. Course-based community: Online courses included opportunities for interaction and community building through collaborative work, discussions, and live office hours.
2. **Faculty**
    a. Faculty meetings: A bi-annual faculty meeting that included time for socialization, dinner, presentations on teaching and learning, and a keynote speaker.
    b. Faculty development: Various faculty development events were held throughout the year to discuss, share, and learn teaching practices.
    c. Course development cohorts: Divisional teaching and learning center, consisting of instructional designers, technologists, and media experts would collaborate with faculty to create online courses that utilize best practices to foster a SoC within the specific course.
    d. Academic program meetings: Online program leadership would hold meetings for the faculty in their program to attend.

To examine the social dimensions of online students engaged in a graduate engineering program, it is necessary to understand the narratives of engaged practitioners (i.e., learners and educators) and analyze the material products of these social practices. This chapter includes two data sources: (a) surveys and (b) document analysis.

SOLUTIONS AND RECOMMENDATIONS

Before implementing changes to increase SoC in the online learning community, the planned changes were aligned with the three primary elements of a CoP. For each group, one interaction site was established to develop and initiate members into the domain and community (i.e., New Student Orientation, Faculty Forward), two were set up to facilitate open knowledge sharing and collaboration, and one was implemented to provide a site of recognition and visibility.

Table 1. Interaction site aligned with a CoP component of development

Practitioner Group	Interaction Site	Strategy	CoP Element	Evidence
Students	Program, Course	New Student Orientation	Domain, Community	Orientation site interaction, end-of-year survey, support cases
Students	Program	Student Community	Community, Practice	Student community interaction, end-of-year survey, support cases
Students	Program, course	Student Advisory Board	Community	Student advisory board interaction, program outcomes based on recommendations
Students	Program	Community Day	Community, Practice	Community day interaction, end-of-year survey
Faculty	Program, course	Faculty Community	Community, Practice	Faculty community interaction, support cases
Faculty	Course	Faculty Forward	Domain, Community	Faculty Forward interaction and completed trainings, course evaluations, support cases
Faculty	Course	Faculty Course Development Cohorts	Community, Practice	Course Development Cohort interaction, course evaluations
Faculty	Program	Faculty Meeting	Community, Practice	Faculty meeting survey, support cases

Data used for analysis includes survey responses to bi-annual faculty meetings, end-of-year graduating student surveys, community interactions and activity on the platform Microsoft Teams, orientation site activity from Blackboard and Canvas, Community Day participation and interactions, and outcomes from the student advisory board meetings.

For Students

For students, the sites of interaction include New Student Orientation, Student Community, Student Advisory Board, and Community Day. These sites of interaction are listed in the order in which they were implemented across the part-time online graduate engineering programs. The implementation order was driven by the stages of development of a CoP, which include orienting new community members with a baseline to interact (e.g., New Student Orientation) and introducing them to the community (e.g., Student Community), establishing formal spaces for visibility or recognition (e.g., Student Advisory Board), and organizing a central event for interaction (e.g., Community Day).

These implemented changes were based on core elements of the CoP framework, Moore's three types of interaction, Bloom's Taxonomy, and adult learning theory (see the Theoretical Framework section). Each provides a structure with overlapping elements: (a) required expertise and interest in the domain; (b) orientation of newcomers to the community; (c) shared resources and knowledge; and (d) shared sites of interaction. The four implemented changes below address the overlapping elements of CoP, Moore's three types of interaction, Bloom's Taxonomy, and adult learning theory.

1. **New Student Orientation**: An online student orientation that welcomes new students to the school and provides guidance on the following aspects:
    a. Getting started as a new student

   b. Academic ethics
   c. Online learning
   d. Preparing for your course
   e. Opioid and sexual harassment resources
2. **Student Community**: An online student community on a collaborative chat platform (Microsoft Teams) that includes over 1,900 active students. The community provides an opportunity for social interaction with all students and within their specific academic program, news about the school, and direct access to teaching and technology support and student services teams.
3. **Student Advisory Board**: A student advisory board provides a voice to a broad and diverse population of online learners. The advisory board advises program leadership on pressing academic and student services issues, respond to communication requests, and meets remotely with the program administration twice per academic year.
4. **Community Day**: A community day is a hybrid (in-person, virtual) conference-style event that seeks to build community among students, faculty, and alumni. Community day offerings include new student orientation, workshops, leadership town halls, resource fairs, technical talks, lab demonstrations, and networking events.

For Faculty

Four primary sites of faculty interaction and community development were implemented: Faculty Meeting, Faculty Course Development Cohorts, Faculty Community, and Faculty Forward. These sites of interaction are listed in the order in which they were implemented. The implementation order for faculty, unlike student implementation, was driven by existing processes and needs. Specifically, the bi-annual faculty meetings and course development cohorts existed but were altered in significant ways, whereas Faculty Community and Faculty Forward were entirely new ventures.

These implemented changes were informed by research from Dolan (2011), who found that part-time faculty noted three specific areas of concern: (a) lack of regular and in-depth communication with other members of the community (i.e., students, staff, other faculty); (b) lack of recognition and visibility in the community (i.e., awards, notes of appreciation); and (c) lack of professional development opportunities.

1. **Faculty Meeting**: A bi-annual faculty event that includes sessions for professional development, opportunities for networking between program faculty and staff, and annual faculty awards. Awards announced at the faculty meetings include New Instructor Award, Outstanding Instructor Award, Sustained Excellence Award, and Exceptional Online Course Design Award. The spring and fall faculty meetings address all three concerns in Dolan's (2011) study.
2. **Faculty Course Development Cohorts**: A systematic course design and development process in which faculty work collaboratively with an instructional designer and have opportunities to discuss course design, teaching, and learning practices with their faculty peers. The course development cohorts address the third concern in Dolan's (2011) study.
3. **Faculty Community**: An online faculty community on Microsoft Teams includes over 600 active faculty members. The community provides an opportunity for social interaction and sharing of teaching practices, news about the school, and direct access to the teaching and technology support team and library liaison. The faculty community addresses the first and second concerns in Dolan's (2011) study.

4. **Faculty Forward**: A faculty development program that provides opportunities for faculty to learn and engage with one another. The Faculty Forward program addresses the third concern in Dolan's (2011) study.

Group collaboration, knowledge sharing, and communication can be increased by providing spaces for interactions between learners and faculty. In this section, the authors discuss the strategies deployed in creating and sustaining a SoC for faculty and students over 10 years, from 2012 to 2022. This period of time is broken into two main phases of the community building strategy.

Phase 1 (2012–2017): Divisional Events, Curriculum and Course Design, and Support

As the school entered into online graduate education, it adopted research-supported best practices in establishing a SoC for faculty and students that included both task-driven interactions to facilitate the teaching and learning goals and socioemotional interactions to develop friendships and the social well-being of its members (Rovai, 2001). These efforts included division-wide events and establishing opportunities for interactions throughout the curriculum.

Faculty Meetings

The online graduate division offered bi-annual faculty meetings that included time for general updates, socialization, dinner, presentations on teaching and learning, and a keynote speaker. Faculty meetings are offered both remotely and in person to allow members who cannot attend in person to participate virtually.

Fall meetings are an opportunity for the administration to offer faculty development workshops designed to introduce new pedagogical opportunities and foster academic discussion among faculty attendees and presenters. Workshop presenters include members, program leadership, divisional teaching and learning centers, and faculty members who want to share knowledge with their peers. Workshop topics at previous fall faculty meetings include "Virtual Reality and Other Technologies for Enhancing Remote Education," and "The Impact of Dyslexia in Engineering: Inclusive Instructional Design Principles and Strategies." Workshops are collaborative and provide ample opportunities for faculty to interact with one another and presenters.

Spring faculty meetings provide an opportunity for individual programs to hold annual program meetings with their faculty members prior to the general faculty meeting. Through individual program meetings, program leaders review program goals, updates, faculty accomplishments, challenges, and strategies for successful teaching and learning. Spring general meetings hold space to recognize faculty excellence in teaching with the announcement of faculty award winners in the following categories:

- Exceptional Online Course Design Award
- Outstanding New Instructor Award
- Outstanding Instructor Award
- Sustained Excellence Award

Faculty who teach remotely often feel isolated from their peers, program leadership, and institutional leadership. The bi-annual faculty meetings not only focus on delivering institutional, programmatic, and pedagogical updates to faculty that are critical to maintaining a high level of instruction, but also facilitate a SoC across the online programs and encourage cross-collaboration and support.

Curriculum Design

With the advancement of technology, programs began re-evaluating their curriculum to determine which courses or experiences could (and should) be offered online. The curriculum design plans followed the current best practices, including balance, rigor, coherence, progression, appropriate, focused, and relevant (William, 2013). The programs also included another tenet: building a SoC throughout the curriculum and within the courses.

For example, the Applied Biomedical Engineering program integrated opportunities for students to be on campus at a specific point in time and provided opportunities for hands-on learning in the labs:

You will also work alongside our colleagues who are scientists, physicians, and engineers at the world-renowned Johns Hopkins Hospital during a unique hybrid two-weekend residency course in Baltimore. Dynamic and life-saving solutions evolve from these biomedical engineering course projects, including a student who redesigned the Ebola protective suit by integrating a cooling system. In this hands-on, immersive lab experience, you will also design and build your own EKG monitor. (Johns Hopkins Engineering for Professionals, 2022a, para. 2)

Other programs, such as Engineering Management, included a live capstone session that brought all students taking the final course of the program together for one day:

The course also includes one Saturday Capstone session in the Baltimore, MD area at the end of the semester. In-person participation with your team is encouraged. Students unable to attend in person can participate online. The Saturday session consists of student teams presenting their capstone technical strategic plan, issues, actions, and execution plans built around an evolving case study. A roundtable discussion will also be held where students have the opportunity to ask probing questions of visiting executives as part of the Capstone Day experience. (Johns Hopkins Engineering for Professionals, 2022b, para. 1)

Course Design and Development

As the division was ramping up its production of online courses, faculty, at that time, did not understand how interactions in an online course would occur. A common question from faculty when they were asked to create and teach an online course was as follows: "How can I interact with my students if I cannot see their facial expressions?"

Research indicates that following a systematic course development process and partnering faculty with instructional designers can greatly impact the overall quality of a course (Dick et al., 2015; Gagné et al., 2004; Halupa, 2019). Thus, to support faculty in implementing research-supported practices in providing opportunities for regular and substantive interaction, a course development process was provided. The course development process included partnering instructional designers with each course, following a systematic course development process (see Figure 1) and a quality review process using the QM (2011) Course Design Rubric.

Figure 1. Course development process

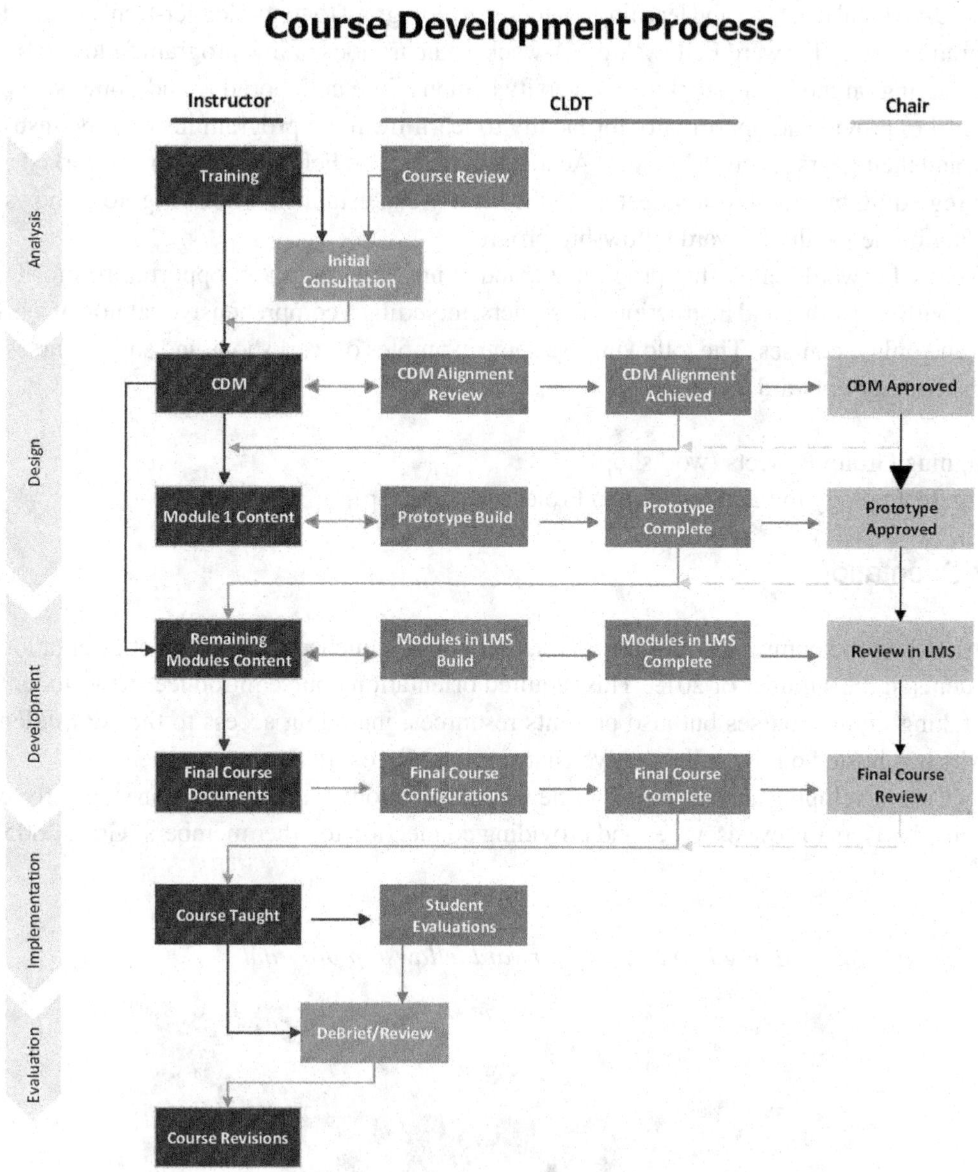

Phase 2: 2018–2022

To improve the SoC, the school moved into Phase 2, which focused on the following outcomes:

1. Improve how students and faculty are oriented to the school;
2. Improve communications with students and faculty; and
3. Expand opportunities for collaboration inside and outside of the course context.

Faculty Development

In 2018, the divisional teaching and learning center received a grant to fund a longer-term faculty development program, Faculty Forward Fellowship, a 4-week asynchronous online program followed by a 3-day in-person training on the Johns Hopkins University campus. The collaborative and immersive program was designed to provide an opportunity for faculty to learn from the program instructors, instructional designers, and their peers (Faculty Forward Academy, 2022). The Fellowship program marked a change in culture regarding faculty development and how faculty can establish a SoC. Figure 2 shows faculty collaborating in the faculty forward fellowship program.

The Faculty Forward Fellowship program expanded into creating more opportunities for faculty to collaborate with each other and instructional designers, including a comprehensive catalog of workshops, webinars, and online courses. The following are some examples of workshops and short courses offered through the Faculty Forward Academy.

- Designing Group Projects (workshop)
- Using Technology to Facilitate Group Projects (workshop)

Student Orientation

As the first step in the community building plan, the school launched a new student orientation course for all students in the summer of 2019. This required orientation course introduces new students to the basics of taking online courses but also presents resources, including access to the EP Student Community, which is hosted on the collaborative chat service Microsoft Teams.

For any CoP, developing an orientation for newcomers is a foundational initiation step to the community, reducing barriers to key resources and providing connection to other members (Gray, 2005). Since

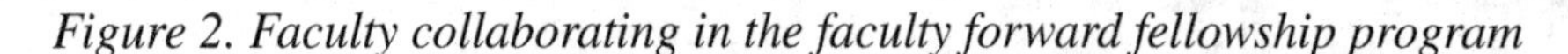

Figure 2. Faculty collaborating in the faculty forward fellowship program

the launch in 2019, over 5,000 students have completed the New Student Orientation course, and 65% of these students interacted monthly in the Teams community within 30 days of completing orientation. The student orientation course included the following topics:

1. Getting Started as a New Student
2. Academic Ethics
3. Online Learning
4. Preparing for Your Course
5. Opioid and Sexual Harassment Resources

Student Community

Following the student orientation, the school created a dedicated student community on Microsoft Teams with conversation channels for each program and news and access to real-time information and assistance from the student services team. Currently, the student community includes over 1,900 active students. These students visit the community for social interaction with all students and within their specific academic program, news about the school, and direct access to the teaching and technology support and student services teams.

After analyzing data over the past 2 years, emerging usage patterns highlight the value of this community, particularly in these areas (listed by interaction frequency):

1. Student assistance through the Student Services channel
    a. Registration process or course availability
    b. Tuition and refund policies and processes
    c. Course schedule
    d. Advisory and degree audit appointments
2. Student interaction and collaboration through the Program channels
    a. Future availability of courses on particular topics
    b. Information on a course, such as an instructor and texts
    c. Program-specific events
    d. Employment and professional development opportunities

While the student orientation connects newcomers to important resources and peers, the Student Community provides a space for ongoing networking and collaboration as students progress through their program.

Faculty Community

The Faculty Community was launched at the same time as the student community and hosted on the same collaborative chat service, Microsoft Teams. Since the majority of the faculty teaching online courses are part-time and working full-time as professionals in their field, the development of a community was viewed as essential. To address this request, the authors structured the community to function at two levels:

Figure 3. Faculty discussing teaching strategies in the Faculty Community Microsoft Teams site

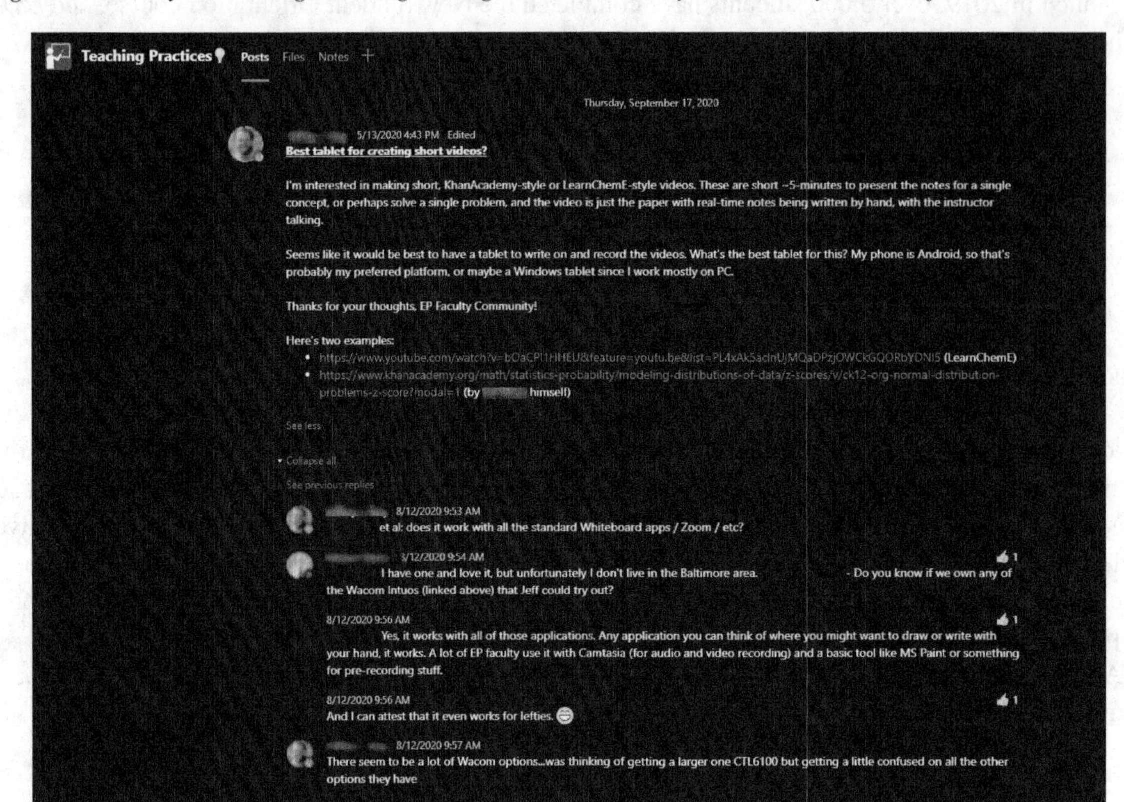

Figure 4. Faculty discussing teaching strategies in their specific program channel

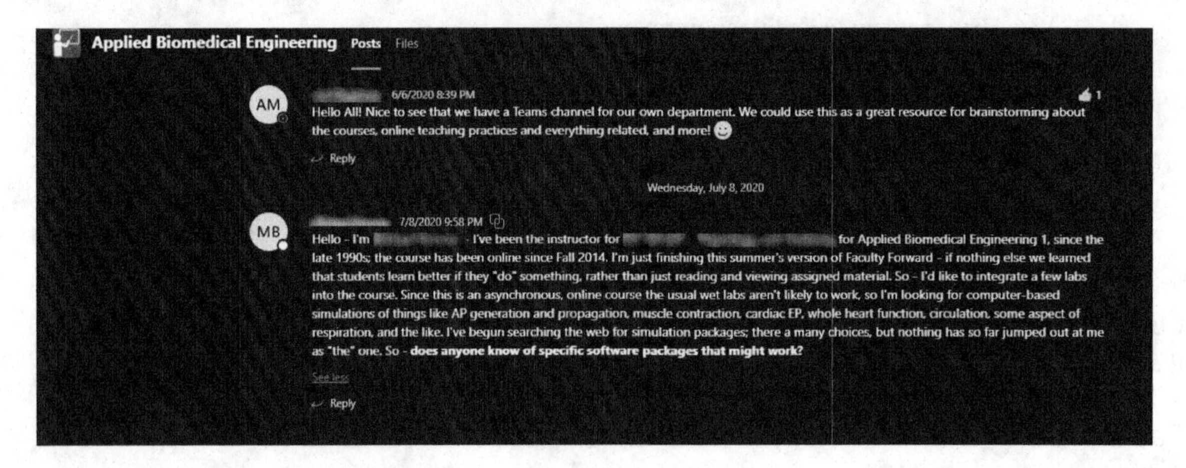

1. **School-level**: Specific channels in the Faculty Community were created to provide opportunities for all faculty to hear about news and policy updates and to share and discuss teaching and learning best practices, as shown in Figure 3.
2. **Program-level**: Specific channels were also created for communication and collaboration in the specific programs, as shown in Figure 4.

Currently, over 98% of the 618 faculty are active in the Faculty Community, posting updates about events in their programs, sharing teaching best practices, and asking questions about domain-specific software and tools.

The development of the community around a common set of tools and platforms accomplished an additional goal, which was helping unify the set of tools and practices used by the larger community consisting of students, faculty, and the support teams for the development and delivery of courses, resulting in better use of the platforms and more frictionless handoff and escalation of issues that emerge in courses.

Student Advisory Board

A student advisory board is a valuable tool for academic programs to gain insights from a broad and diverse population of online learners. In addition, student advisory boards provide students and program administrators with the opportunity to form trusting relationships and build community with one another as well as the students the board represents. In 2020, the school established a student advisory board consisting of approximately 15 current students from 12 master's degree programs who are representative of the school's diverse student body in many ways. The student advisory board:

1. Provides representation for EP students;
2. Reviews proposals and makes recommendations regarding new and current academic and student services initiatives;
3. Shares curricular components of the program experience with the members of the administration team; and
4. Discusses pressing issues in graduate education and the student community.

Student advisory board members are self-nominated and must submit a letter of interest and a recent curriculum vitae to be considered for a 1-year term appointment. Members are selected by administrators through a review process that ensures the board is representative of EP students in areas of diversity (race, gender, disability, age, etc.), time in program, degree program, and student classification (e.g., international, domestic, full-time, part-time, etc.).

The student advisory board meets virtually twice per year, once in the fall and spring term, and on an ad hoc basis. Advisory board members are also active in a dedicated channel of the Microsoft Teams student community, allowing them to interact with each other and the administrators between meetings. Meeting agendas are determined by student advisory board members and members of the administration team, including academic and student affairs, course design and technology support, institutional research, and marketing. Ad hoc meetings and Microsoft Teams discussions are often utilized when a new initiative is being considered or if students are experiencing a programmatic issue that requires swift feedback or response.

Over the last 2 years, the student advisory board has provided feedback on initiatives such as the community day, a student academic success coordinator role, and professional development webinar offerings as well as academic and curricular matters such as class size, the instructor/student/TA relationship, office hours expectations, and tutoring services. As a result of these discussions with the board, the school has moved forward with the hiring process for a student success coordinator who will serve as an administrative advisor to all students, expanded its professional development webinar series to include offerings beyond technical writing, and published an office hours expectation guide for faculty. Through feedback, advocacy, and collaboration, student advisory boards provide an avenue for students to make a lasting impact on their learning communities and help bridge relationships between university administrators and students.

Connect to Campus

Student surveys conducted over 5 years revealed that only an average of 34.2% of EP students felt a SoC in their graduate programs. With the understanding that a SoC is important for retention and degree completion and with the goal of creating more community among online learners, administrators began to explore the idea of an on-campus event. In early 2022, the planning for an annual one-day conference-style event known as Connect to Campus began. Connect to Campus is designed to build community, improve student retention, and enrich student services and consists of several interactive, synchronous offerings available both in-person and via Zoom for remote attendees. The inaugural Connect to Campus event took place in early Fall 2022 and followed the agenda below:

- Welcome – Dean, Vice Dean, Associate Deans
- Program Town Halls – Program Chairs, Vice Chairs, Managers (HYBRID OPTION)
- Tech Talks – Faculty and Alumni (HYBRID OPTION)
- Lunch and Networking
- Campus Tours
- Afternoon Session 1 (students will choose one; HYBRID OPTION)
    - New Student Orientation – Academic Resources, Student Affairs Resources, Campus Support Resources
    - Alumni Panel
- Afternoon Session 2 (students will choose one; HYBRID OPTION)
    - Study Skills Workshop
    - Work, School, Life Balance Workshop
    - Resume Building Workshop
    - Leadership Workshop
    - Student Focus Groups
- Affinity Group Receptions – Students, Faculty, Program Chairs, Vice Chairs, Managers, and Alumni
- Optional Sporting Events (on and off campus)

Figure 5. Students connecting over during Lunch and Networking

Figure 6. Students connecting with department leadership during Tech Talks

Figure 7. Students connecting with administrative staff during a Student Affairs session

Figure 8. Students connecting with the campus during the Campus Tours

FINDINGS

Through data collected from student course evaluations, engagement in the EP Student Community, EP Faculty Community, and faculty development events, the authors show a positive trend in building a SoC within a CoP. The following details demonstrate the positive trend.

Interactions in Online Courses

From 2014 to 2019, students were asked to evaluate their interactions in their online course by responding to the following question: "The other students' comments in the discussion forums contributed to my learning." The average score for this question was 3.83.

In 2020, the question on the survey was revised because it was deemed too narrow since discussion forums are just one form of interaction. The new question asked students, "My interaction with the other students in the course contributed to my learning." The average score from 2020 to 2021 was 3.98. This demonstrates a .15 increase in student evaluation scores for the effectiveness of interactions on their learning in an online course.

Faculty Development

Prior to the creation of the Faculty Forward Academy in 2019, a few faculty development sessions were offered per year. Attendance tracking for those events was not done. However, it is estimated that the authors offered approximately 10 sessions with 75 participants. Since 2020, the Faculty Forward Academy has offered 188 sessions with 1,080 attendees.

Figure 9. Growth of faculty development sessions offered and attendees

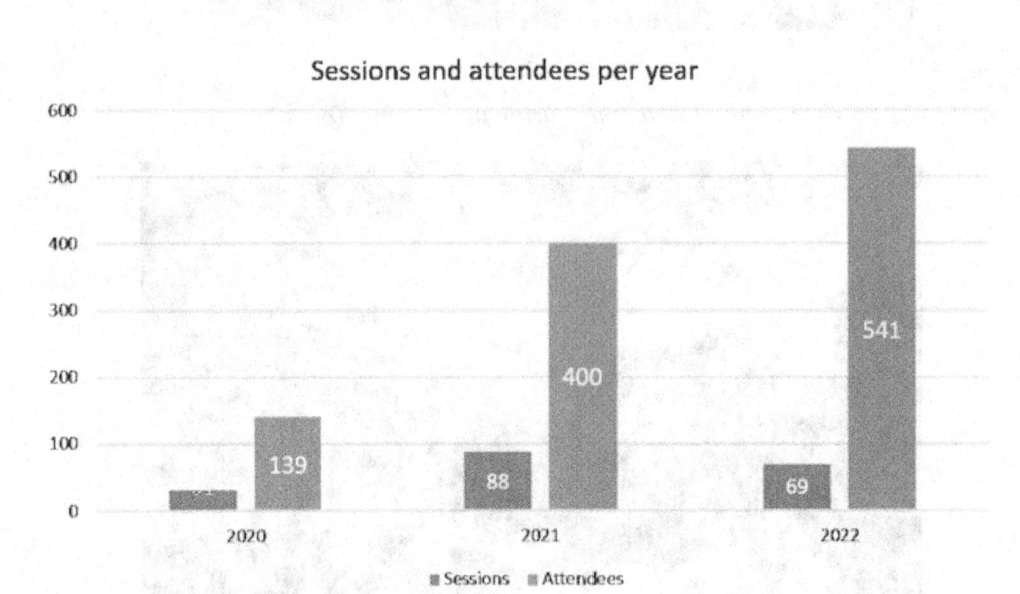

Student Community

The Student Community was launched on a collaborative chat platform, Microsoft Teams, on Monday, January 13, 2020, at the start of the spring semester. Between launch and the time of writing this chapter, the Student Community has grown to 1,934 active members (see Figure 5 for growth over time).

Faculty Community

The Faculty Community was launched on a collaborative chat platform, Microsoft Teams, on January 1, 2020, at the start of the spring semester. Between launch and the writing of this chapter (i.e., 28 months), the Faculty Community has grown to over 600 active members (see Figure 6 for growth over time).

Figure 10. Growth of active students in the Student Community

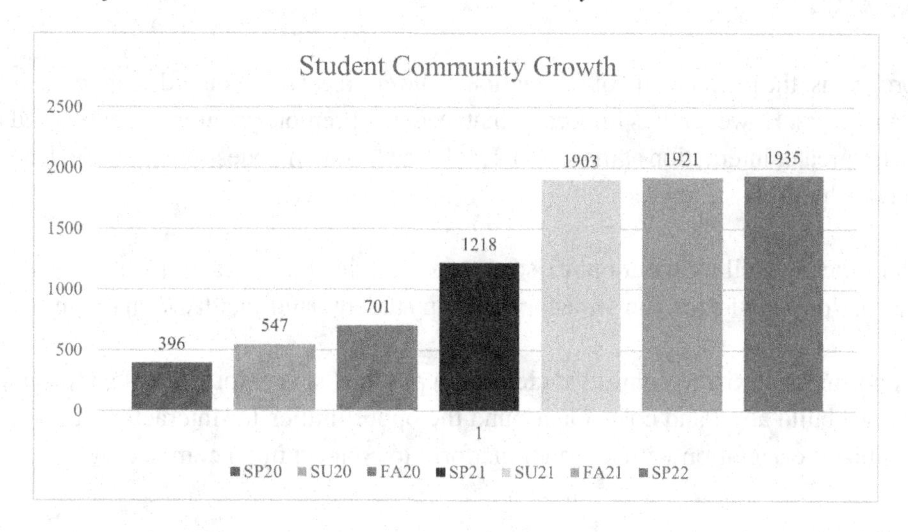

Figure 11. Growth of active faculty in the Faculty Community

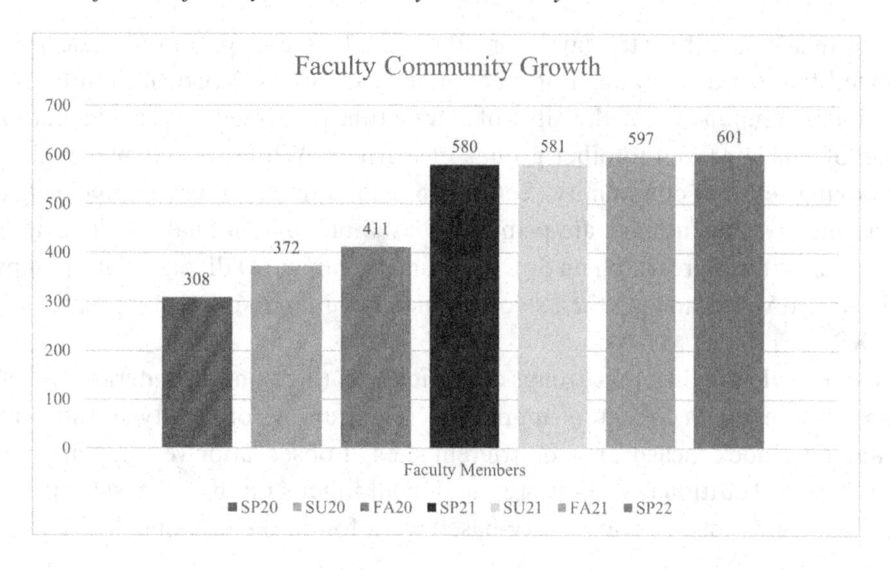

Connect to Campus

During the inaugural event, administrators, faculty, and alumni hosted nearly 200 graduate students in-person and remotely. Following the event, 90% of surveyed attendees indicated that Connect to Campus increased their SoC in just one day. Additionally, 97% of surveyed attendees stated they would attend Connect to Campus in Fall 2023. Surveyed attendees provided feedback that administrators plan to incorporate in future Connect to Campus events, such as increased social networking opportunities, inviting a greater number of faculty and alumni, and building opportunities to interact more with program leadership. Administrators will continue to monitor student survey results to measure if Connect to Campus, along with other SoC building initiatives, increase the percentage of students who feel a SoC in their graduate programs overall.

MAIN TAKEAWAYS

For remote programs, the primary site of interaction is the course, which includes opportunities with the instructor(s) and peers. However, it is important that leaders of remote programs identify and implement opportunities for greater interaction at the school level. The following questions can help leaders identify appropriate and achievable strategies:

1. What does the data tell us about our existing opportunities for interaction?
2. What technology platforms can we adopt to help students and faculty connect in and out of the course?
3. What opportunities exist to connect students in a physical space, for example, the campus?
4. How can we build align and cohesion around the opportunities for interaction, for example, connecting student orientation with a virtual platform to connect upon completion?

FUTURE RESEARCH DIRECTIONS

This chapter provides specific strategies for administrators and program leadership in higher education to facilitate the development of a community of practice (CoP). Future research could examine individual components of the sites of interaction presented in this chapter, and explore the application of this approach to other groups of learners. While this chapter highlights an approach to improving sense of community (SoC) within an online graduate engineering community of students and faculty, the findings are primarily based on content analysis. The authors sought three primary areas of future research on SoC in online learning: (a) diversification of participants, (b) community-based work, and (c) the use of extended or immersive reality, which is commonly referred to as XR.

Future research could use different groups of participants (e.g., online undergraduates in a school of information), different data collection approaches (e.g., surveys of faculty and students using the Sense of Community Index measure), and different sites of observation (e.g., small, public university in the Southeast). Additionally, future research could focus on the successful implementation of pedagogical strategies such as community-based work for online learning. Since the COVID-19 pandemic, there has been renewed interest in this research space, partly due to many experiential learning courses that were typically taught onsite having to move online. Continued research in this area will provide helpful insights and strategies for increasing SoC through various opportunities for interaction within an online course. Finally, with the expansion of immersive reality technologies into education, future research could help researchers and practitioners understand the impact of immersive learning on the SoC.

CONCLUSION

The analysis of the implemented sites of interaction suggests that adding visible spaces for interaction for members of an online learning community increases SoC and supports the development of an active CoP. The planning and selection of opportunities for interaction involve taking inventory of existing platforms and resources that may be relevant to the goal, aligning the inventory with the type of interaction it may facilitate, identifying gaps, and developing a proposal to address these gaps.

ACKNOWLEDGMENT

We would like to thank Dan Horn, Associate Dean for Engineering for Professionals, and Eric Weinstein, Senior Institutional Data Analyst for their assistance with this chapter.

REFERENCES

Allen, I., & Seaman, J. (2014). *Grade change: Tracking online education in the United States*. Babson Survey Research Group and Quahog Research Group, LLC.

Allen, I. E., Seaman, J., Poulin, R., & Straut, T. T. (2016). *Online report card: Tracking online education in the United States*. Babson Survey Research Group and Quahog Research Group, LLC.

Andersen, J. C., Lampley, J. H., & Good, D. W. (2013). Learner satisfaction in online learning: An analysis of the perceived impact of learner-social media and learner-instructor interaction. *Review of Higher Education and Self-Learning*, *6*(21), 81–96.

Bart, M. (2012). *Online student engagement tools and strategies*. Magna Publications.

Berge, Z. L. (1995). Facilitating computer conferencing: Recommendations from the field. *Educational Technology*, *35*(1), 22–30.

Berge, Z. L. (2008). Changing instructor's roles in virtual worlds. *Quarterly Review of Distance Education*, *9*(4), 408–414.

Berry, S. (2017). *Exploring community in an online doctoral program: A digital case study* (Publication No. 10257431) [Doctoral dissertation, University of Southern California]. ProQuest Dissertations & Theses Global.

Brooks, C. D., & Jeong, A. (2006). Effects of pre-structuring discussion threads on group interaction and group performance in computer-supported collaborative argumentation. *Distance Education*, *27*(3), 371–390. doi:10.1080/01587910600940448

Cheng, C., Paré, D., Collimore, L., & Joordens, S. (2011). Assessing the effectiveness of a voluntary online discussion forum on improving students' course performance. *Computers & Education*, *56*(1), 253–261. doi:10.1016/j.compedu.2010.07.024

Collins, R. (2004). *Interaction ritual chains*. Princeton University Press. doi:10.1515/9781400851744

Cox, B., & Cox, B. (2008). Developing interpersonal and group dynamics through asynchronous threaded discussions: The use of discussion board in collaborative learning. *Education*, *128*(4), 553–565.

Cox, M. (2001). Faculty learning communities: Change agents for transforming institutions into learning organizations. In D. Lieberman & C. Wehlburg (Eds.), *To Improve the Academy* (Vol. 19). Anker.

Cranton, P. (1994). *Understanding and promoting transformative learning: A guide for educators of adults*. Jossey-Bass.

Creswell, J. (1998). *Qualitative inquiry and research design: Choosing among five traditions*. Sage.

Darabi, A., & Jin, L. (2013). Improving the quality of online discussion: The effects of strategies designed based on cognitive load theory principles. *Distance Education*, *34*(1), 21–36. doi:10.1080/01587919.2013.770429

Davies, J., & Graff, M. (2005). Performance in e-learning: Online participation and student grades. *British Journal of Educational Technology*, *36*(4), 657–663. doi:10.1111/j.1467-8535.2005.00542.x

De Miranda, M. A. (2004). The grounding of a discipline: Cognition and instruction in technology education. *International Journal of Technology and Design Education*, *14*(1), 61–77. doi:10.1023/B:ITDE.0000007363.44114.3b

Delahunty, J., Verenikina, I., & Jones, P. (2014). Socio-emotional connections: Identity, belonging and learning in online interactions. A literature review. *Technology, Pedagogy and Education*, *23*(2), 243–265. doi:10.1080/1475939X.2013.813405

DeLoach, S. B., & Greenlaw, S. A. (2007). Effectively moderating electronic discussions. *The Journal of Economic Education*, *38*(4), 419–434. doi:10.3200/JECE.38.4.419-434

deNoyelles, A., Zydney, J. M., & Chen, B. (2014). Strategies for creating a community of inquiry through online asynchronous discussions. *Journal of Online Learning and Teaching*, *10*(1), 153–165.

Dick, W., Carey, L., & Carey, J. O. (2015). *The systematic design of instruction* (8th ed.). Pearson.

Dolan, V. (2011). The isolation of online adjunct faculty and its impact on their performance. *International Review of Research in Open and Distance Learning*, *12*(2), 62–77. doi:10.19173/irrodl.v12i2.793

Du, J., Yu, C., & Olinzock, A. (2011). Enhancing collaborative learning: Impact of "question Prompts" design for online discussion. *Delta Pi Epsilon Journal*, *53*(1), 28–41.

Duffy, T. M., & Cunningham, D. J. (1996). Constructivism: implications for the design and delivery of instruction. In D. Jonassen (Ed.), *Handbook of research on educational communications and technology* (1st ed., pp. 1–31). Routledge/Taylor & Francis Group.

Dziuban, C., Moskal, P., Kramer, L., & Thompson, J. (2013). Student satisfaction with online learning in the presence of ambivalence: Looking for the will-o'-the-wisp. *Internet and Higher Education, 17*, 1–8. doi:10.1016/j.iheduc.2012.08.001

Eagan, M. K., & Jaeger, A. J. (2008). Closing the gate: Part-time faculty instruction in gatekeeper courses and first-year persistence. *New Directions for Teaching and Learning, 2008*(115), 39–53. doi:10.1002/tl.324

Ehrenberg, R. G., & Zhang, L. (2005). Do tenured and tenure-track faculty matter? *The Journal of Human Resources, 40*(3), 647–659. doi:10.3368/jhr.XL.3.647

Engage, C. (2022). *Community engaged teaching and learning*. https://www.campusengage.ie/our-work/making-an-impact/community-engaged-teaching-and-learning

Ertmer, P. A., Richardson, J. C., Belland, B., Camin, D., Connolly, P., Coulthard, G., Lei, K., & Mong, C. (2007). Using peer feedback to enhance the quality of student online postings: An exploratory study. *Journal of Computer-Mediated Communication, 12*(2), 412–433. doi:10.1111/j.1083-6101.2007.00331.x

Faculty Forward Academy. (2022). *Fellowship*. https://facultyforward.jhu.edu/fellowship

Fairclough, N. (2001). Critical discourse analysis as a method in social scientific research. In R. Wodak & M. Meyer (Eds.), *Introducing qualitative methods: Methods of critical discourse analysis* (pp. 121–138). SAGE Publications Ltd.

Finegold, A., & Cooke, L. (2006). Exploring the attitudes, experiences and dynamics of interaction in online groups. *The Internet and Higher Education, 9*(3), 201–215. doi:10.1016/j.iheduc.2006.06.003

Foshay, W., Silber, K., & Westgaard, O. (1986). *Instructional Design Competencies: The Standards*. International Board of Standards for Training, Performance and Instruction.

Gagné, R. M., Wager, W. W., Golas, K. C., Keller, J. M., & Russell, J. D. (2004). *Principles of instructional design* (5th ed.). Wadsworth Publishing Company.

Garrison, D. R., Anderson, T., & Archer, W. (2001). Critical thinking, cognitive presence, and computer conferencing in distance education. *American Journal of Distance Education, 15*(1), 7–23. doi:10.1080/08923640109527071

Giles, D., Stommel, W., Paulus, T., Lester, J., & Reed, D. (2015). Microanalysis of online data: The methodological development of "digital CA". *Discourse, Context & Media, 7*, 45-51.

Goggins, J., & Hajdukiewicz, M. (2020). *Community-engaged learning: A building engineering case study*. https://sword.cit.ie/cgi/viewcontent.cgi?article=1069&context=ceri

Gray, B. (2005). Informal learning in an online community of practice. *International Journal of E-Learning & Distance Education, 19*(1), 20–35.

Halupa, C. (2019). Differentiation of roles: Instructional designers and faculty in the creation of online courses. *International Journal of Higher Education, 8*(1), 55–68. doi:10.5430/ijhe.v8n1p55

Han, J., Jiang, Y., Mentzer, N., & Kelley, T. (2022). The role of sense of community and motivation in the collaborative learning: An examination of the first-year design course. *International Journal of Technology and Design Education, 32*(3), 1837–1852. doi:10.100710798-021-09658-6

Hatcher, J. A., & Bringle, R. G. (2010). Reflection: Bridging the gap between service and learning. *College Teaching, 45*(4), 153–158. doi:10.1080/87567559709596221

Herring, S. (1999). Interactional coherence in CMC. *Journal of Computer-Mediated Communication, 4*(4), 0. Advance online publication. doi:10.1111/j.1083-6101.1999.tb00106.x

International Technology and Engineering Educators Association. (2020). *Standards for technological and engineering literacy: Defining the role of technology and engineering in STEM education*. Author.

Jacobs, S., Mishra, C. E., Doherty, E., Nelson, J., Duncan, E., Fraser, E. D., Hodgins, K., Mactaggart, W., & Gillis, D. (2021). Transdisciplinary, community-engaged pedagogy for undergraduate and graduate student engagement in challenging times. *International Journal of Higher Education, 10*(7), 84–95. doi:10.5430/ijhe.v10n7p84

Jacoby, D. (2006). Effects of part-time faculty employment on community college graduation rates. *The Journal of Higher Education, 77*(6), 1081–1103. doi:10.1353/jhe.2006.0050

Jaeger, A. J., & Eagan, M. K. (2011). Examining retention and contingent faculty use in a state system of public higher education. *Educational Policy, 25*(3), 507–537. doi:10.1177/0895904810361723

Järvelä, S., Järvenoja, H., & Veermans, M. (2008). Understanding the dynamics of motivation in socially shared learning. *International Journal of Educational Research, 47*(2), 122–135. doi:10.1016/j.ijer.2007.11.012

Jeong, A. (2004). The combined effects of response time and message content on growth patterns of discussion threads in computer supported collaborative argumentation. *Journal of Distance Education, 19*(1), 36–53.

Jinhong, J., & Gilson, T. A. (2014). Online threaded discussion: Benefits, issues, and strategies. *Kinesiology Review (Champaign, Ill.), 3*(4), 241–246. doi:10.1123/kr.2014-0062

Johns Hopkins Engineering for Professionals. (2022a). *Applied Biomedical Engineering Master's Program Online*. Johns Hopkins Engineering Online. https://ep.jhu.edu/programs/applied-biomedical-engineering

Johns Hopkins Engineering for Professionals. (2022b). *Executive Technical Leadership Online*. Johns Hopkins Engineering Online. https://ep.jhu.edu/programs/engineering-management

Jones, A., & Issroff, K. (2005). Learning technologies: Affective and social issues in computer-supported collaborative learning. *Computers & Education, 44*(4), 395–408. doi:10.1016/j.compedu.2004.04.004

Kay, R. H. (2006). Developing a comprehensive metric for assessing discussion board effectiveness. *British Journal of Educational Technology*, *37*(5), 761–783. doi:10.1111/j.1467-8535.2006.00560.x

Kehrwald, B. (2008). Understanding social presence in text-based online learning environments. *Distance Education*, *29*(1), 89–106. doi:10.1080/01587910802004860

Kim, T. L., Wah, W. K., & Lee, C, T. A. (2007). Asynchronous electronic discussion group: Analysis of postings and perception of inservice teachers. *Turkish Online Journal of Distance Education*, *8*(1), 33–41.

Knowles, M. S. (1984). *The adult learner: A neglected species* (3rd ed.). Gulf Publishing Co.

Krathwohl, D. R. (2002). A revision of Bloom's taxonomy: An overview. *Theory into Practice*, *41*(4), 212–218. doi:10.120715430421tip4104_2

Kupczynski, L., Mundy, M.-A., & Maxwell, G. (2012). Faculty perceptions of cooperative learning and traditional discussion strategies in online courses. *Turkish Online Journal of Distance Education*, *13*(2), 84–95.

Lave, J. (1991). Situating learning in communities of practice. In L. B. Resnick, J. M. Levine, & S. D. Teasley (Eds.), *Perspectives on socially shared cognition* (pp. 63–82). American Psychological Association. doi:10.1037/10096-003

Lee, K. (2007). Online collaborative case study learning. *Journal of College Reading and Learning*, *37*(2), 82–100. doi:10.1080/10790195.2007.10850199

Mager, R. F. (1997). *Preparing Instructional Objectives*. Center for Effective Performance.

Magnuson, C. (2005). Experiential learning and the discussion board: A strategy, a rubric, and management techniques. *Distance Learning*, *2*(2), 15–20.

Marinoni, G., van't Land, H., & Jensen, T. (2020). *The impact of Covid-19 on higher education around the world*. International Association of Universities. https://www.iau-aiu.net/IMG/pdf/iau_covid19_and_he_survey_report_final_may_2020.pdf

Marzano, R. J. (Ed.). (2009). *On excellence in teaching*. Solution Tree Press.

Masters, K., & Oberprieler, G. (2004). Encouraging equitable online participation through curriculum articulation. *Computers & Education*, *42*(4), 319–332. doi:10.1016/j.compedu.2003.09.001

Matters, Q. (2011). *The Quality Matters higher education rubric*. https://www.qualitymatters.org

Matters, Q. (2013). *Rubric and standards*. https://www.qualitymatters.org/rubric

McGinty, A. S., Justice, L., & Rimm-Kaufman, S. E. (2008). Sense of school community for preschool teachers serving at-risk children. *Early Education and Development*, *19*(2), 361–384. doi:10.1080/10409280801964036

McIntosh, S., Braul, B., & Chao, T. (2003). A case study in asynchronous voice conferencing for language instruction. *Educational Media International*, *40*(1-2), 63–74. doi:10.1080/0952398032000092125

McIsaac, M. S., Blocher, J. M., Mahes, V., & Vrasidas, C. (1999). Student and teacher perceptions of interaction in online computer-mediated communication. *Educational Media International*, *36*(2), 121–131. doi:10.1080/0952398990360206

McMillan, D. W., & Chavis, D. M. (1986). Sense of community: A definition and theory. *Journal of Community Psychology*, *14*(1), 6–23. doi:10.1002/1520-6629(198601)14:1<6::AID-JCOP2290140103>3.0.CO;2-I

McNeil, R. C. (2011). A Program Evaluation Model: Using Bloom's Taxonomy to Identify Outcome Indicators in Outcomes-Based Program Evaluations. *Journal of Adult Education*, *40*(2), 24–29.

Mentzer, N. (2014). Team Based Engineering Design Thinking. *Journal of Technology Education*, *25*(2), 52–72. doi:10.21061/jte.v25i2.a.4

Meyer, K. A. (2014). Student engagement in online learning: What works and why. *ASHE Higher Education Report*, *40*(6), 1–114. doi:10.1002/aehe.20018

Milliman, J., Czaplewski, A. J., & Ferguson, J. (2003). Workplace spirituality and employee work attitudes: An exploratory empirical assessment. *Journal of Organizational Change Management*, *16*(4), 426–447. doi:10.1108/09534810310484172

Moore, M. (1989). Editorial: Three types of interaction. *American Journal of Distance Education*, *3*(2), 1–7. doi:10.1080/08923648909526659

Muilenburg, L. Y., & Berge, Z. L. (2002). Designing discussion for the online classroom. In *Designing instruction for technology-enhanced learning* (pp. 100–113). IGI Global. doi:10.4018/978-1-930708-28-0.ch006

Murphy, K. L., & Cifuentes, L. (2001). Using Web tools, collaborating, and learning online. *Distance Education*, *22*(2), 285–305. doi:10.1080/0158791010220207

Online Learning Consortium. (2014). *Quality scorecard for administration of online programs*. https://onlinelearningconsortium.org/consult/quality-scorecard

Online Learning Consortium. (2022). *Quality framework*. https://onlinelearningconsortium.org/about/quality-framework-five-pillars

Oregon State University. (2022). *Ecampus essentials*. https://ecampus.oregonstate.edu/faculty/courses/Best_Practices_Online_Course_Design.pdf

Paquin, J. L. (2006). *How service-learning can enhance the pedagogy and culture of engineering programs at institutions of higher education: A review of the literature*. University of Nebraska Omaha. https://digitalcommons.unomaha.edu/slcedt/19/

Redmon, R. J., & Burger, M. (2004). WEB CT discussion forums: Asynchronous group reflection of the student teaching experience. *Curriculum and Teaching Dialogue*, *6*(2), 157–166.

Reeves, T., & Gomm, P. (2015). Community and contribution: Factors motivating students to participate in an extra-curricular online activity and implications for learning. *E-Learning and Digital Media*, *12*(3–4), 391–409. doi:10.1177/2042753015571828

Reilly, J. R., Vandenhouten, C., Gallagher-Lepak, S., & Ralston-Berg, P. (2012). Faculty development for e-learning: A multi-campus community of practice (COP) approach. *Journal of Asynchronous Learning Networks*, *16*(2), 99–110. doi:10.24059/olj.v16i2.249

Rourke, L., Anderson, T., Garrison, D. R., & Archer, W. (1999). Assessing social presence in asynchronous text-based computer conferencing. *Journal of Distance Education*, *14*(2), 50–71.

Rovai, A. P. (2001). Building classroom community at a distance: A case study. *Educational Technology Research and Development*, *49*(4), 33–48. doi:10.1007/BF02504946

Rovai, A. P. (2002). Building sense of community at a distance. *International Review of Research in Open and Distance Learning*, *3*(1), 1–16. doi:10.19173/irrodl.v3i1.79

Royal, M. A., & Rossi, R. J. (1996). Individual-level correlations of sense of community: Findings from workplace and school. *Journal of Community Psychology*, *24*(4), 395–416. doi:10.1002/(SICI)1520-6629(199610)24:4<395::AID-JCOP8>3.0.CO;2-T

Seaman, J. E., Allen, I. E., & Seaman, J. (2018). *Grade increase: Tracking distance education in the United States*. Babson Survey Research Group. https://onlinelearningsurvey.com/reports/gradeincrease.pdf

Shea, P., Li, C. S., & Pickett, A. (2006). A study of teaching presence and student sense of learning community in fully online and web-enhanced college courses. *The Internet and Higher Education*, *9*(3), 175–190. doi:10.1016/j.iheduc.2006.06.005

Sherer, P. D., Shea, T. P., & Kristensen, E. (2003). Online communities of practice: A catalyst for faculty development. *Innovative Higher Education*, *27*(3), 183–194. doi:10.1023/A:1022355226924

Swan, K. (2002). Building learning communities in online courses: The importance of interaction. *Education Communication and Information*, *2*(1), 23–49. doi:10.1080/1463631022000005016

Swan, K., & Shih, L. (2005). On the nature and development of social presence in online course discussions. *Journal of Asynchronous Learning Networks*, *9*(3), 115–136. https://olj.onlinelearningconsortium.org/index.php/olj/article/view/1788

Walker, C. H. (2016). *The correlation between types of instructor-student communication in online graduate courses and student satisfaction levels in the private university setting* [Doctoral dissertation]. Carson-Newman University. https://classic.cn.edu/libraries/tiny_mce/tiny_mce/plugins/filemanager/files/Dissertations/Christy_Walker.pdf

Weltzer-Ward, L., Baltes, B., & Lynn, L. K. (2009). Assessing quality of critical thought in online discussion. *Campus-Wide Information Systems*, *26*(3), 168–177. doi:10.1108/10650740910967357

Wenger, E. C. (1998). *Communities of practice: Learning, meaning, and identity*. Cambridge University Press. doi:10.1017/CBO9780511803932

Wenger, E. C. (2011). *Communities of practice: A brief introduction*. University of Oregon. http://hdl.handle.net/1794/11736

Wenger, E. C., McDermott, R., & Snyder, W. (2002). *Cultivating communities of practice: A guide to managing knowledge*. Harvard Business School Press.

Wenger, E. C., & Snyder, W. M. (2000). Communities of practice: The organizational frontier. *Harvard Business Review*, *78*(1), 139–145. https://hbr.org/2000/01/communities-of-practice-the-organizational-frontier

Wiliam, D. (2013). *Principled curriculum design*. SSAT (The Schools Network) Limited.

Williams, S., Jaramillo, A., & Pesko, J. (2015). Improving depth of thinking in online discussion boards. *Quarterly Review of Distance Education*, *16*(3), 45–66.

Yarbrough, J. R. (2018). Adapting adult learning theory to support innovative, advanced, online learning—WVMD model. *Research in Higher Education*, *35*, 1–15.

Chapter 7
Teachers' Unique Knowledge to Effectively Integrate Digital Technologies Into Teaching and Learning:
Community Engagement to Build in the Online Space

Abueng R. Molotsi
University of South Africa, South Africa

Leila Goosen
https://orcid.org/0000-0003-4948-2699
University of South Africa, South Africa

ABSTRACT

In order to provide readers with an overview of, and summarize, the content, the purpose of this chapter is stated as to evaluate teachers' unique knowledge, which they need to effectively integrate digital technologies into teaching and learning. This is built on teachers' use of the technological, pedagogical, and content knowledge (TPACK) model to improve their delivery of lessons. It is important to note that the research reported on in this chapter is positioned against the background of community engagement in the online space.

INTRODUCTION

This section will describe the general perspective of the chapter and end by specifically stating the objectives.

DOI: 10.4018/978-1-6684-5190-8.ch007

The paper by Koehler, Mishra and Cain (2013, p. 13) began "with a brief introduction to the complex, ill-structured nature of teaching" and learning. The latter authors described "a teacher knowledge framework for technology integration called technological pedagogical content knowledge (originally TPCK, now known as TPACK, or technology, pedagogy, and content knowledge)." The framework built on the "construct of pedagogical content knowledge (PCK)" by the educational researcher Shulman (1986) "to include technology knowledge. The development of TPACK by teachers is critical to" effectively integrate digital technologies into teaching and learning – this is accomplished by building on the TPACK model to improve the use of teachers' unique knowledge needed to deliver their lessons.

Like the book, this chapter will evaluate key issues and best practices pertaining to a community engagement project aimed at supporting teachers in the online space and remote settings of schools in the Dinaledi cluster of Bojanala district, North-West province, South Africa. Included will be an analysis of various community engagement efforts within (online) university programs. Further, using virtual means as a strategy for global business management will be briefly examined. Additionally, as part of this book, the chapter will *review* best practices for community engagement and considerations for the optimization of these practices, from effective virtual delivery to supporting emergency environmental challenges, such as pandemic conditions.

Target Audience

Similar to the book of which it is to form part of, this chapter could be of benefit to administrators in various settings, community organizers, coordinators, faculty, grant writers, key stakeholders in remote education, program directors, researchers and students.

Recommended Topics

Based on the recommended topics provided for this book, the chapter will particularly pay attention to:

- What is community engagement?
- Best practices for remote community engagement
- Remote community engagement in key settings
- Impacts of the pandemic
- Remote community engagement in higher education
- Remote community engagement in key business models
- Online education in government settings

Objectives

As for the book as a whole, readers of this chapter should be able to:

1. evaluate the critical areas of remote community engagement
2. apply best practices for remote community engagement
3. compare and contrast theories of engagement in a remote environment
4. apply evidence-based strategies for remote community engagement in key settings
5. create personalized engagement strategies in remote settings

6. evaluate pandemic-related learning objectives in remote settings

One area of impact for this chapter in the book is in training those, who work in remote settings, to effectively engage with their communities. For instance, key stakeholders were inspired to create online education programs. Additionally, the value of this chapter to the book can be seen in the impacts it may have on accessing connections. For instance, not all members of a community have equal access to engagement due to barriers. Community engagement via remote means breaks down such barriers, leading to broader participation.

"The purpose of the project introduced in" the earlier chapter by Molotsi and Goosen (2022, p. 1) was "stated as investigating in what ways teachers are using disruptive methodologies in teaching and learning to foster" students' skills by implementing TPACK. Similarly, the purpose of this chapter is stated as to evaluate teachers' unique knowledge, which they need to effectively integrate digital technologies into teaching and learning. Thus, the TPACK model is being used and will be used by the teachers involved in the project to build and improve their delivery of lessons. The outcomes of the chapter in terms of the project on which it is based will be a knowledge base of the extent and ways in which teachers use digital technologies to support learners and deliver lessons in the Dinaledi cluster of Bojanala district, North West province.

BACKGROUND

This section of the chapter will provide broad definitions and discussions on the topic of teachers' unique knowledge needed to effectively integrate digital technologies into teaching and learning, especially against the background of a *community engagement* project. It will also incorporate the views of others (in the form of a literature *review*) into the discussion to support, refute, or demonstrate the authors' position on the topic. Please note that TPACK will be explained and defined in the following section.

What Is Community Engagement?

"Community engagement and citizen participation" had long been important themes for both leaders and citizenry. Often termed 'community engagement', an Australian journal article on political science by Head (2007, p. 441) outlined some of the potential influences "of the citizenry and community groups. Important distinctions" were also drawn with regard to participation on whose terms?

With regard to background information, in a realist *qualitative* case-study, *community engagement* was seen by De Weger, Baan, Bos, Luijkx and Drewes (2022) as key to citizen-centered "and *sustainable* healthcare systems as involving" low-income citizens in the design, "implementation and improvement of services and policies" was "thought to tailor these more closely to communities' own needs". *Sustainable* and inclusive quality education through research-informed practice on information and communication technologies (Goosen, 2018a).

Towards a science of community engagement, Newman (2006, p. 302) criticized procedures, which left "vital processes of community engagement largely to trial and error. Rigorous *qualitative*" assessment of trial investigators was needed "to develop best practices in engagement with local communities." Rigor "in designing and initiating strategies for community engagement" was also crucial.

Best Practices for Remote Community Engagement

According to Dennis (2021a, p. 82), unforeseen "events, such as the global" COVID-19 pandemic, "have the potential to necessitate abrupt closures of the physical campuses of higher education institutions. In these situations," best practices for "emergency remote teaching procedures may be implemented".

To ensure quality in online programs in terms of case studies and best practices, online adjunct faculty members were "best engaged by delivering regular structured activities", which provided structured development and maintained support (Dennis, Halbert, & Fornero, 2022, p. 211). This was particularly important for online adjunct "faculty because of the geographic distance" that "may separate them from one another".

Remote Community Engagement in Key Settings

Against the background of management communication, Dempsey (2010, p. 359) was critiquing "community and community engagement by … illustrating how a campus/community divide" served "as a rich source of critique and … demonstrating the need to reshape community engagement around a critical understanding of community and community" engagement efforts.

Since this was affecting community well-being, in their journal article on investigative medicine, the aim of Holzer, Ellis and Merritt (2014, p. 851) "was to illustrate how community engagement can help to remedy" the shortfalls of community engagement. After briefly describing these shortfalls, the latter authors considered three case examples, which demonstrated "the potential of community engagement" in medical research, as well as why community engagement was needed?

In their journal article in the Stanford Social Innovation Review, Barnes and Schmitz (2016) referred to community centers and other non-profit organizations when explaining why community engagement matters (now more than ever).

Community engagement in flood risk "management in general, and its professionalization, in particular," had its paradoxes. The international journal article on disaster risk reduction by Puzyreva, et al. (2022) examined the micro-level facets of the professionalization of community engagement and "of professionalization for community engagement in flood risk" management, with insights from four European countries.

Community "engagement by different groups of stakeholders depends on numerous factors, ranging from national policies to the local culture of public engagement." In the study by Lioubimtseva and Cunha (2022, p. 257), the focus was "on the level of engagement by" various "groups, most typically involved in local climate adaptation", as well as justice in climate action planning and equity in climate adaptation planning, based on the experiences of small-and mid-sized cities in the United States and France.

A community engagement project using educational technologies for growing innovative e-schools in the 21st century (Goosen, 2015) can be considered, and cross- and trans-disciplinary approaches to *action research* and action learning for e-schools, community engagement, and Information and Communication Technologies for Development (ICT4D) discussed in the chapter by Goosen (2018b), as well as *ethical* data management and research integrity in the context of e-schools and community engagement in Goosen (2018c).

Impacts of the Pandemic

According to Dennis (2021b, p. 329), events, which "are unforeseen in nature and have widespread impact demonstrate the potential to necessitate abrupt changes in both personal and professional parameters of life." Efforts therefore need to be made towards research on developing a post-pandemic paradigm for virtual technologies.

"Technological Pedagogical Content Knowledge … is a crucial necessity for instructors and teachers during the teaching and learning process, particularly when using instructional technologies. The goal of" the study reported on in the journal article on language and linguistic studies by Sahrir, Hamid, Zaini, Hamat and Ismail (2022, p. 1111) was "to learn more about the TPACK" levels of knowledge by investigating the technological pedagogical content knowledge skills of Arabic school trainee teachers in online assessment during the COVID-19 pandemic.

Students' perceptions of e-assessment in the context of the Covid-19 pandemic via a case study of the University of South Africa (UNISA) were the subject matter of a conference paper by Van Heerden and Goosen (2021).

Remote Community Engagement in Higher Education

Although Dennis (2021b, p. 329) acknowledged that university departments, which "are already primarily online are accustomed to operating in" such an environment, there was still a need in higher education for supporting faculty and students during pandemic conditions. One example of the former was described by Goosen and Naidoo (2014), on computer lecturers using their institutional Learning Management System (LMS) for Information and Communication Technology (ICT) education in the cyber world – this is important, as Goosen (2016) indicated that both students in higher education and learners at basic education (school) levels want access to e-learning.

Towards rethinking teaching and learning in the 21st century, institutions "offering Open Distance e-Learning (ODeL) are increasing support levels to" students by involving e-tutors and "providing them with educational technologies and" digital tools (Goosen & Molotsi, 2019, p. 43), as well as opportunities "to select what they feel comfortable using to facilitate e-learning" (Molotsi & Goosen, 2019, p. 37). The latter paper also provided e-tutors' perspectives on the collaborative learning approach as a means to support students of computing.

The journal article on higher education theory and practice by Dennis, DiMatteo-Gibson, Halbert, Gonzalez and Byrd (2020, p. 19) described "the development, implementation, and preliminary evaluation of a research colloquium series for adjunct faculty teaching in a doctoral program offered through the online campus of a midsized university. The study described" by the latter authors provided an analysis towards building faculty community.

Watson (2007) believed that commitments to managing civic and community engagement were a strategic matter, which went "to the heart of the culture and values of any higher education institution (HEI). Consequently, getting this right is a matter of thinking clearly and understanding the strategic context".

Community engagement, from the perspective presented by Wood and Zuber-Skerrit (2013, p. 2) and quoting the Australian Universities Community Engagement Alliance (2008, p. 2), should expand "the role of higher education from a passive producer of knowledge to an active participant in collaborative discoveries". The former authors endorsed "the concept of community engagement" and suggested

Participatory Action Learning and *Action Research* as a methodology for community engagement by faculties of education.

The journal article on leadership, accountability and *ethics* by Dennis, Halbert, DiMatteo-Gibson, Agada and Fornero (2020, p. 30) presented "an analysis of the development and implementation of a faculty evaluation model in the online campus of a midsized three-campus university. Faculty evaluation is an important practice" that "can be utilized to improve" the student experience.

Remote Community Engagement in Key Business Models

The chapters by Bolton, Goosen and Kritzinger (2021) detailed an empirical study into the impact on innovation and productivity towards the post-COVID-19 era with regard to the digital transformation of an automotive enterprise, and Ngugi and Goosen (2021) innovation, entrepreneurship, and sustainability for ICT students towards the post-COVID-19 era, respectively.

Online Education in Government Settings

The reader should note that the community engagement project reported on in this chapter is being carried out in the context of online education conducted in government school settings.

MAIN FOCUS OF THE CHAPTER

Issues, Problems, Gaps, Challenges, Barriers

This section of the chapter will present the authors' perspective on issues, problems, gaps, challenges, barriers, etc., as these relate to the main theme of the book, community engagement in the online space, and arguments supporting the authors' position. It will also compare and contrast with what had been, or is currently being, done as it relates to the specific topic of the chapter, on teachers' unique knowledge needed to effectively integrate digital technologies into teaching and learning by building on the TPACK model.

"To focus a special edition of" the international journal of heritage studies on the issues of heritage and "community engagement might imply a lack of imagination on" the part of Watson and Waterton (2010, p. 1). The latter authors therefore focused on "the turn towards community involvement and a growing recognition on the part of practitioners that community engagement" helped them. Similarly, the focus of this chapter was on evaluating teachers' unique knowledge, which they need to effectively integrate digital technologies into teaching and learning.

Despite a consensus that using ICTs to facilitate multilingual mathematics teaching and learning and that "technologies facilitate English as a Foreign Language (EFL) teaching," university EFL teachers in China were "not using technologies at the optimal level" expected. "To address the *problem* of ineffective technology use," the study by Zhang and Chen (2022) purported to delineate the modeling of dichotomous technology in terms of the roles of TPACK, as well as affective and evaluative attitudes towards technology. Similarly, the authors of this chapter aimed to address the *problem* of teachers' ineffective technology use by evaluating teachers' unique knowledge, which they need to effectively integrate digital technologies into teaching and learning.

In a journal article on consumer marketing responding to a gap identified, the research by Haverila, Haverila and McLaughlin (2022) investigated "the impact of member motives as antecedents of brand community engagement (BCE)." The latter research also addressed "the consequences of BCE, relationship quality (RQ) and satisfaction in" brand communities. Finally, the research explored the development of a brand community engagement model from a service-dominant logic perspective. In the project reported on in this chapter, community engagement is used to foster teachers' use of the TPACK model to improve their delivery of lessons.

According to an Australasian journal article on educational technologies, existing research on TPACK showed "little about in-service secondary school science teachers' TPACK through a quantitative approach." Jang and Tsai (2013) were therefore exploring the TPACK of Taiwanese "secondary school science teachers using a new *contextualized* TPACK model", as well as associations with in-service teachers' TPACK.

Despite a commitment to community engagement, "information about the many types and examples of community engagement" had not previously been compiled (Driscoll, 2008, p. 38). There was, however, much to celebrate. An article on change in the Magazine of Higher Learning also provided a tool for analyzing community-engagement classification in terms of intentions and insights.

Gorelick (2022) reviewed lessons learned from "community engagement strategies utilized in the … African-American Antiplatelet Stroke Prevention Study" (AAASPS) trial and "Studies of Dementia in the Black Aged" (SDBA) observational studies, which had been directed by the latter author. In these studies, community engagement was "a means to help overcome challenges to the delivery of health care and preventative services."

When conducting a community engagement studio to adapt enhanced milieu teaching, in order to address "common barriers to community engagement", the model presented by Quinn, Cotter, Kurin and Brown (2022, p. 1) in an American journal article on speech language pathology included "researcher instruction and coaching on community engagement principles and communicating with nonacademic audiences, addressing researchers' lack of training in convening community members".

Towards breaking barriers, building bridges and increasing community engagement in evaluation and program planning, Roach and Fritz (2022) looked at "community engagement in Wayne County, MI. A strained relationship between" the "Michigan Department of Health and Human Services" (MDHHS) and the community had "created barriers to sustained and effective community engagement". "The local evaluation (LE) was developed to determine the barriers and facilitators to engagement".

"What Is Technological Pedagogical Content Knowledge"?

With regard to contemporary issues in technology and teacher education at the time, Koehler and Mishra (2009, p. 66) indicated that in terms of "pedagogy and content, expert teachers bring TPACK into play any time they teach." Regarding teaching with technology, the latter authors were "involved in considering TPACK as a professional knowledge construct." Their "work on the TPACK *framework*" at the time sought "to extend this tradition of research". Also regarding contemporary issues in technology and teacher education, the study reported on by Archambault and Crippen (2009) sought to examine the "knowledge levels with respect to each of the domains described by the TPACK" model among K-12 *online* distance educators in the United States. When teaching with technology, the TPACK profile of a teacher can be used "as a *framework* for evaluating" and understanding teaching expertise in *online* "*higher education*. Through interviews and non-participant observation," Benson and Ward (2013, p.

153) "created individual TPACK profiles for three professors". A cross-institutional impact study on the *online* training of TPACK skills for higher education scholars used the TPACK "model developed by Mishra and Koehler (2006). Afterwards", Rienties, et al. (2013, p. 481) described "the five learning steps of the *online*" program "and the impact of the" program on participants' TPACK.

In the 21st century, teachers must not only be able to use *educational technologies* for an ICT4D massive open *online* course, but also apply the "content knowledge and teaching strategies known as TPACK." Goosen and Mukasa-Lwanga (2017) showed how such *educational technologies* can be used in distance education to take it beyond the horizon with *qualitative* perspectives. The research by Utomo (2022, p. 1) aimed "to design a learning model to develop" and improve the TPACK of prospective elementary mathematics teachers, since the synthesis of *qualitative* evidence-based "*learning by design* could improve the TPACK of" such teachers. Using the TPACK framework to unite disciplines in *online* learning, Anderson, Barham and Northcote (2013, p. 549) looked at issues related to "using the TPACK framework as a lens through which to" analyze "academic teachers' views and practices about *online* teaching". Educating *online* student teachers to master professional digital competence when the TPACK-framework goes *online*, Tømte, Enochsson, Buskqvist and Kårstein (2015, p. 32) pointed out that "quality *online* instruction cannot rely on content or technological expertise alone; the development of reflection and understanding of the *online* teachers' TPACK profiles may provide" further insight. A preliminary evaluation of a short *online* training workshop for TPACK development by Alsofyani, bin Aris and Eynon (2013, p. 119) evaluated "the acceptance of using short *online* training for TPACK development." Niess, van Zee and Gillow-Wiles (2010, p. 42) looked at knowledge growth "in an *online* graduate course designed for integrating dynamic spreadsheets as teaching and learning tools" when teaching mathematics/science by moving PCK to TPACK through *online* professional development. Developing new schemas for *online* teaching and learning by using "TPACK as a framework for discussion and knowledge development surrounding *online*" "pedagogy, and technology would improve *online* course development and result in more satisfied learners" (Ward & Benson, 2010, p. 483). "The findings of the research" by Kaleli (2021, p. 399) showed the effect of individualized *online* instruction on the TPACK skills and achievement in the lessons "of the groups on the pretest and posttest scores." "TPACK was useful to the" study on learning to teach *online* "during multiple stages of the research process and" directly applicable to measuring the influence of faculty development training on teaching effectiveness through a TPACK lens (Brinkley-Etzkorn, 2018, p. 29). When investigating TPACK in a journal article on educational computing research, Niess (2011, p. 299) indicated that technological pedagogical and content knowledge was "a *framework* for thinking about the … evolution of TPACK as the knowledge" grows in teaching with technology that teachers require, and asked questions on the TPACK development of "teacher preparation with respect to developing TPACK". In the first section of their chapter in a research handbook on frameworks and approaches related to educational technologies, teacher knowledge, and classroom impact, Koehler, Shin and Mishra (2012, p. 16) provided "a brief overview of the TPACK *framework* and" discussed the need for their "*review*. In the second section," the latter authors identified empirical studies at the time that utilized TPACK assessments. They categorized these in terms of how TPACK could be measured. After a systematic literature *review* of TPACK studies in the context of computers and education to analyze how researchers were investigating teacher collaborative discourse in the learning by design process, Yeh, Chan and Hsu (2021, p. 1) identified eleven TPACK studies … to reveal the distribution of TPACK (sub)sets" toward a *framework* that connects *individual* TPACK and collective TPACK. Schmid, Brianza and Petko (2021, p. 1) compared "*individual* TPACK components among groups. Subsequently, unique *profiles* of all TPACK" components were used to investigate the relationships between the self-reported technological pedagogical content knowledge *profiles* of pre-service teachers and digital technology use in lesson plans. Rosenberg and Koehler (2015, p. 187) began their journal article regarding research on technology in education in the context of technological pedagogical content knowledge "with a brief history of prior research on the TPACK *framework*, and then" described "the importance of *context* in TPACK" as a conceptual framework for research, together with "a systematic *review* of TPACK". Unpacking the *contextual* influences of teachers' construction of technological pedagogical content knowledge in the context of computers and education, Koh, Chai and Tay (2014, p. 20) pointed out that *qualitative* TPACK studies tended "to exemplify the *seven* TPACK *constructs*" and described TPACK-in-Action, a *framework,* which "can be used to visualize the interplay between TPACK" *constructs.*

Towards *contextual* understanding of the TPACK *framework*, Swallow and Olofson (2017, p. 228), in a journal article regarding research on technology in education, indicated that the "technological, pedagogical, and content knowledge … *framework* considers the role of technology in teaching. Although TPACK is grounded in *context*, one limitation is the lack of understanding about the interactions between particular contexts, knowledge development, and instruction." The journal article on digital learning in teacher education by Mishra (2019, p. 76) indicated that while giving the TPACK diagram an upgrade and developing "teachers' knowledge types and overall TPACK, it" became clear that *contextual* knowledge needed to be considered as part of the "success of any TPACK development, or a teacher's attempts at technology integration" (Mishra P., 2019, p. 77). The study by Graham, et al. (2009, p. 78) "piloted an *instrument* for measuring" the confidence levels of in-service science teachers "in four of the *seven* TPACK knowledge *constructs*", including "TPACK, TPK, TCK, and TK. The *instrument* was useful in helping SciencePlus program coordinators to see significant" increases in teachers' TPACK confidence. In a journal article on educational technology and society, Koh, Chai and Tsai (2014, p. 185) investigated demographic factors, TPACK *constructs*, and the "perceptions of 354 practicing teachers in Singapore as assessed through" TPACK, as well as "practicing teachers' constructivist-oriented TPACK." In the context of the education sciences, Lyublinskaya and Kaplon-Schilis (2022, p. 79) were unpacking performance indicators for using the TPACK levels rubric and systematically examined the criteria and guidelines for the selection and implementation (Goosen, 2004) "of the rubric in order to" provide an analysis of "differences in the levels of TPACK". A validated *instrument* was also "developed on the basis of the model for the progressive levels of TPACK". Towards the sustainability of changes in teacher training within the TPACK model *framework*, Rodríguez Moreno, Agreda Montoro and Ortiz Colón (2019, p. 2) indicated that teachers' TPACK abilities contributed to their "attitudes towards ICT and their" relationship with TPACK. A systematic review and documentary analysis were "developed around the TPACK model, after analyzing" several studies. Schmid, Brianza and Petko (2020, p. 1) were developing a short assessment instrument "for measuring TPACK (TPACK.xs), and" "to use this instrument to investigate" and compare the factor structure of an integrative and a transformative model "view regarding how the TPACK knowledge domains interact." Regarding preservice teachers' TPACK development, "a brief introduction to TPACK" was provided by Wang, Schmidt-Crawford and Jin (2018, p. 234) to *review* the related conceptual *framework*. The "historical development of TPACK and summaries of prior literature reviews around TPACK" were also discussed. A journal article by Brantley-Dias and Ertmer (2013, p. 103) provided a "*review* of the TPACK construct and" addressed "the development, verification, usefulness, application, and appropriateness of TPACK as a" construct. "As part of this effort," the latter authors traced "the roots of TPACK back to earlier mentions of technological knowledge to address" various questions. The handbook of technological pedagogical content knowledge for educators indicated that there seemed to be "agreement that TPACK can best be developed through '*learning by design*' (Koehler & Mishra, 2008)". Using theoretical perspectives on "developing TPACK and under which conditions", the purpose of the chapter by Voogt, Fisser, Tondeur and van Braak (2016, p. 33) was to advance an understanding of TPACK. A journal article in the context of social education research showed that in "Ghana, the integration of technology into the teaching and learning process seems to be making strides in" higher education. This was, however, not the case in senior high schools. Therefore, the study by Mensah, Poku and Quashigah (2022, p. 80) sought to carry out a TPACK assessment of such schools' Geography teachers. "The consideration that the only goal of games is the achievement of entertainment is still commonly accepted, although there is now" a growing perspective, which "believes in the use of games to promote learning." The exploratory quantitative research based on the TPACK model by Huertas-Abril and García-Molina (2022, p. 554) examined in-service Spanish teachers' attitudes towards digital game-based learning. The study by Maipita, Dongoran, Syah and Hafiz (2022, p. 6) aimed to map the knowledge of pre-service teacher students in the Faculty of Economics Universitas Negeri Medan related to their mastery of TPACK and "the quality of organizational support related to technology-oriented learning".

Against the background of educational research, a short review of TPACK for teacher education by Gur (2015, p. 777) asked what the distribution was "according to the year of the" subjects of articles, which "focused on ICT and TPACK" findings and how TPACK can be integrated. In a conference paper on innovate learning, Petko (2020, p. 1349) expressed the opinion that although TPACK had "been highly inspiring for theory and practice, there" had been some questions regarding TPACK over the previous two years. As an example, Saubern, Henderson, Heinrich and Redmond (2020, p. 1) asked whether it was time to 'reboot' TPACK? The latter authors' opinions, as well as "those of other TPACK researchers" were summarized in four parts, the first of which addressed improving "understanding of the measurement of TPACK and the validation of TPACK *instruments*". According to the TPACK model, in their study, Lavidas, Katsidima, Theodoratou, Komis and Nikolopoulou (2021, p. 396) adapted the TPACK survey *instrument* of Koh and Chai (2014) "to fit into the Greek" educational "context of preschool teachers, as well as investigate their" perceptions about TPACK. The journal article on computers in education by Maor (2017, p. 71) explored the use of the different domains of the TPACK model to develop digital pedagogues as part of a higher education experience. "To maximize students' learning, TPACK was used in the design". Results from a co-word analysis of the academic performance of the term TPACK in Web of Science by Soler-Costa, Moreno-Guerrero, López-Belmonte and Marín-Marín (2021, p. 1) towards sustainability found that technological pedagogical content knowledge and the "evolution of the TPACK concept in the publications" reviewed showed that research on TPACK was on the rise. Valtonen, (2020, p. 2823) provided fresh perspectives on pre-service teachers' own appraisal of their *challenging* and confident TPACK areas, "based on their own descriptions of TPACK" content, using the TPACK areas, which they highlighted and the areas they indicated were worth considering. *Reviewing* a large-scale implementation across 23 virtual exchanges, Rienties, Lewis, O'Dowd, Rets and Rogaten (2020, p. 577) looked at the impact of virtual exchange on TPACK and how the "*growth* in TPACK over time significantly predicted reported gains in" Foreign Language (FL) competence, while pre-existing TPACK knowledge "further strengthened their TPACK skills". Regarding TPACK research with in-service teachers, Hofer and Harris (2012, p. 4704) asked where the TCK was? "While the majority of extant studies" focused on evidence and the "*growth* of TPACK holistically, some" had begun to distinguish the teacher knowledge in TPACK.

The article by Alemán-Saravia and Deroncele-Acosta (2021, p. 104) sought to reveal the key *recommendations* and *emerging trends* related to "the incorporation of ICT in the teaching-learning process based on the findings of a systematic literature review on" the TPACK framework.

SOLUTIONS AND RECOMMENDATIONS

This section of the chapter will discuss e.g., how teachers in the Dinaledi cluster of Bojanala district used information and communication technologies for the development of solutions to technical problems when using digital technologies and dealt with the challenges and issues that they encountered in teaching and learning, including those presented in the preceding section (Goosen, 2018d). The section will also refer to recommendations in this regard.

Methods

Please note that detailed information on the research design and methodology implemented in this project is provided in Molotsi and Goosen (2022). As also described by Archambault and Crippen (2009, p. 74), one "method to begin examining and measuring TPACK among a large group of teachers", who are working together in the Information Technology class (Mentz & Goosen, 2007), "is through quantitative methods, specifically through the use a survey methodology using a carefully developed questionnaire."

The interview of the participant resulting in Transcript 30 (T30) started by providing the teacher with some background information on the project, which is being run in North-West province on technology pedagogical content knowledge. It is all about teaching and learning with technologies when teachers are using different technologies in class. The project is trying to engage, through community engagement with Bojanala district, to support teachers and build their confidence in using online technologies. However, in order for teachers to be assisted, there is a need to know what the current status is. The participant was also reminded that the interview was being recorded.

The interviewer of T22 similarly pointed out that this information was really needed "so that we will be able to know exactly how to assist you. So we are going to analyze all these interviews and come up with a suitable intervention that can be" carried out at these schools.

In terms of the biographical details of the participants featured in this chapter, T18 is a young male between the ages of 20 and 30, who teaches Grades 8, 9 and 10, while T30 is an older female between the ages of 41 and 60, who teaches Grades 10, 11 and 12. T22 offers the subject Technology to both grades 8 and 9.

The reader should note that for the sake of authenticity, participants' own words are presented wherever possible. The interviewer of the different teachers was not necessarily the same person in each case.

How Do Teachers in the Dinaledi Cluster of Bojanala District Enhance Teaching and Learning Using Digital Technologies?

The interviewer also asked T30 the above 'main' research question before proceeding to the secondary numbered research questions that follow. Please note that these questions are not necessarily related to TPACK, which might (not) already be used in all situations.

T30 responded by providing some background information on the context of the study: Schools in the Dinaledi cluster of Bojanala district are situated in a semi-rural area. "So usually we are using technology." In 2021, the grade 12's (students) had "been given … tablets. The problem is connectivity and at the beginning, especially data." At the time of the interview, they had a committee, but even if they sent something on the group, "not all of them can learn because they cannot afford it."

Interviewer: "Which group are you referring to? Is it (a) WhatsApp group?"

T30: "Ja, the WhatsApp group that we are using so that they can get information on their tablets. … And then from there, … we have only one overhead projector, which is really hard to manage. … Sometimes, we are given …files, (but) it is difficult to get it on … time when we need it. … And then, there is a shortage of laptops: we only have five laptops, so they are used departmentally."

1. *How often do you integrate digital technologies into teaching and learning?*

T18: "So far, we don't integrate that much. Grade" 9's "are overlooked because" of the priority of Grades 10-12.

Please note that no detail related to this question was provided for T22.

T30: "It is very seldom between the teacher and the learner, but … I do use it on a daily basis because I" have "a laptop, but teaching and learning that is where we are using it seldom because we've only got one".

2. *What information communication technology infrastructure does your school have?*

T18: All teachers have laptops, two projectors and a smartboard. Grade 12's had been given tablets.

T22: For this academic year, "some of the equipment must (still) be bought."

Interviewer: … "You have already mentioned that you've got … one?"

T30: "Overhead projector."

Interviewer: "Is it a data projector or … overhead?"

T30: "Yes, the one that connects to the laptop."

Interviewer: "To the laptop, it is a data projector. Okay, what else do you have? You said you've only got how many laptops?"

T30: "Five in the school."

Interviewer: "Is there any other technology that you have (in terms of) infrastructure? You don't have … Wi-Fi?"

T30: "No. the school does not have … Wi-Fi."

Interviewer: "The school does not have .. Wi-Fi, and … the network?"

T30: "The network is very poor. Even now I find it difficult to connect."

3. *Which digital technologies do you use to deliver your lessons?*

T18: "We don't integrate technology that much." He uses a chalk board but when he is planning, he uses technologies, such (as) Moodle and You Tube videos aligned with what he was teaching.

T22: "In most cases I use a laptop, where I downloaded the information from YouTube and then take it to the classroom."

T30: "Digital technology that is the one that I just mentioned. … The data projector only."

Interviewer: "And your laptop?"

T30: "And it is very seldom, we can go a month without even using it with the learners. … It is not something that we do frequently. It is used by different teachers. …. It is very seldom we use it."

4. *Explain how you use a digital technology you have just mentioned.*

T18: You Tube (YT) "videos – plan lesson aside and get a lesson on YT in line with what he's prepared and check what is it that he can take from the YT lesson. Moodle – put … your planning on Moodle and questions in a form of a quiz, also upload videos."

Interviewer: … "You were saying you download information (from) YouTube and so forth?"

T22: "Yes."

Please note that no detail related to this question was provided for T30.

What Challenges Do Teachers in the Dinaledi Cluster of Bojanala District Encounter When Using Digital Technologies in Teaching and Learning?

5. *What background training did you receive to use digital technologies in delivering your lessons?*

T18: (A) "Moodle workshop attended in February 2021. In (the) PSFs (subject workshops), (the) subject advisor (was) showing (us) how to use technology."

T22: "From years back, there was this company; I forgot the company's name. It has long been. It was a collaboration. … They … trained us in using … IT. I think it more than five years ago. … And then from there, there was … Adopt A School. … So, Adopt A School is also an initiative … training educators from the school where I was based in terms of … ICT.

T30: "I have had training. It is … knowledge that I got from the university when I (did) ICT."

Interviewer: "Where you studied?"

T30: "Yes. There is no special training that was done maybe by the department (or) by the school. … The only training that we … did was at Unisa."

Interviewer: … "Was it … for … your qualification?"

T30: "Yes."

Interviewer: ... "Which qualification is that?"

T30: … "I was doing (a) Bachelors (of) Arts in education and then there was a compulsory … computer course."

6. *What type of support do you receive from your school with regards to the use of digital technologies in delivering your lessons?*

T18: "Our school has a challenge of not" having "enough classrooms. If you are planning to deliver (your lessons) with a projector, you are allowed to use the projector in a spare space, like a storeroom."

T22: … "from the school itself, we are being mentored by people, who … occupied … management positions and then from there, … like I indicated earlier on, we are given ample chance to say what are the resources that we are going to need for the next academic year. So with the promise that they gave us, in a way it is supported that we get from the school to say some of the items that they have listed as acquisition will be bought at the beginning of the year.

Interviewer: … "you are being supplied with ICT resources?"

T22: "Yes, like, for instance, … because the school is the poor of the poorest, there were people here two weeks ago who say they want to install Wi-Fi. That is another important aspect that they are looking at. … So that we are refraining from using our phones and stuff like that."

T30: "The school has not actually" supported "digital science. I don't know, the teachers are not digitally savvy. We have to … arrange as a teacher on your own, because, remember, with the overhead projector? … So, I even had to buy mine because I have been in the school since last year. Since I am now permanent, I am thinking of buying my own. So actually, when a teacher wants to (do) a digital lesson, maybe to show them. Like I am teaching English, to show some adverts, or whatever. When I am doing … Geography, to show them some graphs. I need to arrange for myself. Maybe things will develop."

7. *How do you obtain solutions to technical problems when using digital technologies?*

T18: In most cases he asks someone, who is knowledgeable. Sometimes he goes to his phone to Google and get a solution.

T22: "You know, as an educator, one needs to improvise in anyway and the teachers should be (as) flexible as possible, so … whenever I come (across) challenges, I (source) some of the information from other schools, from colleagues and stuff like that, so that at the end of the day, I am able to use them in my school. For example, we have subject meetings with other educators within the area. That is where we interact and we get information, in one way or the other. That on its own (assists) in terms of … subject delivery."

Interviewer: "All right, do you have a WhatsApp group?"

T22: "Yes, we do. In that WhatsApp group, we … focus (directly on) the educators within the sub-district."

T30: "Technological problems like I am not sure about the technical problems that you are talking about, because usually it can be an issue of maybe a laptop not reading the data that you are using. We usually change it with another laptop from another department. Lend me your laptop, because it is not reading this? That is the only … technical problems that we sometimes have."

Teachers, who obtain solutions to technical problems when using digital technologies are likely to implement online education.

Table 1 provides an example of the *document analysis checklist* that can be used by e-learners, teachers and managers involved in online learning at e-schools in South Africa (Goosen & Van der Merwe, 2015).

Recommendations

Please note that this chapter reports on some of the first data gathering opportunities for the project. At a later stage, lesson observation checklists will be used towards e.g., determining teachers' reflections and recommendations for further investigation/future research directions as a means to strengthen and promote the use of the TPACK model in the planning and delivery of lessons.

Table 1. Document analysis checklist

Item	Yes/No	Comments
1. Does the school have a professional development framework for digital learning?	No	No comment
2. Does the school have the national ICT policy and/or its own ICT policy?	Yes	Although it is not that much in detail
3. Does the ICT policy of the school include the following?		
3.1. The ICT vision of the school.	Yes	No comment
3.2. Steps for implementation.	Yes	No comment
3.3. Measures that were taken for optimal usage.	Yes	No comment
3.4. Development for staff.	Yes	No comment
3.5. A plan with **future** goals.	No	Not really
3.6. Integration of ICT in teaching and learning.	Yes	It does have
3.7. Roles and responsibilities of various stakeholders.	Yes	It does include it
3.8. Scheduling and management of training.	No	No comment

FUTURE RESEARCH DIRECTIONS

This section of the chapter will discuss future and emerging trends and provide insight about the future of the theme of the book, community engagement in the online space, from the perspective of the chapter focus. The viability of a paradigm, model, implementation issues of proposed programs, etc., may also be included in this section. Future research directions within the domain of the topic of teachers' unique knowledge needed to effectively integrate digital technologies into teaching and learning by building on the TPACK model will finally be suggested.

The data displayed in Table 1 was only provided by a single respondent, as it actually forms part of what will be explored during further research conducted in this project, including checking whether the ICT policy of a particular school contains a plan with future goals.

CONCLUSION

The purpose of this chapter was stated as to evaluate teachers' unique knowledge, which they need to effectively integrate digital technologies into teaching and learning. Built on teachers' use of the TPACK model to improve their delivery of lessons, it was important to note that the research reported on in this chapter is positioned against the background of community engagement in the online space. From surveys, the results presented suggested that this community engagement project helped teachers towards effective teaching and learning in the online space. Similar to what Lobo and Vélez (2022) concluded in their international journal article related to policy, based on the results from surveys, the authors of this chapter concluded that best practices with regard to the community engagement projected were obtained by working together with the teachers to build their TPACK knowledge.

REFERENCES

Alemán-Saravia, A., & Deroncele-Acosta, A. (2021). Technology, Pedagogy and Content (TPACK framework): Systematic Literature Review. *XVI Latin American Conference on Learning Technologies* (pp. 104-111). IEEE. 10.1109/LACLO54177.2021.00069

Alsofyani, M., bin Aris, B., & Eynon, R. (2013). A Preliminary Evaluation of a Short Online Training Workshop for TPACK Development. *International Journal on Teaching and Learning in Higher Education*, *25*(1), 118–128.

Anderson, A., Barham, N., & Northcote, M. (2013). Using the TPACK framework to unite disciplines in online learning. *Australasian Journal of Educational Technology*, *29*(4), 549–565. doi:10.14742/ajet.24

Archambault, L., & Crippen, K. (2009). Examining TPACK among K-12 online distance educators in the United States. *Contemporary Issues in Technology & Teacher Education*, *9*(1), 71–88.

Australian Universities Community Engagement Alliance. (2008). *Position Paper 2008-2010: Universities and Community Engagement.* Retrieved from http://admin.sun.ac.za/ ci/resources/AUCEA_universities_CE.pdf

Barnes, M., & Schmitz, P. (2016). Community engagement matters (now more than ever). *Stanford Social Innovation Review*, *14*(2), 32–39.

Benson, S., & Ward, C. (2013). Teaching with technology: Using TPACK to understand teaching expertise in online higher education. *Journal of Educational Computing Research*, *48*(2), 153–172. doi:10.2190/EC.48.2.c

Bolton, A., Goosen, L., & Kritzinger, E. (2021). An Empirical Study into the Impact on Innovation and Productivity Towards the Post-COVID-19 Era: Digital Transformation of an Automotive Enterprise. In *Handbook of Research on Entrepreneurship, Innovation, Sustainability, and ICTs in the Post-COVID-19 Era* (pp. 133–159). IGI Global. doi:10.4018/978-1-7998-6776-0.ch007

Brantley-Dias, L., & Ertmer, P. (2013). Goldilocks and TPACK: Is the construct 'just right'? *Journal of Research on Technology in Education*, *46*(2), 103–128. doi:10.1080/15391523.2013.10782615

Brinkley-Etzkorn, K. (2018). Learning to teach online: Measuring the influence of faculty development training on teaching effectiveness through a TPACK lens. *The Internet and Higher Education*, *38*, 28–35. doi:10.1016/j.iheduc.2018.04.004

De Weger, E., Baan, C., Bos, C., Luijkx, K., & Drewes, H. (2022). 'They need to ask me first'. Community engagement with low-income citizens. A realist qualitative case-study. *Health Expectations*, *25*(2), 684–696. Advance online publication. doi:10.1111/hex.13415 PMID:35032414

Dempsey, S. (2010). Critiquing community engagement. *Management Communication Quarterly*, *24*(3), 359–390. doi:10.1177/0893318909352247

Dennis, M. (2021a). Best Practices for Emergency Remote Teaching. In Handbook of Research on Emerging Pedagogies for the Future of Education: Trauma-Informed, Care, and Pandemic Pedagogy (pp. 82-100). IGI Global. doi:10.4018/978-1-7998-7275-7.ch005

Dennis, M. (2021b). Supporting Faculty and Students During Pandemic Conditions: An Online Department Chair's Perspective. In Handbook of Research on Developing a Post-Pandemic Paradigm for Virtual Technologies in Higher Education (pp. 329-346). IGI Global.

Dennis, M., DiMatteo-Gibson, D., Halbert, J., Gonzalez, M., & Byrd, I. (2020). Building Faculty Community: Implementation of a Research Colloquium Series. *Journal of Higher Education Theory & Practice*, *20*(6), 19–30.

Dennis, M., Halbert, J., DiMatteo-Gibson, D., Agada, C., & Fornero, S. (2020). Implementation of a faculty evaluation model. *Journal of Leadership, Accountability and Ethics*, *17*(5), 30–35.

Dennis, M., Halbert, J., & Fornero, S. (2022). Structured Development and Support for Online Adjunct Faculty: Case Studies and Best Practices. In Quality in Online Programs (pp. 211-228). Brill.

Driscoll, A. (2008). Carnegie's community-engagement classification: Intentions and insights. *Change: The Magazine of Higher Learning*, *40*(1), 38–41. doi:10.3200/CHNG.40.1.38-41

Goosen, L. (2004). *Criteria and Guidelines for the Selection and Implementation of a First Programming Language in High Schools. 178.* Potchefstroom Campus: North West University.

Goosen, L. (2015). Educational Technologies for Growing Innovative e-Schools in the 21st Century: A Community Engagement Project. *Proceedings of the South Africa International Conference on Educational Technologies* (pp. 49-61). African Academic Research Forum (AARF).

Goosen, L. (2016, February 18). *"We don't need no education"? Yes, they DO want e-learning in Basic and Higher Education!* Retrieved from https://uir.unisa.ac.za/handle/10500/20999

Goosen, L. (2018a). Sustainable and Inclusive Quality Education Through Research Informed Practice on Information and Communication Technologies in Education. In *Proceedings of the 26th Conference of the Southern African Association for Research in Mathematics, Science and Technology Education (SAARMSTE)* (pp. 215-228). University of Botswana.

Goosen, L. (2018b). Trans-Disciplinary Approaches to Action Research for e-Schools, Community Engagement, and ICT4D. In Cross-Disciplinary Approaches to Action Research and Action Learning (pp. 97-110). IGI Global.

Goosen, L. (2018c). Ethical Data Management and Research Integrity in the Context of E-Schools and Community Engagement. In *Ensuring Research Integrity and the Ethical Management of Data* (pp. 14–45). IGI Global. doi:10.4018/978-1-5225-2730-5.ch002

Goosen, L. (2018d). Ethical Information and Communication Technologies for Development Solutions: Research Integrity for Massive Open Online Courses. In Ensuring Research Integrity and the Ethical Management of Data (pp. 155-173). IGI Global.

Goosen, L., & Molotsi, A. (2019). Student Support Towards Rethinking Teaching and Learning in the 21st Century: A Collaborative Approach Involving e-Tutors. In *Proceedings of the South Africa International Conference on Education* (pp. 43-55). AARF.

Goosen, L., & Mukasa-Lwanga, T. (2017). Educational Technologies in Distance Education: Beyond the Horizon with Qualitative Perspectives. In U. I. Ogbonnaya, & S. Simelane-Mnisi (Ed.), *Proceedings of the South Africa International Conference on Educational Technologies* (pp. 41 - 54). African Academic Research Forum.

Goosen, L., & Naidoo, L. (2014). Computer Lecturers Using Their Institutional LMS for ICT Education in the Cyber World. In C. Burger, & K. Naudé (Ed.), *Proceedings of the 43rd Conference of the Southern African Computer Lecturers' Association (SACLA)* (pp. 99-108). Nelson Mandela Metropolitan University.

Goosen, L., & Van der Merwe, R. (2015). e-Learners, Teachers and Managers at e-Schools in South Africa. In *Proceedings of the 10th International Conference on e-Learning* (pp. 127-134). Academic Conferences and Publishing International.

Gorelick, P. (2022). Community Engagement: Lessons Learned From the AAASPS and SDBA. *Stroke*, *53*(3), 654–662. Advance online publication. doi:10.1161/STROKEAHA.121.034554 PMID:35078349

Graham, R., Burgoyne, N., Cantrell, P., Smith, L., St Clair, L., & Harris, R. (2009). Measuring the TPACK confidence of inservice science teachers. *TechTrends*, *53*(5), 70–79. doi:10.100711528-009-0328-0

Gur, H. (2015). A short review of TPACK for teacher education. *Educational Research Review*, *10*(7), 777–789. doi:10.5897/ERR2014.1982

Haverila, K., Haverila, M., & McLaughlin, C. (2022). Development of a brand community engagement model: A service-dominant logic perspective. *Journal of Consumer Marketing*, *39*(2), 166–179. Advance online publication. doi:10.1108/JCM-01-2021-4390

Head, B. (2007). Community engagement: Participation on whose terms? *Australian Journal of Political Science*, *42*(3), 441–454. doi:10.1080/10361140701513570

Hofer, M., & Harris, J. (2012, March). TPACK research with inservice teachers: Where's the TCK? In *Society for Information Technology & Teacher Education International Conference* (pp. 4704-4709). Association for the Advancement of Computing in Education.

Holzer, J., Ellis, L., & Merritt, M. (2014). Why we need community engagement in medical research. *Journal of Investigative Medicine*, *62*(6), 851–855. doi:10.1097/JIM.0000000000000097 PMID:24979468

Huertas-Abril, C., & García-Molina, M. (2022). Spanish Teacher Attitudes Towards Digital Game-Based Learning: An Exploratory Study Based on the TPACK Model. In Handbook of Research on Acquiring 21st Century Literacy Skills Through Game-Based Learning (pp. 554-578). IGI Global.

Jang, S., & Tsai, M. (2013). Exploring the TPACK of Taiwanese secondary school science teachers using a new contextualized TPACK model. *Australasian Journal of Educational Technology*, *29*(4). Advance online publication. doi:10.14742/ajet.282

Kaleli, Y. (2021). The Effect of Individualized Online Instruction on TPACK Skills and Achievement in Piano Lessons. *International Journal of Technology in Education*, *4*(3), 399–412. doi:10.46328/ijte.143

Koehler, M., & Mishra, P. (2008). Introducing TPCK. In A. C. Technology (Ed.), *Handbook of technological pedagogical content knowledge (TPCK) for educators* (pp. 3–29). Routledge.

Koehler, M., & Mishra, P. (2009). What is technological pedagogical content knowledge (TPACK)? *Contemporary Issues in Technology & Teacher Education*, *9*(1), 60–70.

Koehler, M., Mishra, P., & Cain, W. (2013). What is technological pedagogical content knowledge (TPACK)? *Journal of Education*, *193*(3), 13–19. doi:10.1177/002205741319300303

Koehler, M., Shin, T., & Mishra, P. (2012). How do we measure TPACK? Let me count the ways. In *Educational technology, teacher knowledge, and classroom impact: A research handbook on frameworks and approaches* (pp. 16–31). IGI Global. doi:10.4018/978-1-60960-750-0.ch002

Koh, J., & Chai, C. (2014). Teacher clusters and their perceptions of technological pedagogical content knowledge (TPACK) development through ICT lesson design. *Computers & Education*, *70*, 222–232. doi:10.1016/j.compedu.2013.08.017

Koh, J., Chai, C., & Tay, L. (2014). TPACK-in-Action: Unpacking the contextual influences of teachers' construction of technological pedagogical content knowledge (TPACK). *Computers & Education*, *78*, 20–29. doi:10.1016/j.compedu.2014.04.022

Koh, J., Chai, C., & Tsai, C. (2014). Demographic factors, TPACK constructs, and teachers' perceptions of constructivist-oriented TPACK. *Journal of Educational Technology & Society*, *17*(1), 185–196.

Lavidas, K., Katsidima, M., Theodoratou, S., Komis, V., & Nikolopoulou, K. (2021). Preschool teachers' perceptions about TPACK in Greek educational context. *Journal of Computers in Education*, *8*(3), 395–410. doi:10.100740692-021-00184-x

Lioubimtseva, E., & Cunha, C. (2022). Community engagement and equity in climate adaptation planning: experience of small-and mid-sized cities in the United States and in France. In *Justice in climate action planning* (pp. 257–276). Springer. doi:10.1007/978-3-030-73939-3_13

Lobo, I., & Vélez, M. (2022). From strong leadership to active community engagement: Effective resistance to illegal coca crops in Afro-Colombian collective territories. *The International Journal on Drug Policy*, *102*, 103579. Advance online publication. doi:10.1016/j.drugpo.2022.103579 PMID:35121354

Lyublinskaya, I., & Kaplon-Schilis, A. (2022). Analysis of Differences in the Levels of TPACK: Unpacking Performance Indicators in the TPACK Levels Rubric. *Education Sciences*, *12*(2), 79. doi:10.3390/educsci12020079

Maipita, I., Dongoran, F., Syah, D., & Hafiz, G. (2022). TPACK Knowledge Mastery of Pre-Service Teacher Students in the Faculty of Economics Universitas Negeri Medan. *The Age (Melbourne, Vic.)*, *18*(27), 6–43. doi:10.2991/aebmr.k.220104.018

Maor, D. (2017). Using TPACK to develop digital pedagogues: A higher education experience. *Journal of Computers in Education*, *4*(1), 71–86. doi:10.100740692-016-0055-4

Mensah, B., Poku, A., & Quashigah, A. (2022). Technology Integration into the Teaching and Learning of Geography in Senior High Schools in Ghana: A TPACK Assessment. *Social Education Research*, *3*(1), 80–90.

Mentz, E., & Goosen, L. (2007). Are groups working in the Information Technology class? *South African Journal of Education*, *27*(2), 329–343.

Mishra, P. (2019). Considering contextual knowledge: The TPACK diagram gets an upgrade. *Journal of Digital Learning in Teacher Education*, *35*(2), 76–78. doi:10.1080/21532974.2019.1588611

Mishra, P., & Koehler, M. J. (2006). Technology pedagogical content knowledge: A framework for teacher knowledge. *Teachers College Record*, *108*(6), 1017–1054. doi:10.1111/j.1467-9620.2006.00684.x

Molotsi, A., & Goosen, L. (2019). e-Tutors' Perspectives on the Collaborative Learning Approach as a Means to Support Students of Computing Matters of Course! In *Proceedings of the 48th Annual Conference of the Southern African Computer Lecturers' Association* (pp. 37-54). University of South Africa.

Molotsi, A., & Goosen, L. (2022). Teachers Using Disruptive Methodologies in Teaching and Learning to Foster Learner Skills: Technological, Pedagogical, and Content Knowledge. In Handbook of Research on Using Disruptive Methodologies and Game-Based Learning to Foster Transversal Skills (pp. 1-24). IGI Global.

Newman, P. A. (2006). Towards a science of community engagement. *Lancet*, *367*(9507), 302. doi:10.1016/S0140-6736(06)68067-7 PMID:16443036

Ngugi, J., & Goosen, L. (2021). Innovation, Entrepreneurship, and Sustainability for ICT Students Towards the Post-COVID-19 Era. In Handbook of Research on Entrepreneurship, Innovation, Sustainability, and ICTs in the Post-COVID-19 Era (pp. 110-131). IGI Global. doi:http://doi:10.4018/978-1-7998-6776-0.ch006

Niess, M., van Zee, E., & Gillow-Wiles, H. (2010). Knowledge growth in teaching mathematics/science with spreadsheets: Moving PCK to TPACK through online professional development. *Journal of Digital Learning in Teacher Education*, *27*(2), 42–52. doi:10.1080/21532974.2010.10784657

Niess, M. L. (2011). Investigating TPACK: Knowledge growth in teaching with technology. *Journal of Educational Computing Research*, *44*(3), 299–317. doi:10.2190/EC.44.3.c

Petko, D. (2020, June). Quo vadis TPACK? Scouting the road ahead. In *EdMedia+ innovate learning* (pp. 1349-1358). Association for the Advancement of Computing in Education.

Puzyreva, K., Henning, Z., Schelwald, R., Rassman, H., Borgnino, E., de Beus, P., Casartelli, S., & Leon, D. (2022). Professionalization of community engagement in flood risk management: Insights from four European countries. *International Journal of Disaster Risk Reduction*, *71*, 102811. Advance online publication. doi:10.1016/j.ijdrr.2022.102811

Quinn, E., Cotter, K., Kurin, K., & Brown, K. (2022). Conducting a Community Engagement Studio to Adapt Enhanced Milieu Teaching. *American Journal of Speech-Language Pathology*, *31*(3), 1–19. doi:10.1044/2021_AJSLP-21-00100 PMID:35007426

Rienties, B., Brouwer, N., Bohle Carbonell, K., Townsend, D., Rozendal, A., van der Loo, J., Dekker, P., & Lygo-Baker, S. (2013). Online training of TPACK skills of higher education scholars: A cross-institutional impact study. *European Journal of Teacher Education*, *36*(4), 480–495. doi:10.1080/02619768.2013.801073

Rienties, B., Lewis, T., O'Dowd, R., Rets, I., & Rogaten, J. (2020). The impact of virtual exchange on TPACK and foreign language competence: Reviewing a large-scale implementation across 23 virtual exchanges. *Computer Assisted Language Learning*, *35*(3), 577–603. doi:10.1080/09588221.2020.1737546

Roach, M., & Fritz, J. (2022). Breaking barriers and building bridges: Increasing community engagement in program evaluation. *Evaluation and Program Planning*, *90*, 101997. Advance online publication. doi:10.1016/j.evalprogplan.2021.101997 PMID:34503853

Rodríguez Moreno, J., Agreda Montoro, M., & Ortiz Colón, A. M. (2019). Changes in teacher training within the TPACK model framework: A systematic review. *Sustainability*, *11*(7), 1870. Advance online publication. doi:10.3390u11071870

Rosenberg, J. M., & Koehler, M. J. (2015). Context and technological pedagogical content knowledge (TPACK): A systematic review. *Journal of Research on Technology in Education*, *47*(3), 186–210. doi :10.1080/15391523.2015.1052663

Sahrir, M., Hamid, M., Zaini, A., Hamat, Z., & Ismail, T. (2022). Investigating the technological pedagogical content knowledge (TPACK) skill among Arabic school trainee teachers in online assessment during COVID-19 pandemic. *Journal of Language and Linguistic Studies*, *18*(2), 1111–1126.

Saubern, R., Henderson, M., Heinrich, E., & Redmond, P. (2020). TPACK–time to reboot? *Australasian Journal of Educational Technology*, *36*(3), 1–9. doi:10.14742/ajet.6378

Schmid, M., Brianza, E., & Petko, D. (2020). Developing a short assessment instrument for Technological Pedagogical Content Knowledge (TPACK. xs) and comparing the factor structure of an integrative and a transformative model. *Computers & Education*, *157*, 103967. Advance online publication. doi:10.1016/j.compedu.2020.103967

Schmid, M., Brianza, E., & Petko, D. (2021). Self-reported technological pedagogical content knowledge (TPACK) of pre-service teachers in relation to digital technology use in lesson plans. *Computers in Human Behavior*, *115*, 106586. Advance online publication. doi:10.1016/j.chb.2020.106586

Shulman, L. (1986). Those who understand: Knowledge growth in teaching. *Educational Researcher*, *15*(2), 4–14. doi:10.3102/0013189X015002004

Soler-Costa, R., Moreno-Guerrero, A. J., López-Belmonte, J., & Marín-Marín, J. A. (2021). Co-word analysis and academic performance of the term TPACK in web of science. *Sustainability*, *13*(3), 1481. Advance online publication. doi:10.3390u13031481

Swallow, M., & Olofson, M. (2017). Contextual understandings in the TPACK framework. *Journal of Research on Technology in Education*, *49*(3-4), 228–244. doi:10.1080/15391523.2017.1347537

Tømte, C., Enochsson, A., Buskqvist, U., & Kårstein, A. (2015). Educating online student teachers to master professional digital competence: The TPACK-framework goes online. *Computers & Education*, *84*, 26–35. doi:10.1016/j.compedu.2015.01.005

Utomo, E. (2022). The synthesis of qualitative evidence-based learning by design model to improve TPACK of prospective mathematics teacher. In *The 5th International Conference on Combinatorics, Graph Theory, and Network Topology 2021*. IOP Publishing. 10.1088/1742-6596/2157/1/012044

Valtonen, T., Leppänen, U., Hyypiä, M., Sointu, E., Smits, A., & Tondeur, J. (2020). Fresh perspectives on TPACK: Pre-service teachers' own appraisal of their challenging and confident TPACK areas. *Education and Information Technologies*, *25*(4), 2823–2842. doi:10.100710639-019-10092-4

Van Heerden, D., & Goosen, L. (2021). Students' Perceptions of e-Assessment in the Context of Covid-19: The Case of UNISA. In *Proceedings of the 29th Conference of SAARMSTE* (pp. 291-305). SAARMSTE.

Voogt, J., Fisser, P., Tondeur, J., & van Braak, J. (2016). Using theoretical perspectives in developing an understanding of TPACK. In Handbook of technological pedagogical content knowledge (TPACK) for educators (p. 33). Academic Press.

Wang, W., Schmidt-Crawford, D., & Jin, Y. (2018). Preservice teachers' TPACK development: A review of literature. *Journal of Digital Learning in Teacher Education*, *34*(4), 234–258. doi:10.1080/21532974.2018.1498039

Ward, C., & Benson, S. (2010). Developing new schemas for online teaching and learning: TPACK. *Journal of Online Learning and Teaching*, *6*(2), 482–490.

Watson, D. (2007). *Managing civic and community engagement*. McGraw-Hill Education.

Watson, S., & Waterton, E. (2010). Heritage and community engagement. *International Journal of Heritage Studies*, *16*(1-2), 1–3. doi:10.1080/13527250903441655

Wood, L., & Zuber-Skerrit, O. (2013). PALAR as a methodology for community engagement by faculties of education. *South African Journal of Education*, *33*(4), 1–15. doi:10.15700/201412171322

Yeh, Y., Chan, K., & Hsu, Y. (2021). Toward a framework that connects individual TPACK and collective TPACK: A systematic review of TPACK studies investigating teacher collaborative discourse in the learning by design process. *Computers & Education*, *171*, 104238. Advance online publication. doi:10.1016/j.compedu.2021.104238

Zhang, M., & Chen, S. (2022). Modeling dichotomous technology use among university EFL teachers in China: The roles of TPACK, affective and evaluative attitudes towards technology. *Cogent Education*, *9*(1), 2013396. Advance online publication. doi:10.1080/2331186X.2021.2013396

Chapter 8
An Extendable Functioning Waiting Room Solution for Online Education Platforms:
The Beginning of an Online Efficiency Era

Kaan Kırlı
Yeditepe University, Turkey

Cagla Ozen
https://orcid.org/0000-0003-1817-9806
Yeditepe University, Turkey

ABSTRACT

When COVID-19 spread from China, the majority of people witnessed that the form of education changed. This situation has increased the need for online education platforms. Although videoconference platforms such as Zoom and Google Meet became popular to meet the requirement, these platforms have some drawbacks. To illustrate, the majority of students are not active and cannot take a proactive role during the lectures. Furthermore, they stay idle and get bored on the waiting rooms of these platforms. To fill this gap, an extendable functioning waiting room was created by using ASP.NET framework via Visual Studio platform. Proposed waiting room present a solution for enhancing user experience (UX) by increasing user engagement on the online education platform. This chapter would be interesting for not only educators who use online learning platforms but also videoconference platform users and developers.

INTRODUCTION

Before the pandemic period, face-to-face education was popular all around the world. After the Covid-19 was spread from China to all around the world, majority of people have witnessed that the form of education was changed. This situation has increased the need for online education platforms. "Online learning platforms have been developed to provide online learners with an alternative to the conventional ways

DOI: 10.4018/978-1-6684-5190-8.ch008

of receiving information across typically large e-learning networks" (Chansanam et al., 2021, p.349). Videoconferencing platforms such as Zoom and Google Meet were used popularly as an online education platform during pandemic period. These two platforms have characteristics as document sharing, screen sharing and chat features. Besides, these platforms have scalable information systems (IS) infrastructure so that they can easily respond to increased number of users' demands during the pandemic period. With these platforms educational institutions could continue education online. Therefore, these platforms become necessity for online education for the institutions.

When Covid-19 pandemic has risen, the importance of online education platforms has dramatically increased because educators and students have no chance for face-to-face education. Although user experience (UX) of students and teachers on online education platforms and efficiency of the interaction have been discussed in the literature, this is not the case in research and practices on students' engagement problems on these platforms. To enhance students' engagement problems on online education platform, "An Extendable Functioning Waiting Room" for Zoom and Google Meet has been proposed in this study. There are two sub rooms in the main room: a chat room (one-to-one) and a game room. These rooms enable students to socialize with their classmates and spend their time more effectively during they are in waiting mode. In this manner, students were not only interacting with their educators but also with their classmates via online learning platforms. Hence, user engagement on these platforms will be developed. Proposed system was developed by using ASP.NET framework and Visual Studio platform. C# and HTML were used as programming and markup languages.

This system can be extended on other online platforms such as online conferences to enhance conference visitors' UX with the platform.

This chapter is structured as follows; literature review on the topic will be provided in the next section. Then, methodology, focus of the chapter along gap of the research will be discussed. Afterwards, the methodology of the proposed solution for the gap will be discussed. Then, future research directions will be explained. Finally, conclusion will be represented in the last section.

BACKGROUND

Covid-19 pandemic period has increased research and articles on online education and its effects. This section will be based on these studies. In this section, studies related with students' concentration and participation problems on online education, importance of user experience (UX), efficiency of online learning environments for students and educators, useful properties that supports online education platforms will be discussed.

Importance of User Experience (UX)

User experience (UX) describes the emotions, beliefs, preferences, and other reactions that users have before, during, and after interacting with a product or system. User requirements, expectations, and previous UXs all influence user happiness. Furthermore, the authors pointed out that, to maximize the impact, a network environment that learners can modify should be prioritized. They argued that the measurement in current evaluation methods primarily concentrates on two directions of virtual learning syllabus and system feature: Establishing repaired assessment measurement from the perspectives of course quality, instructional impact, educational technology, system usage, and so on, and implementing the assessment using qualitative or quantitive techniques. According to these authors, to score and assess the platform and courses, studies are increasingly concentrating on customers' feedback and comments in the research of UX and user satisfaction. Moreover, most of the customer feedback is centered on system suitability, consistency, reaction time and other factors (Chen et al., 2020).

UX begins with the solution and ensures that it functions in the context of the user (Gaborov & Ivetić, 2022). Besides, the UX technique is vital because it ensures that customers will find the items useable, desired, and valuable. Furthermore, UX focuses on producing goods that are not just functional and beneficial to customers, but also desired and cheap. Users' expectations must be considered by UX designers. Users dislike it when the interface does not behave in the way they expect it to.

One of the most appealing features of online education is its adaptability, which allows students to balance competitive commitments such as homeschooling and busy job calendars as well as students struggling with annoying circumstances. Loneliness and learner isolation are two further factors that contribute to student failure or dissatisfaction with virtual learning knowledges. Moreover, the authors argued that personal interactions that arise spontaneously in a live situation are a goal of online learning groups (Serrano-Solano et al., 2021).

The conceptual model is taken on Aranyi and van Schaik's model and has artifact characteristics linked to UX components, such as perception of instrumental features (pragmatic quality), emotional reactions, and perception of non-instrumental features (hedonic quality). Furthermore, the authors argued that the dimensions, criteria, and metrics in the UX components are added to the new conceptual UX assessment model. This conceptual model also showed a crucial factor for online systems: pragmatic quality, which includes efficiency and effectiveness, and hedonic quality, that involves attractiveness and satisfaction (Yusof et al.,2022).

Efficiency of Online Learning Environments for Students and Educators

Online learning environments are replacing traditional educational situations to combat the problem during this pandemic (Serrano-Solano et al.,2021). For that reason, business software market has turned to develop online education services during COVID-19 (Chen et al.,2020). The internet is the only method for education to break down obstacles to injustice, create chances for young people to acquire twenty first century learners, and allow them to study in ways that suit their needs and personalities. And, to do so, instructors must use technology to enable students to study anything, at any time and from any location (Ayu, 2020). Furthermore, there has been significant progress in education, with the manner of instruction shifting from educator-centered to student-centered (Almahasses et al, 2021). The authors argued that the educator is the source of education in educator-centered education, and students are the beneficiaries of his/her knowledge. Moreover, educators take on the role of "assistant to students who set and implement their own norms" in a student-centered approach.

The pandemic caused many changes in life because of pandemic prevention techniques' effects, and the educational system was no exception. Many countries have adopted the virtual learning method in their schools and universities in during COVID-19 pandemic that has spread the worldwide. The educators and students had to change their learning and teaching techniques to adapt quickly to the pandemic, even if they are not comfortable with digital education. The contrasts between in-person and online education are important, and both educators and students are aware of them. Digital learning is a significant innovation for the sustainability of educational environments (Saha et al., 2022).

However, this new model of education has some drawbacks. Learners may face additional challenges due to a lack of competence with technology. Due to the lack of face-to-face contact with instructors and other students in the e-learning process, some students may find it difficult to appreciate their contents. As a result, some students require the essential hardware and self-regulatory behavior skills to proceed responsibly and successfully use all e-learning tools to efficiently access online content. Furthermore, lecturers were worried about low participation rates in e-learning activities, particularly in forum discussions. They proposed incentives to entice people to participate. They also understood the challenges of working in an online context. One common remark was that it was difficult to get students to participate in the debate. Students are difficult to govern. Distractions are common among students. Since college educators frequently teach that they were educated, they may be unaware of what constitutes good online teaching, especially if they have never taken an online course as a student (Ayu,2020). In addition, the faculty may be expected to move away from educator-centered lectures, which are prevalent in face-to-face instruction. Interactive and engaged lesson application via chat forum tasks, digital simulations, and other instructional technologies, and synchronously or asynchronously webinars are examples of the change from lectures to student-centered teaching approaches. As a result, educators who are more used to educator-centered approaches may find it difficult to transfer to online training that emphasizes student-centered pedagogies (Perrotta & Bohan, 2020).

In the method of embracing remote learning, perceptions of contact are frequently stated as pleasant and disappointing. These findings suggest that individual student personal qualities impact interactions more than distant teaching technologies. Therefore, instructors and students will be happy with interactions only if they completely understand and use the online platform, engage actively, and devote time to communication. Moreover, academic accomplishment is also considered as a source of happiness and discontent, with the influence varying based on the traits and talents of the students. Academic accomplishment requires a high level of concentration. According to the author, the involvement of professors and students is critical to the success of distant learning. Students want to be given the tools they need to participate in classes in a realistic setting that allows for easy participation and active instruction by the instructor. For students, online learning is novel and demanding, but it is apparent that educators are struggling as well (Fatoni et al.,2020).

Online learning has substituted face-to-face learning; students, educators, and institutions no longer interact socially; instead, they interact with technological devices. Educators and students are introduced to a new environment in which they have no opportunity to engage face to face (Quimí and Alexandra, 2022). People become more productive by gaining research skills, self-learning, decision-making, time management, and a higher quality of life through online education, which overcomes space-time obstacles and encourages autonomous learning. The reasons for investigating the obstacles of virtual education focus on current learning, because today, many institutions offer online programs that students from any city or country can connect via their computers and now even their mobile phones, and this has many advantages, not only for university departments, but also for the development of society. Besides, the authors pointed out that online education is described as the use of technical means to develop the teaching-learning process when instructors and learners are not physically present in a classroom. It is regarded as the progression of the new era, in which individuals all over the world may access information at any time and from any location. Educational institutions have attempted to incorporate techniques into their methodologies and educational systems to remain connected to this virtual modality and contribute to students' full training. Many individuals across the world have found that studying online is a good option, and the benefits of online education are perfect for people that study and will not have time to participate a classroom (Quimí and Alexandra, 2022).

A good online teaching activity requires a variety of technologies, including videoconferencing, chats, and shared documents (Serrano-Solano et al.,2021). The need for developing e-learning is to follow a basic principle: students should be able to use technological tools and menus with ease on the offered panel, allowing them to study more efficiently (Ayu, 2020). Online learning (also known as e-learning) is a type of distant education in which the process of learning is mediated by technology and all instruction is provided over the internet (Heng & Sol, 2021, p.5). Using contemporary technology, the adoption of distant learning systems will boost the quality of education. (Liu et al., 2020). As pointed out by Selvaraj et al. (2021), educators should create methods to engage pupils and confidence to speak out when they have questions. Institutions must ensure that they communicate out to all of their students, resolve their problems, and provide guidance. Parents should ensure that their children engage in these activities as well. Selvaraj et al. discuss that while it is undoubtedly insufficient, students must discover methods to make the most use of the resources given to them. Students have a constitutional right to education which should not be denied. While there is a lot of potential in this area of learning, it appears that children need more exposure to these tools, and educators need to develop more efficient methods to use them.

Online learning created a nice way of information distribution that was not restricted by time or location, allowing students to receive education at any time and from any location. Learners say that the online environment makes it easier to incorporate education into their hectic schedules (Yusnilita, 2020). The educators and students can obtain immediate feedback on the quality of the course, delivery, and experience inside a face-to-face classroom setting. In a classroom context, an educator may observe students' body language, and these nonverbal cues allow the instructor to quickly change their teaching style to better align the students' needs. When compared to other existing channels, extra questioning and personalized concentration in the classroom allows for a more thorough evaluation of the student's comprehension of topics being taught. Investigating and evaluating how online classes should be constructed and arranged while taking students' and professors' viewpoints into consideration should be an essential part of creating online learning approaches. Educational boards have enforced the delivery of online programs at the college / university levels in light of current COVID-19 pandemic (Rathor, 2022).

USEFUL PROPERTIES OF ONLINE EDUCATION PLATFORMS

In this part, two popular online education platforms' (Zoom and Google Meet) main properties will be listed:

On Zoom

On Zoom desktop application, there are four main useful features: Screen Sharing, Document Sharing, Raising Hand, and Whiteboard.

Screen Sharing

On screen sharing feature, educators can teach the lesson by presenting their own screens. It can be a website, a schema, a table and so on. Besides, students' can present their projects and/or assignments via this feature.

Usage steps are as the following:

Step 1: "Share Screen" button should be clicked on the bottom of the screen (see Figure 1).

Step 2: If the user clicks arrow button at the right top, main screen can be shared by one participant or multiple participants (see Figure 2).

Step 3: After "Share Screen" button is clicked, "Basic Choice" menu is shown. In this example, "Getting started with ASP.NET 4.7." is selected and "Share" button is clicked (see Figure 3).

Step 4: shared screen will be presented (see Figure 4).

Figure 1. "Share Screen" Button

Figure 2. "Share Screen" Settings Menu

Figure 3. "Share Screen" Basic Choice

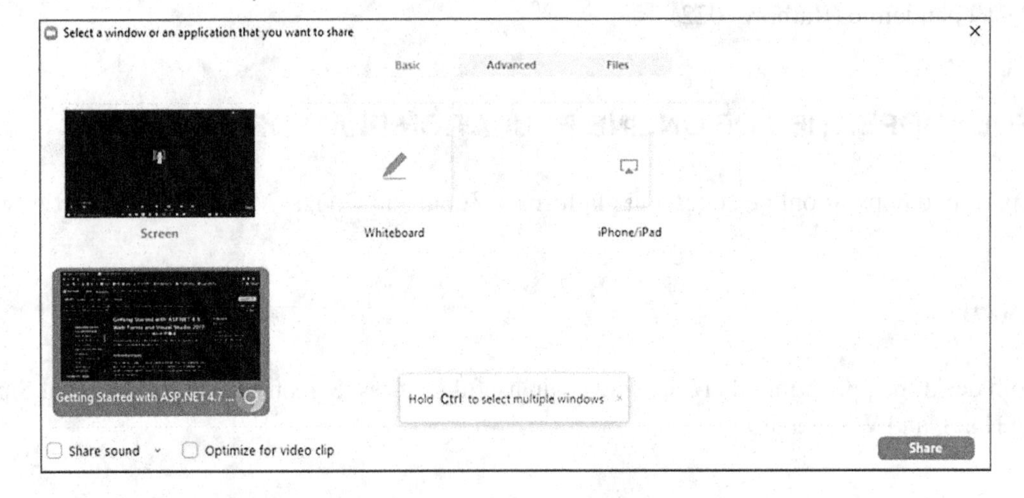

Figure 4. Zoom Screen Sharing

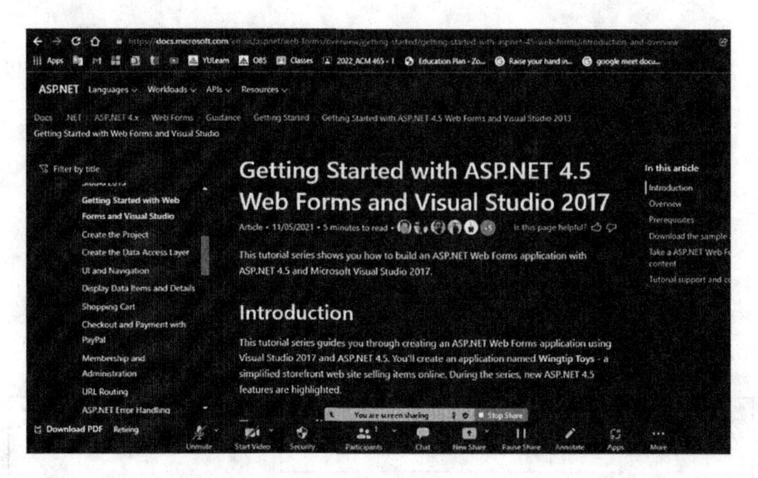

Document Sharing

On document sharing feature, important documents (such as curriculums, course slides, assignments and project instructions, midterm, and final exams) can be shared by educators with their students. Students' can also share their documents with their educators via this feature.

Usage steps are at the below:

Step 1: "Chat" button should be clicked at the bottom of the screen (see Figure 5).

Step 2: "Chat" menu screen will be opened. On this screen "Document Sharing" icon should be clicked (see Figure 6).

Step 3: "Document Sharing" options will be appeared. If the user's document's path is on the computer, "Your Computer" selection should be clicked (see Figure 7).

Step 4: "Open" window is shown (see Figure 8).

Step 5: "Zoom share screen 2.png" document is found by using document path and by selecting the document. Then, "Open" button should be clicked (see Figure 9).

Step 6: "Zoom share screen 2.png" document will be sent to "Chat" part successfully. Besides, "File sent successfully" notification can be seen on the "Chat" screen (see figure 10 and 11).

Figure 5. "Chat" Button on Zoom

Figure 6. "Chat" Menu Part and "Document Sharing" Icon Plotted with Blue Color

Figure 7. "Document Sharing" Type Options

Figure 8. "Document Sharing" Computer Menu

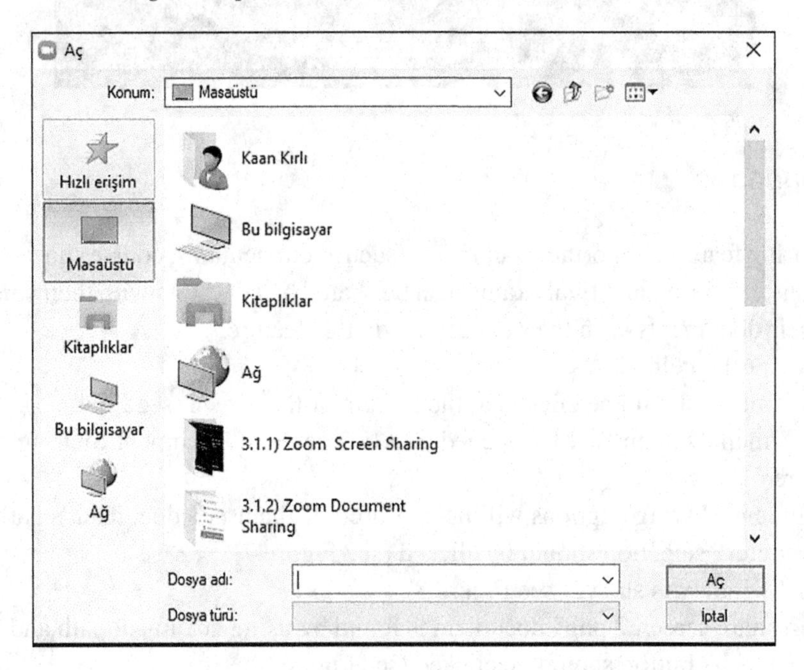

Raising Hand

Raising hand feature can be used on education with two ways: attendance and class-participation. First, this feature can be used for checking students' attendance. Students' absence can be controlled by an educator via "Raise Hand" icon button. Secondly, students can participate class discussions by raising hand feature. By simply clicking on "Raise Hand" icon button, a student can participate to the discussion. Then, the student's voice can be turned on by the educator.

Steps for this feature are as follows:

Step 1: "Reactions" button should be clicked (see Figure 12).

Step 2: "Reactions" menu is shown (see Figure 13). "Raise Hand" icon button should be clicked to inform educator about hand raise attempt.

Step 3: Hand icons are shown on the user's screen and on "Participants" part (see Figure 14).

Step 4: To lower the hand, "Lower Hand" icon button should be clicked (see Figure 15).

Figure 9. "Document Sharing" File Open

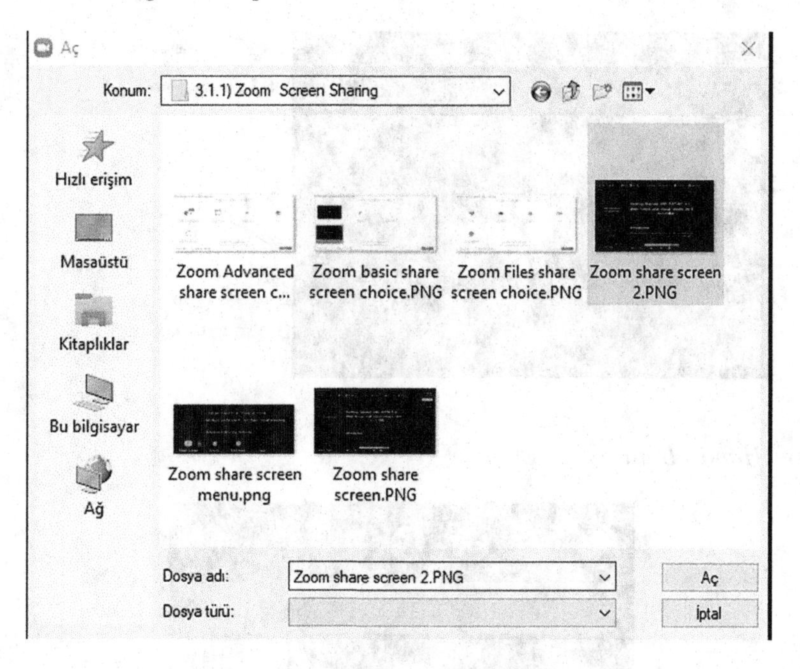

Figure 10. Document Is Sent

Figure 11. "File sent successfully" Notification

Figure 12. "Reactions" Button

Figure 13. "Reactions" Menu and "Raise Hand" Button

Figure 14. "Raise Hand" Feature Plotted with Red Color

Figure 15. "Lower Hand" Button

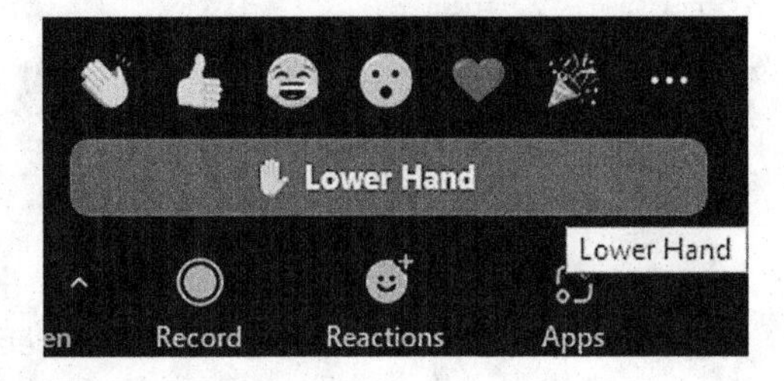

Whiteboard

By using "whiteboard" feature, educators can use their screen like a traditional whiteboard. Furthermore, educator can easily plot a table, a schema and can write a text. There is also a "save" option of this whiteboard. Besides, used whiteboard can be saved as a document and sent to the students by the educator.

The following steps should be performed to use this feature:

Step 1: "Share Screen" button should be clicked at the bottom of the screen (see Figure 16).

Step 2: "Whiteboard" choice should be selected from basic choice tab and "Share" button should be clicked (see Figure 17).

Step 3: A plotted whiteboard can be shown (see figure 18).

On Google Meet

On Google Meet, there are three main features: Screen Sharing, Raising Hand, and Sharing Whiteboard File.

Figure 16. "Share Screen" button

Figure 17. "Whiteboard" on Basic Choice

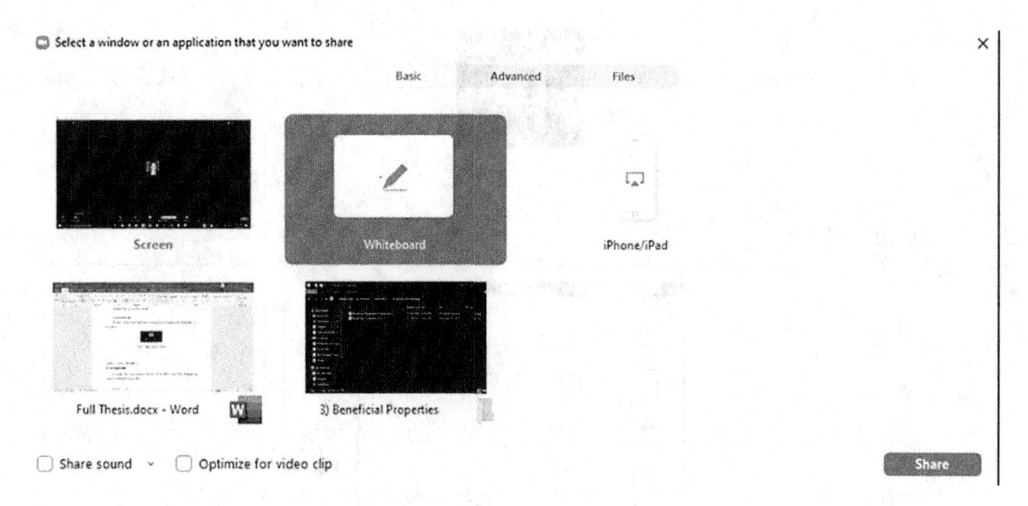

Figure 18. Whiteboard on Zoom

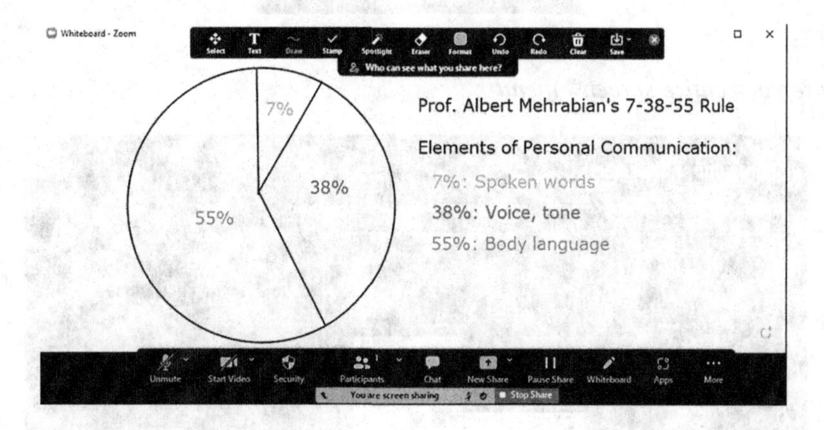

Screen Sharing

An educator can use this screen sharing feature to demonstrate their own displays. For instance, the educator may prefer to present a webpage, a schema, or data. Students' may use this feature to show their own projects and/or assignments if their educators give them permission to do this. Furthermore, students may prefer to demonstrate any facts related to the class topic for being active and their class participation grades.

The following are the steps for using this feature:

Step 1: "Share Screen" button should be clicked on the bottom of the screen (see Figure 19).

Step 2: "Present" menu is seen. In this menu, there are three sharing options: "Your entire screen", "A window", and "A tab" (see Figure 20).

Step 3: If the user clicks on "Your entire screen" button, "Share your entire screen" menu will be seen (see figure 21). The main screen should be selected, and "Share" button should be clicked so that the entire screen will be shown to other users.

Figure 19. "Share Screen" Button

Figure 20. "Present" Menu

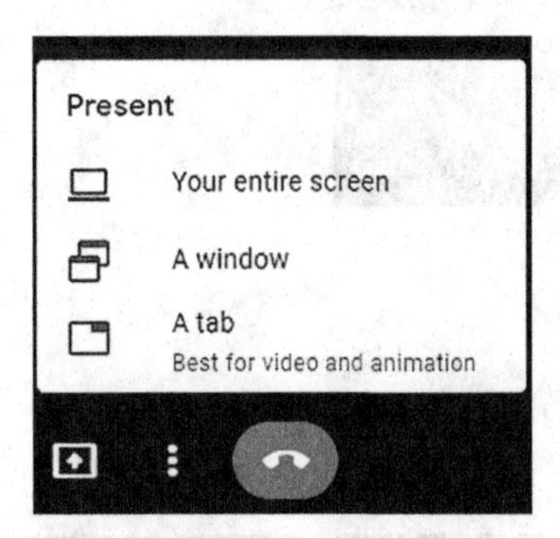

Figure 21. "Share your entire screen" menu

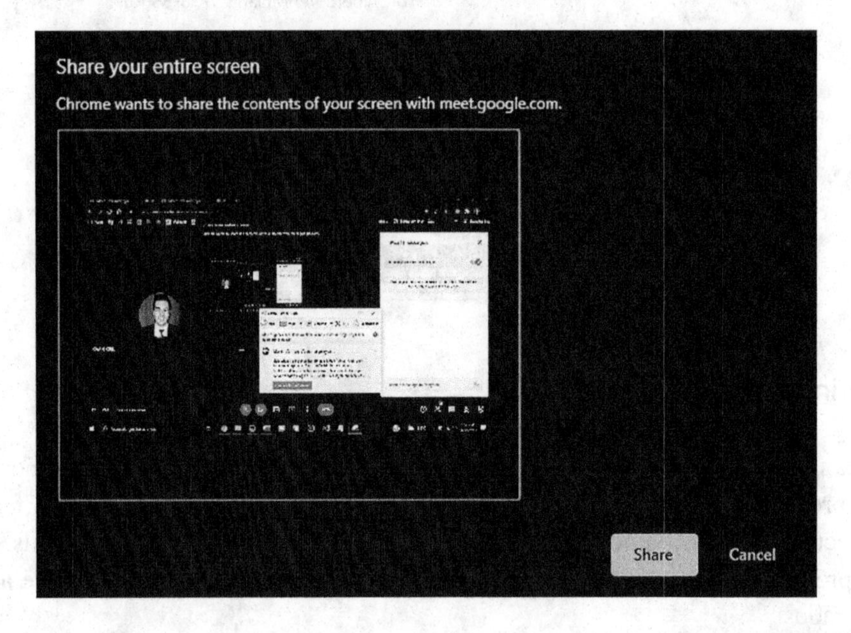

Step 4: If the user clicks on "A window" button, "Share an application window" menu will be seen (see Figure 22). The application screen should be selected, and "Share" button should be clicked.

Step 5: Application window is shown to the other users (see Figure 23).

Step 6: If the user clicks on "A tab" button, "Share a Chrome tab" menu will be seen (see Figure 24). The target tab should be selected, and "Share" button should be clicked. The tab will be shared with other users (see Figure 25 and 26).

Figure 22. "Share an application window" Menu

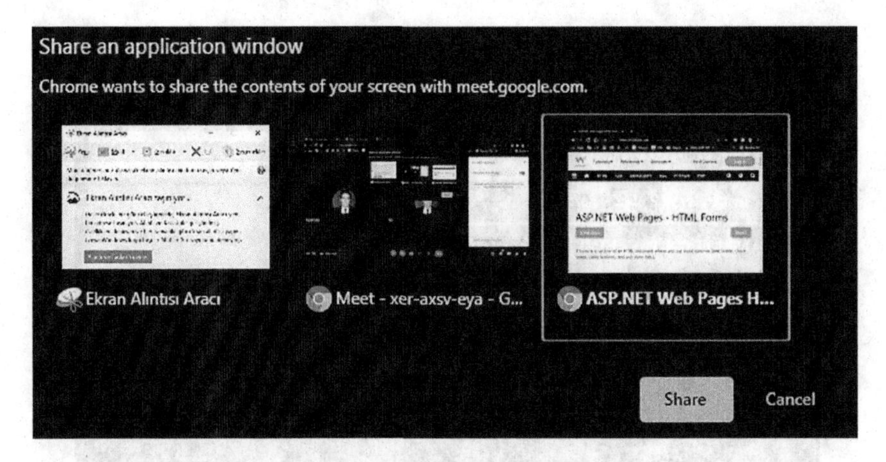

Figure 23. Application Window

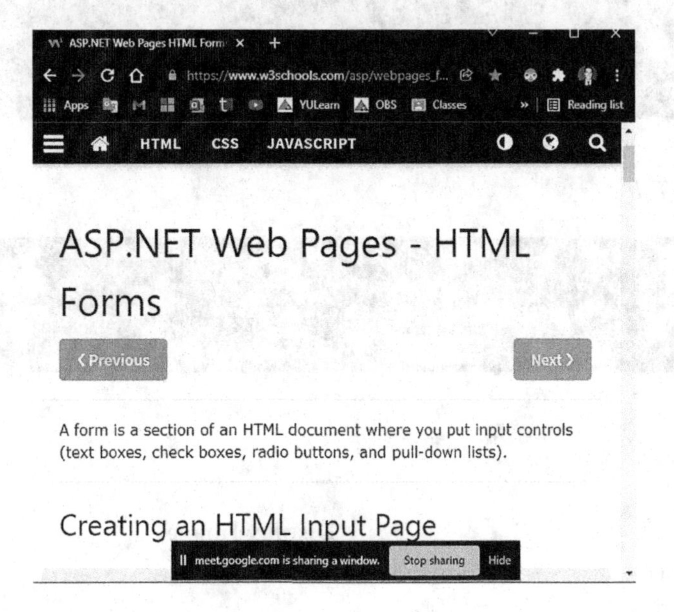

Raising Hand

Raising hand feature can be used on education with two ways: attendance and class-participation. First, this feature can be used for checking students' attendance. Students' absence can be controlled by an educator via "Raise Hand" icon button. Secondly, students can participate class discussions by raising hand feature. By simply clicking on "Raise Hand" icon button, a student can participate to the discussion. Then, the student's voice can be turned on by the educator.

The general steps to use this feature are as the following:

Step 1: To raise hand, "Hand raise" icon should be clicked at the bottom of the screen (see Figure 27).

Step 2: "Raising Hand" icon is shown on bottom left of the screen (see Figure 28)

Step 3: To lower hand, same icon should be clicked again on bottom left of the screen.

Figure 24. "Share a Chrome tab" menu

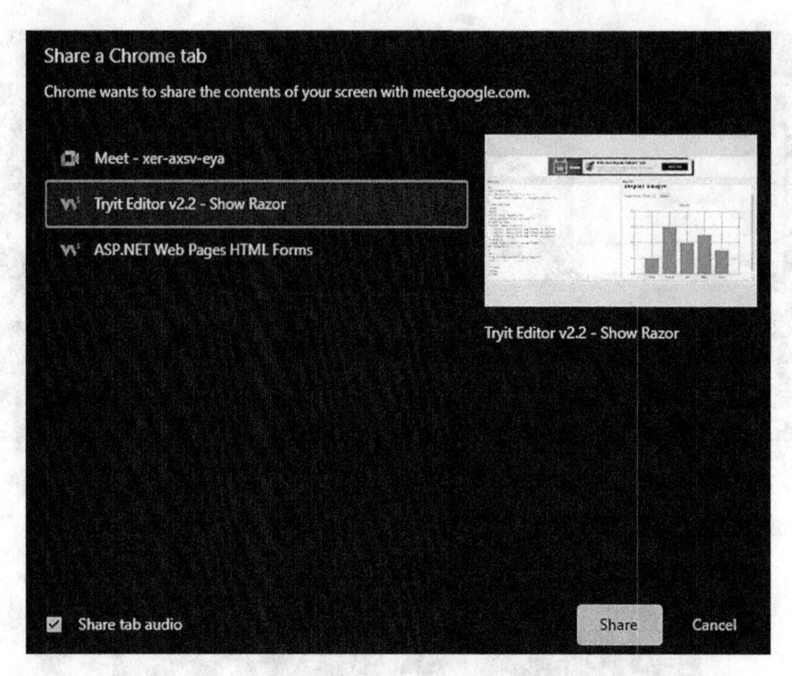

Figure 25. Sharing A Tab

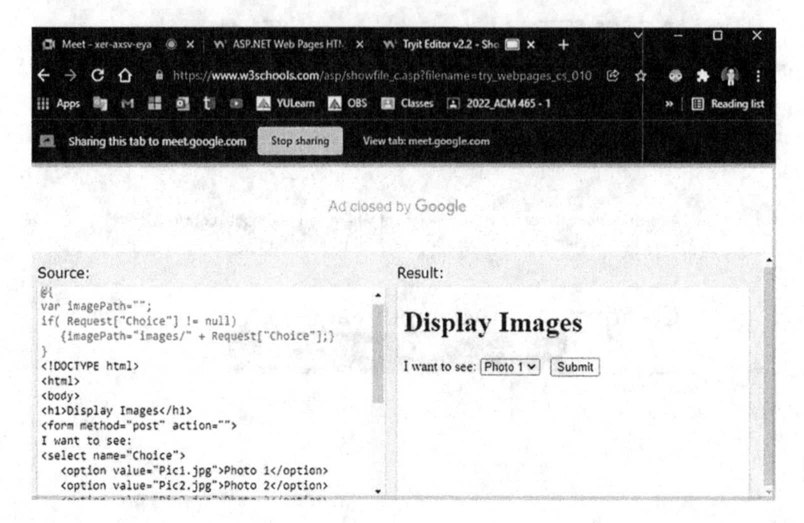

Sharing Whiteboard File

By using "whiteboard" feature, educators can use their screen like a traditional whiteboard. Furthermore, educator can easily plot a table, a schema and can write a text. There is also a "save" option of this white-board. Besides, used whiteboard can be saved as a document and sent to the students by the educator.

The following are the guidance for the usage of this feature:

Step 1: "Activities" icon should be clicked at the bottom right of the screen (see Figure 29).

Figure 26. Sharing A Tab on Meet Room

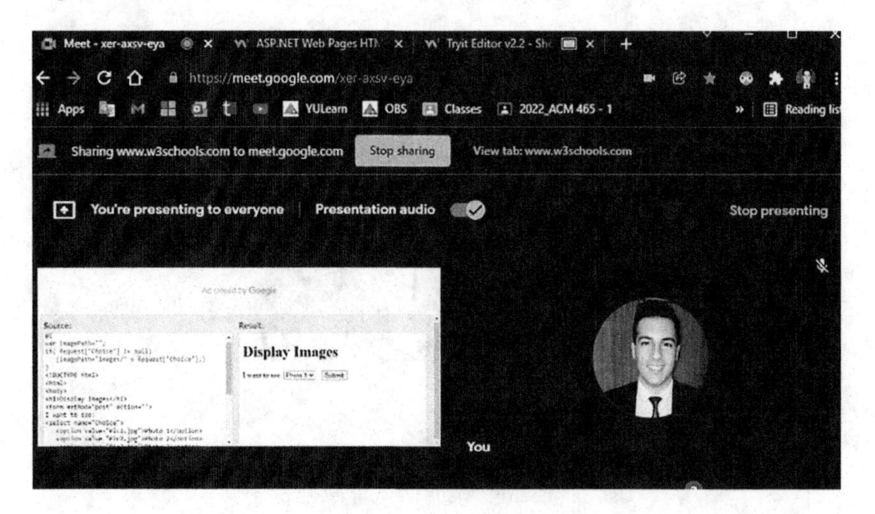

Figure 27. "Hand raise" Icon Plotted with Green Color

Figure 28. "Raising hand" icon on the screen

Figure 29. "Activities" Icon on The Bottom Right

Figure 30. "Activities" Menu

Figure 31. "Whiteboarding" Menu

Step 2: "Activities" menu will be shown and "Whiteboarding" icon button should be clicked (see Figure 30).

Step 3: "Whiteboarding" menu will be shown and "Start a new whiteboard" button should be clicked (see Figure 31).

Step 4: New whiteboard is shown and any subject can be charted and explained. If the subject has been already existed, "Choose from Drive" button should be clicked. By using "Share" button, whiteboard can be shared via "anyone with the link" (see Figure 32 and 33).

Figure 32. Whiteboard Feature

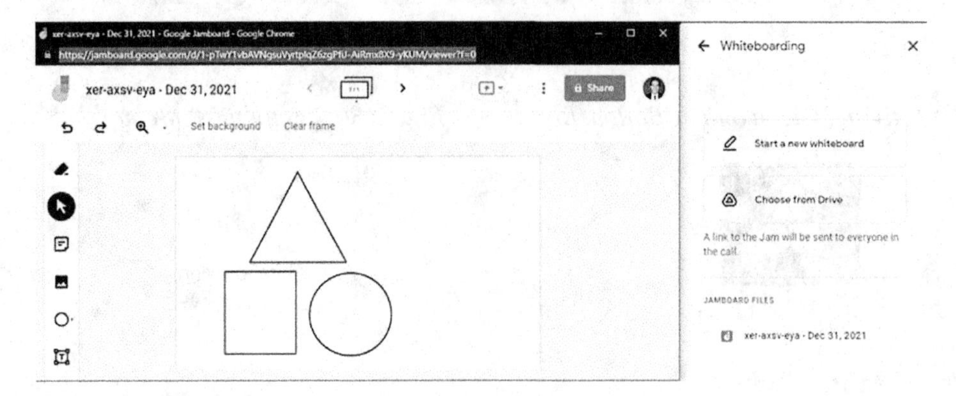

Figure 33. Sharing Whiteboard via Link

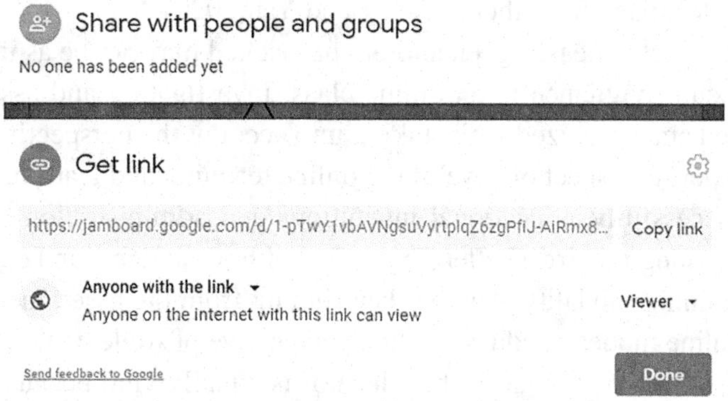

Share with people and groups
No one has been added yet
Get link
https://jamboard.google.com/d/1-pTwY1vbAVNgsuVyrtplqZ6zgPfIJ-AIRmx8... Copy link
Anyone with the link
Anyone on the internet with this link can view
Viewer
Send feedback to Google
Done

MAIN FOCUS OF ONLINE EDUCATION PLATFORMS

Issues, Controversies and Problems of Online Education Platforms

COVID-19 has prompted educational institutions all across the world to adopt innovative techniques on short notice. Most institutions have moved to an online method during this period, employing Microsoft Teams, Blackboard, Zoom, and other online platforms. Educational institutions in impacted areas are looking for temporary ways to keep teaching going, but it's crucial to remember that the quality of learning is dependent on digital availability and efficiency (Muthuprasad et al., 2020). Moreover, the possibility of future crisis events demanding the use of remote classrooms is considerable. As a result, now is the time to plan online learning activities in which students may actively engage. This implies that the present remote learning program has to be more structured and organized (Fatoni et al., 2020).

The COVID-19 pandemic provides a chance to revisit and examine the necessary circumstances for online learning, as previous experiences with urgent teaching and online teaching do not appear to have adequately prepared higher education institutions. What renders the concept of online education under COVID-19 particularly interesting is the suddenness with which all participants are forced to participate. Furthermore, between the earlier and present studies, there are variations in the volume and manner of engagement in participants. Such variations may have an influence on students' expectations of online learning, necessitating a detailed examination of their experiences and needs during this trying period. In addition, when studying at home, students reported having difficulty concentrating on online learning, which was judged undesirable. Due to a lack of space, learners had to put up with noise from their families while online studying, yet as the preceding remarks indicated, they struggled to stay focused in class. The lack of room has had an influence on students' online learning in this aspect. The e-resources that students have access to were an unusual source of focus difficulty. When they were widely available online, learners find it an excuse to pay less attention in class or even put off studying because they could review the contents afterwards. The authors clear that minimal student-to-student interaction was noted as a problem in online learning by the comments. Students found it difficult to make friendships in online classrooms and felt uneasy having talks online. This might be due to the fact that when the microphone or camera were turned off, students couldn't see or hear one other on e-learning platforms. Also, students may be unwilling to participate in online conversations if they cannot see or hear their peers. Apart from that, online learning has been discovered to have issues with student-to-educator engagement. Students in particular indicated difficulty engaging in immediate interactions with educators or obtaining immediate feedback from them (Yeung and Yau,2020).

In a classroom, things that are easily seen and/or approached may not be as important as things that are more subtle and require vigilance in an online class. Investigating and assessing whether online classes should be created and organized while taking into account the perspectives of students and professors should be an important aspect of developing online teaching and learning techniques. To utilize the techniques most successfully, educational institutions and administration who will be the future distributors of online learning require a better knowledge of how students and educators view and react to online classes as a training modality. Another key finding from the assessment of past research was that while evaluating online modes of education, the perspective of students was given priority over the perspective of educators. The viewpoint of the educator is equally vital because if they, as educators, are dissatisfied with the online method, the educational foundation itself would be weakened. This new advent of online classrooms has been equally difficult for professors, who are also trying to acquire this new teaching approach. As a result, the importance of this study is to find out how educators and students feel about online classrooms in contrast to face-to-face lectures. Additionally, online classes were reported to be useful in terms of time savings, but both educators and students found that they were less effective and less structured than the classroom learning modes. The author clears that he comforts level of educators and students with online class design, structure, level of interaction between students and faculty, the quality and amount of class content, technical support, and overall online class delivery experience impacts the success or failure of online education. It is important that awareness is increased about the convenience and accessibility of online channels to increase the adoption of these by students and educators (Nambiar, 2020).

The most common source of dissatisfaction with virtual learning was network instability, which students described as having disrupted their lessons. The importance of networks in the online learning environment is one of the most essential areas for development. Poor communication only with educator, difficulty to collaborate successfully with classmates, and a shortage of possible feedback to also be exchanged between students were all sources of dissatisfaction with online learning interactions. Educators and students will be satisfied with interactions only if they can understand the capabilities of online platforms and utilize them effectively, participate actively, and spend enough time to communication. Moreover, educators and students both play vital roles in enhancing the efficiency of online learning. Students want to be given the tools they need to participate in class in a realistic atmosphere that allows for active teaching and engagement. To promote remote learning, both students and instructors should be aware of the challenges they face, and solutions should be found (Shim & Lee, 2020).

Attendance and participation in online courses are problematic, making online education channels hard to embrace. Furthermore, online education, particularly remote learning, has become easier because to the internet and various technology. A personal computer, laptop, or indeed a smartphone can be used for online teaching and learning. The facility's most important feature is that it may be accessed at any location and at any time. On the other side, students are unhappy with online learning and instruction. As a result of a number of challenges faced by students, including technical problems, uncertainty clarification, difficulty understanding practical courses, and so on (Rathor, 2022).

A classroom is indeed a lively learning setting where professors have direct contact with students, but online programs do not provide this possibility. Less student-educator connection, bad internet access, a lack of student involvement, students' low interest, less utility for evaluation, less student participation, and a lack of suitable instructions were all clear issues for the educators. Moreover, during the epidemic,

Figure 34. Zoom's "The meeting host will let you in soon" waiting mode

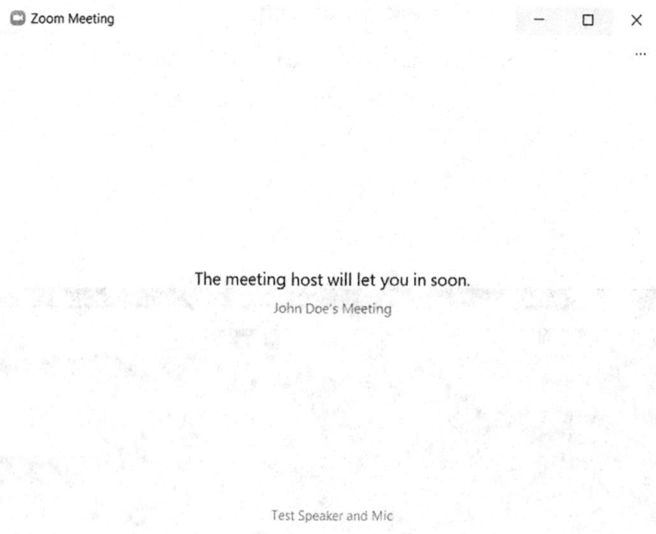

e-teaching was seen to be efficient, time-saving, and easy to distribute knowledge, but it was not suited for monitoring or performing practical lessons, and it was less effective for assessing students (Saha et al., 2022)

"Due to the fact that mental load on the user is considerable during the learning process, further load, caused by inappropriately designed interface, is not desirable. From ergonomics point of view, mental load is much more demanding on the user than the motoric load, and that is why the key demand on the learning interface is intuitiveness and simplicity." (Weinschenk, S.,2011, as cited in Oveslova, H.,2015, p.221)

Especially, since Covid-19 came and online education platforms were started to use, the mind's mental load was become more annoying. Hence, for students, as a social gap, the waiting modes on Zoom and Google Meet cannot be satisfied because of no social activity, no proactive role, and the lack of user experience (UX).

On Zoom and Google Meet, when students want to enter room by using "join" button, "waiting mode" is activated by educator as a host, and until students are accepted to room by educator, they may be bored, and their participation and concentration may be decreased after the lesson is started. Thus, lack of efficiency for students may be seen because of this circumstance.

To demonstrate the problem clearly, the waiting modes of Zoom and Google Meet are at the below:

To solve this huge problem, a solution is needed.

A proper instructional environment and system that support academic accomplishment must be developed to help student learning and achievement (Shim & Lee, 2020).

THE METHODOLOGY, SOLUTIONS AND RECOMMENDATIONS

To enhance socializing among students, to increase their participation to the online lectures, to improve efficiency of online educational environments for the students and to improve user engagement on these

Figure 35. Google Meet's "Asking to join" waiting mode

Figure 36. Design Part

environments, proposed solution of this study is to create an Extendable Functioning Waiting Room (EFWR). This room will be a valuable feature for Zoom, Google Meet and other online education platforms. With this effort, most of the challenges and gaps of online educational environments that were discussed in the previous sections, will be handled.

EFWR was developed with ASP.NET Web Form. "ASP.NET is a framework for building web apps and services with .NET and C#." (ASP.NET, n.d.). ASP.NET Web Form was developed on Visual Studio 2019 Community platform. In reference to Visual Studio (2022), Visual Studio which was developed by Microsoft, is an integrated development environment (IDE) and code editor for software developers and teams.

On Functional Waiting Main Room, students can chat with each other until they are accepted by the educator as a host. Also, there are two sub rooms of Functional Waiting Main Room: One-to-One Room and Game Room. On One-to-One Room, two participants are able to chat with each other. On Game Room, participants can play game while they are waiting for host.

Development steps are as at below:

Step 1: On the design part of ASP.NET Web Form, labels, textboxes, images, buttons, image buttons and list boxes are implemented, and all text components are labelled (see Figure 36).

Step 2: On the source (implementation) part, all HTML and ASP codes are set automatically as the authors made on design. Also, some parameters (width, height, color, font-size etc.) can be adjusted from here (see Figure 37 and 38).

Figure 37. Source (HTML and ASP) (Part 1)

```
FunctionalWaitingMainRoom.aspx   EndCall.aspx   Accepted.aspx   GameRoom.aspx   OneToOneRoom.aspx
İstemci Nesneleri & Olayları   (Olay Yok)
1   <%@ Page Language="C#" AutoEventWireup="true" CodeBehind="FunctionalWaitingMainRoom.aspx.cs"
2       Inherits="ACM498GUIProject.FunctionalWaitingMainRoom" %>
3
4   <!DOCTYPE html>
5   <html xmlns="http://www.w3.org/1999/xhtml">
6   <head runat="server">
7   <meta http-equiv="Content-Type" content="text/html; charset=utf-8"/>
8       <title>Functional Waiting Room</title>
9   </head>
10  <body>
11      <form id="form1" runat="server">
12          <asp:Panel ID="Panel1" runat="server" BackColor="RoyalBlue" ForeColor="White"
13              Height="322px" Width="904px" style="margin-left: 3px">
14              <asp:Label ID="Label3" runat="server" Font-Size="Large"
15                  Text="Welcome to Functional Waiting Room! You can choose a sub room:"></asp:Label>
16               <asp:Button ID="Button6" runat="server" OnClick="Button6_Click" ForeColor="White" BackColor="Blue"
17                  Text="One-to-One Room" Font-Bold="true"/>
18              <asp:Button ID="Button7" runat="server" ForeColor="White" BackColor="ForestGreen" Text="Game Room"
19                  OnClick="Button7_Click" Font-Bold="true"/>
20              <asp:Panel ID="Panel2" runat="server" Height="290px" style="margin-left: 697px;background-color:black" Width="208px">
21                  <asp:Label ID="Label1" runat="server" ForeColor="White" Text="Participants"></asp:Label>
22                  <br/>
23                  <asp:ListBox ID="ListBox1" runat="server" Height="76px" Width="206px">
24                      <asp:ListItem>Mary Jackson</asp:ListItem>
90 %   Sorun bulunamadı   Sat: 8   Krkt: 30   BŞL
Tasarım | Böl | Kaynak
```

Figure 38. Source (HTML and ASP) (Part 2)

```
26                      <asp:ListItem>David Anderson</asp:ListItem>
27                  </asp:ListBox>
28                  <asp:Label ID="Label2" runat="server" ForeColor="White" Text="Chat"></asp:Label>
29                  <asp:ListBox ID="ListBox2" runat="server" Height="114px" style="margin-left: 0px; margin-top: 10px" Width="209px">
30                      <asp:ListItem>David: Hey guys, how are you?</asp:ListItem>
31                      <asp:ListItem>Mary: Hi David!</asp:ListItem>
32                      <asp:ListItem>Joe: I&#39;m excited for this lesson!</asp:ListItem>
33                  </asp:ListBox>
34                  <asp:TextBox ID="TextBox1" runat="server" Height="24px" Width="127px"></asp:TextBox>
35                  <asp:Button ID="Button5" runat="server" Height="27px" ForeColor="White" BackColor="DarkBlue"
36                      OnClick="Button5_Click" Text="Enter" Width="69px" Font-Bold="true"/>
37              </asp:Panel>
38               </asp:Panel>
39          <div style="height: 40px; width: 907px; text-align:center; background-color:black">
40              <asp:ImageButton ID="ImageButton1" runat="server" ImageUrl="~/Img/soundopened.png" Height="37px"
41                  Width="40px" OnClick="ImageButton1_Click"/>
42              <asp:ImageButton ID="ImageButton2" runat="server" ImageUrl="~/Img/cameraopen.jpg" Height="37px"
43                  Width="39px" OnClick="ImageButton2_Click"/>
44              <asp:ImageButton ID="ImageButton3" runat="server" ImageUrl="~/Img/sharescreen.png" Height="37px"
45                  Width="40px" OnClick="ImageButton3_Click"/>
46              <asp:ImageButton ID="ImageButton4" runat="server" ImageUrl="~/Img/end-call.png" Height="37px"
47                  Width="40px" OnClick="ImageButton4_Click" />
48          </div>
49      </form>
50  </body>
51  </html>
```

Step 3: When it is clicked any button or image button on the design part, "aspx.cs" page is opened and all buttons' methods are shown. Inside some methods, page transitions can be seen (see Figure 39 and 40).

Step 4: To run the project, "IIS Express (Google Chrome)" button should be clicked (see Figure 41), and output results on the browser can be seen:

After the program is run, Functional Waiting Main Room can be seen (see Figure 42). On this GUI, main menu, chat, video, sound and screen sharing parts are available. Sub rooms can be chosen.

After "One-To-One Room" button is clicked, on the One-To-One Room, two participants are able to fill the quotas and chat with each other. Also, video camera is able to be used (see Figure 43).

Figure 39. "aspx.cs" (C#) Part

```
FunctionalWaitingMainRoom.aspx.cs   FunctionalWaitingMainRoom.aspx   EndCall.aspx
ACM498GUIProject                    ACM498GUIProject.FunctionalWaitingMainRoom
1   using System;
2   using System.Collections.Generic;
3   using System.Linq;
4   using System.Web;
5   using System.Web.UI;
6   using System.Web.UI.WebControls;
7
8   namespace ACM498GUIProject
9   {
        2 başvuru
10      public partial class FunctionalWaitingMainRoom : System.Web.UI.Page
11      {
            0 başvuru
12          protected void Page_Load(object sender, EventArgs e)
13          {
14
15          }
            0 başvuru
16          protected void ImageButton1_Click(object sender, ImageClickEventArgs e)
17          {
18              ImageButton1.ImageUrl = "~/Img/soundclosed.png";
19          }
            0 başvuru
20          protected void ImageButton2_Click(object sender, ImageClickEventArgs e)
21          {
22              ImageButton2.ImageUrl = "~/Img/cameraclosed.jpg";
23          }
```

Figure 40. "aspx.cs" (C#) Part

```
FunctionalWaitingMainRoom.aspx.cs   FunctionalWaitingMainRoom.aspx   EndCall.asp
ACM498GUIProject                    ACM498GUIProject.FunctionalWaitingM
            0 başvuru
24          protected void ImageButton3_Click(object sender, ImageClickEventArgs e)
25          {
26
27          }
            0 başvuru
28          protected void ImageButton4_Click(object sender, ImageClickEventArgs e)
29          {
30              Response.Redirect("EndCall.aspx");
31          }
32
            0 başvuru
33          protected void Button5_Click(object sender, EventArgs e)
34          {
35
36          }
37
            0 başvuru
38          protected void Button6_Click(object sender, EventArgs e)
39          {
40              Response.Redirect("OneToOneRoom.aspx");
41          }
42
            0 başvuru
43          protected void Button7_Click(object sender, EventArgs e)
44          {
45              Response.Redirect("GameRoom.aspx");
46          }
47      }
48  }
```

Figure 41. "IIS Express" Run Button

Figure 42. Functional Waiting Main Room

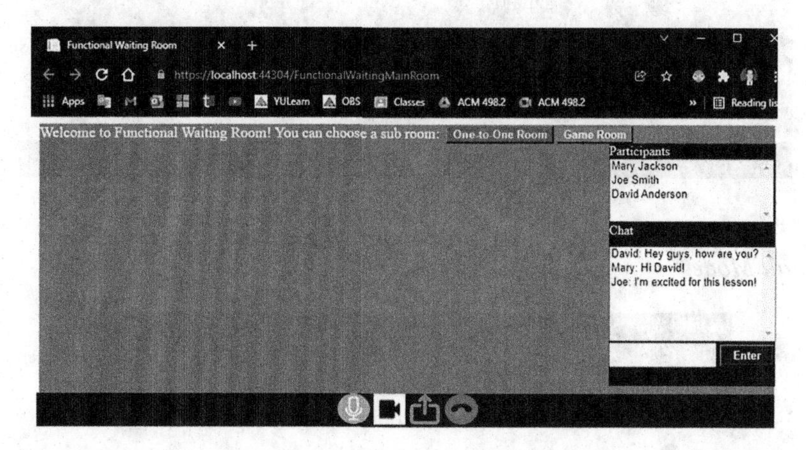

Figure 43. One-To-One Room

On the Game Room, after "Game Room" button is clicked, it is able to be played game with all participants (see Figure 44).

When end call (red button) icon is clicked, the user leaves from the EFWR (see Figure 45).

If the educator as a host has accepted his/her students, a notification will be sent to these students and they will move to the main classroom automatically (see Figure 46).

Figure 44. Game Room

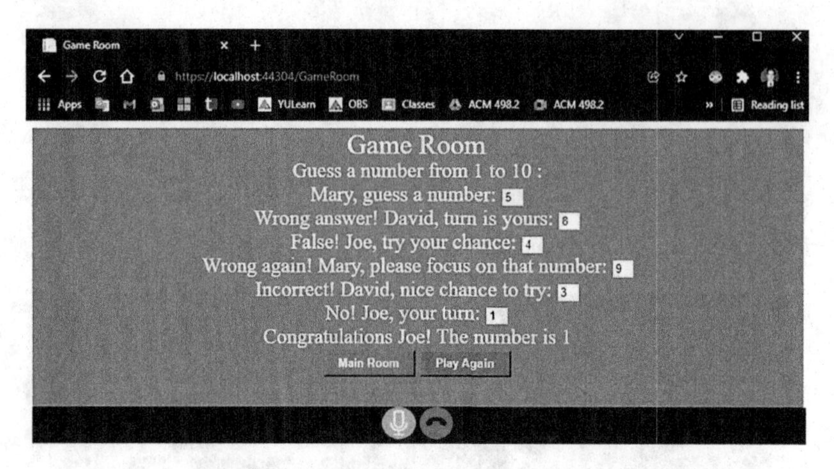

Figure 45. End Call Mode

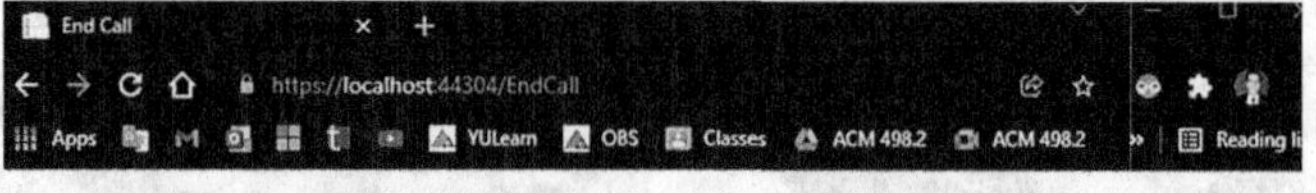

Figure 46. "Accepted" Mode

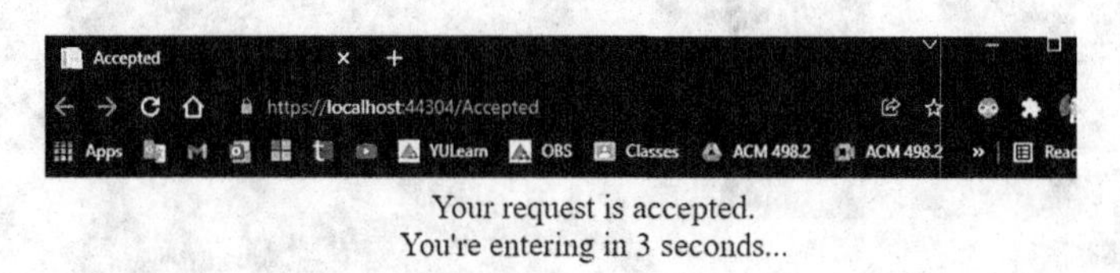

FUTURE RESEARCH DIRECTIONS

To develop the proposed system, there are twelve aspects for future research directions: Other Potential Application Areas, Emerging Web Technologies, New Generation Coding, Disruptive Technologies, User Experience (UX), Gamification, Break Periods, Different Types of Companies, Psychology, Advertisement, Sociology and New Games.

Other Potential Application Areas: First of all, not only in educational area, but also other areas can benefit from this solution. For instance, in any business meeting, participants can use this EFWR.

Emerging Web Technologies: Emerging web technologies such as blockchain integration and cloud computing can be implemented to this solution.

New Generation Coding: In terms of future coding, ASP.NET Core MVC can be applied. Besides, low code - no code platforms can be used for highly developed EFWRs.

Disruptive Technologies: In terms of disruptive technologies, this EFWR model can be used on the Metaverse World, particularly for online job and education meetings by using Augmented Reality (AR),

Virtual Reality (VR) and/or Mixed Reality (MR) technologies. Moreover, Artificial Intelligence (AI) and Internet of Things (IOT) topics can be integrated on this model. Filtering (just like e-commerce websites' feature) can be added into waiting rooms as well.

User Experience (UX): New user-oriented designs and trends in the future can be applied into EFWR.

Gamification: In terms of gamification, new and current game mechanics can be applied into EFWR for online educational platforms and other sectors' online platforms. In addition, this research has a potential to affect user engagement.

Break Periods: For giving break periods on any of these online platforms like EFWRs, functional break mode rooms can be developed.

Different Types of Companies: In terms of different types of companies, they are able to create and use this EFWR by modifying some features before their own job conferences and educational purposes.

Psychology: The EFWR's effects on participants' psychology can be studied and tested. Psychological researches on community engagement may have positive effects as well.

Advertisement: Advertisement aspect can be applied into the EFWR model and its effects can be studied to empower user engagement.

Sociology: This aspect can be applied into the EFWR to enrich user engagement.

New games: Finally, new games (just like bowling, rock paper scissors, okey game and so on) can be developed and tested to enrich the environment of the EFWR.

Above-mentioned future aspects showed that this project has a potential to extend in the future.

CONCLUSION

The potential for online courses is huge. In the future, if online teaching and learning improves in terms of innovation, it may prove to be a suitable option to classroom learning (Rathor,2022).

While promoting online education, college and university administration should focus on creating an organized and user-friendly platform for online mode of education that is accessible to all without putting financial strain on students and professors. Besides, proper technological training for educators on how to conduct online classes should be prioritized, as it has been discovered to be a prerequisite for effective online class adoption. Even though the sample size is insufficient to generalize to the wider online higher education community, the data can shed light on the common challenges that professors and students experience in online programs (Nambiar,2020).

Students are being affected by social gap because of the pandemic period. This situation leads to the students' lack of participation during their lectures and deterioration of their concentration. To fill this gap, instead of" waiting mode", an effective solution can be developed between "Join" button and main room.

To fill the gap, a solution that is the Extendible Functioning Waiting Room Project was developed. Thanks to Waiting Main Room, One-To-One Room and Game Room, students can expand good time with each other until they are accepted by a host educator.

The main aim of this study's effort is to improve students' experience and engagement with the online learning environments via the Functional Waiting Room. This functional room can be developed for Zoom, Google Meet and other online educational platforms not only to support education but also business purposes.

ACKNOWLEDGMENT

This research received no specific grant from any funding agency in the public, commercial, or not-for-profit sectors.

REFERENCES

W3schools.com. (2021). Available from: https://www.w3schools.com/asp/webpages_forms.asp

W3schools.com Tryit Editor v2.2-Show razor. (2021). Available from: https://www.w3schools.com/asp/showfile_c.asp?filename=try_we bpages_cs_010

Almahasees, Z., Mohsen, K., & Amin, M. O. (2021, May). Faculty's and students' perceptions of online learning during COVID-19. *Frontiers in Education.*, *6*(3), 638470. Advance online publication. doi:10.3389/feduc.2021.638470

Ayu, M. (2020). Online learning: Leading e-learning at higher education. *The Journal of English Literacy and Education*, *7*(1), 47–54. doi:10.36706/jele.v7i1.11515

Chansanam, W., Tuamsuk, K., Poonpon, K., & Ngootip, T. (2021). Development of online learning platform for Thai university students. *International Journal of Information and Education Technology (IJIET)*, *11*(8), 348–355. http://www.ijiet.org/vol11/1534-IJIET-1860.pdf. doi:10.18178/ijiet.2021.11.8.1534

Chen, T., Peng, L., Jing, B., Wu, C., Yang, J., & Cong, G. (2020). The Impact of the COVID-19 Pandemic on User Experience with Online Education Platforms in China. *MDPI Sustainability*, *12*(18), 4–17. doi:10.3390u12187329

Fatoni, N. A., Nurkhayati, E., Nurdiawati, E., Fidziah, G. P., Adha, S., Irawan, A. P., . . . Azizi, E. (2020). University Students Online Learning System During Covid-19 Pandemic: Advantages, Constraints and Solutions. *Systematic Reviews in Pharmacy, 11*(7), 570-576. Retrieved from: https://www.sysrevpharm.org/articles/university-students-online-learning-system-during- covid19-pandemic-advantages-constraints-and-solutions.pdf

Gaborov, M., & Ivetić, D. (2022). The importance of integrating Thinking Design, User Experience and Agile methodologies to increase profitability. *jATES. Journal of Applied Technical and Educational Sciences*, *12*(1), 1–17. doi:10.24368/jates286

Google Chrome. (n.d.). Available from: https://www.google.com/chrome/

Google Drive. (n.d.). *Extendable Functioning Waiting Room Project.* Available from: https://drive.google.com/drive/folders/1hdmd8E5VC8TbID0ASD4U _Qj7Oql9iVwx?usp=sharing

Google Jamboard. (n.d.). Available from: https://jamboard.google.com/d/1-pTwY1vbAVNgsuVyrtplqZ6zgPfi J-AiRmx8X9-yKUM/edit?usp=sharing

Google Meet Help. (2021). *Raise your hand in Google Meet*. Available from: https://support.google.com/meet/answer/10159750?hl=en&co=GENIE.Platform%3DDesktop

Google Meet. (n.d.). Available from: https://meet.google.com/

Heng, K., & Sol, K. (2021). Online learning during COVID-19: Key challenges and suggestions to enhance effectiveness. *Cambodian Journal of Educational Research, 1*(1), 3-16. Retrieved from https://www.academia.edu/51722743/The_Roles_of_Parental_to_Promote_Inclusive_Education_During_COVID_19?bulkDownload=thisPaper-topRelated-sameAuthor-citingThis-citedByThis-secondOrderCitations&from=cover_page

Liu, Z.-Y., Lomovtseva, N., & Korobeynikova, E. (2020). Online Learning Platforms: Reconstructing Modern Higher Education. *International Journal of Emerging Technologies in Learning*, *15*(13), 4–21. doi:10.3991/ijet.v15i13.14645

Microsoft ASP.NET. (n.d.). Available from: https://dotnet.microsoft.com/en-us/apps/aspnet

Microsoft Visual Studio. (2022). Available from: https://visualstudio.microsoft.com/

Microsoft Visual Studio Community 2019 v.16.11.7. (2021). Available from: https://docs.microsoft.com/tr-tr/visualstudio/releases/2019/release-notes#16.11.7

Muthuprasad, T., Aiswarya, S., Aditya, K. S., & Jha, G. K. (2021). Students' perception and preference for online education in India during COVID -19 pandemic. *Social Sciences & Humanities Open*, *3*(1), 100101. doi:10.1016/j.ssaho.2020.100101 PMID:34173507

Nambiar, D. (2020). The impact of online learning during COVID-19: Students' and educators' perspective. *International Journal of Indian Psychology*, *8*(2), 783–793. https://ijip.in/?s=10.25215%2F0802.094

Nerds Chalk. (2021). *Google Meet Hand Raise Not Available? Here's Why and What to do*. Available from: https://nerdschalk.com/google-meet-hand-raise-not-available-heres-why-and-what-to-do/

Ovesleová, H. (2015). E-Learning Platforms and Lacking Motivation in Students: Concept of Adaptable UI for Online Courses. In A. Marcus (Ed.), Lecture Notes in Computer Science: Vol. 9188. *Design, User Experience, and Usability: Interactive Experience Design. DUXU 2015*. Springer. doi:10.1007/978-3-319-20889-3_21

Perrotta, K., & Bohan, C. H. (2020). A Reflective Study of Online Faculty Teaching Experiences in Higher Education. *Journal of Effective Teaching in Higher Education.*, *3*(1), 50–66. doi:10.36021/jethe.v3i1.9

Quimí, J., & Alexandra, J. (2022). *Face-to-face vs online learning advantages and disadvantages* [Master's thesis]. Universidad Estatal Península de Santa Elena. Retrieved from https://repositorio.upse.edu.ec/handle/46000/6928

Rathor, D. (2022). Students' and Educators' Perspectives on the Impact of Online Education during Covid- 19. *Social Science Journal for Advanced Research, 2*(1), 1-5. Retrieved from https://www.ssjar.org/index.php/ojs/article/view/29

Right Attitudes. (2008). Available from: https://www.rightattitudes.com/2008/10/04/7-38-55-rule- personal-communication/

Saha, S. M., Pranty, S. A., Rana, M. J., Islam, M. J., & Hossain, M. E. (2022). Teaching during a pandemic: Do university educators prefer online teaching? *Heliyon*, *8*(1), e08663. doi:10.1016/j.heliyon.2021.e08663 PMID:35028450

Selvaraj, A., Vishnu, R., Ka, N., Benson, N., & Mathew, A. J. (2021). Effect of pandemic based online education on teaching and learning system. *International Journal of Educational Development*, *85*, 102444. doi:10.1016/j.ijedudev.2021.102444 PMID:34518732

Serrano-Solano, B., Föll, M. C., Gallardo-Alba, C., Erxleben, A., Rasche, H., Hiltermann, S., ... Grüning, B. A. (2021). Fostering accessible online education using Galaxy as an e-learning platform. *PLoS Computational Biology*, *17*(5), e1008923. doi:10.1371/journal.pcbi.1008923 PMID:33983944

Shim, T. E., & Lee, S. Y. (2020). College students' experience of emergency remote teaching due to COVID-19. *Children and Youth Services Review*, *119*, 105578. doi:10.1016/j.childyouth.2020.105578 PMID:33071405

Shutterstock. (n.d.a). *Gülümseyen kendine güvenen Latin genç kız okul öğrencisi*. Available from: https://www.shutterstock.com/tr/image-photo/smiling-confident-latin-teen-girl-school-1740966191

Shutterstock. (n.d.b). *Young happy Hispanic Indian Latin student*. Available from: https://www.shutterstock.com/tr/image-photo/young-happy-hispanic-indian-latin-student-2007080243

Yeung, M. W., & Yau, A. H. (2022). A thematic analysis of higher education students' perceptions of online learning in Hong Kong under COVID-19: Challenges, strategies and support. *Education and Information Technologies*, *27*(1), 181–208. doi:10.100710639-021-10656-3 PMID:34421326

Yusnilita, N. (2020). The impact of online learning: Student's views. *ETERNAL*, *11*(1). Advance online publication. doi:10.26877/eternal.v11i1.6069

Yusof, N., Hashim, N. L., & Hussain, A. (2022). A Conceptual User Experience Evaluation Model on Online Systems. *International Journal of Advanced Computer Science and Applications*, *13*(1). Advance online publication. doi:10.14569/IJACSA.2022.0130153

Zoom Video Communications, Inc. (2021). Available from: https://zoom.us/

ADDITIONAL READING

Bulturbayevich, M. B., Rahmat, A., & Murodullayevich, M. N. (2021). Improving Educator-Student Collaboration And Educational Effectiveness By Overcoming Learning Challenges. *Aksara: Jurnal Ilmu Pendidikan Nonformal*, *7*(1), 153–160. doi:10.37905/aksara.7.1.153-160.2021

Kumar, G., Singh, G., Bhatnagar, V., Gupta, R., & Upadhyay, S. K. (2020). Outcome of online teaching-learning over traditional education during covid-19 pandemic. *International Journal (Toronto, Ont.)*, *9*(5), 77014–77711. doi:10.30534/ijatcse/2020/113952020

KEY TERMS AND DEFINITIONS

ASP.NET: Open source and web-application framework which was created by Microsoft.

C#: Programming language that was developed by Microsoft.

Concentration: The power of focusing on anyone or anything.

COVID-19: Known as Coronavirus (SARS-CoV-2) disease that started from some animals that infected some people at Wuhan/China in 2019, and spread to all around the world.

Efficiency: The ratio analysis of input and output.

Google Meet: Online meeting platform that was created by Google.

HTML: Hyper Text Markup Language.

Online Education Platform: A platform where educators and students can perform education and training with the help of a website and internet connection.

Pandemic: Epidemic of an infectious disease that spreads throughout more than one continent or worldwide.

Participation: The activity of joining in something.

User Engagement: Participating an organization for the user's welfare.

User Experience (UX): A person's emotions and behaviors about a product, a system, or a service.

Visual Studio: Coding platform of Microsoft for web sites, desktop applications and mobile applications.

Waiting Room: A virtual room where students can productively spend time before educator accept them on online education platforms.

Zoom: Online meeting platform that was created by Zoom.

Chapter 9
Creating a Community in the Course Room Through Discussion Posts

Tricia Mazurowski
Adler University, USA

ABSTRACT

Online education has become increasingly popular due to the convenience of learning in an online environment. Online learning allows adult learners to continue their studies remotely. Discussion boards are used in an online learning environment in lieu of lectures from an instructor. Creating a community in an online learning environment through the use of discussion boards is important to engage adult learners. Discussion boards allow students to engage with and learn from their peers and from the instructor. This chapter will explore how to create a community in the course room through discussion posts to increase student engagement in an online learning environment. Online educational programs often experience a higher attrition rate since students may feel isolated from their peers and their instructors. Increasing engagement in the course room through discussion posts is important to ensure that adult learners are engaged in the course room. Interaction with peers and the instructor is instrumental in increasing engagement through online learning.

BENEFITS OF ONLINE DISCUSSION BOARDS

Due to advances in technology and the convenience of online learning, virtual classrooms are becoming increasingly popular. There are several benefits to creating interactive online discussion boards in the classroom to enhance student interaction and learning.

Social interactions in the online environment promote group dynamics within the learning community. Online discussion boards create an opportunity for peer-peer and instructor-student learning. Learning is accomplished by reading peers' discussion post and responses to other peers along with instructor feedback to students (Aloni & Harrington, 2018). Students are also able to post their response in the discussion board at their convenience. Students are also able to practice writing more succinctly to pre-

DOI: 10.4018/978-1-6684-5190-8.ch009

pare for more formal writing assignments. Forming a community in an online learning environment also increases participation for introverted students who may find it difficult to participate in a face-to-face learning environment (Aloni & Harrington, 2018). Students also have time to focus on their answers in the discussion board which facilitates the ability to critically reflect on their answers. Students also have time to respond to their peers while providing evidence from the literature to support their thoughts, which enhances critical thinking (Aloni & Harrington, 2018). Interaction with the instructor and peers in an online forum helps to create positive group dynamics in an online learning environment.

FORMING COMMUNITIES IN AN ONLINE LEARNING ENVIRONMENT

Online learning has continued to become an increasingly popular way for adult learners to complete their course work remotely. Although this method of learning is convenient, many online programs have a higher attrition rate than face-to-face learning environments due to several factors including both personal issues and issues related to difficulty adjusting to an online learning environment (Chen & Bogachenko, 2022). One of the factors that contributes to a higher attrition rate for online learning programs is the feeling of isolation from peers and instructors (Chen & Bogachenko, 2022). Creating a networked structure such as a community is an effective way to learn in the 21st century (Jan & Vlachopoulos, 2018). Discussion boards are used in online course rooms in lieu of traditional face-to-face lectures. Discussion posts allow students to engage in an online discussion with their peers and instructor using an online learning management system. Discussion boards also offer a flexible form of learning for both students and instructors due to the asynchronous nature of online discussion boards. Online discussions may also be used to enhance a sense of community and engagement for the class. This may be accomplished by the instructor engaging the students by responding to their discussions, asking questions to promote critical thinking, and by providing real-world examples for the students to consider and apply to their careers. It is noted that instructors who engage in a social presence in an online discussion board will impact student engagement in a course.

Interactions among the instructor and students in an online discussion board will have a substantial impact on students' learning and willingness to engage in the online community (Celik, 2013). The social exchange theory proposes that there is a cost analysis in relationships, where the individual should see the benefits as outweighing the risks. This theory may be used to explore cohesion in an online learning environment. Students must consider the benefits of being a member of an online learning community as exceeding the risks of belonging to the online learning community. Positive communication among members of an online discussion board will influence participation and contribution to learning in the online environment (Celik, 2013). As students continue to engage with their peers and the instructor in the course, they will continue to find value in participating in an online discussion board. Fostering engagement in an online community also encourages each student to participate. Students who develop positive relationships within the online community will have a better learning experience in a virtual classroom. The benefits of utilizing online discussion boards allow for a positive interaction and support in a virtual learning environment.

Researchers suggest that online discussion boards offer a supportive environment for students to grow and learn through interactions with the instructor and peers (Grant & Lee, 2014). While online discussion boards offer an opportunity for students to learn from the experience and knowledge of their peers, instructors must also be engaged in the discussion to provide guidance that allows students to find meaningful reflection in the discussion to create a community in an online learning environment (Grant & Lee, 2014). Instructor must log into the course frequently and be active participants in the discussion board to create a community in the discussion board by encouraging participation and self-reflection. Instructor involvement in the discussion board is pivotal to create positive group dynamics.

CREATING POSITIVE GROUP DYNAMICS IN THE ONLINE LEARNING ENVIRONMENT THROUGH ONLINE DISCUSSION BOARDS

Group dynamics will develop in online discussion boards through interaction between the students and the instructor and peer-peer interaction. It is important to understand the group cohesion and social interactions in the online learning environment to understand the value of collaborative learning. Online discussion boards are used by instructors in virtual classroom to enhance writing and critical thinking skills, as well as to create social interactions between students and instructors in the classroom (Aloni & Harrington, 2018; Celik, 2013). Positive interactions in the online discussion board create supportive behaviors from the group. Students are also more willing to engage in the discussion when the interactions are supportive and positive. Social presence in the discussion board allows students to feel more connected with their peers and faculty, which is an important indicator of student satisfaction (Bloomberg, 2013). As students and instructors get to know one another through their interaction in the discussion boards, this is more likely to foster group cohesiveness. It is important for the instructor to continually monitor and participate in the online discussion without dominating the discussion to create positive group dynamics through interactions in the course (Aloni & Harrington, 2018). Instructors who respond to students' discussion posts are more likely to engage the students in the online learning experience by creating an environment that is both interactive and supportive of learning and critical thinking. The social network analysis theory will be explored to offer an understanding of how interpersonal interactions in an online learning environment create a positive learning community in online discussion boards.

SOCIAL NETWORK ANALYSIS AND COMMUNITY OF PRACTICE IN AN ONLINE DISCUSSION BOARD

Social network analysis provides an exploration of interpersonal interactions in an online community. Social network analysis is used to explore how individuals interact in online environments to create connectivity and community (Ruane & Lee, 2016). Social networks allow students in an online learning environment to develop peer-peer interactions to develop relationships. Online discussion boards offer a venue for support and community. Social network analysis has also demonstrated that peer-peer interactions in an online discussion board have a positive impact on student learning (Ruane & Lee, 2016). Communities of practices are defined as entities that have an identity which consists of the group members but is different from the group members (Ruane & Lee, 2016). Community identity serves to create community cohesion in an online learning environment.

Community identity consists of the classroom culture. A group will have an individual identity which defines the culture of the group. Communities of practices are critical for social interaction, support, individual thinking, and knowledge-building (Ruane & Lee, 2016). The openness of an online learning environment also allows for more interaction in the course where social interactions allow learning to occur. The process of creating a community through online discussion boards in the course room begins with designing a course for an online community.

DESIGNING COURSE TO CREATE A SENSE OF COMMUNITY AND ENGAGEMENT IN AN ONLINE ENVIRONMENT

Curriculum designers must ensure that the course design they select encourages active engagement in the course through a variety of activities including discussion boards, videos, reading assignments, group activities, and synchronous sessions where the class meets together in an online meeting platform. A good learning design is expected to enhance learning outcomes by incorporating a variety of online learning activities to engage a variety of learning styles. In the Activity-Centered Analysis and Design (ACAD) method, there are three structures of learning design. The first is the set design (the space and platform where the learning occurs), the second is the social design of a course (groups, roles, communities), and the third is the intended design which helps to create the actual activity of learning (Jan & & Vlachopoulos, 2018). Good course room learning designs should not be static but should have the ability to be altered if needed as a course progresses to completion. Curriculum developers must also include innovative and engaging pedagogical practices for courses developed in an online learning environment.

Incorporating Bloom's Taxonomy when designing courses for an online environment provides specific and measurable goals that are used throughout the course to keep the educational needs of the learner in mind. Adult learners will come to the course room with diverse professional and educational experiences; therefore, multiple methods of learning are required to engage adult learners in the online environment (Bloomberg, 2022). The type of interactions in an online learning environment includes learner-learner interactions, learner-instructor interactions, and learner-content interactions (Mukuni et al., 2021). Course designers must focus on each of these interactions when considering pedagogical practices for online courses. Adult learners are self-directed and autonomous; therefore, subject-centered learning is not engaging for the adult learner. Adult learners seek to apply what is learned in the course room to real-world experiences that they may apply to their careers (Bloomberg, 2022). The instructor must go beyond delivering instruction but provide an experience that cultivates a learning journey for the students where they are engaged in their own learning. This occurs when a learning community is built where the instructor is present in the course and serves as a mentor to the adult learner (Bloomberg, 2022). Instructor presence in the online learning environment serves as a source to create increased learner engagement in the course room. Team-based learning is a method of course design that involves a framework designed to promote active and collaborative learning.

Team-Based Learning in Online Course Design

Although team-based learning is often used in a face-to-face learning environment, this framework may be used successfully to design courses in an online learning environment. This adaption of applying team-based learning to the online environment increased due to the coronavirus pandemic beginning in 2020 (Parrish et al, 2021). The global pandemic significantly increased the need for online course delivery. Although there has been an increase in online learning, there remains concerns about limited interactions between other students and instructors (Parrish et al., 2021). Limited interactions in an online learning environment may create a lower sense of belonging; therefore, it is important to design courses in an online environment to increase a sense of community and belonging. Online instruction using team-based learning frameworks is likely to provide students with the benefits of face-to-face learning (Parrish et al., 2021). Team-based learning may be utilized in both asynchronous and synchronous course to build a community in an online course room to increase student engagement.

Team-Based Learning in Asynchronous Online Learning Environments

Asynchronous learning involves students accessing course materials and completing discussion boards and assignments on their own time. Asynchronous learning involves watching pre-recorded video lectures, completing independent writing assignments and research, completing tests and quizzes online, and participating in online discussion boards (Coursera, 2022). A benefit of asynchronous learning is that it is flexible, which allows adult learners to complete discussion boards and assignments on their own time. Although there are benefits to asynchronous learning, some challenges faced by students is that it is difficult to form a sense of community and there is also a lower sense of belonging (Parrish et al., 2021). Team-based learning may be incorporated to provide a more positive experience for students in an asynchronous online learning environment. This may involve assigning learners to teams where they are provided an opportunity to respond to assigned discussions and activities where students can respond to other teams and review the instructor's response to each of the teams to gain a deeper understanding of the topic (Parrish et al., 2021). Palsolé and Awalt (2008) affirmed that team-based learning in asynchronous courses improved student engagement and retention in the course (as cited in Parrish et al., 2021).

Team-Based Learning in Synchronous Online Learning Environments

Synchronous learning occurs when students and the instructor meet at the same time, but not all in the same place. Synchronous learning also occurs in an online environment using technology to meet virtually. Synchronous learning is often referred to as distance learning (Edglossary, 2013). Online learners who engaged in synchronous sessions were more likely to have a positive learning experience due to the increased engagement with their peers and the instructor. Real-time communication with peers and the instructor, as well as immediate feedback from the instructor were also noted as positive experiences for online learners (Parrish et al., 2021). Synchronous sessions built into the course design provide an engaging experience for online learners, which creates the opportunity to build a sense of community in an online classroom environment through interaction in an online meeting using technology such as Zoom or Teams to facilitate interaction between the instructor and the peers to create engagement in an online course to enhance learning.

CREATING ENGAGEMENT IN THE ONLINE COURSE ROOM ENVIRONMENT

Creating a sense of community in online discussion boards is necessary to increase engagement for adult learners in an online community. It is important for instructors to both create and maintain engagement for adult learners in an online community. Instructors should empower adult learners to engage in self-directed learning. Instructors should also engage in self-reflection to reduce implicit bias in the course room (Bloomberg, 2022). Pedagogy ought to be adaptive while keeping the needs of the learner in mind. Mukuni et al. (2021) defined the three types of interactions that influence engagement in the online course room. Learner-learner interaction is one way in which students develop a sense of community in an online learning environment. Learner-learner engagement occurs when students interact with and receive feedback in the discussion boards from their peers in an online course room environment (Mukuni et al., 2021).

Another element of online course room interaction includes learner-instructor interaction, which occurs when the instructor provides feedback to students in discussion boards and through grading rubrics in the course. Instructors may increase student engagement in the course room by responding to each student's post in discussion boards by adding information to the students' post and asking questions to promote critical thinking. Instructors may also increase engagement in an online course room by posting videos to provide real-world examples of how to apply the learning objectives in the course to the workplace. Instructors may also provide additional articles or online videos to students so that they may explore how the learning objectives in the course may apply to real-world scenarios in their respective careers. Students who report having a higher degree of interaction with their instructors also report higher satisfaction and engagement in the course (Mukuni et al., 2021). It is also important for instructors to support and encourage each learner.

The third type of interaction in an online course room relates to the course room content. This involves the content in the course that the student uses to promote learning. The learning content in the course is developed to allow students to acquire knowledge which may be applied to the workplace environment (Mukuni et al., 2021). The content in an online learning environment may involve students reading from textbooks or peer-reviewed journals, viewing videos, and completing assignments, projects, and independent research. Projects and assignments are likely to involve individual and group projects. Student engagement with the course content is also an indicator of satisfaction with the learning experience. Assignments involving the ability to problem solve and apply principles to the workplace are more likely to increase student engagement (Mukuni et al, 2021). The theoretical framework of Community of Inquiry also provides a way to evaluate the online learning process to develop a sense of community and engagement in the course room.

COMMUNITY OF INQUIRY OF THEORETICAL MODEL

The Community of Inquiry (CoI) is a theoretical model grounded in social constructivism used to explore the effectiveness of online learning. The three significant presences of online learning examined using this theoretical model include teaching presence, social presence, and cognitive presence (Honig & Salmon, 2021). The thought is that each of the three presences yield a robust learning experience in an online learning environment. The CoI framework has also been associated with such outcomes as student satisfaction with online discussion boards (Honig & Salmon, 2021). The teaching presence of this model include the design of the course, instructor facilitation of the course, and instructor response and feedback to students. The social presence of this theoretical framework includes expression, communication, and group cohesion. Finally, the cognitive presence includes "the ability to construct and confirm meaning through sustained reflection and discourse in a critical

Table 1. Three Significant Presences

Teaching presence	Teaching presence involves cognitive and social processes to engage students in meaningful learning (Honig & Salmon, 2021).
Social presence	Social presence involves learners' ability to communicate with a sense of purpose and to build a sense of community in an online environment (Honig & Salmon, 2021).
Cognitive presence	Cognitive presence involves forming meaning through self-reflection and inquiry (Honig & Salmon, 2021).

community of inquiry" (Garrison et al., 2001 p. 11 as cited in Honig & Salmon, 2021). Table 1 below outlines the definition of each of the presences.

Learner presence in an online community is a valuable tool to measure the effectiveness of online learning. The three-tiered approach developed in the Community of Inquiry theoretical model serves as a tool to understand the importance of creating an online learning environment that is engaging from the perspective of both instructors and online learners. Faculty teaching in an online learning environment must also ensure that they practice inclusive teaching practices to develop a sense of community and belonging in an online learning environment.

The CoI approach is very important in the online learning environment where there is a need to remove distance in the classroom, so that everyone can work together for a common goal. The CoI model was built on the foundation that learning occurs within a community (Hauser & Darrow, 2013). Almost 30% of students currently enrolled in a higher education program are taking at least one course online through a massive open online course (MOOC) (Beckett, 2019). The increase in online courses has paved the way for synchronous and asynchronous learning with discussion boards being at the heart of many online courses. This is especially true of courses in humanities and social sciences, where instructors promote a thoughtful discussion between instructors and peers to extend learning (Beckett, 2019). The promotes an engaging experience for students where they can learn from their online community in the course. There is a growing body of research on communities of inquiry to ensure that there is sustained communication and ongoing learning. As noted in Table 1, communities of inquiry are used to measure teaching presence, social presence, and cognitive presence. CoI researchers are interested in sustaining students' cognitive presence in online discussion boards. This goal is achieved through four phases.

The four phases include a triggering event, exploration, integration, and resolution (Beckett, 2019). A triggering event creates a sense of being aware of the unknown. Exploration occurs in the online learning environment when information is exchanged in a discussion board. Integration has been achieved when different thoughts and ideas come together. Resolution has been achieved when the ideas come together to form new knowledge (Beckett, 2019). CoI researchers also use discussion board to lead students to a higher level of Bloom's Taxonomy. Instructors should utilize discussion boards to help students form knowledge and understanding of concepts explored in assignments to a synthesis and critical evaluation of these concepts (Beckett, 2019). Instructors must be active participants in discussion boards to promote learning and the development of an inclusive and respectful community in the online learning space. Each presence of the Community of Inquiry model will be analyzed to understand how each presence solidifies the formation of a robust community in an online learning environment.

Role of Teaching Presence in Forming a Community in an Online Learning Environment

Teaching presence is one of three components of the Community of Inquiry theoretical model. According to CoI, teaching presence is comprised of direct course room instruction, facilitating debate, and course design (Silva et. al., 2021). Instructors use design and organization to develop the learning environment. This includes establishing course curriculum and establishing instructional methods and media methods in the course (Kilis & Yildirim, 2019). Instructors facilitate debate in an online discussion board by encouraging a healthy debate as students come to resolutions. Instructors facilitate course room instruction through constructive feedback, reinforcing student contributions in the discussion, and setting a positive learning environment through support and acknowledgement of each student (Kilis & Yildiriim, 2019). A higher teaching presence in the online learning environment is correlated with increased student engagement, transfer of knowledge, and student participation. Teaching presence also includes an instructor's ability to clearly communicate course requirements and facilitating a healthy debate in the online discussion board (Silva et. al., 2021). Instructors may create an interactive online learning community through course organization to support student learning needs. Creating an interactive environment in an online course room includes the use of multiple instructional media resources to increase student engagement. Instructional resources may include videos, Podcasts, blogs, interactive online exercise, recorded lectures, or visual elements to enhance learning (Silva et al., 2021). Instructors may create teaching presence in the online environment by ensuring that they are present in the course discussions by providing timely responses to students' discussions and inquiries. Teaching presence in an online learning environment correlates with increased social presence in the course room to form a sense of community.

Role of Social Presence in Forming a Community in an Online Learning Environment

Social presence in an online learning environment is a contributor to student satisfaction with online discussion boards. Social presence is also a context from the Community of Inquiry framework. Social presence consists of open communication and group cohesion in an online discussion board. Social presence in an online discussion board is influenced by course content, instructional strategies, student comfort with technology, personality styles such as introversion and extroversion, and learning styles (Nasir, 2020). Social presence in online learning is a popular topic among researchers since social presence is correlated with increased student satisfaction in the online classroom. In a course where there is a prominent level of social presence, learners are able to express themselves in a manner that is conducive to promote their learning (Garcia-O'Neil, 2016). Curriculum designers and instructors must consider factors to promote social presences in an online course room to foster both learning and the formation of an online community.

There are several strategies that instructors may incorporate to increase social presence in the classroom to develop a sense of community. One strategy that instructors may use to enhance social presence online is to encourage students to include their feelings and individual experiences when responding to their peers in the discussion board (Garcia-O'Neil, 2016). Sharing emotions and individual experiences allows students to connect with their instructor and peers on a personal level. Sharing individual experiences in the discussion board also enhances peer-peer learning because students are able to explore how the concepts and theories explored in the discussion board may be applied to real-world situations.

Instructor Strategies to Promote Social Presence in an Online Learning Environment

Instructors play a pivotal role in managing the online classroom, enhancing social presence, and forming a sense of community. Instructors may enhance their online visibility by including video feedback on discussion boards and posting videos in the course to explain the assignments and learning objectives. Video feedback enhances social presence in the online course because students can see their instructor. Instructors may speak with their emotions, provide real-world examples of how theories in the course may be applied, and speaking with students in a conversational manner using video feedback (Garcia-O'Neil, 2016). It is important that media posted in an online classroom is high quality, so that students can view social cues (e.g., eye contact, facial expressions) to enhance learning. Any photos that are used in an online course should also be high quality to enhance a richer social presence (Garcia-O'Neil, 2016). Multiple forms of media should also be used to ensure the course is welcoming and inclusive. For example, visual media, written media, and videos using closed caption. Students should be encouraged to respond to the instructor's welcome message with their personal written or video introduction to form a community in an online learning environment (Garcia-O'Neil, 2016). Applying these strategies will enhance the social presence of the instructor and the students in the online classroom to enhance learning. Cognitive presence in an online learning environment is also necessary to enhance student learning and development.

Role of Cognitive Presence in Forming a Community in an Online Learning Environment

Cognitive presence refers to finding meaning through knowledge and self-reflection (Honig & Salmon, 2021). Constructing meaning through communication and self-reflection is a key element of critical thinking. Cognitive presence plays a pivotal role in high-level learning in an online learning environment. Cognitive presence, according to the CoI framework, begins with experience something that is unknown until new knowledge transfer (Akti-Aslan & Turgut, 2021). The second phase of cognitive presence occurs when learners share their knowledge and experience with their peers and instructor in the discussion board. The second phase also involves sharing knowledge and experience by asking questions to promote a learning dialog (Akti-Aslan & Turgut, 2021). As a result, learners will gain new knowledge and understanding. In the third phase of cognitive presence, learners are expected to synthesize knowledge and make connections to develop solutions (Akti-Aslan & Turgut, 2021). Finally, new ideas and solutions are created. Instructors enhance cognitive presence in an online learning environment through their interactions with learners in the course room.

Akti-Aslan and Turgut (2021) recommend frequent instructor-student interaction in an online course room to facilitate cognitive presence. Instructor-student interaction also develops a sense of community in an online learning environment. Encouraging multiple perspectives in an online discussion board enhances cognitive presence by promoting an open dialog where knowledge and experience is shared in the course. Self-reflection exercises are also conducive to cognitive presence in an online course. Assignments such as self-assessments and peer reviews allow students to perform a self-reflection of their participation and performance in the course (Akti-Aslan & Turgut, 2021). Instructors may also enhance cognitive presence in an online discussion board by responding to students with questions that encourage critical thinking to promote an environment conducive to knowledge sharing.

INSTRUCTOR INVOLVEMENT IN CREATING AN ONLINE LEARNING COMMUNITY

Instructor involvement is essential to creating a sense of community in an online learning environment. Instructors share their knowledge and expertise with students as subject matter experts to facilitate synthesis of knowledge and critical thinking. Instructor presence in the virtual classroom bridges the distance between the instructor and the students by offering various methods of facilitating online learning (Orcutt & Dringus, 2017). The course climate created by the instructor defines the learning experience for students and the development of an online community. Instructors may create a climate of connectedness in the online learning environment by developing a shared sense of belonging in the online community. Teaching presence in an online learning environment involves the instructor demonstrating support and availability to students. Teaching presence influences both social interactions and cognition in an online learning environment (Orcutti & Dringus, 2017). Teaching presence is an integral part of the Community of Inquiry theoretical model. Active interactions in an online learning environment must be carefully conducted by the instructor to facilitate engagement and a sense of community in the course. The level of presence in the course and instructor visibility will influence student interaction and engagement. Instructors can enhance presence in an online course by responding to student discussions with constructive feedback, theoretical examples, incorporating additional resources that contribute to the discussion, or asking questions to promote knowledge transfer and critical thinking. Strategies to increase a sense of community in an online course involve a sequence of action to establish a presence.

An instructor can begin to create a sense of community in an online learning environment by establishing rapport with students. Creating a welcome post and welcome video at the beginning of the course allows students to feel a part of the online community. This also provides an opportunity for the instructor to share some personal information to introduce themselves to the students (Orcutti & Dringus, 2017). Initiating interactions with students from the beginning of the class also establishes a teaching presence in the course. This will also set the tone for expectations for the remainder of the course. Instructors should provide constructive and thoughtful feedback on each student's discussions and assignments to enhance learning. Setting instructor expectations at the beginning of the course will set students up for success. Instructors must also respond to students in a timely manner to provide support and assistance (Orcutti & Dringus, 2017). Instructors must also develop an inclusive pedagogy to form a sense of community in an online environment.

INCLUSIVE TEACHING PRACTICES TO DEVELOP A SENSE OF COMMUNITY IN AN ONLINE LEARNING ENVIRONMENT

As online learning continues to increase due to the flexibility of such programs, the course room will become more global and more diverse; therefore, it is important for online instructors to examine the accessibility of the content in the course room to ensure the course room is inclusive for all students (Cash et. al., 2021). The Higher Education Opportunity Act of 2008 was developed to create a framework for online learning that provides flexibility in the way course content is delivered, provides accommodations to support challenges to ensure academic achievement for all students with disabilities or whose first language is not English (Cash et al., 2021). The Higher Education Opportunity Act provides a framework to remove barriers and to include a variety of adult learning styles to accommodate students' needs. One example of ways to remove barriers and improve accessibility in an online course room include providing captions for video recordings. Incorporating captions for all video recordings removes barriers for student with hearing impairments or those students who speak English as a second language (Cash et al., 2021). Using text alternatives for graphic images also removes barriers for those who may not have access to images on their computer. In addition to ensuring the online course room materials are more accessible to students, instructors must also incorporate inclusive teaching practices to a diverse group of students to increase a sense of belonging in the course room.

Today's online learning environment has become increasingly more diverse regarding cultural diversity, racial diversity, generational diversity, socioeconomic status, and ability levels (Williams et al., 2020). Despite the increased diversity of the student population, there remains a gap in diversity between educators and the students they teach. Due to this gap in ethnic/cultural diversity, higher education is tasked with developing opportunities for both instructors and students to develop an increased cultural awareness and cultural competence (Williams et al., 2020). Educators are now expected to have a teaching pedagogy that includes a well-developed sense of cultural awareness and teaching from a social justice perspective. Higher education institutions must incorporate inclusive pedagogy in their professional development plans to increase cultural awareness and cultural competence of teaching faculty (Williams et al., 2020). This will allow faculty to promote inclusive course rooms, support marginalized groups in the course room, and foster difficult conversations (Williams et al., 2020).

The Creating Inclusive Communities (CIC) framework is a pedagogical approach to increase faculty awareness to promote diversity and inclusion in the classroom. This model was developed to allow faculty to promote conversations around ethnic and racial differences. CIC emphasized principles including critical consciousness and self-reflection to promote an inclusive community within an online course room (Williams et al., 2020). Course work must also include a variety of activities and online discussions to promote a sense of inclusion in an online learning environment. Activities include a varied method of content delivery including videos, photos, graphics, reading assignments, and discussion boards. The online content must be diversified so that it is welcoming and speaks to a variety of learners. Including resources from various and diverse perspectives enhances the feeling of belonging and inclusion in an online learning environment.

Why Is Inclusivity Important When Fostering a Sense of Community in Online Discussion Boards?

Educators strive to create an online learning environment that is welcoming and inclusive, but why is inclusivity important? Inclusivity ensures that students feel that their opinions and ideas are important, and they are equal participants with their peers in the course (Williams et al., 2020). Language should be inclusive and respectful of all. Instructors must begin by recognizing and acknowledging their own biases. Although this may be difficult, it is important to acknowledge biases to reframe thinking to create an inclusive community. Using universal phrases is important. For example, American slang may not be understood or well-received by students not located in the United States. Using gender-neutral phrases is also important. Instructors must be cognizant of gender stereotypes when giving examples to eliminate gender bias. The ability to create and maintain an inclusive community in an online learning environment is for instructors to learn approaches to address complexities and remove barriers to student participation in an online discussion board (Awang-Hashim et al., 2019).

Inclusion in the classroom involves a teaching pedagogy that fosters human rights and values diversity. An inclusive pedagogy involves designing, assessing, and delivering curricula that is relevant to all students in the classroom (Awang-Hashim et al., 2019). Instructors and curriculum designers must design discussion board prompts to embrace diversity and individual differences that enhances learning in the online classroom. This includes four dimensions to enhance student learning and inclusive teaching practices. The dimensions include an institutional commitment to support diversity and inclusion, diversity in the curriculum, an inclusive teaching environment, and assessing instructors and curriculum to ensure there is an inclusive teaching pedagogy (Awang-Hashim et al., 2019). Diverse and inclusive curriculum and teaching pedagogies are important strategies for fostering a sense of community in an online discussion board because students are more likely to feel included and that they are equal to their peers.

Value of Inclusive Teaching to Create Connectedness in Online Learning Environments

Inclusive teaching involves the practice of increasing student participation from all. Inclusive teaching also involves providing information that applies to a broad range of learning needs. Instructors must have a well-developed sense of cultural intelligence to enhance the learning experience for students (Williams et al., 2020). Instructors must encourage an online learning environment where there is freedom to learn, and students feel comfortable sharing their prior knowledge and experiences to create new learning (Figueroa, 2014). This teaching practices allows students to explore their learning styles, as well as create connections in the online classroom. These practices include the ability to foster an inclusive classroom, manage difficult conversations in the discussion board, and provide support for all voices in the classroom (Williams et al, 2020). Instructors must utilize these inclusive teaching practices in addition to enhancing teaching presence to foster a sense of community through the online discussion boards.

FOSTERING A SENSE OF COMMUNITY THROUGH ONLINE ASYNCHRONOUS DISCUSSION BOARDS

Online asynchronous discussion boards serve as a tool to promote an active dialog among participants from diverse backgrounds. Students and instructors participate in online discussion boards to provide feedback, knowledge, and support. Online discussion boards also allow students to develop a sense of community while working one one's own time. Instructors may create a sense of community in an online learning environment by creating a safe space to communicate with respect in an online environment, include group activities to foster teamwork, and monitor student engagement in the discussion boards. It is also important for online instructors to evaluate teaching presence in the course, social presence, and cognitive presence (Payne, 2021). There are several strategies that instructors may use to create a sense of community and engagement through online discussion boards.

One strategy that instructors may use to create a sense of community through online discussion boards is to use media to create engagement in the course. Media also helps to pique students' interest in the course content. The media may include instructor videos, YouTube or Vimeo videos, TED Talk videos, audio recordings, or graphic art images (Payne, 2021). Another strategy online instructors may use to create a sense of community in an online discussion board is to use storytelling or real-world examples of how the learning objectives may be applied in the workplace so that student have a practical application of learning. Socratic questioning allows for thoughtful questioning so that students can engage in critical thinking. Socratic questioning includes the following examples from Payne (2021):

- Clarification questions (e.g., "What do you mean by that statement?")
- Questions about an initial question (e.g., "Does this question lead to other important issues?")
- Assumption questions (e.g., "What is the author assuming here?")
- Reasoning questions (e.g., "Can you explain your reasoning for your response to X?")
- Source questions (e.g., "Has your opinion been influenced by someone or something else?")
- Consequence questions (e.g., "What alternatives might you consider?"
- Viewpoint questions (e.g., "How are student X's and student Y's responses similar and different to your response?")

Instructors may also put out introductory posts in the discussion board each week to summarize the learning objectives, apply the learning objectives and theories to real-world scenarios, and explore any salient topics from the literature to engage the students in the discussion (Payne, 2021). Instructors should also refer to students by name in their response to create a sense of community in the discussion board. It is also important to empower students by acknowledging and honoring their unique experiences and promote an atmosphere of collaborative learning.

COLLABORATIVE LEARNING IN ONLINE DISCUSSION BOARDS

Collaborative learning in the online learning environment involves student-student collaboration and student-instructor collaboration. This approach allows students to learn from both the instructor and their peers to increase their understanding of a subject. Instructors may promote collaborative learning by posting in the online discussion board to encourage student interaction. Collaborative learning is also encouraged by organizing students into groups to work together to solve a problem. Peer review assignments may also be used in an online learning environment to allow students to provide feedback to their peers and to learn from their peers (Simkovich, 2022). Problem-based learning allows students to independently work to solve a problem and report their findings to their peers in the online discussion board. In this method of collaborative learning, students conduct independent research to identify a solution to the problem (Simkovich, 2022). The goal is to provide a collaborative environment where students respond to their peers' solutions to engage each student in the online discussion. Case studies may be used in online discussion boards to expose students to real-world examples that may be applied to the workplace to explore how to apply solutions to real-world problems (Simkovich, 2022). Instructors can assign students to distinct roles to explore solutions to the case study to create collaboration. Instructors may promote a sense of community through collaborative learning by establishing rules and clear expectations in the online discussion board, encouraging diverse perspectives, encouraging students to use critical thinking to promote high-quality and interactive discussions, and using a modern approach to collaboration such as different formats including videos, graphics, or interaction through online meeting platforms (Simkovich, 2022).

The research on student engagement in online discussion boards has revealed the need to ensure teaching presence in the online course room, social presence to allow students to build a sense of community in the discussion board space, and cognitive presence to allow student to develop meaningful learning ((Honig & Salmon, 2021). Each of these concepts assist in creating a sense of community in the online discussion board through interaction, support, and collaborative learning. Students who experience teaching presence, social presence, and cognitive presence are more likely to be satisfied with their experience in an online discussion board. Instructors must be active participants in an online discussion board to create a supportive sense of community that fosters both instructor and student engagement to enhance the learning experience in an online environment.

REFERENCES

Aktı Aslan, S., & Turgut, Y. E. (2021). Effectiveness of community of inquiry based online course: Cognitive, social and teaching presence. *Journal of Pedagogical Research*, *5*(3), 187–197. doi:10.33902/JPR.2021371365

Aloni, M., & Harrington, C. (2018). Research based practices for improving the effectiveness of asynchronous online discussion boards. *Scholarship of Teaching and Learning in Psychology*, *4*(4), 271–289. doi:10.1037tl0000121

Awang-Hashim, R., & Valdez, N. (2019). Strategizing inclusivity in teaching diverse learners in higher education. *Malaysian Journal of Learning and Instruction*, *16*(1), 105–128. doi:10.32890/mjli2019.16.1.5

Beckett, K.S. (2019). Dewey Online: A Critical Examination of the Communities of Inquiry Approach to Online Discussions. *Philosophical Studies in Education, 50*, 46-58.

Bloomberg, L. D. (2022). Designing and delivering effective online instruction, how to engage the adult learner. *Adult Learning*, *34*(1), 55–56. doi:10.1177/10451595211069079

Cash, C., Cox, T., & Hahs-Vaughn, D. (2021). Distance educators' attitudes and actions towards inclusive teaching practices. *The Journal of Scholarship of Teaching and Learning*, *21*(2). Advance online publication. doi:10.14434/josotl.v21i2.27949

Çelik, S. (2013). Unspoken social dynamics in an online discussion group: The disconnect between attitudes and overt behavior of English language teaching graduate students. *Educational Technology Research and Development*, *61*(4), 665–683. doi:10.100711423-013-9288-3

Chen, J. C., & Bogachenko, T. (2022). Online community building in distance education: The case of social presence in the blackboard discussion board versus multimodal VoiceThread interaction. *Journal of Educational Technology & Society*, *25*(2), 62–75. https://www.researchgate.net/publication/359278705_Online_Community_Building_in_Distance_Education_The_Case_of_Social_Presence_in_the_Blackboard_Discussion_Board_versus_Multimodal_VoiceThread_Interaction

Doucette, B., Sanabria, A., Sheplak, A., & Aydin, H. (2021). The perceptions of culturally diverse graduate students on multicultural education: Implication for inclusion and Diversity Awareness in higher education. *European Journal of Educational Research*, *10*(3), 1259–1273. doi:10.12973/eu-jer.10.3.1259

Douglas, T., James, A., Earwaker, L., Mather, C., & Murray, S. (2020). Online discussion boards: Improving practice and student engagement by Harnessing facilitator perceptions. *Journal of University Teaching & Learning Practice*, *17*(3), 86–100. doi:10.53761/1.17.3.7

Figueroa, I. (2014). The value of connectedness in inclusive teaching. *New Directions for Teaching and Learning*, *2014*(140), 45–49. doi:10.1002/tl.20112

Garcia-O'Neill, E. (2021, May 12). *Social presence in online learning: 7 things instructional designers can do to improve it*. eLearning Industry. Retrieved November 13, 2022, from https://elearningindustry.com/social-presence-in-online-learning-7-things-instructional-designers-can-improve

Grant, K., & Lee, V. (2014). Teacher educators wrestling with issues of diversity in online courses. *Qualitative Report*. Advance online publication. doi:10.46743/2160-3715/2014.1275

Hauser, L., & Darrow, R. (2013). Cultivating a Doctoral Community of Inquiry and Practice: Designing and Facilitating Discussion Board Online Learning Communities. *Education Leadership Review, 14*(3).

Honig, C. A., & Salmon, D. (2021). Learner presence matters: A learner-centered exploration into the community of Inquiry Framework. *Online Learning*, *25*(2). Advance online publication. doi:10.24059/olj.v25i2.2237

Jan, S. K., & Vlachopoulos, P. (2018). Influence of learning design of the formation of online communities of learning. *The International Review of Research in Open and Distributed Learning*, *19*(4). Advance online publication. doi:10.19173/irrodl.v19i4.3620

Kebble, P. G. (2017). Assessing online asynchronous communication strategies designed to enhance large student cohort engagement and foster a community of learning. *Journal of Education and Training Studies*, *5*(8), 92. doi:10.11114/jets.v5i8.2539

Kilis, S., & Yıldırım, Z. (2019). Posting patterns of students' social presence, cognitive presence, and teaching presence in online learning. *Online Learning*, *23*(2). Advance online publication. doi:10.24059/olj.v23i2.1460

Nasir, M. K. (2020). The influence of social presence on students' satisfaction toward online course. *Open Praxis*, *12*(4), 485. doi:10.5944/openpraxis.12.4.1141

Orcutt, J. M., & Dringus, L. P. (2017). Beyond being there: Practices that establish presence, engage students and influence intellectual curiosity in a structured online learning environment. *Online Learning*, *21*(3). Advance online publication. doi:10.24059/olj.v21i3.1231

Payne, A. L. (2021). A resource for E-moderators on fostering participatory engagement within discussion boards for online students in Higher Education. *Student Success*, *12*(1), 93–101. doi:10.5204sj.1865

Ruane, R., & Lee, V. (2016). Analysis of Discussion Board Interaction in an online peer-mentoring site. *Online Learning*, *20*(4). Advance online publication. doi:10.24059/olj.v20i4.1052

Sabbott. (2013, August 29). *Synchronous learning definition*. The Glossary of Education Reform. Retrieved July 29, 2022, from https://www.edglossary.org/synchronous-learning/

Silva, L., Shuttlesworth, M., & Ice, P. (2021). Moderating relationships: Non-designer instructor's teaching presence and distance learners' cognitive presence. *Online Learning*, *25*(2). Advance online publication. doi:10.24059/olj.v25i2.2222

Simkovich, J. (2022, March 1). *Collaborative learning strategies for professors in 2022*. Ment. io. Retrieved July 30, 2022, from https://www.ment.io/collaborative-learning-strategies-for-professors/

What is asynchronous learning? (n.d.). Coursera. Retrieved July 29, 2022, from https://www.coursera.org/articles/what-is-asynchronous-learning

Williams, S. A. S., Hanssen, D. V., Rinke, C. R., & Kinlaw, C. R. (2019). Promoting race pedagogy in Higher Education: Creating an inclusive community. *Journal of Educational & Psychological Consultation*, *30*(3), 369–393. doi:10.1080/10474412.2019.1669451

Chapter 10
Thirty Days in 2020 and How They Future-Proofed a University

Michael Graham
National Louis University, USA

Bettyjo Bouchey
National Louis University, USA

ABSTRACT

The COVID-19 pandemic was a catalyst for change in myriad ways. While nearly all the societal changes were negative, the crisis also brought about positive adaptations in response to this crisis. National Louis University's (NLU's) initial response to the crisis, beginning in February 2020 and through subsequent decisive actions, kept the university vibrant throughout the pandemic. Perhaps more importantly, these actions created a foundation for long-term institutional success. The steps included technological enhancements and intentional activities focused on ensuring that institutional personnel would remain connected to each other and to the students that NLU serves. Many of these changes are now a permanent part of the institution as the world moves past the COVID-19 era. Utilizing NLU as an exemplar, recommendations will be given designed to help college and university leadership evolve in this new hybrid reality. Ultimately, the institutions that can effectively adapt to this hybrid existence will be the ones most capable of continued success in an uncertain future.

BACKGROUND

National Louis University (NLU), an urban, broad-access, Hispanic-serving institution in Chicago, Illinois, has for many years had a number of remote campuses in different states. In 2014, the decision was made to embrace mobility for its faculty and staff in order to provide equitable coverage at campuses beyond the main building in downtown Chicago. All full-time faculty and staff were provisioned with laptops, all campus phones could freely place calls in any state, and critical technology for business functions (e.g.,

DOI: 10.4018/978-1-6684-5190-8.ch010

the student information system (SIS) and the learning management system (LMS)) could be accessed from anywhere. In parallel, a decision was made to provide transparency to students by requiring every course to be housed within the LMS and include at least syllabi information, a gradebook, and key assignment submission areas. Further balancing the needs of a disparate faculty and staff population, NLU became an early adopter of Zoom in 2015 to provide a rich web-conference meeting format, no matter someone's location. In part to manage these critical infrastructure investments, the institution hired a chief information officer (CIO) who had experience with managing technology infrastructure at a large higher education institution and with business continuity planning and implementation.

Around the same time and through 2019, NLU also prioritized investment in online education infrastructure. NLU successfully converted over 70% (50) of its traditionally face-to-face programs into the online environment between 2015 and 2019, a feat accompanied by key hires in instructional design, the formal adoption of Quality Matters to inform top-notch course design, a ramp-up of online teaching professional development, and the early formalization of a role focused on online learning. This person would become dean of online education in 2021.

Fortuitously, NLU had data-driven decision-making frameworks it had developed in response to the university president's earlier decision to establish a business intelligence unit. This unit used different sets of software and services to help the institution analyze all these new data points and create actionable reports and dashboards. This was done first with a focus on enrollment metrics and later to mine student performance data from the learning management system and other essential software and systems utilized across the university.

DAYS 0 TO 30 OF FULLY REMOTE

While the institution was clearly positioned for the future of work, it had not yet focused any effort on how all this technological infrastructure would enhance or detract from the culture of the institution nor on how it would impact working relationships and university operations overall. Little did NLU know that these steps would position them for the most unprecedented disruption in most people's lifetimes—the COVID-19 global pandemic that reached the United States in early 2020. By mid-February 2020, the first suspected case of the virus came to NLU with an international student hailing from China. With little guidance offered by the Chicago Department of Public Health or the Centers for Disease Control to colleges and universities at that time, a crisis team was formed of executive leaders in the institution (the Emergency Response Team or ERT), co-led by a recent hire, the vice president (VP) of student affairs, who was formally trained in crisis management for higher education.

By March 3, NLU was evaluating its mobile working infrastructure for its ability to extend it to the entire campus community, perhaps for an extended period of time given the dialog around stay-at-home orders and possible shutdowns. What ensued was not unlike what other institutions faced at the time, though NLU had a very solid foundation of mobile working and online learning to leverage. Fortunately, key systems such as the student information system, learning management system, and the institution's web conferencing platform were all cloud-based services that would scale to support all faculty, staff, and students accessing them at a distance. It was quickly identified, however, that telephony capabilities would need to be modified quickly to be deployed through voice-over-internet protocol (VOIP). This was completed in record-time, by March 10, by the agile information technology (IT) team.

On March 13, the ERT and the president's cabinet met to confirm the need to shut down on March 20 in response to sweeping viral contagion in Chicago and around the world. By March 16, faculty and staff were testing out their new home office working environments and working with IT to troubleshoot access and needs to ensure there would be little to no disruption of university operations. At the same time, a group of academic stakeholders were ramping up efforts to support remote teaching and learning to ensure students did not encounter a disruption in their learning. By March 18, faculty had been rapidly trained in synchronous online teaching via Zoom, and a set of communications had been created for students to help them with their expected seamless transition to learning from home (nearly 30% of the student population at NLU was already fully online and would experience no interruption in their learning). NLU's Continue Working (for employees), Continue Teaching (for faculty), and Continue Learning (for students) websites were rapidly developed and deployed in the same week to provide a one-stop-shop place of information and prevent miscommunication.

That same week, the CIO and a group of key leaders across the institution who were focused on this massive shift to remote working, teaching, and learning launched the University Online Operations Group (UOG). The UOG had broad participation from every administrative, student-facing/serving, and academic unit across the institution. At its peak, nearly 50 individuals attended the UOG meetings. Its original purpose was to ensure that there was a specific place that brought everyone together to help resolve challenges and issues with remote operations. It was also a place where questions from the university could be collected and relayed to the ERT. The CIO and VP of student affairs were in both groups; they acted as the conduit between them to relay and respond to information. This group also eventually became a connection and morale booster that kept everyone focused on supporting students and on their own mental well-being though activities that brought levity to an incredibly challenging situation (e.g., Zoom background challenges, a daily joke, sharing good news, and simply having a few moments to vent). It also increased operational efficiency in solving challenges because all stakeholders were in the same meeting for ease of resolution and action planning on the fly. For example, this group devised calling strategies to check on students' well-being and to help support enrollment and retention initiatives for the university, and they also identified a need to survey students to measure their satisfaction with their new teaching and learning environments. In the past, both initiatives may have taken weeks or months to come to consensus on, but in this way of being, both were decided upon in one meeting and deployed the following week.

By March 23, an on-demand, video-based support system was launched to help students and faculty navigate their new teaching and learning environment (the NLU Concierge), and the 24x7 Helpdesk was bolstered to handle the additional load. Real-time training for faculty who wanted to add information to their course in the LMS and those faced with instructional challenges could also schedule consultations with the best online faculty across the institution. They had volunteered to be available to their colleagues 1 hour per week for troubleshooting and thought partnership.

While NLU weathered the same storm that all institutions were navigating at the time, the resulting knowledge and embrace of remote working culture, as well as flexible working and learning environments that were refined and improved throughout the pandemic, gave way to a renaissance in institutional culture. Armed with the foresight and experience gained from the pandemic, NLU embodied the best attributes of a hybrid university (Selingo et al., 2021) to buffer against new, temporary shutdowns related to COVID-19 variants. It has just as importantly built culture and community in a multimodal way (i.e., in person and at a distance) to further strengthen student support and enable the institution to attract and retain the best talent, not all of whom live in Chicago. Moreover, the institution has, almost by accident, future proofed the institution through its newfound skill at managing across and through physical boundaries into cyberspace. It has built a culture that is diverse, engaged, and uniformly focused on the student experience.

This chapter details actionable and realistic recommendations to institutions that are interested in cultivating thoughtful hybrid operations that leverage the best of in-person and remote working and build new pathways of engagement with students, faculty, and staff. A hybrid operation with a healthy culture can and should draw upon tenets of a customized culture (Grant & Judy, 2017), where an institution can take steps to (a) truly understand the existing culture and, in doing so, identifying where change might be needed; (b) develop a plan to align culture with student success; and (c) empower university stakeholders to become durable stewards of the new culture. This chapter will discuss key attributes of how NLU developed their bustling work culture that specifically relate to how to align to a student success culture and how to empower stakeholders to caretake the new culture into the future, including:

- leveraging and ramping up technological infrastructure to support a hybrid university with action orientation and a focus on student success;
- the formation of teams infused with an agile methodology mindset to self-manage and quickly resolve challenges;
- investment in student support services meant to engage students in a multimodal, highly responsive way;
- intentional strategies used to cultivate and nurture relationships and community with a deliberate focus on human connections, whether in-person or at a distance; and
- decisive, authentic leadership dedicated to the morale and support of all university stakeholders.

Organizations that are truly interested in modern-day risk management should be investing more, not retreating from technologies and ways of being that support a hybrid university. Those intentional strategies must be put in place to ensure all organizational stakeholders embrace technology and find ways to establish connection and meaning, in person and at a distance. This is a strong call to action for universities and colleges (as well as private industry) across the United States to embrace what is truly meant by those who reference "the new normal," which does not involve myopic retreats into the comfort of traditional, residential operations. The new normal is here. It is a sustainable, flexible system of technologies and ways of being that will create newer levels of engagement and satisfaction among organizational stakeholders, offer competitive advantage, mitigate risk, and provide the best of both worlds (in person and at a distance).

CURRENT STATE

Institutions across the world are struggling with establishing their new normal after the pandemic. Thus, it is not surprising that a growing body of literature looks at the challenges of developing an effective hybrid workplace. The literature identifies central themes: leadership matters now more than ever, deliberate culture creation and cultivation is critical, and reinstating connections between the workplace and organization's community is central to supporting a new hybrid work environment (Babapour Chafi et al., 2022; Bednar, 2021; Brower, 2021; Cohen, 2021; Evans, 2022; Fayard et al., 2021; Gratton, 2021; O. C. Tanner, 2022; Rigby, 2021; Selingo et al., 2021). Still, there is no all-encompassing solution to the issue of developing culture in the postpandemic workplace. In many ways, the return to campus by institutions of higher education presents a more formidable challenge to the institutional culture than the pandemic itself. COVID-19 was a crisis that forced institutions to move at a significantly faster pace, increased deliberate communication, increased stakeholder engagement, and (because they were worried about what the future held, both personally and professionally) allowed leaders to shepherd others through change. It resulted in ubiquitous work-from-home realities; a new blending of personal and professional lives; and, as the United States reached a postpandemic phase, the need to attend to a hybrid work environment that toggled work-from-home and on-campus working hours. With the new challenge of creating culture and reinventing work processes to span both modes of working, the hybrid university represents the best of online and on-campus experiences. But, leaders have little experience with managing this new reality—they have no blueprint. It is time to think anew about how a professional culture can exist in a postpandemic university that spans geography, time zones, and trauma.

The postpandemic reality for institutions has begun in earnest. Although many have weathered it completely, albeit with short shutdowns or the reinstatement of mask-mandates, leaders fear the so-called great resignation and are reimagining their cultures. An approach that looks at lessons learned is critical. In many cases, the ways of being during the pandemic may have shown the strength or weakness of the culture as well. Many institutions unintentionally built strength in agility and facility during that time—artifacts of culture that should be executed in the future.

As Schein (2004, p. 107) indicated:

In their responses to such crises what some of their deeper assumptions really were. In this sense an important piece of an organization's culture can be genuinely latent. No one really knows what response it [organization] will make to a severe crisis, yet the nature of that response will reflect deep elements of the culture.

Because the elements of culture may be good, bad, or even neutral, there are decisions that leadership can make now to reinstitute a culture that incorporates the best of the pandemic operations. Moreover, leaders should be deliberate and decide what is wanted and needed in this new postpandemic culture. It is critical to understand whether the existing business model can be sustained in a largely hybrid environment or whether it depends on in-person operations (few operations actually do). From a student-first perspective, if they are the product of the on-demand culture promulgated by industry disrupters such as Amazon and Netflix, then it is also important to understand the concept of experience liquidity (Thayer, 2021) and how it might impact which services are brought back to a solely on-campus offering and which ones are offered multimodally over the long term. Moreover, leaders should reflect upon the actions taken during the pandemic with a keen eye to the entrepreneurial actions taken by their team that can be recognized and formalized in this postpandemic way of being. As NLU navigated its return to campus, leaders were reminded of the key attributes of their existing culture that the university was able to capitalize upon to support stakeholders during the shutdown:

- Leveraging and ramping up technological infrastructure to support a hybrid university with an action orientation. Today, NLU boasts a robust technological environment that supports over 11,000 university stakeholders (students, faculty, and staff) whether they are working on campus or at home. Recent technological upgrades have been added to ensure that any future disruption can be handled without interruption to teaching, learning, or working. While the university had maintained currency in networking technology, it decided to upgrade the network backbone, increase the purchase of advanced network security, and accelerate its transfer of systems from on-premise to the cloud to maximize the advantages of this technology. Although NLU as a multicampus environment had invested in Zoom videoconferencing capabilities in 2015 to augment communication between campuses, the pandemic exacerbated the need for conferencing capabilities. In order to ensure the best possible human connection through the pandemic, the university tripled its investment in Zoom licensing to ensure that the ability to talk to one another was a ubiquitous part of every faculty and staff member's daily work experience, no matter their location. Additionally, in preparation for a hybrid academic experience as COVID-19 continues to impact faculty and student attendance on campus, the university made a large investment in classroom technology. This provides enhanced audio and video in the classroom, which maximizes the sense of community for on-campus and at-home students, faculty, and staff.

- The formation of agile teams infused with a mindset to self-manage and quickly resolve challenges. NLU had always had an entrepreneurial culture where individual teams often had latitude, within regulatory confines, to create processes and procedures that suited the function and its stakeholders. Now, functional areas have a new appreciation of their interdependencies. As a result, NLU has embarked on a service blueprinting project (Shostack, 1984) to map, evaluate, and improve student service across the institution. Service blueprinting may help differentiate the institution in a more hybrid future and remove barriers to student entry and success. This work will be an ongoing and a deliberate part of service improvement for the institution, regardless of how its hybrid identity evolves.
- Investment in student support services meant to engage students multimodally and in highly responsive ways. Armed with its previous strengths in online student support and the experiences during the shutdowns of 2020 and 2021, NLU created a strategic plan initiative to perform a full-scale evaluation of student support for online students to ensure equity and parity among on-campus student offerings, and respond to all students' needs. The dean of online education and the dean of students cosponsored a peer- and self-evaluation of each service through a combined assessment that utilized the Online Learning Consortium's (OLC) Online Student Support Scorecard (OLC, n.d.), the Quality Matters (QM) Learner Success rubric (QM, n.d.), and best-practices gleaned from their own experiences and those of peers. The scores were combined and used to generate a student survey that informed a year-long plan of improvement. Its goal was near to full multimodality of all student support services by the end of the 2022–2023 academic year.
- Intentional strategies to cultivate and nurture relationships and community that had a deliberate focus on human connections, whether in person or at a distance. Because NLU had geographically spread as part of its growth and had so much strength in online programming, it inherently had ways to build enduring relationships. Even so, the institution made special efforts to honor the needs of its faculty, staff, and students as the campus reopened to be sure that comfort levels were high and that flexibility was still integral to scheduling for employees and students. Most campus events intended for faculty and staff are now multimodal and as more and more events return to campus, efforts are made to ensure that those staying at a distance are as engaged as those gathering in person.
- Decisive, authentic leadership based on the morale and support of all university stakeholders. The president's cabinet and the academic cabinet regularly discuss how faculty, staff, and students are feeling as the return to campus continues. A reason for that is to keep in mind the balance between the needs of the stakeholders and the goals of the institution. The tension between the two is managed through open, honest dialog where all voices can be heard. For example, the president hosts a

monthly town hall where she takes questions and concerns directly from the audience and answers them with honesty and authenticity.

Each attribute was either intentionally or unintentionally woven into NLU's way of being after the pandemic, though perhaps with a different lens or level of urgency. It cannot be overstated that culture is built through a series of emergent strategies that culminate in a substantive impact.

CONSIDERATIONS FOR EFFECTIVE COMMUNITY ENGAGEMENT

Creating and cultivating an intentional organizational structure not only takes time, but it also touches each precious resource of the organization, physical and human. With intention and reflection, leaders of institutions of higher education can bring the best of this crisis into a reimagined university that is hyper-focused on its students, faculty, and staff. The customized culture of a hybrid university ultimately is one that is doggedly focused on student success and where each stakeholder is emboldened to hold fast to the culture.

Technological Infrastructure Meant for Action and Student Support

While many institutions were able to ramp up their technological infrastructure in response to the pandemic, some relied on emergency funding to procure these new products and services. With this type of funding likely coming to a close (Office of Postsecondary Education, 2021), it is critical for IT departments to critically evaluate the use and efficacy of not only systems brought online during the pandemic but also legacy systems that were implemented for situations that may no longer exist. If an institution is ready to fully commit to a student-first mentality, every system, software application, and process should be evaluated through the lens of student service and support so financial and human resources can be directed toward those that are strictly doing what is needed.

As a means of continued focus on student success, NLU was quick to augment its existing data reporting on student engagement (see Figure 1) to ensure engagement in the classroom. It also took an even more proactive stance by reaching out to students who were not engaging and found the need to further formalize data usage across the university after the pandemic. NLU no longer took for granted that a student who was out would be coming back, so a more unified and assertive system of student outreach was needed. Existing models of intrusive advising used strictly for online students previously (Earl, 1988; Heisserer & Parette, 2002; Upcraft & Kramer, 1995) have been more universally instituted, though with modifications for scale, to connect with students who have not attended class regularly, have fallen behind, or have exhibited other at-risk academic behavior, regardless of their modality.

In addition, student onboarding has been reevaluated and a dashboard has been created through data mined from the customer relationship management (CRM) system. The dashboard charts the student journey from application to registration to completion of an improved version of student orientation and eventual attendance in their first class (see Figure 2).

Institutional leaders should evaluate their existing technology and data infrastructure to assess their needs so they can move from a data-informed organization to one that is data driven: one where data mined from systems across the university are used to chart individualized action meant to improve student service or success.

Figure 1. Student Engagement Dashboard Snapshot

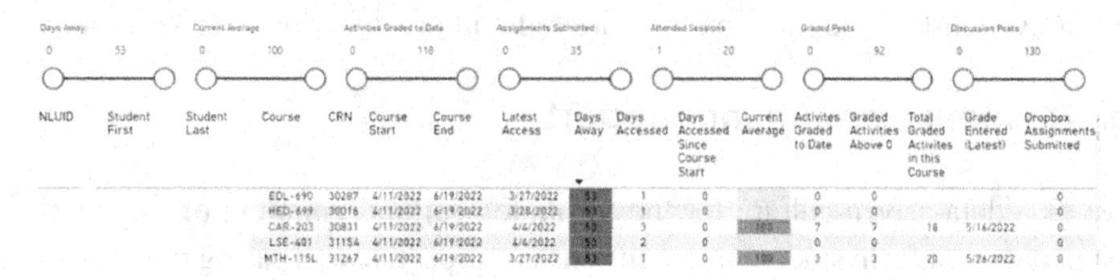

Institutional leaders should also evaluate their students' ability to effectively engage with the institution through technology. For example, it may be necessary to continue to provide internet hotspots and laptops to students to ensure they can access services at a distance. This is also a way to close digital divides among underserved communities. The hybrid university not only ensures that its services can be accessed in multiple means and modes but also that its stakeholders can access them. This holistic approach helps the institution fulfill its mission and extends an ethos of care to its stakeholders, who know that hotspots and laptops provided by the university can also be used in other aspects of the recipients' lives.

Agile Teams Focused on Rapid Problem Solving

Many institutions increased their agility and responsiveness during the pandemic (Garrett et al., 2020), but questions remain about whether that sustained energy and action orientation can be maintained. Borrowing from lean methodology (Lean Enterprise Institute, n.d.) and the tenets of agile ways of being (Sackolick, 2020), institutions of higher education should be actively looking for ways to shed their historical hierarchical structures and adopt a flatter, self-managing organization that leverages initiative and the intrinsic motivation of faculty and staff.

While this may sound easier said than done, actions can be taken to encourage teams to look back on how they operated during the pandemic and reimagine their team culture and processes so they are more flexible and responsive. For example, a series of team retreats to reflect on pandemic operations and participation in postmortems or lessons learned sessions (Project Management Institute, 2021) could glean (a) what went right during those times, (b) what went wrong, and (c) what should be changed to leverage the benefits and implement lessons learned. From here, an action plan for change can be developed to document the operation, as well as a timeline to implement lasting changes that incorporate the vestiges from the pandemic that served as a platform for rapid problem solving and self-management. These active teams should embrace the mission of the university and support it through future-proofing their operation to weather any storm. Essentially, the focus for university teams should now be to maintain

Figure 2. Student Onboarding Dashboard Snapshot

554
Registered, Orientation Not Completed

Registered in Past 3 Days...	Registered 3-29 Days Ago...	Registered 30-89 Days Ago...	Registered 90+ Days Ago...
4	66	362	122

% of Open Applications Registered, with Orientation not Completed
20.0%
55.6%
0.0%
100.0%

Roster View

a proactive stance in university operations as well as student success, whereas before it may have only been the latter, or even just on the function they offered without attention to student needs.

Multimodal, Responsive Student Support

The imperative of a data-driven culture for the technological underpinning of a university cannot be overstated. Using data to inform student interventions is a proactive and caring piece of student service, and it is inextricably tied to multimodality as well. Gone are the days of offering certain services only on campus to residential students. The new reality is experience liquidity (Thayer, 2021): now that students (but also faculty and staff) have experienced their learning and services in different modes, they may expect this flexibility to continue. Drawing on lessons learned from industry disrupters such as Amazon and Netflix, this new on-demand culture and the idea that one can obtain nearly any service in the mode one prefers at the time one needs it should be integrated into university operations to ensure that most, if not all, student services can be universally accessed in person and at a distance. Leaders who take the time to reflect on their pandemic operations can easily visualize how key functions such as onboarding, financial aid, learning support, and career development were offered to students during that time. Although those solutions may not have been elegant, a recommitment to continuing and improving these offerings demonstrates an understanding of not only the contemporary student but also of equity of service between residential and online students. Simply ensuring that web conferencing is a part of each offering is a first step, though recent advances in artificial intelligence and machine learning allow chat features, asynchronous services, and residential services such as housing to be paired with technology to offer more modes to all students. Relatedly, ensuring that hours of operation continue to include nights and weekends can be paired with multimodal student support to meet the possible need for more flexibility in staff schedules.

Key services can be evaluated for their veracity and efficacy through frameworks from trusted organizations, such as the OLC Online Student Support Scorecard (OLC, n.d.) and the QM Learning Success rubric (QM, n.d.). These frameworks can be used in concert with sets of team-based lessons learned to identify areas of student support that should continue to be multimodal and areas that may need to be bolstered in the future. Again, this relentless pursuit of student service excellence can be seen not only through the lens of which services to offer but how those services are offered. Although the investment to offer all student services in a multimodal way may seem daunting or cost prohibitive, the trend toward student consumerism when selecting a university might direct students from one's institution to another if high-quality supports are not made available.

Strong Relationships Across Modes

As institutions build a new culture for the future, relationships across modes present a new challenge. The culture of an organization is delicate and durable. Fears and concerns about returning to campus are a reality that all leaders must face with grace and care across in-person and at-a-distance interactions. Deliberate work behaviors should be forged in all interactions to establish, reinforce, and cultivate deep relationships. For example:

- Being responsive to outreach, no matter the channel. Emails, phone calls, texts, and chats should be answered within a reasonable timeframe that the organization establishes and holds people accountable to. Communication protocols can be documented by their use (e.g., when to call versus use internal messaging) and how to engage someone in the communication (e.g., when using internal messaging or phone, asking the recipient whether they are available to talk).

- Intentional use of salutations and stakeholder names in communication, whether in writing or verbal, is an essential cultural artifact that demonstrates an ethos of care. Ensuring that pronunciation is accurate and preferred pronouns and names are used can also telegraph a culture that values each individual.
- Taking the time to acknowledge meeting attendees, both in person and at a distance, and making space for interpersonal chat at the beginning or end of a meeting is fun and, more importantly, a way to cultivate belonging and shared fellowship among colleagues.
- Establishing multiple means of engaging university stakeholders through multimodal town halls, question-and-answer sessions, asynchronous comment boxes, and small-group meetups demonstrates the institution's willingness to meet someone where they are and in the mode they are most comfortable with.
- Ensuring that all voices are heard through compassionate, progressive elevation of quieter colleagues (e.g., mentioning their name in one's remarks) and creating forums that ask each person to contribute in specific and meaningful ways is a way to create community. This also ensures that multiple and important perspectives are brought into each conversation.
- Equitably engaging the in-person and virtual audiences in a meeting setting is critical. The first step is being sure to make eye contact with those in the physical room and those on the web-conferencing system. This is closely followed by ensuring that all questions and activities associated with the meeting have both a physical and a virtual engagement strategy associated with them.
- Reimagining the hallway connections that often serve as quick problem-solving sessions in-person. Using emoticons in written communication, chat, and messaging and pinning the video of speakers to read their body language are key ways to look for subtle cues in a digital space that are more easily read in-person.

Decisive, Authentic Leadership

Once these aspects of culture are aligned around student success, brave and authentic leaders should reinforce a culture that holds each stakeholder accountable to the new culture and empowers each individual to be the best versions of themselves. Embracing the hybrid nature of the university and its commitment to student service excellence should not come at the cost of faculty and staff morale—ideally, morale would be improved through this renewed energy and vigor. The feats accomplished during the pandemic should be acknowledged and honored, and institutional leadership should be willing to reevaluate legacy systems of management for alignment with the new reality. For example, remote work should be embraced as not only a means of faculty and staff satisfaction but also as a way of combating the so-called great resignation and opening up new opportunities to recruit talent in new regions (that may even have lower costs of living). Sick time should be reevaluated and perhaps collapsed into a single category of paid time off to provide more flexibility for faculty and staff and to telegraph trust that each stakeholder is working toward a unified student success culture, which involves taking care of oneself too.

In being deliberate, leaders should also look at both the positive aspects and the challenges of working at home and being in the office. In order to effectively develop the hybrid culture that many institutions are working toward, a variety of issues should be addressed by the institution, namely: (a) actively living the culture, (b) evaluating the quality and meaning of work on campus and at a distance, (c) intentionally telegraphing and cultivating connection across modes, (d) focusing on well-being, and (e) ensuring equity across modes of the institution.

Actively Living the Culture

Prior to the pandemic, most institutions had probably discussed or said "living the culture," but it often lacked a real focus because everyone was on campus and an inherent connection existed through proximity. Now, to bridge the gap between working on campus and at home, leaders should explicitly highlight how employees can live the culture in any modality. A key aspect of that is ensuring that student success is achieved no matter their modality or location.

Evaluating Quality and Meaning.

Leaders should review the work that is performed in each department. They should analyze what is gained and what is lost by moving some employees to fully remote (at a distance) and keeping some on campus or by developing schedules where employees are on campus some days and remote on others. In essence, the work product of each department should be evaluated to indicate whether its quality can be maintained no matter the mode of delivery or, if quality is compromised, how that aspect of the work product can be optimized in its needed mode. For example, some functions are difficult to complete at a distance, while for others the mode does not make a significant difference in quality or the effort needed to perform them.

Institutional leaders must also critically evaluate the meaningfulness of on-campus work time, asking themselves whether on-campus work has substantive value. This is particularly important for employees who have been able to be successful for 2 years working completely remote. If there are specific needs to be in the office, such as work performed by student-facing departments, that is self-evident; but if there is no reason for an employee to return to the office other than to be able to be watched by their supervisor, returning to work can reduce morale. Institutions must think anew about what happens when employees come to campus. Experiences such as team-building exercises, all-organization gatherings, or development of other cultural connections within teams and the institution should be the focus of requiring remote employees to come to campus, not an affinity for proximal working environments.

Telegraphing and Cultivating Connection

Institutional leaders who hope to embrace the hybrid future and create enduring community engagement must ensure that all faculty and staff understand the culture they are part of. Intentionally telegraphing and cultivating connection can be accomplished through shared language and habits and through celebration. The message needs to be reinforced regularly by all members of the institution. It should be expressed at large and small meetings and gatherings so all members feel connected to the culture, its focus on student success, and the team spirit of moving in the same direction. Special care should be given in selecting new leaders who have the capacity to engage teams both remotely and in person; existing leaders should be scaffolded if they exhibit gaps in this area as well. Moreover, leaders should make sure that all faculty and staff understand how the institution is doing in terms of achieving its goals through its culture, not just through the activity of individual contributors.

Focusing on Well-Being

Leaders should increase efforts to focus on the well-being of faculty and staff so they feel valued and recognized for their contributions. They should also ensure all institutional stakeholders, including students, have a voice in its future. As the institution moves into this new hybrid environment, it should review its mental, social, and physical well-being policies and related coverages and provide an equitable set of services meant to address the whole person, not just their work selves. The institution should also offer proactive wellness programs, such as fitness and meditation challenges, that promote connection to other employees and provide support regardless of modality and location

Leveling the Playing Field

Perhaps one of the most deliberate steps that institutions should take is to ensure that there is equitable treatment of faculty, staff, and students, regardless of their modality, in all policies and practices. This may seem elementary, but it may be more nuanced than one might think. For example, one way that staff may get promoted in an organization is because of their willingness to be involved in a variety of on-campus work projects. Employees who are on campus and capable of developing strong interpersonal relationships with superiors and leaders throughout the institution may be seen as more capable than coworkers who are primarily remote. Care should be taken to ensure that all employees are evaluated equitably based on metrics developed from work that is performed toward quantifiable standards that are largely mode-agnostic. Performance and promotion standards should only be based on actual work product and should be strictly guarded against subjective interference from affinities developed through physical proximity. While this is difficult, simply stating this policy in promotion policies is an effective way to bring it into the consciousness of leaders tasked with promotion decisions. Leaders should hold themselves accountable for consistently questioning their adherence to the difficult practice of measuring employee performance strictly on the work product and not on factors related to personal relationships facilitated by employees who are on campus.

Orientation may become even more critical in this new environment, where new employees do not benefit from the direct observation of routines and norms that can be modeled by existing employees who are in the office daily. Particular focus should be paid to highlighting the culture of the institution, the expectations for new employees as a part of that culture, ways that new employees can be successful, and resources to aid employees in their journey to becoming an active part of the institution.

Furthermore, new cultural norms for hybrid work should be shared with all employees. As faculty and staff are afforded more flexibility in their work environments, remote work should be treated as seriously as on campus work so they are working in a largely distraction-free, professional location. Policies should be developed for home-office requirements and work time, as needed, to ensure expectations are clear for all.

CONCLUSION

While several factors are involved in what and how to move a university into a hybrid environment with high community engagement and an invigorated culture, the following questions are designed to stimulate thinking by institutional leaders making key decisions during their journey toward creating this new future. While these questions are not exhaustive, they should aid in helping leaders analyze whether departments within the institution are able to sustain a hybrid culture over time. Questions include:

- Is the unit aligned to the new institutional culture?
- Is the department student-facing, and is there a large population of students on campus requiring assistance in person?
- Is the employee base capable of providing coverage in multiple modalities?
- Was leadership able to maintain a connection to all campus constituencies during the pandemic?
- Is the institution experiencing high staff turnover?
- Has the institution managed to maintain high student persistence?
- Is there technology available for students and employees to connect to the university ubiquitously?
- Has the institution been able to provide the same level of service throughout the pandemic?
- What type of work is performed (office/clerical, academic, strategic/creative)?
- What level of institutional collaboration is required for employees of the unit to be successful?
- Are there any other factors affecting the institutional environment that might prevent a more hybrid environment?

These questions can be used as a means of reflection that can guide possible action around how work is structured, which includes the possibility of a more hybrid approach, overall.

Culture is no longer a concept that one can assume is transferred from employee to employee or one that is simply maintained naturally. Instead, institutional leaders should take great care to be deliberate in developing ways to ensure that the institution's culture is imagined anew as it is faced with a new, hybrid future and that this new culture is collectively created, shared, and well-understood. Only through such intentional effort can the hybrid university of the future truly create a near obsession with student service, like service leaders in other sectors such as Hilton Hotels and Disney, and evolve through this crucible of change and disruption that the pandemic brought to the world.

Positive cultural habits and ways of being that were created and operationalized during the pandemic should continue to be reinforced to meet the demands of the contemporary student and those of faculty and staff adjusting to their new realities. Leadership at all levels matters now more than before because personal connections across modes (on campus and at a distance) require more effort and a conscious shift in behavior. Designing a culture centered around student service in which all contributions are valued, no matter where they originate (on campus or otherwise), is the truest manifestation of equity and a way to orient oneself to the mission of the institution, rather than traditional ways of being.

As institutions imagine their new futures and walk through the deliberate steps necessary to ensure it is a positive one, constant attention to recalibration should be kept in mind as well. Armed with the knowledge that disruption can take the most unimaginable shapes and forms, an institutional leader can future proof their college or university through an impeccable focus on student, faculty, and staff service and an eye toward embracing change as it arrives (and perhaps even before that). The bravest leaders are not breathing a sigh of relief as they reach a postpandemic moment: They are taking a collective breath and then learning everything they can from it to ensure sustainability.

REFERENCES

Babapour Chafi, M., Hultberg, A., & Bozic Yams, N. (2022). Post-pandemic office work: Perceived challenges and opportunities for a sustainable work environment. *Sustainability*, *14*(1), 294. doi:10.3390u14010294

Bednar, J. (2021). Best of both worlds: Hybrid work schedules are fast becoming the new norm. *BusinessWest*, *38*(6), 25–28.

Brower, T. (2021, February 7). *How to sustain company culture in a hybrid work model.* Forbes. https://www.forbes.com/sites/tracybrower/2021/02/07/how-to-sustain-company-culture-in-a-hybrid-work-model/?sh=565f883610
09

Cohen, A. (2021, December 30). *What really happens when workers are given a flexible hybrid schedule?* Bloomberg. https://www.bloomberg.com/news/articles/2021-12-30/the-flexible-hybrid-work-schedule-that-employees-actually-want

Earl, W. R. (1988). Intrusive advising of freshmen in academic difficulty. *NACADA Journal*, *8*(2), 27–33. doi:10.12930/0271-9517-8.2.27

Evans, E. (2022). Cracking the hybrid work culture conundrum: How to create a strong culture across a workforce you may never even see. *Strategic HR Review*, *21*(2), 46–49. doi:10.1108/SHR-12-2021-0065

Fayard, A. L., Weeks, J., Khan, M., & Hines, F. (2021). Designing the hybrid office: From workplace to "culture space." *Harvard Business Review*, *99*(2), 114–123.

Garrett, R., Legon, R., Fredericksen, E. E., & Simunich, B. (2020). *CHLOE 5: The pivot to remote teaching in spring 2020 and its impact.* Quality Matters. https://www.qualitymatters.org/qa-resources/resource-center/articles-resources/CHLOE-5-report-2020

Grant, M., & Judy, C. (2017, February 15). *Viewpoint: 3 steps to cultivating a customized culture.* Society for Human Resource Management. https://www.shrm.org/resourcesandtools/hr-topics/employee-relations/pages/viewpoint-3-steps-to-cultivating-a-customized-
culture.aspx

Gratton, L. (2021). Four principles to ensure hybrid work is productive work. *MIT Sloan Management Review*, *62*(2), 11A–16A.

Heisserer, D. L., & Parette, P. (2002, March). Advising at-risk students in college and university settings. *College Student Journal*, *36*(1), 69–84.

Lean Enterprise Institute. (n.d.). *What is lean?* https://www.lean.org/explore-lean/what-is-lean/

Office of Postsecondary Education. (2021). *Higher education emergency fund III: Frequently asked questions.* United States Department of Education. https://www2.ed.gov/about/offices/list/ope/arpfaq.pdf

Online Learning Consortium. (n.d.). *OLC quality scorecard: Online student support.* https://onlinelearningconsortium.org/consult/olc-quality-scorecard-student-support/

Project Management Institute. (2021). *A guide to the project management body of knowledge* (7th ed.).

Quality Matters. (n.d.). *QM program review annotated criterion.* https://www.qualitymatters.org/sites/default/files/program-review-docs-pdfs/Annotated-Program-Criteria.pdf

Rigby, A. (2021, August 12). *A manager's guide for creating a hybrid work schedule.* Trello. https://blog.trello.com/creating-a-hybrid-work-schedule

Sackolick, I. (2020, February 25). *What is agile methodology? Modern software development explained.* InfoWorld. https://www.infoworld.com/article/3237508/what-is-agile-methodology-modern-software-development-explained.html

Schein, E. (2004). *Organizational culture and leadership.* John Wiley & Sons.

Selingo, J., Clark, C., Noone, D., & Wittmayer, A. (2021) *The hybrid campus: Three major shifts for the post COVID university.* The Deloitte Center for Higher Education Excellence. https://www2.deloitte.com/content/dam/insights/articles/6756_CGI-Higher-ed-COVID/DI_CGI-Higher-ed-COVID.pdf

Shostack, L. (1984, January). Designing services that deliver. *Harvard Business Review Magazine.* https://hbr.org/1984/01/designing-services-that-deliver

Tanner, O. C. (2021). *Hybrid workplace: The future of work is a combination of workplaces.* https://www.octanner.com/global-culture-report/2022/hybrid-workplace.html

Thayer, B. (2021). *Planning for higher ed's digital-first, hybrid future: A call to action for college and university cabinet leaders.* Education Advisory Board. https://eab.com/research/strategy/whitepaper/plan-digital-first-hybrid-future-higher-ed/

Upcraft, M. L., & Kramer, G. (1995). Intrusive advising as discussed in the first-year academic advising: Patterns in the present, pathways to the future. Academic Advising; Barton College.

Chapter 11
From Loneliness to Belonging Post Pandemic:
How to Create an Engaged Online Community and Beyond

Andrea D. Carter
https://orcid.org/0000-0002-5712-6611
Adler University, USA

Lilya Shienko
Adler University, USA

ABSTRACT

The effects of isolation, loneliness, and ostracism on social engagement and online communities has been acknowledged throughout the pandemic by researchers. This chapter contextualizes and normalizes the behaviours associated with isolation and loneliness and draws attention to belonging methodology. Belonging indicators and practices are revealed and described so as to break the patterns many are struggling with. Finally, the chapter provides context for engagement, collaboration, and creating community post pandemic.

INTRODUCTION

It is too early to know the real impact and trauma the world has endured throughout the global pandemic. While research soars and in-person situational elements collide, the variables of living conditions and social interactions have become delicate and complicated. Social isolation forced people into online platforms disrupting social norms and behaviors to curb the pandemic's spread (Abrams et al., 2021; Hales et al., 2021). Institutions navigated what was deemed acceptable behavior during the immediate crisis, yet our present reality demands navigating online environments in a different capacity. The topography

DOI: 10.4018/978-1-6684-5190-8.ch011

has changed, requiring considerate understanding and a myriad of solutions to encourage social beings to re-engage, collaborate, and renew belongingness.

Volatile economic activity affecting the globe further complicates recovery. It also demonstrates a clear need to understand and support the emotional, psychological, and physiological ramifications of thwarting social animals from social exchanges (Public Health Reports, 2021). We must also acknowledge that before the pandemic social exchanges were different. McGinty et al. (2020) and Luchettie et al. (2020) argue that contrary to 'experts' predictions, loneliness during the early months of the pandemic was not as worrisome because, at that time, people perceived a shared sense of community and everyone was in the situation together (Hales et al., 2021). Research indicates that the perception of togetherness was distinctively different at the end of the pandemic, where many felt ostracized and socially rejected from relational and collective loneliness patterns (Hales et al., 2021). Most of the population was sent home and told to "isolate" to slow the spread. However, distinct groups had different rules and regulations, and inequities and disproportionate burdens were realized as social fragmentations (Abrams et al., 2021).

Additionally, as the lines faded between work and home and school and home, researchers began to report the toll of isolation and loneliness leading to depression, anxiety, and negative behavioral patterning (Kovacs et al.,2021; Shelvin et al., 2020; Entringer & Gosling, 2021; Wilkialis et al., 2021; and Zhou et al., 2021). Moreover, by remaining distant from others, exclusion and the ability to ignore and devalue the importance of others created the perfect storm of negative social experiences. Isolation, ostracism, and loneliness are all negative social experiences that threaten several basic needs: belonging, self-esteem, control, meaningful existence, and certainty (Williams, 2009; Hales & Williams, 2020). Which begets the question, "As the world returns to a new normal, how will humans learn to reconnect with others to lessen the effects of isolation and loneliness?"

This chapter looks to contextualize the effects of isolation and loneliness on human behavior, social engagement, and online communities. Understanding these factors will lead the reader to the key indicators of belonging, including comfort, contribution, connection, psychological safety, and wellbeing. Methodology to understand how to create belonging spaces will be explored and contextualized, leading the reader to create authentic belonging behavior and reimagining collective applications and spaces. Finally, the reader will learn mechanisms to reemerge into a culturally responsive community in remote spaces and beyond.

The Effects of Isolation and Loneliness: Thwarting Human Social Engagement

Humans need to be social, feel part of a group and belong (Besse et al., 2021). This need is profoundly physiological and part of the human evolutionary process that keeps us alive, similar to the biological warning signals akin to hunger and thirst (Cacioppo et al., 2014; Mund et al., 2020). Therefore, as the institutions that structure social lives were disrupted due to the pandemic, modern society did not entirely understand this threat. The pandemic's public health and economic consequences will be felt for years to come; however, to ignite processes and strategies for recovery, we must first understand how trauma has permeated society.

Isolation and loneliness are undesirable human experiences because they lead to feelings of discontent within one's social experience (Hortulanus et al., 2009). Moreover, to be ostracized, even when there are reasonable and abundant external explanations, makes people feel bad. While ostracization and loneliness have similarities, loneliness is the sensation that remains once someone is alone against their desire (Hales et al., 2021). *Loneliness* is an aversive emotional state perceived as an inadequacy in the quality or quantity of one's social relationships (Entringer & Gosling., 2021; Lui et al., 2020; Mund et al., 2020). While loneliness is a subjective condition, it is not solely the experience of being isolated but rather a discrepancy between an individual's preferred and actual social relationships and networks (Besse et al., 2021; Entringer & Gosling, 2021; Kovacs et al., 2021; Peplau & Perlman, 1982). This discrepancy is relevant in the global pandemic because people were ordered to stay home and socially distance themselves from others, ultimately affecting the quality and quantity of relationships and human interaction. Entringer and Gosling (2021) indicate that while intimacy and closeness can be created in different ways, in-person contact is only sometimes necessary for the quality of relationships to be present. However, Kerr (2021) argues that student loneliness was on the rise even before the unprecedented social isolation of COVID-19 (Ewell, 2010; Kerr, 2021). Kerr (2021) reveals that 39% of U.S. high school seniors report loneliness (Twenge et al., 2019), and 49% of college students feel lonely (Kerr, 2021). One-third of all adults in the U.S. are lonely (Kerr, 2021).

Moreover, it warrants an understanding that while loneliness is a human experience occurring at some point in a human being's lifespan, the isolation and ostracization observed during the pandemic were excessive. The most significant cause for concern is the post-pandemic extremism that people are experiencing from prolonged periods of isolation (Ackerman & Peterson, 2020).

When humans cannot interact as they have always known, physiological, psychological, and cognitive repercussions come to pass. Interestingly, Cacioppo and Hawkley (2009) demonstrate through their early research on the Evolutionary Theory of Loneliness (EVT) that when primitive humans found themselves outside social bounds, they became more sensitive to threats, which likely supported survival in primitive times. Today, however, loneliness's self-preservation response brings a hypersensitivity toward perceived social threats (Cacioppo et al., 2016). There now exists an urgency to control and aggressively defend against exclusionary behaviors while at the same time engaging in them (Besse et al., 2021). Association with social media and loneliness also perpetuate the inability to have in-person relationships, demonstrating the imbalances being experienced post-pandemic (Bonsaksen et al., 2021). It ultimately leads to an unconscious mistrust of others or anxiety towards people and social situations. Cacioppo and Patrick (2008) argue that lonely people experience their world as a threatening place producing behaviors that confirm negative expectations and inflame a perpetual cycle of loneliness. Brain imaging supports this notion. Findings demonstrate that lonely participants over non-lonely participants were quicker to select images of social rejection, indicative of social threat, rather than non-social threats, like images of snakes (Cacioppo et al., 2016).

While most research has examined loneliness through the lens of the individual, it is also essential to explore its cultural implications. Specifically, considerations of how lonely individuals are impacted by individualistic versus collectivistic societies.

The Impact of Culture on Loneliness: Individualistic Verses Collectivistic Societies

Oyserman and Lee (2008) argue that culture impacts individuals because different societies have different experiences. Therefore, individualistic versus collectivistic societies provide differing insights into acceptable psychological and societal processes (Kim et al., 1994; Triandis, 1995; Bond et al., 2004). Moreover, culture provides predominant constructs through values, self-concept, relationality, and cognitive processes, which differ significantly between individualistic and collectivistic societies. Markus and Kitayama (1991); Oyserman and Lee (2008); and Triandis (1995) outline that individualism regards the core unit of an individual as the central focus enabling both a hierarchical, idiocentric, independent and separated experience of self.

Alternatively, collectivism regards the core unit as a society existing to promote individuals' well-being, emphasizing equality, allocentrism, interdependency, and relational self-awareness (Markus & Kitayama, 1991; Oyserman & Lee, 2008; Triandis, 1995; Westerhof et al., 2000). The dominating culture will dictate defining elements and factors that influence how individuals think about themselves and others (Hoefstede, 1980; Triandis, 1995; Inglehart, 2000; Inglehart & Baker, 2000; Oyserman & Lee, 2008). Moreover, from a cultural and psychological perspective, the constructs of individualism and collectivism provide fundamental differences in the relationship between an individual and the society in which an individual resides. Within individualistic cultures, an individual must fit in and assimilate into the culture's values, norms, beliefs, and behaviors to be accepted. Conversely, belonging and inclusion are empirically and conceptually provided in collectivistic societies simply for being born into that culture (Blondel & Inoguchi, 2006).

What is essential here is to recognize that people can reside in an individualistic society and have collectivistic values and vice versa. That said, from a social identity perspective, the groups and memberships individuals ascribe to will evoke different values within the contexts with which they identify (Abrams et al., 2021). The values that an individual and a culture subscribe to will manifest in daily interactions and norms, outlining those who belong and those who are deemed out-grouped or ostracized—understanding that 'people's social identities support how they define themselves and how they navigate the social world (Abrams & Hogg, 1988).

Further, cross-national literary studies differ in how the self is defined and how interpersonal relationships are conceptualized and are based on the frameworks of an individualistic versus a collectivistic society and culture. These differences between groups and cultures feed into who is accepted, what differences are judged, and what behaviors influence inclusive or exclusive etiquette (Oyserman & Lee, 2008). Additionally, Markus, Mullally, and Kitayama (1997) argue that these descriptors are always individual and collectivistic; however, culture cannot be separated or held as a constant like an independent variable. "Culture is only realized through individuals, and individuals can lead meaningful lives only through acculturation" (Shweder & Sullivan, 1993; Westerhof et al., 2000).

These cultural elements are essential to understand because when belongingness needs are not sufficiently met, as in the global pandemic, individuals are likely to experience a heightened perception of loneliness (Besse et al., 2021). Understanding the complex interconnected challenges involves a dedication to work towards a collaborative and complimentary initiative inclusive of different social identities. It is, therefore, essential to explore conditions that are helpful to generate harmony, empathy, and cohesion rather than rivalry that leads to shaming, naming, and blaming (Abrams et al., 2021).

What is Belonging?

Scholars have long used a broad range of related constructs to explain the meaning of belonging in a theoretical framework. There is no doubt that Alfred Adler's (1931) contributions in relation to belonging theory were less well known than Abraham Maslow's (1968) contributions, but Adler's (1931) contributions are unique and significant. Adler articulated that fundamental human motivation is based on the human need to belong (Ferguson, 1989). Early in Adler's research, he argued that for people to feel belonging, they must perceive both equality and the ability to contribute to the community (Ferguson, 2010). Interestingly, Ferguson (2010) points out that Adler also indicated that children who do not feel belonging strive to be extraordinary rather than contribute. They mistakenly believe that being special will bring a feeling of belonging. These "mistaken goals" are readily evident in children and adults (Ferguson, 2010). Interestingly, post-pandemic, these indicators of not belonging are significant and manifest through aggressive tendencies that are now on the rise (Krings et al., 2021).

However, let us look back to belongingness foundations. Adler's baseline of belonging is connected with Maslow's Hierarchy of Needs (1968) and supports interrupting pain cycles. Maslow (1968) recorded belonging within the middle of his motivational hierarchy, ranking belonging above food, safety, and 'other basic needs, and below esteem, which depicts respect and acknowledgment from others (Maslow, 1968). The evolution of Adlerian psychology and Maslow's Hierarchy of Needs contribute to belonging as a basic psychological need that, if fulfilled, results in favorable outcomes, such as; motivation, focus, engagement, and more significant health responses (Cacioppo, 2018; Eisenberg, 2012; Pressman et al., 2019; Rizzolatti & Fogassi, 2014; Tuck et al., 2017).

Baumeister and Leary (1995) define belonging as: "a pervasive drive to form and maintain at least a minimum quantity of lasting, positive, and significant interpersonal relationships" (p. 500). Moreover, Baumeister and Leary (1995) articulate that for belonging to be present, individuals must sense and perceive that they are accepted, valued, included, and encouraged by others (p. 499). When these elements are present, lonely individuals do not feel the pressure to perform (Krings et al., 2021). When the group behaves in a belonging manner, the individual can recognize they can be themselves and experience quality interactions.

Interestingly, scholarly research from the 1980s to 1990s depicts belonging as a natural formation. An "abundance of evidence" suggests that social bonds of belongingness form quickly (Bond, 1980; Clark et al., 1987). However, Beecher (1990), Dreikurs (2000), Stein (2002), and Zoltán (2011) argue that belonging has deteriorated in modern society. They articulate that we live in a discouraging time with an unprecedented aggressive individualistic expression where competition and rivalry override respect for human dignity, emotional warmth, and the opportunity to experience the feeling of belonging to a community. So what has changed? Furthermore, perhaps more importantly, how do we revive the social bonds of belongingness to form quickly and navigate the reparations of loneliness from the pandemic?

THE BELONGING-FIRST MODEL AND METHODOLOGY

Ferguson (2010) details that Alfred Adler understood the inextricably interwoven direct correlation between individual and societal wellbeing. One's ability to fully belong relies on the ability to feel equal, be a valued contributor, and have a social interest of commitment to the welfare of the human community (Ferguson, 2006). Adler, in all respects, connected the wellbeing of the individual with the wellbeing of the community. He highlighted that mental health increases when individuals feel they belong and all individuals in a community feel they belong (Ferguson, 2010). In addition to belonging acting as a motivator for engagement and performance, belonging also provides the blueprint for interindividual and intergroup consonance.

With that in mind, Carter (2022) sought to create metrics and indicators for belonging to offer foundational frameworks to communities and individuals within social institutions. A primary grounded theory mixed methods research design involved measuring and identifying cohesive terminology to describe and measure belonging. With a linguistic search to define five themes representing belonging, a quantitative survey was validated with expert selection and degrees of relevance and clarity for each survey

Figure 1. Belonging-First Model: Reciprocated Responsibility of Belonging Behaviours
Note: This model shows that the responsibility of belonging is reciprocated equally between the culture and environment and the individual. It also denotes the importance of all five key indicators that produce the experience of belonging as an individual and a culture. Adapted from "Belonging within the Workplace: Mixed Methods Constructivist Grounded Theory Study For Instrument Validation and Behavioural Indicators For Performance & Governance," by A.D. Carter (2022). [Master's thesis, Adler University]. ProQuest Dissertations and Theses Global. p.137. https://www.proquest.com/docview/2716586005/FECBB62CCA204025PQ/1. Copyright 2022. Reprinted with permission.

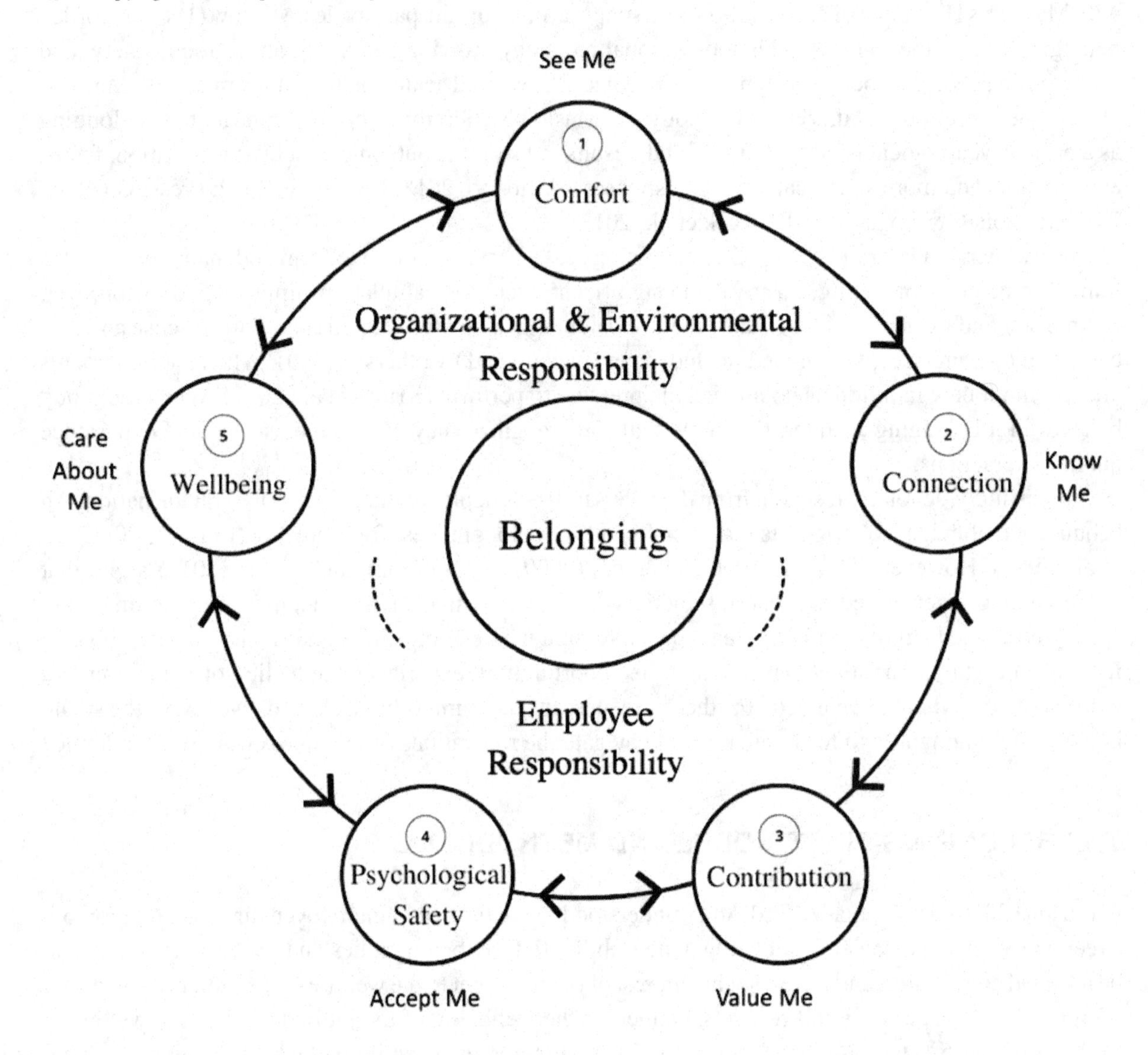

item measuring belonging. The results produced five key indicators that identify belonging; they are 1) comfort, 2) contribution, 3) connection, 4) psychological safety, and 5) wellbeing (Carter, 2022). From this primary grounded theory research, Carter (2022) created the Belonging-First model, methodology, and the Workplace Belonging Scale, which is validated to score and measure belonging in the workplace.

The Belonging-First model outlines the primary needs that must be present for individuals and cultures to act responsibly for creating a belonging environment where everyone can belong. From a foundational perspective, comfort represents the need to be seen, connection represents the need to be known, contribution represents the need to be valued, psychological safety represents the need to be accepted, and wellbeing represents the need to be cared about (Carter, 2022). The Belonging-First methodology indicates that when an individual joins an institution, it is in the institution's best interest to provide actions and behaviors that both indicate and enable the experience of belonging from the onset. This provides the contextual key indicators to the individual that the environment supports the individual and wants them to succeed. This sentiment then motivates the individual to engage with others in their environment and model the actions and behaviors they are experiencing. When these actions and behaviors are modeled for the individual from the onset, the individual then reciprocates these actions and behaviors with others in the institution and upholds a belonging-first outlook and the resulting benefits (Carter, 2022).

The Belonging-First model is upheld through Albert Bandura's 1986 Social Cognitive Theory (SCT). Bandura (1989) asserts that there is a dynamic and reciprocal interaction between a person (an individual with a set of learned experiences), the environment or institution in which they are part of (external social context), and the behavioral responses of those within the environment or institution (the responses to the stimuli to perform and achieve goals). With reciprocal determinism at its core, behavioral capability, observational learning, and reinforced behavioral responses enable positive expectations and anticipated consequences, with these five key indicators upholding the influence of successful modeling. Further, with these five key indicators accounted for, a person's confidence or self-efficacy has the propensity to increase, enabling the individual to perform optimally and succeed. The more individuals who perform optimally and succeed, the more institutional performance and success augments. These crucial factors are often understood through engagement and collaboration measurements.

Engagement and Collaboration in Online Communities

With the rise of reliance on communication due to COVID-19, we can see how technology provides us not only knowledge but also breaks barriers in online communities where individuals can not only learn but also communicate "anytime, anyplace, and to anyone" (Morris, 2010; Lorenzo, 2001; Robinson & Hullinger, 2008). These online social institutions enable the creation of groups, allow individuals to dialogue and interact online, bridge diverse perspectives and people together, and essentially provide an opportunity to create lifelong connections (i.e., see Brindley et al., 2009; Kassop, 2003; Sun & Chen, 2016). For instance, when we study online educational institutions, it is questionable if online learning and interaction have the same outcomes as in-person communication or if it is even better (Hannay & Newvine, 2006; Harasim, 2000; Kassop, 2003; Peat & Helland, 2004). Research studies look at online 'learners' outcomes, test scores, grades, attitudes about online learning, and overall satisfaction (Robinson & Hullinger, 2008). Many studies conclude that rather than replicating in-person learning and communication, research should stress the unique benefits of communicating in the online environment and garner 'users' feedback on online engagement (Müller et al., 2021; Singh et al., 2021; Stojan et al., 2022).

If individuals are not engaged, they will not have the motivation to learn or interact with others online, which results in reduced productivity and is often accompanied by lower attendance ratings (Astin, 1984; Robinson & Hullinger, 2008). Objectively, to measure engagement, communication, and online learning, all parties of the online community should be evaluated. This binds the concept of engagement with collaboration (Huang et al., 2018; Robinson & Hullinger, 2008). Engagement can be measured as "physical energy that [individuals] expand on [particularly in collaborative] activities" (Robinson & Hullinger, 2008, p.101). According to studies such as Deslauriers et al. (2019), Kuh (2003), and Robinson & Hullinger (2008), engagement is generally expressed in online learners' effort to study, receive feedback, analyze, and solve problems. In the Robinson and Hullinger (2008) study, researchers measured online student engagement, utilizing effective educational practices such as "taking into consideration level of academic challenge, active and collaborative learning, student-faculty interaction and enriching educational experiences (Robinson & Hullinger, 2008, p.102-103).

In this particular study, students reported their experience in online environments through an online questionnaire. The aspect that was found most fascinating, which can be related essentially to any online community, was the stress on collaboration. Active and collaborative learning reflected the students' efforts to engage with peers, in-class discussions, and collaborate with students on projects, as well as with students from other classes (Robinson & Hullinger, 2008). It is here that the link between engagement, collaboration, and belonging comes to light. Interaction between instructors and students has shown the most frequent collaboration with indicators of belonging. Specifically, promoting an inclusive learning environment and culture promotes learning in the online environment (Garrison, 2006; Palloff & Pratt, 2010). By viewing the online class as an institution or community, the environment's culture promoted collaboration, which motivated individuals to engage frequently in collaborative behaviors (Robinson & Hullinger, 2008; Ouzts, 2006; Wang et al., 2003). Nearly "80% of students collaborated at least sometimes, 40% worked together very often", and 65% acknowledged some benefits to their learning and growth by engaging in collaboration (Robinson & Hullinger, 2008, p.105).

How Belonging Supports Engagement and Collaboration

Research studies such as by Thomas et al. (2014) show how a sense of belonging in the online environment enhances participation, success, and retention in online educational settings and how the online environment provides more opportunities for belonging to individuals from underrepresented groups. A sense of belonging allows individuals to have a more personal connection to the content learned and to peers. Everyone experiences belonging differently in the online space. However, when the facilitator uses the five indicators of belonging to create an environment where users feel they belong, there is a higher propensity for engagement and collaboration.

Theoretically, suppose we triangulate the engagement and collaboration knowledge with the Belonging-First Model and methodology and apply it to online and in-person environments. In that case, the expectation is an increase in collaboration, higher engagement with online and in-person users, and an enriched student experience through the social state of the environmental space (Robison & Hullinger, 2008; Seifert, 2016; Carter, 2022). The more users work together, the higher chance they will get to know each other better, share ideas, and intentionally engage in open discussions that lead to growth and success.

An online example of this sentiment can be seen with video calls. The main goal of the standard online environment is to create a community where face-to-face interactions can be replaced primarily by video calls. While standard video calls are commonly used, perceiving that they increase engagement, belonging indicators allow the user's comfort to be considered a driver of engagement. With this perspective, users are relieved of the pressure to keep cameras on at all times. When users have the freedom to turn their cameras on or off, this permission drives a level of comfort and connection to the online experience that can otherwise cause apprehension.

Moreover, enabling users to contribute by participating in community discussions once their comfort levels are accounted for supports an individual's psychological safety and wellbeing (Carter, 2022). By not having the pressure to be seen on video, the user relaxes into a more comfortable state of mind and instead serves as an active agent in their online community (Thomas et al., 2014).

Using Comfort and Connection Indicators as Foundations for Engagement

McDougall (2015) articulates that comfort contributes to an individual being their authentic self and individually gaining a sense of belonging; Carter (2022) speaks to the importance of the environment and institution providing belonging methodology in the commencement of individual interaction. For example, one of the five key indicators required to perceive belonging, Comfort, requires the leader and those in the culture to *see each individual for their authentic self*. To obtain the perception of comfort and being seen, these initial actions and behaviors were found to be instrumental: learning individuals' names, pronouns, culture, and what they value at the onset of joining the group. With this initial step, the research indicates that individuals feel both seen and accepted (Carter, 2022). Moreover, an articulated tactic is to ensure that a student's name is pronounced accurately because a person's name is a primary indicator of identity. To initially *see a person as they are* and *create comfort for that person's authentic self*, learning their name demonstrates to the individual that you see them and want them to feel comfortable in the new environment (Carter, 2022).

By modeling the importance of someone's name new to the institution or group, the leader models the importance of accepting individuals as they are and for who they are. When this is modeled as an essential factor of the institution, others within the institution will understand the importance of learning each individual's name correctly. Interestingly, when comfort is present in the Robinson and Hullinger (2008) research, students were not reluctant to ask for help when necessary, and peer tutoring was more commonly sought after (Robinson & Hullinger, 2008).

Connection, also characterized as being *known*, is the second key indicator of belonging. Based on Carter's (2022) primary research, alignment, values, and trust were key characteristics that demonstrate connection within an environment. A specific tactic to bring forward connection and being known, is to create standards of engagement so that all members are aligned with the values of the group, which in turn leads to the development of trust. Initially, standards of engagement can outline the actions and behaviors that all individuals value, such as; 1) articulating that the space (whether online or in-person) is meant to support multiple norms and truths, 2) that each individual can listen with the intention of hearing, 3) to not interrupt someone when they are speaking, 4) to provide the opportunity to be heard, specifically by those who are more quiet and introverted, 5) to start initial meetings by reminding participants of the values within the standards of engagement and their importance (Carter, 2022).

When these elements of connection are triangulated, a community's engagement and collaboration enable individuals of different perspectives and backgrounds to utilize the online space to work together to solve problems that mirror real-life issues (Samson, 2015). Elements of connection allow users to enrich their technical skills and knowledge and learn more when they actively discuss the material with peers, connecting content to people and valuing the elements of collaboration (Robinson & Hullinger, 2008). The more individuals become used to connecting through sharing in the online environment, the more conversations occur, which increases the sense of belonging. As individuals see and know each other, they grow in their understanding of each other, and more dialogue occurs (Peacock et al., 2020; Peacock & Cowan, 2019).

When standards of engagement through belonging are valued, dialogue is healthy. Everyone gets a turn, all parties are respected, different points of view are considered, and individuals learn to become more accepting of the modeling they receive within the group. These initial characteristics and tactics of comfort and connection set the group norms or online environment guidelines, and the culture embraces and promotes belongingness.

Building on Engagement with Characteristics from The Contribution Indicator

To create interaction and engagement online, online community designers bode well by creating projects in which users are empowered to contribute, and their data can be analyzed in terms of interaction and performance. For example, researchers such as McBrien et al. (2009) analyzed ways that online features such as asynchronous learning environments affect 'users' online experience by enabling mechanisms for authentic contribution. Tactics such as dialogue, structures for contribution, and learner autonomy when contributing; can reduce the distance in group interaction and promote belonging. For example, the dialogue elicited comfort from online users and engagement. The majority (91%) of the comments were positive regarding dialogue in terms of quality and quantity experienced in the online environment, and users even reported that the online experience "felt like being in a [traditional] classroom" (McBrien et al., p.8). They also reported being so relaxed; learning information was easy. As a result, a tip for those creating belonging in the online environment is to use tactics that create comfort and connection and then provide mechanisms and structures to make it easy for users to contribute – belongingness will begin to form naturally after that. By encouraging others to contribute by speaking out more, enabling participants to contribute specific topics for discussion, and outlining how users can contribute constructive feedback with further questions - even the most silent or shy users are provided the opportunity to contribute to the dialogue.

From the initial primary research (Carter, 2022), specific tactics were discovered that lead to environments where contributions are valued. First, articulate that contributions within the environment are valued and describe and model how these are to be done. Modeling the expectations for contribution creates comfort on how and what to do while reinforcing the value of connection to all members. Contribution characteristics emphasize the importance of relationship building through collaborating, dialoguing, and developing mutually beneficial partnerships (Carter, 2022). McBrien et al. (2009) align with this perspective, indicating that when everyone in the online class had an opportunity to talk, individuals could openly express their points of view. As a result, users reported discussions "became more involved (p.8)."

The structure is also essential for online users; hence we propose having a technical step-by-step instruction that explains how to use the online environment along with recorded/live video training that can help create that comfort and connection users need for engagement. Moreover, online community facilitators should also pay attention to different 'members' needs in the online environment – how experienced are users with online group interactions? To solve this, a facilitator can ignite the contribution element by starting conversations that engage different topics of discussion, asking for perspectives and reasoning, and allowing community users to contribute their topics and opportunities to be proactive in their online experience.

Another option to create contribution and conversational order, which reinforces belonging, is to articulate the structure of engagement by using gestures like a hand or another emoji as code to request permission to speak. Moreover, guidance for contributions, specific feedback for growth, and providing the authority to lead and support are tactics that are helpful as online users often report that virtual group work can be challenging when the structure of contributions is not initially established (Palloff & Pratt, 2010; Sutia et al., 2019). Designers can also create online forums with frequently asked questions to help new users adjust by reading 'others' problems/solutions and sharing their own. These tactics support the contribution indicator and help users in the online environment by highlighting what does not work, demonstrating the contributions that drive comfort and connection and enabling users to adjust (a collaborative/interactive and proactive approach to learning). The facilitator, in this case, should be a fly on the wall, so users can focus more on sharing and conversing amongst themselves and developing peer interpersonal relationships that enhance belonging (Baran & Correia, 2009; Carter, 2022).

ACCOUNTING FOR PSYCHOLOGICAL SAFETY & WELLBEING WITHIN ONLINE ENVIRONMENTS

Initially, when starting an online environment, it is essential to understand the correlation between belonging and engagement. While we have shared this importance earlier in this chapter, psychological safety and wellbeing are often factors that are produced once comfort, connection, and contribution have been accounted for. However, the ability to achieve psychological safety and wellbeing of users is also determined by the initial standards of engagement and what is modeled for the user by the facilitator and peers. Psychological safety exists when specific characteristics and tactics are in place, such as; 1) the environment must have open and approachable people and content, 2) users must be able to share personal perspectives that are cared about and considered, and 3) respect and empathy for all users and their unique identities must be present, and discretion, privacy, and confidentiality must be standardized with structural guidelines and governed rules (Carter, 2022). When these elements are not in place, the ability to ask for help, dialogue, or form interpersonal relationships becomes difficult.

Hence having additional features online, such as icebreaker games, bonding activities/social activities, and discussion groups, with clearly articulated and normalized structures of engagement so that members can increase comfort, connection and contributions can be a huge help (McBrien et al., 2009; Linders, 2018; Carter, 2022). We also advise designers and facilitators to create smaller groups online that change, develop and enable users to grow interpersonal relationships. For example, if partners or group members change with each assignment in online schools, students will have opportunities to collaborate with various people in the class. When standards of engagement are created, and the initial three indicators of belonging (comfort, connection, and contribution) are actively modeled and communicated, students are supported to reciprocate and collaborate in an environment whereby everyone is included. This way, no one will be left out; after all, our goal is to increase collaboration and engagement and heighten a sense of belonging.

In other communities, dividing Zoom meetings into small rooms for discussion can replicate this phenomenon and enhance contributions by enhancing comfort and connection that develops more easily in smaller groups of people. For this – having regular meetings or online communication is beneficial for the community's growth as it prepares users to adjust to the online environment while promoting the transfer of new ideas across users. When users gain a sense of autonomy over their online environment, they feel more freedom to communicate openly. It can be predicted that more conversations will flourish into deeper connections and higher productivity (McBrien et al., 2009).

As mentioned, belonging occurs when students feel comfortable expressing their authentic selves online, which class facilitators and fellow members reinforce through communication in online spaces (Hudson et al., 2020). Studies show that as a student's sense of value and belonging in the online environment grows, motivation to learn increases (Morrow & Ackeermann, 2012; Over, 2016). Therefore, it is essential for instructors to show interest in students and to continuously share values, expectations, feedback on progress, and words of encouragement through positive reinforcement, which can be replicable to any online community (Hudson et al., 2020). Online facilitators can create an online student community where students can have a safe space to discuss issues concerning their learning, their interests, academic expectations, and personal progress, as well as online opportunities to socialize with other students. This can start by establishing clear aligned properties of engagement that reinforce the belonging indicators. Separating the online space into chat rooms with categories of topics is one practical way to go about this. As mentioned earlier, hosting themed events will provide a space for students to interact and bond with others.

Essential Themes and Tactics to Support the Transition from Loneliness to Belonging

We see five essential tactics leaders and facilitators need to know to move from perspectives of loneliness into an environment of belonging:

1. An opportunity for interaction and engagement must be established through tactics of comfort.
2. Enforcement of a learning culture must be established through aligned connections with people, values, materials, and governance.
3. Enablement of unique perspectives must be brought forth through individual and group contributions.
4. Fortification of psychologically safe environments where open, empathetic, respectful dialogue and support structures are embedded into the culture.
5. The standards that normalize differences, unique identities, and individual strengths and weaknesses must be upheld with reinforced communications and permissions.

To understand how to make these considerations, let us now consider practical strategies for making the online environment and educational platforms (i.e., LMS) incentivize the sense of belonging for online users (Linders, 2018).

First, at the commencement of an online environment opening, we suggest meetings adopt a methodology that highlights user commonality and incorporates aligning the environmental values with user values, interests, cultures, and individual intersections of identity by enabling users to share what they feel comfortable with. Despite universities transitioning back to classroom settings, many classrooms still operate through online learning or a blended option. Since the start of COVID-19, it has been evident that students experienced stress from the newly found sense of isolation in online spaces (i.e., Cowie & Myers, 2021; Flynn & Noonan, 2020; Hamza et al., 2021; Husky et al., 2020). While many may not have associated isolation and loneliness as the primary reasons for feeling anxious or overwhelmed in their return, the presence of such feelings led us to the question - how can we help online space users at large experience a greater sense of belonging in online spaces?

Interactive games fundamentally act as connectors that help to highlight the commonalities of those in the group. In order for games to be considered psychologically safe and considerate of user wellbeing, the following characteristics are essential:

- Before starting a game, communicate that the game is voluntary.

- Normalize upfront that opting out from a game is an act of self-care and self-ownership, not an act of shame
- Make it known that those who watch are learning and acquiring education and knowledge - help them establish this by directing them to listen with curiosity and compassion.
- Keep the game light; do not risk highlighting afflictions that you are not qualified to handle
- Pre-vet and pre-test your game prior to launching it online
- Let players know the game is not about winning but illustrating a point about the work or learning. The lesson does not need them to be great players.
- The facilitator is required to listen to people with curiosity and compassion.
- Facilitate the game so that the experience is both friendly and rewarding
- Learn to prep and debrief the games

Next, consider sharing humorous content at the beginning of every meeting. Laughter occurs across all cultures, in various situations and intersections of identity, by uniting people with positive emotions (Warren et al., 2021). Additionally, humor helps people cope with loss, difficulties, and challenges in life; therefore, it is a tactic worth considering to support moving from loneliness to belonging (Warren et al., 2021). For this to be embedded with psychological safety and wellbeing, the facilitator must pre-frame humorous content by acting as the online space leader, facilitator, and partner to users in the online space (Hudson et al., 2020).

Warren et al. (2021) demonstrate the research on what makes things funny. For humor to be appreciated as a unifier, it must omit the following:

- Superiority - feeling that you are better than someone or something else by sharing a story or joke that promotes seeing someone else get hurt (emotionally, psychologically, or physically)
- Violate appraisal - create the perception that something threatens beliefs or values

For humor to be appreciated as a unifier, it must include the following:

- Successful comedy - include content that makes people laugh and feel amused.
- Benign appraisal - create the perception that a stimulus or situation does not present a serious problem or shame/blame an individual
- Surprise - includes a perception that diverges from the expected outcome.

Lastly, to support psychological safety and wellbeing, the facilitator needs to be mindful of tone, facial expressions, and words of support and encouragement. In educational settings, students perceive belonging when they receive 'others' care, an element that is also characterized under the wellbeing indicator of belonging.

Belonging Methodology for Educators & Facilitators

Inclusive leadership positively moderates the potentially harmful experience between ethnic-cultural diversity and a belonging climate. While many assume that high diversity yields an inclusive climate where everyone can belong, these assumptions are flawed. In order to create belonging communities and institutional membership where everyone feels supported to engage and learn freely, leaders, educators, and facilitators need to support the environment with crucial behaviors. To deepen this concept, we look to optimal distinctiveness theory, which highlights that two needs must be accounted for so that an individual feels included and is supported to engage: belongingness and uniqueness (Shore et al., 2011). Belongingness in this context involves using the five indicators to support individuals in recognizing their similarities with and validation by others. Uniqueness refers to individuals having their intersections of identity respected and normalized compared to others. With this in mind, for individuals to experience belonging within an online group, educators and facilitators need to ensure that all members are treated as insiders while having the opportunity to express and preserve their unique identities (Ashikali et al., 2021). When inclusive leadership is at the forefront, educators and facilitators can effectively balance belonging and uniqueness by encouraging the exchange of diverse viewpoints and stimulating the discussion of diverse perspectives.

The teacher-student relationship increases satisfaction when teachers are supportive and helpful and make individuals feel like they belong. The words of satisfied students support this point, "The instructor reached out to me when they thought I needed help in the course, and listened to me and helped me when I did not understand the concept" (Hudson et al., 2020, p.4). By creating intentional spaces where belonging is modeled, valued, and upheld with the indicators, the responsibility of belonging becomes shared. Specifically, in the example of needing online support, when educators foster inclusive leadership that fosters belonging indicators, the environment, climate, and all members are empowered to support each other because they are all considered "insiders." This distinction enables the community of users to share perspectives and tools of support, which then heightens engagement and collaboration. By preframing these concepts at the commencement of an online community, the inclusive educator and facilitator support when necessary and empower the community the rest of the time.

To instill these values, the educator or facilitator must be conscious of how they provide direction and feedback by adopting a positive tone and governing values of belonging. To make the online learning environment more inclusive, guidelines and terms of engagement and expectations need to be communicated frequently. By breaking barriers and discussing what inclusion and belonging mean in the classroom and the online meeting space (i.e., using tools so that all students can benefit from the learning), the culture of the online environment will change (Hudson et al., 2020). Students should be encouraged to recreate comfort, connection, and contributions to collaborate and help each other learn using an individual's unique capabilities. Providing office hours is an option; however, other mechanisms for support should exist. Having spaces where students can come together to interact, study, question, and socialize during off-school or off-community hours is also essential. As a result, having peer-guided online study groups online is also a great idea to foster a sense of belonging.

SPECIFIC TOOLS THAT ENHANCE BELONGING AND ENGAGEMENT ONLINE

There are many tools that educators can utilize in online platforms to innovate and try something new that helps students learn and provides a sense of belonging that can be replicated in other online communities. For example, collaborative online tools such as Kahoot, Padlet, or Flipgrid can provide students and instructors with opportunities to collaborate and express themselves and their relation to the material (Hudson et al., 2020). In the online workspace, even in person, using tools such as Mentimeter, AhaSlides, PollEverywhere, or Slido can enable community and belonging perceptions while supporting a person's anonymity and reducing anxiety. Considering multiple options for communication and learning is so important, hence asking students what tools they are most comfortable with is an indirect method of engagement. The most effective engagement tool is showing students they are cared for by instructors, staff, or fellow students' attitudes in class – by accommodating belongingness and uniqueness. Asking for feedback with these tools allows students the comfort of using what they know and can connect easily with. Moreover, feedback is another contribution mechanism that enables individuals to experience inner circle behavior.

Regarding belonging as a tool to enhance engagement, another consideration is to use flexible and hybrid models that support all five belonging indicator touch points. Thomas et al. (2014) bring the example of having workshops in person for a week as a mechanism to change interactions. Alternatively, allow community users to decide to meet in person on their own time and work on projects together. Providing channels for discussion groups can also be very engaging, as online users who struggle with conversing with strangers will have structure and guidelines to feel comfortable starting a topic of discussion. The community leader can start the dialogue with a topic or question, and users can post their answers and interact with others. If a user wants, they can also form discussion groups where they can participate more than once a week. This will allow users to interact more frequently and explore new learning styles, which can be very engaging as it "puts a face to the names" in class (Thomas et al., 2014, p.72). It is also crucial for community leaders to encourage users to express their feelings regarding the material discussed or feelings as they navigate through the online environment (i.e., the user is very active in discussions and upset that there are not enough responses generated in the discussion). This isolation can be approached through open discussion, understanding, and creating new strategies to involve more users in discussions.

The key is to continuously work from the 'users' feedback rather than garner it only once a year when the class ends; that is the way to build an active relationship and new strategies for learning and connection between online members. A sense of belonging comes from being part of the online space; collaboration helps bring the parts of the space together where each feels like their part is meaningful. Engagement results from collaboration and belonging, the formula for successful lifelong learning (Thomas et al., 2014). Users are happy to learn and engage in dialogue with others; they need guidance and clear inclusionary tactics, no matter what age group - a consideration many group leaders fail to recognize, especially in higher learning environments (Thomas et al., 2014). To avoid this, users can provide daily questions to respond to, change up group members after a while so that others get a chance to work with not only select part of the group and create activities for collaboration that are both formal and informal (Thomas et al., 2014). Also, where applicable and users feel comfortable, providing some personal information, such as areas where members live (only with their permission), can help connect members who live in the same time zones to meet outside of their online community. To start conversations with new individuals, users can share what they know about a topic before the start of a lecture. This easy open-ended question can generate many views. This can help the other members build on future conversations. This is called "embedding collaboration into assessment to promote social interaction" and belonging or utilizing the created sense of belonging to promote engagement and collaboration; it works both ways (Thomas et al., 2014, p.74).

Other Online Engagement and Belonging Considerations

When looking to increase engagement and belonging, a rule of thumb is to allow users to interact with as many mechanisms and applications as they feel comfortable. Often, there are 1-3 features of interactions available with LMS or online platforms where open conversations can occur without being forced. Features such as a "course café space" or "lounge" can help online users have unmonitored discussions with the whole group. Users can also share their social media if comfortable and create breakout groups where it makes sense (Thomas et al., 2014). While these virtual spaces all work to create a sense of belonging: by supporting the action for users to connect, share topics that are specific to the group, or provide experiences of learning and growing together, one does need to ensure subgroups or breakout groups do not inadvertently create exclusion or elitist behaviors.

Moreover, real-time online interaction makes collaboration easier worldwide; however, only some people's schedules allow this. Recording meetings is also helpful for users who cannot make it to meetings as they can gain a sense of "really being there," (Thomas et al., 2014, p.75). Group leaders can include the users that are missing by embedding them into the conversation (i.e.," can you say it again for the external [members]") (Thomas et al., 2014, p.75). Technology can also help to personalize the engagement with features such as screen sharing and sharing images or information garnered on the topic in the chat box. Providing personal photos when introducing the self at the beginning of the first community meetings is also helpful as it shows how involved the parties want to share something about themselves (Thomas et al., 2014). This is the first step towards belonging, engagement, and collaboration, showing a willingness to do so.

Other Opportunities to Consider for Engagement and Belonging

What further opportunities can one learn from engaging with the key indicators and tactics of belonging? This chapter is intended for a broad audience as we all play a part in helping others belong in learning environments – online or otherwise. Understanding the model for belonging that incorporates the five indicators and methodology is a starting place to move from sentiments of loneliness to belonging. From an organizational perspective, not all indicators can be addressed at once, and having the ability to score belonging is an action step that larger institutions may consider so they know where to start (Carter, 2022). Fundamentally, change happens with each small action taken once the belonging indicators are understood. That means starting with essential comfort elements can open the gateway for belonging and engagement benefits. Additionally, sharing what the indicators of belonging are at the commencement of any new group allows users and members to consider that the responsibility of belonging resides with the community and each member. By stating these broader concepts, users also take responsibility for greeting other members, making efforts to interact with others, and making discussions and online webinars more engaging (Ways to create belonging in a virtual classroom, n.d.)!

As a group - help establish a sense of community by developing norms and rules regarding behavior that is expected/accepted/not accepted in the online environment, such as respect and fair treatment or intolerance of bullying of any kind. Use the belonging indicators as a methodology to start with. All individuals' voices deserve to be heard and valued (as long as they do not harm the community) (Ways to create belonging in a virtual classroom, n.d.).

Finally, the timing of meetings needs to be regulated by online users, not only group leaders, to promote more autonomy and decision-making as a group (Ways to create belonging in a virtual classroom, n.d.). It would also be interesting to see more features built into online environments that allow users to converse in real-time (i.e., a chat/conference option or a separate community with threads such as Yammer). Having more spaces online to converse would promote interaction and easier collaboration which leads to a sense of belonging (Ways to create belonging in a virtual classroom, n.d.). Culture in these spaces or lounges should also promote values of acceptance by setting ground rules in spaces such as group sessions, teamwork project meetings, and discussion responses. Assessments (if applicable in an online environment such as educational settings) should also focus on user communication and collaboration – how can users improve their teamwork? There can be metrics for this too. Users can also evaluate themselves and their experiences and objectively compare their performance. Users can also help find different ways to deliver content to the community, as not everyone will enjoy using the same online communication modes. There should be many opportunities for community members to express such as creating videos, submitting ideas for discussions, suggesting new tools and resources, and exploring topics in the online space (Ways to create belonging in a virtual classroom, n.d.).

We see a lot of technical opportunities in this field utilizing design thinking when shaping the new online environments (think: who are our users and what are their needs), as the future of learning, working, and communicating is online and designers and researchers alike have much room of refining spaces such as LMS systems to be more collaborative and have more open communication centers for users. When all parties, including the designers and researchers, understand the meaning of belonging, "when [users] feel accepted, valued, and understood as part of the group. They [will] feel that they fit in," (Ways to create belonging in a virtual classroom, n.d., p.1) then the design will genuinely reflect the perspective of the user needs. Interviewing users before redesigning can be very beneficial and a huge opportunity for the field of education to grow.

FINAL LIMITATIONS AND A CALL TO ACTION

The only limitations thus far are that online users still report they do not feel they belong; it will take a communal effort to make this change, and the change will only happen over one day or by more than one member. Creating an environment where people belong is a continual process. It requires leaders to introduce it and members to uphold it. All individuals within a group are reciprocally responsible for creating a space where everyone belongs.

When these expectations are established early on and continually modeled and reinforced, willingness to listen and embrace new perspectives becomes easier for the whole community. As a result, for those reading this chapter – we hope we have inspired you to consider how you can bring belongingness into your community, and the content has motivated you to act now and start your journey today.

REFERENCES

Abrams, D., & Hogg, M. A. (1988). Comments on the motivational status of self-esteem in social identity and intergroup discrimination. *European Journal of Social Psychology, 18*(4), 317–334. doi:10.1002/ejsp.2420180403

Abrams, D., Lalot, F., & Hogg, M. A. (2021). Intergroup and intragroup dimensions of COVID-19: A social identity perspective on social fragmentation and unity. *Group Processes & Intergroup Relations, 24*(2), 201–209. doi:10.1177/1368430220983440

Ackerman, G., & Peterson, H. (2020). Terrorism and COVID-19. *Perspectives on Terrorism, 14*(3), 59–73. doi:10.2307/26918300

Adler, A. (1931). The meaning of life. *Lancet, 217*(5605), 225–228. Advance online publication. doi:10.1016/S0140-6736(00)87829-0

Ashikali, T., Groeneveld, S., & Kuipers, B. (2021). The role of inclusive leadership in supporting an inclusive climate in diverse public sector teams. *Review of Public Personnel Administration, 41*(3), 497–519. doi:10.1177/0734371X19899722

Astin, A. W. (1984). Student involvement: A developmental theory for higher education. *Journal of College Student Personnel, 25*(4), 297–308.

Bandura, A. (1989). Human agency in social cognitive theory. *The American Psychologist, 44*(9), 1175–1184. doi:10.1037/0003-066X.44.9.1175 PMID:2782727

Baran, E., & Correia, A. P. (2009). Student-led facilitation strategies in online discussions. *Distance Education, 30*(3), 339–361. doi:10.1080/01587910903236510

Baumeister, R. F., & Leary, M. R. (1995). The need to belong: Desire for interpersonal attachments as a fundamental human motivation. *Psychological Bulletin, 117*(3), 497–529. doi:10.1037/0033-2909.117.3.497 PMID:7777651

Beecher, W. (1990). Beyond Success and Failure. Ways to Self-Reliance and Maturity. Richardson, TX: Willard & Marguerite Beecher Foundation.

Besse, R., Whitaker, W. K., & Brannon, L. A. (2021). Loneliness among college students: The influence of targeted messages on befriending. *Psychological Reports*, 1–24. doi:10.1177/0033294121993067 PMID:33593152

Blondel, J., & Inoguchi, T. (2006). *Political Cultures in Asia and Europe: Citizens, States and Societal Values* (1st ed.). Routledge. doi:10.4324/9780203966907

Bond, R., & Smith, P. B. (1980). Culture and conformity: A meta-analysis of studies using Asch's (1952b, 1956) line judgment task. *Psychological Bulletin, 119*(1), 111–137. doi:10.1037/0033-2909.119.1.111

Bond, M. H., Leung, K., Au, A., Tong, K.-K., de Carrasquel, S. R., Murakami, F., Yamaguchi, S., Bierbrauer, G., Singelis, T. M., Broer, M., Boen, F., Lambert, S. M., Ferreira, M. C., Noels, K. A., van Bavel, J., Safdar, S., Zhang, J., Chen, L., Solcova, I., ... Lewis, J. R. (2004). Culture-Level Dimensions of Social Axioms and Their Correlates across 41 Cultures. *Journal of Cross-Cultural Psychology, 35*(5), 548–570. doi:10.1177/0022022104268388

Bonsaksen, T., Ruffolo, M., Leung, J., Price, D., Thygesen, H., Schoultz, M., & Geridal, A. O. (2021). Loneliness and its association with social media use during the COVID-19 outbreak. *Social Media + Society, 7*(3), 1–10. doi:10.1177/20563051211033821

Brindley, J. E., Blaschke, L. M., & Walti, C. (2009). Creating effective collaborative learning groups in an online environment. *International Review of Research in Open and Distributed Learning, 10*(3). Advance online publication. doi:10.19173/irrodl.v10i3.675

Cacioppo, J. T., & Patrick, W. (2008). *Loneliness: Human nature and the need for social connection*. WW Norton & Company.

Cacioppo, J. T., & Hawkley, L. C. (2009). Perceived social isolation and cognition. *Trends in Cognitive Sciences*, *13*(10), 447–454. doi:10.1016/j.tics.2009.06.005 PMID:19726219

Cacioppo, S., Capitanio, J. P., & Cacioppo, J. T. (2014). Toward a neurology of loneliness. *Psychological Bulletin*, *140*(6), 1464–1504. doi:10.1037/a0037618 PMID:25222636

Cacioppo, S., Bangee, M., Balogh, S., Cardenas-Iniguez, C., Qualter, P., & Cacioppo, J. T. (2016). Loneliness and implicit attention to social threat: A high-performance electrical neuroimaging study. *Cognitive Neuroscience*, *7*(1-4), 1–4, 138–159. doi:10.1080/17588928.2015.1070136 PMID:26274315

Cacioppo, J. T. (2018). Loneliness in the Modern Age: An Evolutionary Theory of Loneliness (ETL). Advances in Experimental Social Psychology, 58C, 127.

Clark, M. S., Oullette, R., Powell, M. C., & Milberg, S. (1987). Recipient's mood, relationship type, and helping. *Journal of Personality and Social Psychology*, *53*(1), 94–103. doi:10.1037/0022-3514.53.1.94 PMID:3612495

Clark, J. L., Algoe, S. B., & Green, M. C. (2018). Social network sites and wellbeing: The role of social connection. *Association for Psychological Science*, *27*(1), 32–37. doi:10.1177/096372147730833 PMID:30407887

Carter, A. (2022). *Belonging within the workplace* (Publication No. 29393403) [Adler University]. ProQuest. https://www.proquest.com/openview/df3bede74f97b31ba91c846f316515e6/1.pdf?pq-origsite=gscholar&cbl=18750&diss=y

Cowie, H., & Myers, C. A. (2021). The impact of the COVID-19 pandemic on the mental health and wellbeing of children and young people. *Children & Society*, *35*(1), 62–74. doi:10.1111/chso.12430 PMID:33362362

Daymont, T., Blau, G., & Campbell, D. (2011). Deciding between traditional and online formats: Exploring the role of learning advantages, flexibility, and compensatory adaptation. *Journal of Behavioral and Applied Management*, *12*(2), 156.

Deslauriers, L., McCarty, L. S., Miller, K., Callaghan, K., & Kestin, G. (2019). Measuring actual learning versus feeling of learning in response to being actively engaged in the classroom. *Proceedings of the National Academy of Sciences of the United States of America*, *116*(39), 19251–19257. doi:10.1073/pnas.1821936116 PMID:31484770

Dreikurs, R. (2000). *Social Equality: The Challenge of Today*. Adler School of Professional Psychology.

Entringer, T. M., & Gosling, S. D. (2021). Loneliness during a nationwide lockdown and the moderating effect of extroversion. *Social Psychological & Personality Science*, •••, 1–12.

Eisenberg, N. I. (2012). Broken hearts and broken bones: A neural perspective on the similarities between social and physical pain. *Current Directions in Psychological Science*, *21*(1), 42–47. doi:10.1177/0963721411429455

Ewell, P. T. (2010). The U.S. national survey of student engagement (NSSE). In *Public policy for academic quality* (pp. 83–97). Springer. doi:10.1007/978-90-481-3754-1_5

Ferguson. (1989). Adler's Motivational Theory: An Historical Perspective on Belonging and the Fundamental Human Striving. *Individual Psychology, 45*(3).

Ferguson, E. D. (2006). A Long-Awaited Book on Adlerian Psychotherapy for the Modern Reader. *PsycCRITIQUES, 51*(8).

Ferguson, E. D. (2010). Adler's Innovative Contributions Regarding the Need to Belong. *Journal of Individual Psychology, 66*(1). https://web.p.ebscohost.com/ehost/pdfviewer/pdfviewer?vid=1&sid=2d4830a9-1a93-4502-917b-ee9a8991535a%40redis

Flynn, S., & Noonan, G. (2020). Mind the gap: Academic staff experiences of remote teaching during the Covid 19 emergency. *All Ireland Journal of Higher Education, 12*(3).

Garrison, D. R. (2006). Online collaboration principles. *Journal of Asynchronous Learning Networks, 10*(1), 25–34.

Hales, A. H., Wood, N. R., & Williams, K. D. (2021). Navigating COVID-19: Insights from research on social ostracism. *Group Processes & Intergroup Relations, 24*(2), 306–310. doi:10.1177/1368430220981408

Hamza, C. A., Ewing, L., Heath, N. L., & Goldstein, A. L. (2021). When social isolation is nothing new: A longitudinal study on psychological distress during COVID-19 among university students with and without preexisting mental health concerns. *Canadian Psychology, 62*(1), 20–30. doi:10.1037/cap0000255

Hannay, M., & Newvine, T. (2006). Perceptions of distance learning: A comparison of online and traditional learning. *Journal of Online Learning and Teaching, 2*(1), 1–11.

Harasim, L. (2000). Shift happens: Online education as a new paradigm in learning. *The Internet and Higher Education, 3*(1-2), 41–61. doi:10.1016/S1096-7516(00)00032-4

Hofstede, G. (1980). Culture and organizations. *International Studies of Management & Organization, 10*(4), 15–41. doi:10.1080/00208825.1980.11656300

Hortulanus, R., Machielse, A., & Meeuwesen, L. (2009). *Social isolation in modern society*. Routledge.

Huang, K. Y., Kwon, S. C., Cheng, S., Kamboukos, D., Shelley, D., Brotman, L. M., Kaplan, S. A., Olugbenga, O., & Hoagwood, K. (2018). Unpacking partnership, engagement, and collaboration research to inform implementation strategies development: Theoretical frameworks and emerging methodologies. *Frontiers in Public Health, 6*, 190. doi:10.3389/fpubh.2018.00190 PMID:30050895

Hudson, E., Hamlin, E., & Cummings, J. (2020, September 11). *How to cultivate belonging in online learning*. GOA. Retrieved May 27, 2022, from https://globalonlineacademy.org/insights/articles/how-to-cultivate-belonging-in-online-learning

Husky, M. M., Kovess-Masfety, V., & Swendsen, J. D. (2020). Stress and anxiety among university students in France during Covid-19 mandatory confinement. *Comprehensive Psychiatry, 102*, 152191. doi:10.1016/j.comppsych.2020.152191 PMID:32688023

Inglehart, R. (2000). Culture and democracy. *Culture matters: How values shape human progress*, 80-97. Retrieved from https://www.academia.edu/download/55039033/Culture_Matters_How_Values_Shape_Human_Progress_-_SAMUEL_HUNTINGTON.pdf#page=115

Inglehart, R., & Baker, W. E. (2000). Modernization, cultural change, and the persistence of traditional values. *American Sociological Review*, *65*(1), 19–51. doi:10.2307/2657288

Kassop, M. (2003). Ten ways online education matches, or surpasses, face-to-face learning. *The Technology Source, 3.*

Kerr, N. A. (2021). Teaching in a lonely world: Educating students about the nature of loneliness and promoting social connection in the classroom. *Teaching of Psychology*, 1–11. doi:10.1177/00986283211043787

Kim, U., Triandis, H. C., Kâğitçibaşi, Ç., Choi, S.-C., & Yoon, G. (Eds.). (1994). *Individualism and collectivism: Theory, method, and applications*. Sage Publications, Inc.

Kovacs, B., Caplan, N., Grob, S., & King, M. (2021). Social networks and loneliness during the COVID-19 pandemic. *Socius: Sociological Research for a Dynamic World, 7*, 1–16. doi:10.1177/2378023120985254

Krings, V. C., Steeden, B., Abrams, D., & Hogg, M. A. (2021). Social attitudes and behavior in the COIVD-19 pandemic: Evidence and prospects from research on group processes and intergroup relations. *Group Processes & Intergroup Relations, 24*(4), 195–200. doi:10.1177/1368430220986673

Kuh, G. D. (2003). What we're learning about student engagement from NSSE: Benchmarks for effective educational practices. *Change: The Magazine of Higher Learning, 35*(2), 24-32.

Linders, B. (2018, May 11). *Psychological safety in training games*. InfoQ. https://infoq.com/articles/psychological-safety-games

Liu, H., Zhang, M., Yang, Q., & Yu, B. (2020). Gender differences in the influence of social isolation and loneliness on depressive symptoms in college students: A longitudinal study. *Social Psychiatry and Psychiatric Epidemiology*, *55*(2), 251–257. doi:10.100700127-019-01726-6 PMID:31115597

Lorenzo, G. (2001). Learning Anytime, Anywhere... for Everybody? Making Online Learning Accessible. *Distance Education Report*, *5*(11), 1–3.

Luchetti, M., Hyun Lee, J., Aschwanden, D., Esker, A., Strickhouser, J. E., Terracciano, A., & Sutin, A. R. (2020). *The trajectory of loneliness in response to COVID-19.* American Psychological Association. https://www.apa.org/pubs/journals/releases/amp-amp0000690.pdf

Maslow, A. H. (1968). *Toward a psychology of being*. Van Nostrand.

Markus, H. R., & Kitayama, S. (1991). Culture and the self: Implications for cognition, emotion, and motivation. *Psychological Review*, *98*(2), 224–253. doi:10.1037/0033-295X.98.2.224

Markus, H. R., Mullally, P. R., & Kitayama, S. (1997). *Selfways: Diversity in modes of cultural participation.* The Conceptual Self. In *Context: Culture, Experience, Self-Understanding.* University Cambridge Press.

McBrien, J. L., Cheng, R., & Jones, P. (2009). Virtual spaces: Employing a synchronous online classroom to facilitate student engagement in online learning. *International Review of Research in Open and Distributed Learning, 10*(3).

McDougall, J. (2015). The quest for authenticity: A study of an online discussion forum and the needs of adult learners. *Australian Journal of Adult Learning, 55*(1), 94–113.

McGinty, E. E., Presskreischer, R., Anderson, K. E., Han, H., & Barry, C. L. (2020). Psychological distress and COVID-19–related stressors reported in a longitudinal cohort of U.S. adults in April and July 2020. *Journal of the American Medical Association, 324*(24), 2555–2557. doi:10.1001/jama.2020.21231 PMID:33226420

Morris, T. A. (2010). Anytime/anywhere online learning: does it remove barriers for adult learners? In *Online education and adult learning: New frontiers for teaching practices* (pp. 115–123). IGI Global. doi:10.4018/978-1-60566-830-7.ch009

Morrow, J., & Ackermann, M. (2012). Intention to persist and retention of first-year students: The importance of motivation and sense of belonging. *College Student Journal, 46*(3), 483–491.

Müller, A. M., Goh, C., Lim, L. Z., & Gao, X. (2021). Covid-19 emergency elearning and beyond: Experiences and perspectives of university educators. *Education Sciences, 11*(1), 19. doi:10.3390/educsci11010019

Mund, M., Freuding, M. M., Möbius, K., Horn, N., & Neyer, F. J. (2020). The Stability and Change of Loneliness Across the Life Span: A Meta-Analysis of Longitudinal Studies. *Personality and Social Psychology Review, 24*(1), 24–52. doi:10.1177/1088868319850738 PMID:31179872

Ouzts, K. (2006). Sense of community in online courses. *Quarterly Review of Distance Education, 7*(3).

Over, H. (2016). The origins of belonging: social motivation in infants and young children. *Philosophical Transactions of the Royal Society B: Biological Sciences, 371*(1686).

Oyserman, D., & Lee, S. W. (2008). Does culture influence what and how we think? Effects of priming individualism and collectivism. *Psychological Bulletin, 134*(2), 311–342. doi:10.1037/0033-2909.134.2.311 PMID:18298274

Palloff, R. M., & Pratt, K. (2010). *Collaborating online: Learning together in community* (Vol. 32). John Wiley & Sons.

Peacock, S., & Cowan, J. (2019). Promoting sense of belonging in online learning communities of inquiry in accredited courses. *Online Learning, 23*(2), 67–81. doi:10.24059/olj.v23i2.1488

Peacock, S., Cowan, J., Irvine, L., & Williams, J. (2020). An exploration into the importance of a sense of belonging for online learners. *International Review of Research in Open and Distributed Learning, 21*(2), 18–35. doi:10.19173/irrodl.v20i5.4539

Peat, J. A., & Helland, K. R. (2004). *The Competitive Advantage of Online versus Traditional Education*. Online Submission.

Peplau, L. A., & Perlman, D. (1982). Perspectives on loneliness. In L. A. Peplau & D. Perlman (Eds.), *Loneliness* (pp. 1–18). Wiley.

Pressman, S. D., Jenkins, B., & Moskowitz, J. (2019). Positive Affect and Health: What do we know and where next should we go? *Annual Review of Psychology*, *70*(1), 627–650. doi:10.1146/annurev-psych-010418-102955 PMID:30260746

Public Health Reports. (2021). COVID-19 pandemic underscores the need to address social isolation and loneliness. *Association of Schools and Programs of Public Health, 136*(6), 653-655.

Rizzolatti, G., & Fogassi, L. (2014). The mirror mechanism: Recent findings and perspectives. *Philosophical Transactions of the Royal Society of London. Series B, Biological Sciences*, *369*(1644). doi:10.1098/rstb.2013.0420 PMID:24778385

Robinson, C. C., & Hullinger, H. (2008). New benchmarks in higher education: Student engagement in online learning. *Journal of Education for Business*, *84*(2), 101–109. doi:10.3200/JOEB.84.2.101-109

Samson, P. L. (2015). Fostering student engagement: Creative problem-solving in small group facilitations. *Collected Essays on Learning and Teaching, 8*, 153-164

Schweder, R. A., & Sullivan, M. A. (1993). Cultural psychology: Who needs it? *Annual Review of Psychology*, *44*(1), 497–523. doi:10.1146/annurev.ps.44.020193.002433

Seifert, T. (2016). Involvement, collaboration and engagement: Social networks through a pedagogical lens. *Journal of Learning Design*, *9*(2), 31–45. doi:10.5204/jld.v9i2.272

Shevlin, M., McBride, O., Murphy, J., Miller, J. G., Hartman, T. K., Levita, L., Mason, L., Martinez, A. P., McKay, R., Stocks, T., Bennett, K. M., Hyland, P., Karatzias, T., & Bentall, R. P. (2020). Anxiety, depression, traumatic stress and COVID-19-related anxiety in the U.K. general population during the COVID-19 pandemic. *BJPsych Open*, *6*(6), e125. doi:10.1192/bjo.2020.109 PMID:33070797

Shore, L. M., Randel, A. E., Chung, B. G., Dean, M. A., Ehrhart, K. H., & Singh, G. (2011). Inclusion and diversity in work groups: A review and model for future research. *Journal of Management*, *37*(4), 1262–1289. doi:10.1177/0149206310385943

Singh, J., Evans, E., Reed, A., Karch, L., Qualey, K., Singh, L., & Wiersma, H. (2021). Online, Hybrid, and Face-to-Face Learning Through the Eyes of Faculty, Students, Administrators, and Instructional Designers: Lessons Learned and Directions for the Post-Vaccine and Post-Pandemic/COVID-19 World. *Journal of Educational Technology Systems*.

Stein, H. T. (2002). A Psychology for Democracy*: Presentation at the IAIP Congress*. Alfred Adler Institutes of San Fransisco and Northwestern Washington.

Stojan, J., Haas, M., Thammasitboon, S., Lander, L., Evans, S., Pawlik, C., ... Daniel, M. (2022). Online learning developments in undergraduate medical education in response to the COVID-19 pandemic: A BEME systematic review: BEME Guide No. 69. *Medical Teacher*, *44*(2), 109–129.

Sun, A., & Chen, X. (2016). Online education and its effective practice: A research review. *Journal of Information Technology Education*, *15*, 15. doi:10.28945/3502

Sutia, C., Wulan, A. R., & Solihat, R. (2019, February). Students' response to project learning with online guidance through google classroom on biology projects. *Journal of Physics: Conference Series, 1157*(2), 022084. doi:10.1088/1742-6596/1157/2/022084

Thomas, L., Herbert, J., & Teras, M. (2014). *A sense of belonging to enhance participation, success and retention in online programs*. https://ro.uow.edu.au/asdpapers/475/

Triandis, H. C. (1995). A theoretical framework for the study of diversity. In M. M. Chemers, S. Oskamp, & M. A. Costanzo (Eds.), *Diversity in organizations: New perspectives for a changing workplace* (pp. 11–13). Sage Publications Inc. doi:10.4135/9781452243405.n2

Twenge, J. M., Spitzberg, B. H., & Campbell, W. K. (2019). Less in-person social interaction with peers among U.S. adolescents in the 21st century and links to loneliness. *Journal of Social and Personal Relationships, 36*(6), 1892–1913. doi:10.1177/0265407519836170

Tuck, N. L., Adams, K. S., Pressman, S. D., & Consedine, N. S. (2017). Greater ability to express positive emotion is associated with lower projected cardiovascular disease risk. *Journal of Behavioral Medicine, 40*(6), 855–864. doi:10.100710865-017-9852-0 PMID:28455831

Wang, M., Sierra, C., & Folger, T. (2003). Building a dynamic online learning community among adult learners. *Educational Media International, 40*(1-2), 49–62. doi:10.1080/0952398032000092116

Warren, C., Barsky, A., & McGraw, A. P. (2021). What makes things funny? An integrative review of the antecedents of laughter and amusement. *Personality and Social Psychology Review, 25*(1), 41–65. doi:10.1177/1088868320961909 PMID:33342368

Ways to create belonging in a virtual classroom (n.d.). UVA School of Nursing. Retrieved December 3, 2023, from https://www.nursing.virginia.edu/diversity/inclusive-teaching/belonging/

Westerhof, G. J., Dittmann-Kohli, F., & Katzko, M. W. (2000). Individualism and Collectivism in the Personal Meaning System of Elderly Adults: The United States and Congo/Zaire as an Example. *Journal of Cross-Cultural Psychology, 31*(6), 649–676. doi:10.1177/0022022100031006001

Wilkialis, L., Rodrigues, N. B., Cha, D. S., Siegel, A., Majeed, A., Lui, L. M. W., Tamura, J. K., Gill, B., Teopiz, K., & McIntyre, R. S. (2021). Social isolation, loneliness and generalized anxiety: Implications and associations during the COVID-19 quarantine. *Brain Sciences, 11*(12), 1620. doi:10.3390/brainsci11121620 PMID:34942920

Williams, K. D. (2009). Ostracism: Effects of being excluded and ignored. In M. P. Zanna (Ed.), Advances in experimental social psychology (Vol. 41, pp. 275–314). Academic Press.

Zhou, A., Sedikides, C., Mo, T., Li, W., Hong, E. K., & Wildschut, T. (2021). The restorative power of nostalgia: Thwarting loneliness by raising happiness. *Social Psychological & Personality Science*, 1–13.

Zoltán, A. (2011). Alfred Adler's Individual Psychology-towards an Integrative psychosocial foundation of the education in the 21st Century. *Reconnect-Electronic Journal of Social, Environmental and Cultural Studies, 3*(1), 4–19.

Section 3

Applying Principles of Remote Community Engagement

This section provides examples and guidance for the effective application of best practices for leadership with relevance to remote community engagement.

Chapter 12
Undergraduate Perspectives on Community-Engaged Service During COVID-19:
Exploring the Differences Between In-Person and Remote Tutoring Experiences

Craig Allen Talmage
https://orcid.org/0000-0002-0824-8364
Hobart & William Smith Colleges, USA

Kathleen Flowers
Hobart & William Smith Colleges, USA

Peter Budmen
Hobart & William Smith Colleges, USA

Alexander Cottrell
Hobart & William Smith Colleges, USA

Jonathan Garcia
Hobart & William Smith Colleges, USA

Jasmine Webb-Pellegrin
Hobart & William Smith Colleges, USA

ABSTRACT

This chapter chronicles a rapid pivot in community engagement from in-person tutoring to remote (also called virtual) tutoring and a gradual shift back to in-person tutoring during the COVID-19 pandemic. The chapter provides reflections from both staff members in a community engagement and service-learning office and college students who served as tutors of local children and youth during the COVID-19 pandemic. Readers are encouraged to reflect upon two main areas of interest in this chapter. First, dynamics between parents and tutors altered during COVID-19 as children and youth received virtual tutoring from home. Second, the importance of training and professional development for tutors regarding how to use different technologies and engagement strategies for virtual tutoring is imperative. Lessons learned from the Tutor Corps program are shared via reflections from both staff and tutors in hopes that others will share their experiences in rapidly innovating their community engagement work during COVID-19.

DOI: 10.4018/978-1-6684-5190-8.ch012

INTRODUCTION

Across the world, the paradigms and practices of community engagement were shocked by the disruption of the global pandemic caused by COVID-19. The resilience of programs, partnerships, initiatives, and other forces that aimed to deliver needed services and bolster well-being, especially for oft-marginalized and/or underserved community members were tested starting in 2020 with many adaptations continuing thereafter. In many communities (small and large), universities and colleges across the globe were thrusted to reexamine their positions as anchor institutions or hubs for community engagement, challenging and shifting modalities, mentalities, strategies, and structures.

During COVID-19, communities and universities cataloged stories of resilience and innovation highlighting ways to bolster well-being and address community needs; however, many of those stories have not yet been featured as of this writing. Many innovations at the program-level focused on shifting to remote modalities to deliver courses, hold meetings, and promote social interactions (Talmage et al., 2020; 2021). At the institutional partnerships level, many universities, cities, libraries, and other service-providers crafted websites and online tools to help community members find necessary resources (Talmage et al., 2021). During COVID-19, community engagement offices across higher education institutions activated to rapidly pivot how they remotely deliver services, engage with partners, support students, and assess impact.

This chapter shares the rapid adaptation of a more than 30-year-old tutoring program at Hobart and William Smith Colleges (HWS) in Geneva, New York held across seven physical locations in the Finger Lakes region of New York. The program is called HWS Corps' America Reads (also called Tutor Corps or HWS Corps). During COVID-19, the tutor program shifted to remote (i.e., Zoom) tutoring sessions. After the shift back to in-person instruction at HWS, portions of the remote tutoring remain as part of the program, specifically for STEM education for middle schoolers. This chapter chronicles this journey, especially reflections on the actions taken by community engagement staff and the perspectives of college students serving as tutors.

Overall, this chapter informs future local collaborative community engagement efforts beyond COVID-19 between higher education institutions and their local communities. In terms of the overall themes in this volume, readers of this chapter will be able to evaluate the critical areas of remote community engagement and apply best practices for remote community engagement. Readers who are looking to begin or refine their remote engagement will gain insights into effectively engaging local partners for support, parents, and community members looking for services. In particular, this chapter provides reflections from students and staff members to help identify the barriers to college-level student involvement and the motivations of college-level students to serve in remote community engagement initiatives, hopefully informing how to broaden participation and optimize recruitment strategies. The chapter will also highlight the strengths and challenges shared by students who participated in both modalities (remote vs. in-person) before, during, and after the shift to remote tutoring service-delivery.

RAPID PIVOTS IN UNIVERSITY COMMUNITY ENGAGEMENT AND TUTORING DURING COVID-19

The COVID-19 pandemic did not just impact the physical health and well-being of communities, but also the physical/social distancing, prevention measures, and sanitation protocols had profound impacts on mental/emotional health and social well-being of communities across the world. The pandemic disrupted supply chains impacting the quality and access to economic and social services, such as health care, grocery/food, and education. Universities shifted their modes of delivering education to their students and shifted their scientific research to prevent the spread of and treat COVID-19, but they also had to shift their programs that engaged with their broader local and international communities (Gilmore et al., 2020; Perrotta, 2021; Talmage et al., 2020; 2021). Importantly, universities are also connected to the culture and cultural well-being of communities, specifically in regard to arts, nature, and leisure activities; cultural well-being was also disrupted during COVID-19 (Mughal et al., 2022).

Much of community engagement involves service-learning, which has primarily occurred in face-to-face forms. Traditionally, university service-learning activities expressed in local communities include distribution of food to those in need, working with older adults or individuals with disabilities, or tutoring children and youth. In COVID-19, physical distancing to prevent the spread of COVID-19 required innovation in continuing these services, and most programs shifted to remote service delivery (Morton & Rosenfield, 2021; Weisman, 2021). While university responses such as switching to video conference technologies to hold classes showcase top-down responses, many more democratic, grassroots responses arose to address COVID-19 (Ohmer et al., 2021; Talmage et al, 2021). Ohmer and colleagues (2021) note the importance of this shift, so that universities can move beyond helping in their communities to actually engage in problem-solving modes in their community engagement during COVID-19. Furthermore, meaningful remote community engagement can provide supportive, stable, and meaningful benefits to students engaged in service-learning (Burton & Winter, 2021)

Many tutoring programs, in particular, had to shift to remote forms of engagement during COVID-19 (Burton & Winter, 2021; Chambers, 2020; Seru, 2021). While longer-term impacts and lessons learned will be published about the impacts of COVID-19 on tutoring programs, a few scholars have shared initial reflections and findings over the past three years. For the purpose of this piece, literature during COVID-19 on university tutoring programs shifts to remote modalities in general are discussed. At this time, this literature mostly focuses on students' experiences as peer-to-peer tutors at a university rather than being students who tutor children and youth in the community, which is the focus of this article. Still, emergent themes are offered from others and in this piece as they speak specifically to the student experience.

Remote tutoring appears to have been a highly sought-after service during COVID-19 (Yusuf, 2021). Prior to COVID-19, tutoring support often took place in the home with caregivers or siblings (Chabbot & Sinclair, 2020; Yang et al. 2021) and local and neighborhood levels (Chabbot & Sinclair, 2020), such as at community centers and schools. At local schools, students may engage with paid or volunteer tutors from local colleges and universities. Thus, the studies of remote tutoring during COVID-19 connected to the university setting supply considerations for future short-term/emergency or long-term remote tutoring programs.

From these students, scholars have noted that tutoring, even using remote modalities, should not only focus on knowledge acquisition, but also spur curiosity (Cao et al., 2021). The frequency of interactions between remote tutors and tutees matter (Mali & Lim, 2021). Remote tutoring requires substantial amounts of emotional connection and social behaviors, such as trust, empathy, collaboration, closeness, and security (Cao et al., 2021; Pérez-Jorge et al., 2020; Tragodara, 2021). Remote tutoring can and should provide rapport- and relationship-building opportunities, self-directed learning, and considerations of time limits when managing discussions (Kim et al., 2021).

The connections between emotion and motivation among those being tutored are essential considerations in effective remote tutoring (Tragodara, 2021). In this light, tutors are challenged to remotely mediate emotions during sessions as well as leverage optimism when dealing with tough motivational situations (Tragodara, 2021). Van Maaren et al. (2022) found that remote tutoring in general provided sufficient social integration and that quality of technology and individual well-being were not strong barriers; however, specific areas with issues arose (e.g., support for subject-level tutoring and first-year student social integration).

Schools explored different modalities and technologies for remote tutoring during COVID-19. The lessons learned from their usage are important considerations to be successful, as remote teaching and learning is expensive and requires buy-in from multiple stakeholders and proper transitions from one modality to other (Aboagye, 2021). Furthermore, the users of these technologies, specifically the tutors, should receive training regarding the remote tutoring technologies, and they should also be provided proper technological devices for remote tutoring (Johns & Mills, 2021; Omodan & Ige, 2021). Pérez-Jorge et al. (202) contend that synchronous models are the most effective models compared to asynchronous models, but they need to be enhanced. Johns and Mills (2021) recommended both synchronous and asynchronous modalities be provided during COVID-19 to increase flexibility and accessibility.

Okro et al. (2021) added other notable considerations to the discussion of remote tutoring. First, remote tutoring requires matching between tutors and tutees beyond topical knowledge; matching also must include digital capabilities. Second, remote tutors require more and regular professional development. Third, communication procedures and tutoring processes must be more standardized to manage remote tutoring. Fourth, remote tutoring requires different strategies and operations and collaboration across different departmental and university levels. Finally, processes for tutoring and collecting data on tutoring must be agile and aligned with student needs and situations.

This chapter pays attention to the aforementioned elements when discussing the story of HWS Corp's America Reads (also called Tutor Corps or HWS Corps). For simplicity, Tutor Corps will be the name used throughout the chapter. Different from the articles cited, this story focuses on college students tutoring local children and youth rather than peer-tutoring or professional-tutoring in a university setting. In particular, the story from staff and reflections of students will connect to perceptions of tutoring effectiveness, technologies, and modalities, training and development, matching of tutors, and other notable processes and outcomes.

RAPID PIVOTS IN TUTOR CORPS DURING COVID-19

Tutor Corps started in 1988 as a writing and reading workshop by students and faculty at HWS with the goal of increasing literacy among Geneva youth. The Center for Community Engagement and Service Learning (CCESL) at HWS runs Tutor Corps, which generally recruits around 100 undergraduate students each year from HWS to serve as tutors. Tutors are paid, and many qualify for and receive federal work study support. The tutoring is free to local parents and children and youth. Tutors meet with K-12th grade students 45 minutes twice a week. Tutors are matched with local students based on the expertise of each tutor, scheduling, student needs, and common interests. Any local student who has a teacher recommendation is given priority in the matching process.

During COVID-19, the Boys and Girls Club of Geneva (called the Club) remained a substantial partner for the Tutor Corps program. The Tutor Corps connected with Club members (local students) after school through Zoom. The HWS tutors helped students with homework and practicing math and reading skills, while also mentoring them through online games and relationship-building exercises. Tutors were also supported by civic leaders and coordinators, experienced student leaders within the office, each semester. To shift to remote modalities, CCESL staff had to create a website and leverage Google forms for parents to sign their students up, in which there was a Spanish version for both as well. Safety protocols had to be refined and implemented, since HWS was technically hosting minors on campus, although through laptops; All tutors took the HWS Minors on Campus[1] training course, completed background checks and participated in extensive virtual training sessions, so they could record and save all tutor sessions for future reference. One HWS Educational Studies professor and two community partners provided a professional development opportunity and collectively watched a showing of Beyond Measure[2], an educational documentary showing how education in America can be re-imagined.

Remote tutoring was a dramatic switch even after the on-going evolution of the 30-year in-person program. Much of the shift and resources were inspired by the community engagement work around remote tutoring at Wake Forest University[3] and Summer 2020 webinar series put on by Campus Compact[4]. The shift to remote modalities allowed CCESL, HWS, and Boys and Girls Club of Geneva to support local students while also continuing to provide employment opportunities for HWS students. While this chapter will supply information on impacts perceived by parents and children, the primary focus of the chapter is on student perspectives as tutors prior to, during, and after COVID-19 necessitated remote tutoring and reflections of CCESL staff. A series of student voices are presented collected from informal interviews with and email solicitations to tutors during Spring 2022. These responses highlight lessons and insights, on-going questions and issues, and barriers and limitations to remote community engagement.

Today, portions of the remote engagement remain. In response to COVID-19, CCESL curated multiple resources with the local city and organization (Talmage et al., 2021). As of this writing, CCESL still maintains an on-going list of virtual community engagement opportunities[5]. Additionally, this chapter will share decision-making regarding keeping the STEM-focused tutoring program as a remote tutoring program for local middle school students. The chapter continues by giving history behind Tutor Corps, chronicling how it changed during COVID-19, sharing student reflections, and concluding with insights for future research and practice.

BRIEF HISTORY

The HWS Student Literacy Corps began in 1988 by two students with little funding, but a big idea to improve local student literacy (HWS, 2014). It started as a more general literacy program, and then evolved to align with America Reads, a federal program in 1997 started under the Clinton Administration (Edmonson, 1999). HWS has continued to support literacy in local schools and afterschool programs throughout the decades, allowing paid and trained tutors to work one-on-one with elementary students who need extra help in expanding their literacy skills and exploring the best parts of learning to read well (HWS, 2022). America Reads became HWS Corps in 2018 when grant alignment became possible with the 21st Century after-school grant in partnership with the Geneva City School District and the Boys and Girls Club of Geneva. Notable differences are literacy tutoring is the focus for HWS Corps at America Reads sites and homework help at HWS Corps at Boys and Girls Club sites (HWS, 2019). For this chapter, the two sites are discussed together as the chronicle of changes during COVID-19 is presented.

SPRING 2020 - SEMESTER INTERRUPTED / EMERGENCY REMOTE LEARNING

During the spring of 2020, like many colleges in the US, HWS sent students home out of an abundance of caution as the COVID-19 virus spread across the world and the United States. Students departed for Spring Break in late March 2020, only to receive an email days later that they would be unable to return to in-person courses. For all, it was a challenging time of disruption, confusion, and abrupt change to countless traditions found at a private, residential liberal arts college. The staff at CCESL immediately recognized a need to thoughtfully, but also rapidly, pivot community engaged programs to bolster the resilience of their college and community programs.

Together, the CCESL staff and Tutor Corps' tutors crafted online resources and processes, while young children and college students alike adjusted to remote learning to conclude the academic year. Under the direction of CCESL staff, tutors worked hard to adjust to their new reality, while juggling numerous other pandemic induced circumstances (e.g., health concerns for loved ones, travel interruptions most especially for international students, Wi-Fi access to continue studies, etc.) and developed resources, including curated and unique YouTube content, playlists, and videos[6]. For one example, CCESL worked with an introductory psychology class on developing fun activities and web resources for children. The tutors put together 70 videos with fun facts, writing tips, topical overviews, arts and crafts, cooking and baking, book suggestions, music lessons, and stories about the college experience. These resources were shared with teachers, administrators, and parents, as CCESL staff were able to identify contact information and provided an opportunity for tutors to earn allocated federal work study funding to help pay for tuition and school expenses (HWS, 2020a).

FALL 2020 TO SPRING 2021 - LEARNING IN MASKS AND BEFORE FULL U.S. VACCINE ROLLOUT

With summer to prepare for a fully remote tutoring year, the CCESL staff adjusted strategies and resources to utilize online resources, specifically Google Classroom and Zoom, to provide tutoring services to children and their families, teachers and administrators. After an inordinate amount of time and energy spent researching virtual/remote tutoring programs, implementing safety protocols, and recruiting participation through communication sent by schools and CCESL staff outreach to parent groups on social media, CCESL staff engaged 21 schools in the Finger Lakes region of New York, supported 95 students from pre-kindergarten to 12th grade, and employed 64 HWS tutors. The majority of tutors (57.89%) worked with students at grade 5 or lower. Grades 6 to 8 had 15 students served (15.78%). 25 high schoolers (26.32%) were tutored as well. The largest grade levels served, proportionally, were 2nd (20.00%), 3rd (11.58%), and 10th (10.53%) grades. All of the schools who had tutors remained in-person or switched to a hybrid schedule, but the tutors (and also in most cases volunteers or parents) were not physically allowed into the school. The majority (70.52%) of the children and youth being tutored were enrolled in local public schools, but when CCESL staff realized the program could support additional children and more tutors, they ramped up recruitment (to reach the 95 [100%] above). The recruitment increase allowed for the broadest territorial outreach in the program's thirty-four-year history and provided an income source for HWS tutors.

The matching process utilized by Tutor Corps allowed for input from parents and teachers, and in ideal situations, input from *both* parents and teachers, so that CCESL staff could most effectively identify how the child's needs, age, interests, among others could be paired with a tutor who had corresponding experiences to best foster a developmental learning environment.[7] Previously, during in-person sessions, tutors relied upon teacher input and opening activities facilitated by experienced tutor coordinators to pair up tutor/student matches and set a positive tone for a productive semester, so remote-only tutoring was much different as it relied on information from both parents or teachers. One tutor reflected on the "perceived equity gap" where the children she was supporting were reading above-grade-level and had consistent access to personal devices and reliable Wi-Fi, whereas previously tutors were paired only with children who needed additional reading support to get caught up to grade-level reading expectations. After contact information for tutors were shared, tutors and students connected and schedules were set up to ensure a twice weekly commitment.

CCESL staff recognized they needed to find ways to train the tutors to be resilient in the face of the unknown. Tutor training occurred virtually on Zoom and was recorded for later viewing by tutors and staff. Training heavily focused on diving deeper into the America Reads curriculum, which aligns with igniting or sustaining a love of reading is based on Vygotsky's Zone of Proximal Development[8] (Hughes et al., 2011; San Tham, 2020). Training placed emphasis on having fun and focusing on what the students are doing well. Fittingly, one tutor shared that to adapt to tutoring during COVID-19 that they "engaged the students in conversation and got to know them, so they would focus on the work we had to do." The tutors were prompted to look further at the America Reads curriculum and tools even prior to engaging their students.

An additional challenge for the tutors that was shared during the training was that they had to address both parent and teacher goals for their students. The tutors might need to leverage the parents as intermediaries to communicate with teachers if there were questions about their students' needs and goals. Tutors were encouraged to "take it all in stride" as there would be limited info on students and their individual issues. Finally, the training discussed how to deal with online resources that are free but require login emails or signup forms, which would require parental or CCESL help. Additionally, the training emphasized sharing additional resources to support tutors and students with CCESL staff to be vetted and then sent to other tutors.

The virtual training and ongoing mentoring from CCESL staff contain additional advice to navigate the remote tutoring sessions in line with the Zone of Proximal Development. For example, when a book is too easy, a student can lose interest quickly. If a book is too difficult, then the child can become frustrated. Finding the 'sweet spot' between challenge and comfort, can scaffold towards successful reading habits and build fluency, comprehension, and vocabulary. Determining the reading level of the child can be difficult when sessions are in person with timely input from a teacher, so working to identify a reading level remotely was especially challenging. CCESL staff encouraged tutors to ask parents if they knew the reading level of their child, so that books could be identified. Tutors also asked their students what their favorite book or author was, and then they could research that reading level accordingly and select appropriate future books. If the student did not have a favorite author or book, the tutor asked about interests and researched resources for engagement accordingly.

Additional off-line resources were supplied to the tutors. A tutor manual was shared with tutors, although the activities were not yet adapted to the virtual realm. Still, they were adaptable to virtual exchanges and utilized as a way to show new tutors and remind returning tutors what the 'ideal' was to attain during a tutor-student interaction. The tutor manual highlighted opening 'get to know you' activities, a contract of mutual expectation (which prompted a conversation as opposed to a hard copy signed contract as in previous in-person semesters), and numerous activities that had been developed to engage the child in reading (e.g., write a letter to the author, text-to-text, text-to-self, text-to-world, among others[9]). Online resources were added to a newly developed online tutor portal including lesson plans, learning tools, and fun games and exercises were curated and added, drawing from sites such as www.readworks.org, www.starfall.org, www.storyjumper.com, among others. Offline in the CCESL office on campus, tutors were welcome to visit the Book Nook and select books, if they felt comfortable and able to do so.

The sessions began in earnest. As with in-person tutoring, tutors were encouraged by the CCESL staff and resources to over-prepare! Tutors were sensitive to the numerous struggles that their student(s) were facing. Tutors shared how challenging it was to engage their students on Zoom after the novelty wore off and that they worried about the chaos of some of the situations they would witness while tutoring. CCESL staff reminded tutors that their presence, although virtual, was an important and consistent opportunity to highlight positive role modeling and learning in a new way amidst a tumultuous time. Tutor Jasmine shared, "I would let the student pick a game, sometimes it had nothing to do with reading, but it would calm her down which would help when we read later," and CCESL staff reiterated that handling a situation with creativity and flexibility was of utmost importance.

CCESL staff added additional professional development opportunities throughout the semester that highlighted potential adverse childhood experiences and the impact of stressors on learning. CCESL hosted a local school district social worker and HWS professor to talk about how the pandemic affects well-being. They also discussed how U.S. residents rely on schools to supply food, health screenings, and emotional support in the form of counselors for students and families in addition to social support and educational services provided by teachers and support staff. All of these networks and tangible services were impacted during COVID-19. Congruently, remote learning requires resources, and well-resourced communities, which varied across the Tutor Corps sites, had adequate ability to provide high quality internet access and individual/personal devices through which remote learning was made possible. Others did not. Another professional development opportunity included a local social worker, who was also a parent. She offered multiple strategies for the tutors to better engage with students and families.

Tutors who had experience with previous in-person tutoring sessions, noted the difference in family engagement, as parents and tutors never or rarely meet during a typical day in the previous Tutor Corps program; that program occurred after school and students were transported home on buses. The remote sessions also occurred during non-traditional times including evenings or weekends when the child was home from school. (As a reminder, K-5 grades held school in-person, and 6-12 grade students throughout the region were on a hybrid schedule). This new schedule bringing tutors and families together virtually was a welcome aspect to this three-decade old program.

In this same light, parents were provided a virtual tutoring handbook. The handbook contained an overview of the virtual tutoring program, general guidelines (see Table 1), zoom instructions, additional resources for learning, frequently asked questions, and contact information. An additional highlight allowed for sessions to continue after the semester concluded, which allowed for extra weeks of tutoring and a 'change of pace' for tutors to work from home environments as opposed to their residence halls or dedicated library spaces.

These new connections and encounters between parents and tutors had many positive aspects, however, frustration was felt by the tutors when parents would forget to login a young child or an older child would forget to login his/herself. One tutor noted "It was tough when a parent wouldn't communicate with you… you'd wait for 20 minutes after texting a reminder to parent…It was a struggle!" One tutor shared, "It was easier to interact with parents during COVID because most of them were home and working online, but now [Spring 2022] it is hard since they are all busy trying to adjust going back to work in-person."

Table 1. General guidelines for parents during virtual/remote tutoring

Advice Area	Items
Communication	● It is important to communicate with your tutor and with us regularly. ● Once a tutor is matched to your student, we ask that you find a time to talk in more detail about the academic needs of your child and the best times to meet. ● Make sure to check in with your tutor at the beginning of every session. ● Our office will send out bi-weekly surveys. Please complete this survey to provide feedback on the program and your individual tutoring sessions. ● Please communicate with the tutor directly about canceling any sessions. After two missed sessions without communication, your child will be removed from the program.
Safety	● We are working together to ensure the safety of our students. ● A parent/guardian must check in with the student at the beginning of every session. This is to ensure that an adult is present during the tutoring session. ● Please be present in the household for the duration of the session. ● All tutors will take the HWS Minors on Campus Online Training, as well as have had background checks. ● HWS will record all tutoring sessions and the recordings will be stored on HWS's online storage platform for the safety of both HWS tutors and local students. HWS tutor coordinators will be overseeing this process. These recordings will not be made public, although a parent can request to view their child's tutoring sessions.
Expectations	● Your child is expected to receive tutoring for 45 minutes twice a week. Tutors are not allowed to tutor for more than 45 minutes. This ensures equity for all students. ● Tutors are not allowed to meet their tutees in person, as this is set up for virtual tutoring only. ● Communicate with your tutor if you cannot make a session. In the event that your student misses two sessions without communication, they will forgo the opportunity to be in the program. Likewise, if an HWS tutor misses two sessions without notice, they will also forgo the opportunity to be in the program. ● You are expected to complete biweekly surveys through email about the program and how your child is doing with your tutor.
General Tips	● If your child has a work packet or specific homework, tutors are happy to work through it with the student. Please share the packet if you are able. You can do this by forwarding electronic copies to the student or by taking photos of the packet. ● The tutor/students can decide how to design each 45-minute session. ● Please make sure your child has a quiet space for the tutoring session. If you don't have access to a space conducive to learning, consider going to the Geneva Public Library or other quiet space for your tutoring sessions. ● Reminder: This program is **FREE** to all local parents and students.

When parents were able to consistently maintain appointments for younger children and older children were able to log-on for their scheduled tutoring sessions, and tutors were able to prompt a sincere connection with their partner students, the experience became more meaningful for both tutors and youth. One reflection from a parent provides a strong testimony of the powerful impact of remote tutoring:

My two children (1st and 3rd grade) participated in virtual tutoring during the 2020-21 school year when they were learning from home as part of a home learning pod. They both looked forward to their tutoring sessions, which were often the highlight of their weeks! Their tutors found creative ways to keep them engaged and excited about reading, and served as caring mentors as role models for them. The sessions were often the topic of our dinner conversations as they excitedly told us about what activities they had completed with their tutors that particular day. Since all sessions happened virtually throughout the year, it was a very special moment when my daughter had a chance to meet her tutor (Hannah) outside on the campus quad at the end of the spring 2021 semester. Hannah gave her a framed picture that was a screenshot of the two of them during one of their sessions. That picture was placed right on the mantle in our living room where it stayed for many months before being moved to a special place in my daughter's room. Both of my kids still mention their virtual tutors often. When I recently saw Hannah on campus, I just had to give her a big hug and offer words of appreciation to let her know how much the time and effort that she devoted to my daughter had meant to me. The students who worked with my children via virtual tutoring during 2020-21 had such a positive impact on my children (and our entire family!) during a very challenging time. We are so grateful to have been a part of that program!

Although there were many barriers to learning and increased stress from numerous uncertainties brought on by the pandemic, technology became a manner through which new connections could be created and maintained; however, technology was not without issues for tutors, parents, and students. One tutor shared:

Technology was a big challenge during remote tutoring…the features available weren't conducive for math tutoring. Before this, I could easily write something down and turn it around for the tutee to see but through zoom I would have to hold up the paper to the camera for them to read, which was challenging for both of us. The challenges were definitely greater during remote tutoring.

Another tutor shared that with technology, "Everything was more flexible but very chaotic. Sometimes the parents didn't know how to fix a problem in the computer, and I had to walk them through the steps. I had to learn a little bit of everything." Technology was also limited regarding the ability to keep students engaged. One tutor wrote, "The biggest challenge was engaging with the student. Most students were home and, in their room, so they weren't paying as much attention as they do if they were in person." Another wrote, "Keeping kids focused and on task is the biggest challenge. It was easier to do that in person." Despite frustrations, tutors appreciated being shown how to use the Zoom whiteboard, but felt it did not substitute for pencil and paper.

In addition to employment, HWS tutors were also enrolled in service-learning classes for graded credit through which remote tutoring was a popular selection for students to connect course content and experiential, community engaged learning. In one particular elementary school classroom, a teacher indicated an interest in creating a pen pal program where 3rd graders and HWS students would write to each other weekly with help from specific prompts. At the conclusion of the semester, a few of the pen pals were able to virtually meet their assigned writers, which tutors reflected was a special experience that will surely be remembered by tutors and students on both sides of the writing and receiving of the letters. One tutor reflected, "Hearing back from them [students] is always something I look forward to because I love hearing how they respond to the questions I ask them" and when asked about the final virtual meeting he saw it as a "truly an amazing experience to finally connect with these students and finally know what my pen pals look and sound like" (HWS, 2021a, para 5).

Tutor Corps' Virtual STEM (Algebra 1) was able to begin because of Tutor Corps' switch to remote tutoring for the 2020-2021 school year. The work of determining major logistics and barriers had been developed in partnership with CCESL, Geneva City School District (GCSD), and other community partners. However, Tutor Corps' Virtual STEM (Algebra 1) program was specifically targeted towards eighth and ninth graders taking Algebra 1. GCSD wanted to restructure their curriculum to have all eighth graders taking the Algebra 1 Regents Exam by the 2022-2023 school year. Virtual STEM tutors reflected, "This strategic reorganization of the math curriculum in Geneva had been made in the hopes that students who pass Algebra I in middle school will have greater rates of success in subsequent math courses, and more opportunities to take Calculus and other higher-level math courses during their high school careers."

Obstacles in staffing, both at CCESL and within the GSCD school district, left the Tutor Corps' Virtual STEM (Algebra 1) program wanting more. CCESL staff anticipated having all tutors matched with two students, but communication barriers and outreach prevented the program from working at full capacity. CCESL staff were hopeful that through constant and clear communication with GCSD that they could target students that would benefit from the program in the following year.

FALL 2021 AND SPRING 2022 - RETURNING TO IN-PERSON

As HWS prepared for a predominantly in-person return to instruction on campus during the fall 2021-2022 academic year (with the exception of Virtual STEM tutoring), CCESL staff raced to find placements for their tutors in local schools during a time when state guidance for K-12 governance seemed to change on a daily basis. Tutor Corps was able to return to in-person sessions at the Boys and Girls Club, and expanded outreach to two surrounding rural districts and a private Montessori-based school. Through Tutor Corps, CCESL employed 24 tutors at the Boys and Girls Club and 37 in school-based settings. Again, the STEM program remained remote during the academic year with 9 tutors and 16 students.

Unfortunately, leadership changes and pandemic uncertainty impacted the large tutor placement numbers at the closest local public school in the same city as the college. If CCESL staff had known that the local school district was not going to be able to take any tutors in-person, the remote tutoring program may have been maintained. Having 34 fewer tutors enrolled than the previous year was disconcerting to CCESL staff, considering national growing awareness of the pandemic learning losses and a sensitivity among CCESL staff towards ensuring employment opportunities for HWS tutors seeking federal work study opportunities.

Challenges arose throughout the return to in-person tutoring. Through an unfortunate combination of haste and subsequent miscommunication, one of the partnership sites wanted to host tutors, but then also were already hosting experienced teachers during an after-school tutoring program. When HWS tutors arrived for a simultaneous school tutoring effort, confusion and awkwardness arose. Among all parties involved, it was subsequently clarified that what was most important to the partnership was that the needs of the children and youth in this particular community were already being met. These particular HWS tutors were transferred to another district to assist with arts instruction, a much needed and missed experience throughout the COVID-19 lockdown. Throughout the numerous interactions with teachers, parents, and administrators at the 21 schools, it was clear to the CCESL staff how much stress and pressure everyone was under. It is essential to note with gratitude the trust with which the schools showed to CCESL regarding their (flex)ability to manage the tutoring program in different ways.

A common theme CCESL staff heard from tutors was that *my student does not want to read or they are having a tough time focusing*. Looking at the experience students are coming out of, COVID-19, remote learning, and other adverse experiences (e.g., traumas), CCESL staff urged tutors to identify what their students needed most. Relationship building, play, and creativity were all encouraged and prioritized throughout the Spring 2022 semester. Professional development was provided to tutors around the power of play and the equal importance that play has on a child's learning experience.[10]

In reflecting on the pre-, during, and post-COVID tutoring experiences, many tutors expressed a profound necessity to reimagine the traditional notions of tutoring; one tutor commented, "One of my students hated reading, I'd split up the time to do more entertaining things (Starfall) and then would build back in reading." Echoing that same sentiment, many tutors were surprised by the different roles they were required to take on such as that of teacher (behavior management), counselor (emotional regulation), and instructor (what is developmentally appropriate) when positioned online without the ease of access to a professionally trained teacher. Tutors shared they worked more closely with teachers and families alike to discuss the student's needs, progress and success more so than when tutoring in person. Once again, the CCESL team is proud of the skills born out of creativity and necessity. Flexibility, understanding/empathy, and creativity were all stretched in ways that might not have organically happened with in-person tutoring.

Diversity, equity, and inclusion are core values of the students at HWS, a few of the tutors appreciated that pandemic tutoring was open to all students and not just students who were below grade level noting the difference between the ability to read and the ability to comprehend. This same student also recognized the vast inequalities that existed with remote-style tutoring such as access to reliable high-speed internet and a private safe place to learn while remote. The local community had numerous ways for children to receive free books: At well-child care visits at pediatricians' offices[11], little free libraries located throughout local communities, and purchases at book fairs with financial support from a local literacy nonprofit[12]. However, early readers can still face barriers that impact their success with and their affinity for literacy as some books are not accessible in a child's language, interest level, or affirmed outside of school-based settings.

REFLECTING BACK AND LOOKING FORWARD IN PRACTICE

Students are not born hating reading, but they rather can develop a dislike for it. The author, James Patterson tweeted, "There's no such thing as a kid who hates reading. There are kids who love reading, and kids who are reading the wrong books" (JP_Books, 2014). Reading at its core is a highly personal and intimate experience. It is personal in that all readers have varying levels of comfort with their individual reading ability, and intimate in that literature allows the reader to experience, explore, and empathize a parallel world. While the remote model of tutoring allowed for one-on-one connection and a more 'private' tutoring experience, the need and limitations of technology made 'traditional' reading from a book more difficult. As a result, tutors utilized websites and materials made available to them and through their own research to interact and promote learning remotely. One of the largest frustrations tutors expressed was the untimely, inconvenient, or often missed tutoring sessions. College students often share with CCESL staff that they feel their schedules are tight, and CCESL staff heard over and over again how frustrating it would be when students would miss, or they would have to contact parents to help students get on their zoom session. One tutor shared, "Remote tutoring was worse, sometimes the kid wouldn't even log in and if they did, they didn't want to do anything. Now kids are more excited to be in school because they can see their friends and play with other kids." Another shared:

Remote was difficult in that students were distracted at times. It was also hard to connect with them and harder to see if they were having difficulty with what I was teaching. It was better than not having tutoring at all though because it gave the students an opportunity to improve their skills.

For all parties engaged with Tutor Corps, the return to in-person learning also brought with it a greater consideration of the trauma and effects of living through a global pandemic for students and instructors alike. It is incredible to consider that K-6 students went from being walked from class to class within the school to navigating a new schedule, joining Zoom links, and participating in a largely independently driven learning experience which required self-discipline, motivation, and commitment. In this regard, CCESL staff are optimistic and hopeful of the skills students gained during their time with remote learning.

Anecdotally, when surveyed at the end of Spring 2022 semester Tutor Corps meeting, all but two of the tutors said they would prefer to remain with in-person tutoring, even if provided the opportunity to go remote in Fall 2022. One tutor noted that students are more excited to be back in-person as it allows them to play with their friends. As CCESL staff look to address the needs of HWS students, CCESL staff are also giving consideration to how CCESL staff can best support tutors who concluded their high school years amidst turmoil. CCESL staff are also reflecting on how to support future children and youth who are inundated with personal stressors and traumatic events that continue to play out nationally; it seems to CCESL staff that the zone of proximal development appears to be a moving target.

One of the ways the Tutor Corps are hoping to address this concern is through a revamp of the CCESL book collection. To CCESL staff, it is now time to weed out the selection of old, outdated, and culturally insensitive literature and compile more inclusive books with representative characters, plot lines, and themes of diverse backgrounds. The CCESL staff recognize the power of appropriate cultural representation and creating mirrors and windows for students. When students can see themselves in literature, especially represented in a positive manner, they may then be able to see themselves, windows, in a future aspiration that had not previously been considered. Additionally, Tutor Corps has recently been gifted a set of iPads for use with the tutoring program which have been set up with educational apps, vetted by the CCESL Civic Leader for Literacy (who also is a future Masters of Arts in Teaching student). The iPads also have programs that can aid and augment the tutoring experience of Tutor Corps students. Reflecting overall, the technology advances born out of the pandemic allowed Tutor Corps to utilize tools that had not been, nor would have been, fully explored prior to COVID-19.

When it comes to supporting tutors in their academic, personal, and professional lives, the Tutor Corps experience has aided many tutors in finding a sense of purpose and belonging. Based on CCESL's end of year survey results of tutors, 22% and 71% of tutors who responded agreed or strongly agreed that by serving in Tutor Corps they felt more connected to their local community in which they served. Similarly, 44% and 47% agreed and strongly agreed with the statement "I feel more connected and more friendly with my HWS peers as a result of serving with Tutor Corps." One tutor wrote, "I appreciate it [community engagement] more because of the people that I work with and the students." Another reflected, "It [remote tutoring] was creative and better then if we didn't have it at all. It helped me to be more creative in how I helped a student with a subject." The act of service, a shared experience and sense of camaraderie born of it contributes highly to the retention and success of the Tutor Corps program. This notion is resonant among tutors surveyed given that 66% of them tutored two or more semesters. In similar fashion to prioritizing play and relationships with the students, the CCESL staff aims to continue to work hard to build community among the tutors.

Tutoring is CCESL's flagship program, due to the long-standing partnership with area schools and districts and the mutually beneficial impact it has on children (e.g., reading confidence, role modeling, exposure to the college experience, among others) as well as a beneficial impact on tutors (e.g., civic development, career exploration, community connection, earning federal work study funding to assist with education/living expenses, among others). For example, HWS's annual federal work study expenditures dedicated to America Reads far exceeds the 7% federal minimum requirement (U.S. DOE, n.d.) and is among the indicators that signal HWS's commitment to the community as captured in reporting for the successful 2010 and 2020 Carnegie Community Engagement Classifications (HWS, 2020b) and the Washington Monthly College Guide and Rankings (HWS, 2021b).

Although important to track, record, and celebrate reading gains and tutor commitment, CCESL staff have witnessed many valuable anecdotal reflections. In partnership with the HWS Educational Studies Department and Geneva 2030, CCESL staff look forward to reviewing available data and implementing collaborative programming and supportive interventions with CCESL's school partners, children and families. CCESL staff recognize that there is much work to do and that the impact of COVID-19 will be far reaching. Whether through education gains disrupted or personal goals, families and communities impacted, CCESL staff are optimistic that, through relationships that have been maintained or created since the spring of 2020, Tutor Corps will continue to support HWS students with opportunities to connect with local youth. The focus for CCESL staff will continue to support literacy and academic success, while also showing gratitude for how technology kept up connectedness during a time of social/physical distancing.

Overall, CCESL staff and tutors have a more robust appreciation for the way that community building occurs through in-person connections. One tutor shared:

I've always had a passion for tutoring. I'm nearly a senior in college now and I've been tutoring since sophomore year of high school...I'm passionate about becoming an educator. Once tutoring went remote, I felt as though it took the fun out of tutoring. It felt like students were more focused to just get the answer rather than being able to sit down with them and work through a problem.

Another student shared that tutoring in-person is more fun, "there's so much more activities that we can do." In another reflection, a tutor noted, "It was easier to connect with the students in person. It was easier to teach and easier to see if the students were struggling." This final reflection really shows depth in summarizing the tutor experience.

Remote tutoring was nice in the sense that it ensured that I was safe from possible covid exposure, however, the remote tutoring was very challenging. For one, it wasn't personable at all. I was used to having regular tutee's coming every week and I would really get to know them outside of tutoring, whereas when it was over zoom, not as many would request tutoring appointments and if they did, they tended to keep their camera off which made it hard to get to know them. In a way tutoring over zoom felt like the student would show up to get their questions answered but then they would leave right away so it's safe to say that the appointments were much quicker. Along with that, I noticed that there were less repeat tutee's when tutoring was done through zoom as opposed to in-person, which just shows that students are less likely to request a tutor when it's remote. Another challenge with tutoring remotely was showing each other our work. I would have the tutee hold up their paper close to the camera so I could read the question and get a sense as to what the question was asking but it was usually really hard to read it. Then, once I saw the question, I would use the whiteboard feature to write some notes or key points down, but this was very sloppy because I had to write with my mouse since I didn't have an iPad and the type feature wouldn't display the symbols or equations we needed. So, it is definitely nice to write something down on a piece of paper and show it directly to the student for in-person meetings. Like I mentioned previously, in-person meetings are way more personable and they allow me, the tutor, to get to know those who I work with, which is really something that I enjoy over the several years that I've been a tutor. Overall, I feel as though in-person tutoring appointments are way more beneficial to the students who utilize these services.

While the majority of tutoring has shifted back to in-person, the Virtual STEM program remains actively online for now. With a dedicated CCESL staff liaison to the Tutor Corps' Virtual STEM program, questions and other communications are streamlined for more timely responses. Talks for the 2022 school year are progressing, but it is assumed the program will remain virtual for the time being. As these tutoring efforts (both in person and virtual/remote) progress and evolve CCESL staff are looking for even more ways to incorporate playing and learning. Together, two Virtual STEM tutors wrote:

The HWS Corps Virtual STEM program STEM (Algebra I) Tutoring Program faced was the low-levels of tutor initiative, in that a number tutors not only neglected to inform the program coordinator that students were missing meetings, but also turned down requests to tutor a second student (despite being informed during the hiring process that they would be responsible for tutoring two students).

CCESL staff with student leaders aim to bolster the STEM tutoring program by introducing more focused community building and social interaction efforts for both tutors and tutees. Through these low-stakes interactions, tutors and tutees will be able to develop and grow relationships that are not hindered by the routine of tutoring sessions. Alex Cottrell, CCESL Program Coordinator who also simultaneously worked for the Boys and Girls Club, remarked:

While I was not around to witness the virtual tutoring efforts, the relationships formed by tutors and tutees as they were able to play in the different program areas (e.g., gym, art, game room, and computer lab) became so strong. This trust and dependence allowed the children to forge relationships and friendships because of their ability to play and learn at the same time.

CONCLUDING REMARKS

This chapter showcased the story of a rapid pivot to remote tutoring and shifts back to in-person tutoring during COVID-19. Importantly, the chapter provides not only the reflections of staff members in a community engagement and service-learning office, but also student perspectives on community engagement during COVID-19. Readers may resonate with the experience and reflections shared, but two primary areas are worth future examination. First, parent-tutor interactions dramatically shifted during remote tutoring. The lessons learned regarding the benefits and challenges of these interactions need to be better chronicled and examined across institutions and partnerships. Second, lessons were learned regarding the importance of training tutors how to use different technologies as well as ongoing professional development around community engagement for remote tutoring. These lessons line up well with Okro et al.'s (2021) considerations noted earlier; however, they did not particularly discuss parent-tutor interactions. Thus, these lessons deserve more attention in the literature and discussions among practitioners. In sum, this chapter provides only one story but students, staff members, and a professor who had the opportunity, resources, and time, especially, to craft a chapter for others to read.

Hopefully, others will do the same, so lessons learned during COVID-19 that can create a better future for community engagement are not lost.

REFERENCES

Aboagye, E. (2021). Transitioning from face-to-face to online instruction in the COVID-19 era: Challenges of tutors at colleges of education in Ghana. *Social Education Research, 2*(1), 9–19. doi:10.37256er.212021545

Books, J. P. (2014, Jan 9). *There's no such thing as a kid who hates reading. There are kids who love reading, and kids who are reading the wrong books* [Twitter post]. Retrieved 9 June 2022 from https://twitter.com/jp_books/status/421344811224817664?lang=en-GB

Burton, C., & Winter, M. A. (2021). Benefits of service-learning for students during the COVID-19 crisis: Two case studies. *Scholarship of Teaching and Learning in Psychology*. Advance online publication. doi:10.1037tl0000292

Cao, J., Yang, T., Lai, I. K. W., & Wu, J. (2021). Is online education more welcomed during COVID-19? An empirical study of social impact theory on online tutoring platforms. *International Journal of Electrical Engineering Education*. Advance online publication. doi:10.1177/0020720920984001

Chabbott, C., & Sinclair, M. (2020). SDG 4 and the COVID-19 emergency: Textbooks, tutoring, and teachers. *Prospects, 49*(1), 51–57. doi:10.100711125-020-09485-y PMID:32836424

Chambers, D. (2020). An interdisciplinary approach to language learning through community engagement. *13th International Conference Innovation in Language Learning Virtual Edition*. Retrieved 30 May 2022 from https://conference.pixel-online.net/ICT4LL/files/ict4ll/ed0013/FP/6968-EMO4902-FP-ICT4LL13.pdf

Edmondson, J. (1999). *America Reads: A critical policy analysis*. The Pennsylvania State University. Dissertation. Retrieved 4 June 2022 from https://www.proquest.com/docview/304521148?pq-origsite=gscholar&fromopenview=true

Gilmore, B., Ndejjo, R., Tchetchia, A., De Claro, V., Mago, E., Lopes, C., & Bhattacharyya, S. (2020). Community engagement for COVID-19 prevention and control: A rapid evidence synthesis. *BMJ Global Health, 5*(10), e003188. doi:10.1136/bmjgh-2020-003188 PMID:33051285

Hobart & William Smith Colleges. (2014, May 29). America Reads 25th Anniversary [video]. *YouTube*. Retrieved 4 June 2022 from https://www.youtube.com/watch?v=fgvRVDMzilo&t=4s

Hobart & William Smith Colleges. (2019, May 18). The legacy of America Reads at HWS. *HWS Update*. Hobart & William Smith Colleges. Retrieved 4 June 2022 from https://www2.hws.edu/the-legacy-of-america-reads-at-HWS-2/

Hobart & William Smith Colleges. (2020a, April 17). HWS virtual tutoring for Geneva students. *HWS Update*. Hobart & William Smith Colleges. Retrieved 5 June 2022 from https://www2.hws.edu/hws-virtual-tutoring-for-Geneva-students/

Hobart & William Smith Colleges. (2020b, February 7). *HWS earns Carnegie Community Engagement Designation*. Retrieved 10 June 2022 from https://www2.HWS.edu/HWS-earns-carnegie-community-engagement-designation/

Hobart & William Smith Colleges. (2021a, January 15). Service-learning and social science in action. *HWS Update*. Hobart & William Smith Colleges. Retrieved 6 June 2022 from https://www2.HWS.edu/service-learning-and-social-science-in-action/

Hobart & William Smith Colleges. (2021b, September 9). HWS ranked no. 3 in the nation for service. *HWS Update*. Hobart & William Smith Colleges. Retrieved 10 June 2022 from https://www2.HWS.edu/HWS-ranked-no-3-in-nation-for-service/

Hobart & William Smith Colleges. (2022, May 9). The grassroots initiative that grew into America Reads. *HWS Update*. Hobart & William Smith Colleges. Retrieved 4 June 2022 from https://www2.HWS.edu/the-grassroots-initiative-that-grew-into-america-reads/

Hughes, E. M., Brooker, H., Gambrell, L. B., & Foster, V. (2011). Tutoring and mentoring: The results of an America Reads program on struggling readers' motivation and achievement. *Literacy promises: Association of Literacy Educators and Researchers Yearbook, 33*, 205-218. Retrieved 6 June 2022 from https://files.eric.ed.gov/fulltext/ED527009.pdf#page=217

Johns, C., & Mills, M. (2021). Online mathematics tutoring during the COVID-19 pandemic: Recommendations for best practices. *PRIMUS (Terre Haute, Ind.), 31*(1), 99–117. doi:10.1080/10511970.2020.1818336

Mali, D., & Lim, H. (2021). How do students perceive face-to-face/blended learning as a result of the Covid-19 pandemic? *International Journal of Management Education, 19*(3), 100552. https://doi.org/10.1016/j.ijme.2021.100552

Morton, P., & Rosenfeld, D. (2021). Reconceptualizing service-learning during the COVID-19 pandemic: reflections and recommendations. In C. Tosone (Ed.), Shared trauma, shared resilience during a pandemic (pp. 331–339). Springer. https://doi.org/10.1007/978-3-030-61442-3_34.

Mughal, R., Thomson, L. J., Daykin, N., & Chatterjee, H. J. (2022). Rapid evidence review of community engagement and resources in the UK during the COVID-19 Pandemic: How can community assets redress health inequities? *International Journal of Environmental Research and Public Health, 19*(7), 4086. https://doi.org/10.3390/ijerph19074086

Ohmer, M., Carrie, F., Lina, D., Aliya, D., & Alicia, M. (2021). University-community engagement during a pandemic: Moving beyond "helping" to public problem solving. *Metropolitan Universities, 32*(3), 81-91. doi:10.18060/25329

Okoro, C. S., Takawira, O., & Baur, P. (2021). An assessment of tutoring performance, challenges and support during COVID-19: A qualitative study in a South African university. *Journal of University Teaching & Learning Practice, 18*(8). https://doi.org/10.53761/1.18.8.4

Omodan, B. I., & Ige, O. A. (2021). Re-constructing the tutors-tutees relationships for better academic performance in universities amidst COVID-19 new normal. *Mediterranean Journal of Social Sciences*, *12*(2), 30–40. https://doi.org/10.36941/mjss-2021-0010

Pérez-Jorge, D., Rodríguez-Jiménez, M. D. C., Ariño-Mateo, E., & Barragán-Medero, F. (2020). The effect of COVID-19 in university tutoring models. *Sustainability*, *12*(20), 8631. https://doi.org/10.3390/su12208631

Perrotta, D. (2021). Universities and COVID-19 in Argentina: From community engagement to regulation. *Studies in Higher Education*, *46*(1), 30–43. https://doi.org/10.1080/03075079.2020.1859679

San Tham, S. Y. (2020). *Exploring the Self-Reflection of America Reads Tutors* [Doctoral dissertation]. University of Kansas. Retrieved 6 June 2022 from https://www.proquest.com/docview/2426575455

Seru, E. J. (2021). Critical, interdisciplinary, and collaborative approaches to virtual community-engaged learning during the COVID-19 pandemic and social unrest in the Twin Cities. *Journal of Higher Education Outreach & Engagement*, *25*(3), 79–90. https://openjournals.libs.uga.edu/jheoe/article/download/2560/2689

Talmage, C. A., Annear, C., Equinozzi, K., Flowers, K., Hammett, G., Jackson, A., ... Turino, C. (2021). Rapid community innovation: A small urban liberal arts community response to COVID-19. *International Journal of Community Well-Being*, *4*(3), 323–337. https://doi.org/10.1007/s42413-020-00074-7

Talmage, C. A., Baker, A. L., Guest, M. A., & Knopf, R. C. (2020). Responding to social isolation among older adults through lifelong learning: Lessons and questions during COVID-19. *Local Development & Society*, *1*(1), 26-33. doi:10.1080/26883597.2020.1794757

Tragodara, K. S. C. (2021, March). *Virtual tutoring from the comprehensive training model to Engineering students during the COVID-19 pandemic*. Conference paper at the 2021 IEEE World Conference on Engineering Education (EDUNINE). https://ieeexplore.ieee.org/abstract/document/9429115

United States Department of Education. (n.d.). *Programs: Federal Work-Study (FWS) Program*. Retrieved 10 June 2022 from https://www2.ed.gov/programs/fws/index.html#:~:text=Institutions%20must%20use%20at%20least%207%20percent%20of,family%20literacy%20activities%3B%20or%20emergency%20preparedness%20and%20response

Van Maaren, J., Jensen, M., & Foster, A. (2022). Tutoring in higher education during COVID-19: Lessons from a private university's transition to remote learning. *Journal of College Reading and Learning*, *52*(1), 3–22. https://doi.org/10.1080/10790195.2021.2007175

Weisman, M. (2021). Remote community engagement in the time of COVID-19, a surging racial justice movement, wildfires, and an election year. *Higher Learning Research Communications, 11*, 88-95. https://scholarworks.waldenu.edu/cgi/viewcontent.cgi?article=1225&context=hlrc

Yang, Q., Gu, J., & Hong, J. C. (2021). Parental social comparison related to tutoring anxiety, and guided approaches to assisting their children's home online learning during the COVID-19 lockdown. *Frontiers in Psychology*. doi:10.3389/fpsyg.2021.708221

Yusuf, N. (2021). The effect of online tutoring applications on student learning outcomes during the COVID-19 Pandemic. *ITALIENISCH, 11*(2), 81-88. doi:10.1115/italienisch.v11i2.100

KEY TERMS AND DEFINITIONS

America Reads: Clinton Administration program that utilized multi-year grants to support tutor coordinators, reading specialists, and volunteer reading tutors in increasing child literacy (originally called America's Reading Corps).

Asynchronous Tutoring Modalities: Learning strategies that leverage technology to allow students to access tutoring materials on their own time and over time.

Community Engagement: College and university strategies that foster relationships with local stakeholders, beyond students, faculty, and staff, to address local issues and bolster local assets.

Remote and Virtual Tutoring: Tutoring undertaken asynchronously or synchronously using online technologies (e.g., virtual conferencing [Zoom]) to connect with tutees.

STEM Education: Teaching programs focused on Science, Technology, Engineering, and Mathematics, which emphasize hands-on, problem-solving, applied, and 'real-world' learning.

Synchronous Tutoring Modalities: Learning strategies that leverage technology to have students and tutors interact in 'real-time' to utilize tutoring materials.

Vygotsky's Zone of Proximal Development: The gap between what tutees can do with assistance or without assistance from adults or adept peers.

ENDNOTES

1. HWS Minors on Campus Course: https://www.hws.edu/offices/hr/minors_training.aspx.
2. Beyond Measure Documentary: https://beyondtheracetonowhere.org/beyond-measure/.
3. Wake Forest University Community Engagement Virtual Tutoring Resources https://communityengagement.wfu.edu/virtual-engagement/virtual-tutoring/
4. Campus Compact Summer 2020 webinar series: https://compact.org/webinarseries/summer-2020-webinar-series/#1591641159658-25234e00-e978.
5. CCESL Virtual Opportunities: https://docs.google.com/document/d/1iPDmdPsWjz1eTu48OnhBe78V3cbRa70tIEM3vHSZoXs/edit?usp=sharing.
6. For multiple examples see stories found in Talmage et al. (2021).
7. For more on the matching process, visit https://www2.hws.edu/hws-corps-virtual-tutoring/.
8. For a simple overview of Vygotsky's Zome of Proximal Development that CCESL Staff used, please look at https://www.simplypsychology.org/Zone-of-Proximal-Development.html.
9. To learn more about these terms and activities, please visit https://www.theedadvocate.org/how-to-implement-the-text-to-t

ext-text-to-self-text-to-world-teaching-strategy-in-your-cla
ssroom/.

[10] Two Instagram posts about this professional development opportunity: https://www.instagram.com/p/CcRbtz8LEfE/ and https://www.instagram.com/p/CavK1wJvfJr/.

[11] For more on well-child care visits, check out: https://www.healthychildren.org/English/family-life/health-management/Pages/Well-Child-Care-A-Check-Up-for-Success.aspx.

[12] The program is called Geneva Reads. It is a program that aims to remove the financial barriers to reading; see https://www.genevareads.org/.

Chapter 13
Remote Engagement Through Cohort Mentors

Donna DiMatteo-Gibson
Adler University, USA

ABSTRACT

Working virtually can be a lonely experience. Ways to engage employees is so important to uncover especially with so many organizations working virtually. One strategy that the researcher utilized to cultivate engagement was through the use of cohort mentors. This chapter will explore research on mentorship and engagement. The discussion will also present the research that has been done at an online university that embraced engagement through the use of cohort mentors. Doctoral students can benefit extensively from individually administered mentorship programs. This mentorship can be particularly beneficial for doctoral students completing their degrees in the online format. A doctoral cohort mentor model was implemented at the online campus of a mid-sized university in 2019. The model was expanded to include formal advising. Lessons learned and future directions will be discussed.

BACKGROUND

Working virtually can be a lonely experience. This can be particularly challenging when completing an online graduate degree. Gao and Sai (2020) discussed how virtual working could involve both feelings of isolation and loneliness. Many researchers have investigated how to improve and promote the engagement of online students through a variety of methodologies. The importance of success when working or learning in a virtual environment has also recognized that there are unique challenges that can occur when not physically being around other classmates, colleagues, and professors. A focus on how to engage students has been a critical focus in academia, as engagement has been linked to student success. Determining ways to engage employees and students is important, especially with so many schools and organizations working virtually. One strategy that the researcher utilized to cultivate engagement was through the use of cohort mentors. In this chapter, the author will explore research on mentorship, technology, and engagement within academic settings. The author will also review the model that was both researched and applied. The discussion will present the research that was completed at an online

DOI: 10.4018/978-1-6684-5190-8.ch013

university and that embraced engagement through the use of cohort mentors. In particular, the author will examine how this can branch out to what is considered socially responsible academic affairs. Finally, the author will discuss findings and recommendations that can be applied in other universities where mentorship is a tool used to promote student engagement.

Doctoral students can benefit extensively from individually administered mentor programs. This mentorship can be particularly beneficial for doctoral students completing their degrees in the online format. Halbert et al. (2022) implemented a doctoral cohort mentor model at an online campus of a mid-sized university in 2019. The researchers expanded the model to include formal advising. They identified lessons learned, future directions, and recommendations through the completion of this research. In the next section, the author will explore the research on student engagement, mentorship, technology, and virtual learning.

LITERATURE REVIEW

Building Engagement Through Virtual Learning

A significant body of research has examined how there are complex needs of online doctoral students (Akojie et al., 2019; Gibson et al., 2019; Kumar & Coe, 2017). Online students need to feel engaged with their learning, connected with peers and faculty and motivated to learn. Many of the challenges online doctoral students experience are directly related to the institution and also due to the virtual educational environment. Students may feel disconnected when learning virtually, as they are not able to physically engage with their classmates and professors in a face-to-face environment. This could impact their motivation.

It is possible to build engagement in several ways when working virtually with students. One option is to utilize mentors to support and provide that connection with students. Mentors can make a significant impact in encouraging this engagement. Also, one of the benefits of mentors is they can be an essential asset to building engagement. Understanding the challenges online doctoral students face is one area that mentors can focus and assist with (Kumar & Johnson, 2017; Kumar & Johnson, 2019). Also, this information not only can help to build engagement, but also to inform policy and practice, which prioritizes the student experience and supports the educational journey of students. In addition, this can support the development, evaluation, and expansion of an online mentoring program (Kumar, Kumar, & Taylor, 2020). Mentorship has been found to encourage recruitment and retention efforts of staff in rural workplaces (Rohatinsky et al., 2020).

Motivation and Online Education

Motivation is important to investigate, also because it impacts engagement when working and learning virtually. Jameson et al. (2021) found that online faculty utilized specific strategies and activities to improve the motivation of online graduate students. When faculty applied differing strategies and connected with online student needs, this was very beneficial. Also, they found that what they considered high-achieving faculty were great at communication and building connections with students. They reached out to individual students often and provided encouragement, mentorship, and identified realistic goals for success. This has important applicability to engagement of online students. Jameson and Torres (2019) discussed the importance of mentorship in an online learning environment in terms of meeting goals and being successful working in a graduate program. The necessity of creating a positive relationship with the student where they felt supported by a mentor was critical for motivation and overall success. They also highlighted specifically how internal locus of control and motivation were both keys to success and engagement overall.

Virtual Mentorship and Engagement

It is important to examine key research studies in regard to virtual mentorship and engagement. Qua et al. (2021) conducted research at Case Western Reserve University's School of Medicine and Comprehensive Cancer Center. Their goal was to engage high school students in biomedical research and encourage the pursuit of careers in clinical care. The researchers obtained feedback about student and faculty perceptions of the virtual by using both surveys and focus groups. Interestingly, ratings for aspects of the program that were to remain virtual after COVID-19 were highly rated for highly interactive activities and faculty mentorship opportunities. The findings identify activities that sustain student interest and mentorship and as well as interactive discussion-based activities enhanced the virtual learning experience.

Hart (2016) discussed a model that was used to provide remote support. They utilized a model that focused on communication, building relationships, and improving developmental opportunities. Their focus was within virtual work teams. The results of the study have implications for leaders and members who work virtually and for human resource development professionals seeking strategies to create and improve informal developmental relationships through the medium of virtual work team communication. The findings support the ability of building engagement with mentors working virtually and with different work teams.

Pfund et al. (2021) also focused on developing new mentor strategies for research trainees that were working remotely. They found these strategies improved productivity and helped the trainees focus on their goals. They utilized a three-step process where mentors and mentees collaborate to enhance their work effectiveness. Training and development research has investigated alternative mentoring programs because traditional one-on-one mentoring programs can fail to work as a number of employees work remotely. The mentoring approaches have allowed participants to join in monthly meetings either in-person or online.

Ely (2021) discussed mentorship and compared and contrasted traditional mentorship versus virtual mentorship. They found that virtual mentorships may take more time and effort to build, but they provided new access to mentors that is available at the global level. A key application of this model is the importance of spending time of mentors to cultivate this engagement. Creating a successful virtual mentorship starts by identifying the right mentors and setting achievable mentee goals with the virtual mentor. The important tool is that both goals and the strategies to attain them are documented in a virtual mentoring plan. The plan provides information for both the mentor and mentee to keep the virtual mentorship on track. It is also a professional method, no matter where in the world they may work and live, which is fantastic when mentoring at a global scale.

With the increase in the use of virtual teams, organizations have increasingly begun to use virtual mentors to build engagement as well as to develop individuals through the use of technology. Neely et al. (2017) shared a model for understanding the virtual mentoring process. It illustrated how valuable mentors are for supporting virtual workers.

De Janasz et al. (2008) researched e-mentoring with business students. They focused on both strengths and weaknesses associated with this type of mentoring and developed a model of e-mentoring identifying specific antecedents and outcomes. The researchers found that perceived similarity in terms of attitudes and values is positively related to effective e-mentoring. They also found that demographic similarity specifically with gender and race is not related to effective e-mentoring. Very important observations were made in terms of effectiveness of e-mentoring and how this may lead to proteges' enhanced academic performance, professional network, as well as job opportunities.

Remote mentoring serves to play an important role in building community and engagement among online faculty and doctoral students (Kumar, Johnson, Dogan, & Coe, 2019). Studies have focused on mentoring and engagement of individuals. Formal mentoring has grown substantially in recent years and it encompasses both traditional and virtual mentoring. Interestingly, mentoring has demonstrated benefits for individuals in the areas of learning, career planning, and development of leadership skills. Ghosh et al. (2019) investigated mentoring programs and found they impact psychological capital and employee engagement. They utilized a mixed-methods design and gained data from more than three hundred franchisee locations across the United States. They found key attributes were critical in terms of mentorship and engagement and that the frequency of contact between the mentors and mentees was a key factor for engaging in mutually beneficial learning and mentorship.

Garg et al. (2021) also investigated mentoring and engagement. Their focus was on reverse mentoring. They found that reverse mentoring can lead to positive organizational outcomes. They explored the role of work engagement and found that reverse mentoring and job crafting are positively related to work engagement, which, in turn, increases performance and decreases work withdrawal behaviors. Work engagement fully mediates the relationship between reverse mentoring and withdrawal behavior and partially mediates the relationship between reverse mentoring and work performance. The authors connect research on reverse mentoring and work engagement. The study provides evidence to support practitioners in implementing resources for reverse mentoring and job crafting to increase work engagement among employees. The American Psychological Association (APA) (2016) also discussed both mentoring and engagement. They focused on strategic mentoring and assesses mentoring, employee retention, engagement, and succession planning, and found benefits that a strategic mentoring program can bring to credit unions.

Zhang et al. (2020) also investigated virtual mentoring and engagement specifically with doctoral students. The purpose of their study was to evaluate the potential differences between part-time and full-time doctoral students in how technology may serve as a communication and organizational tool for engagement and support of the program. This study provided insights into part-time doctoral students' scholarly development and provided suggestions for designing doctoral programs. Further, additional multifaceted mentoring approaches including peer mentoring and e-mentoring was evaluated. Significant differences found in terms of the opportunities to do research related to grants with faculty, support for scholarly work in addition to the advisor's support, involvement in the teaching/supervision activities, and goals for scholarly development. Recommendations for mentors include appreciating the differences between part-time and full-time doctoral students. They also supported the use of an individual development plan to mentor doctoral students to enhance the effectiveness of mentoring and engagement. Interestingly, the focus on research helped to build the engagement of students in the doctoral program who were working with mentors.

Mentoring, Communication, and Technology

Baran (2016) studied faculty mentoring at a university professional development program and focused on the roles of technology and communication. They utilized a faculty technology mentoring model and mentor's use of blog posts, case reports, and interviews. Baran found critical success factors necessary for effective mentoring and overcoming any mentoring challenges. Motivation, evaluating the mentoring relationship, exploring communication channels, and utilizing support were all critical success factors. Lasater et al. (2014) found that technology was critical for virtual mentoring to be successful. They found that distance does not appear to limit the connecting potential leading to a meaningful mentoring relationship. Also, this identified opportunities that local mentoring relationships may not possess. Lastly, they emphasized that students in distance programs would benefit from distance mentoring relationships. Zhang et al. (2021) explored how a technology-based learning model would be useful for students in a doctoral program. They illustrated similar student concerns in terms of students not feeling supported as well as feeling disconnected from their programs. They also found support that faculty mentoring can help students feel both connected and supported.

Hamilton and Scandura (2003) discussed how virtual mentoring is also referred to e-mentoring. The authors explored the concept of e-mentoring, that is, the potential benefits and challenges in utilizing technology to promote the mentoring relationship and support engagement. They addressed the influence of e-mentoring as a way of providing support for career building, skill acquisition, and coaching. They concluded with action steps to promote an e-mentoring program. Bierema and Merriam (2002) also examined e-mentoring. They provided a thorough analysis of how technology is impacting mentoring. They proposed that e-mentoring can change the conditions under which mentoring is sought and offered. This was very relevant to the mentoring work in this study. E-_mentoring could make mentoring relationships more available to groups that have previously had limited access to mentoring. Benefits, barriers, and opportunities related to virtual mentoring was examined.

A particularly important point is mentoring students using technology and ensuring teachers are meeting the unique needs of our students. Sullivan and Moore (2013) reinforced how technical communication contributes to the mentoring success of women engineers. Changes in technical communication were key to the success of the mentoring program. One option was to effectively schedule work while appreciating the unique challenges of the mentees. Another critical success factor was utilizing good mentoring practices by having an infrastructure of communication.

Application of Research to the Development of the Model

The literature review above was fundamental to the work that was completed to develop and implement a doctoral cohort mentor model at an online graduate-level university where engagement and social responsibility are essential to the mission and vision of the school. This doctoral cohort mentor model aimed at accomplishing the following goals:

1. Building community within individual cohorts by creating opportunities for regular interaction between students and mentors.
2. Cultivating relationships with students by engaging in regular individual meetings.
3. Facilitating engagement within individual cohorts by hosting regular synchronous meetings.
4. Developing students by sharing resources and networking opportunities on a regular basis.

5. To reduce the potential for student isolation by providing regular communication through diverse means.
6. Improving the student experience by providing meaningful opportunities for connection.
7. Improving the mentor experience by providing opportunities to share experiences and provide direction to assigned mentees.

In the following sections, the author will examine whether these goals were met and how they impacted lessons learned and future recommendations.

METHODOLOGY

A qualitative approach was used to gather data and understanding on how the doctoral mentor model impacted virtual engagement of the online students. A focus group was used as an essential tool to gather information for this qualitative investigation. More specifically, the focus group approach was used to gain a detailed understanding of engagement of online students and how the mentor experience impacted student engagement in the online setting.

Part of the doctoral mentoring model involved training the faculty mentors to provide the knowledge, tools, information, and resources that mentors needed to apply the model effectively. This was an important component to the effectiveness of the faculty mentor model. The training occurred via an online platform in two sessions, namely in February 2021and in June 2022. It focused on the following topics: Student outreach, individual/group meetings, specificity of course plans/elective options, and more information on the social justice practicum (SJP).

Student outreach was discussed in terms of how often the mentor should apply it and what means of communication the mentor should utilize. Also, when outreach would take place was also identified. An evaluation of what should occur in both individual and group meetings was discussed. Finally, knowledge of the program, upcoming dissertation process, as well as SJP requirement were explored, so that the mentors had the necessary knowledge to respond to student questions and promote active engagement.

It was essential that the training also detailed the dissertation process at the online university. This was a necessary component to be able to respond to student questions and concerns. The training walked through best practices in student engagement, information about doctoral program planning, and virtual mentorship. This training was followed up with a focus group to gather feedback on the mentoring experience. The focus group then evaluated the experience of faculty mentors and how this model was utilized to promote student engagement.

The focus group addressed five questions, which were focused on assessing the impact of the mentor model on faculty, students, and the institution, and on any positive reflections and suggestions for improvement. The questions were:

1. What impacts, if any, does the cohort mentor model have on faculty?
2. What impacts, if any, does the cohort mentor model have on students, in your view?
3. What impacts, if any, does the cohort mentor model have on the institution?
4. Please share any positive reflections regarding your experience with the cohort mentor model.

Please share any suggestions for improvements to the cohort mentor model.

Key Content Areas Needed of Virtual Mentors

One area that was very important to cover was the content of the dissertation course sequence that would be coming at a later date in the online students' virtual learning experience. The specifics on what is required to start the dissertation were identified via specific course numbers. Also, the composition of the dissertation committee was described. It was communicated that successful completion of all required coursework, and a SJP was essential for moving forward in their research. Also, successful completion of a doctoral qualifying exam had to occur in order to be able to collect data for doctoral research. The necessary knowledge of the entire requirements of the online program was essential for mentors to possess to ensure that questions and concerns were responded within both individual and group mentoring sessions.

Each phase of the dissertation process was reviewed in the training. This was important to allow mentors identify opportunities for students while they were actively completing their coursework to proactively prepare for their upcoming doctoral research. Mentors linked the dissertation with their courses as well as future career goals.

In Phase I of the dissertation, there is a focus on completing at least 3 credits of the dissertation course (IOP-801). In the first module, students work on completing a prospectus. The prospectus outlines the statement of the problem and shares information on the research methodology, theoretical underpinnings, potential participants, and a general overview of the methodology. In Module 2, students focus on developing their proposal. This is where students finalize their first chapter and begin to work on their literature review. Collecting scholarly research is the focus. The importance of having a strong faculty mentor sets the stage for success when students are working on their doctoral research.

Phase II of the dissertation involves focusing on the completion of the first three chapters of the dissertation along with submitting the IRB application to the Online Institutional Review Board. Successful completion of the dissertation proposal must occur before going into the third phase. The fourth and fifth phases of the dissertation focus on data collection and analysis. The student must complete chapters 4 and 5 of their dissertation and summarize their findings as well as link their findings to existing research. Successful completion of dissertation proposal defense must occur before completing the dissertation process. Students who have not successfully defended their dissertation may register for additional credits of IOP 801. This knowledge was conveyed to mentees, so that students could prepare in advance while completing their coursework. This also promoted virtual engagement, which was a key goal of the mentorship model. Resources that are available were shared with mentors, so they could best assist students. Through this research, the identification of resources needed was also defined.

Resources Needed

Important resources are needed to facilitate engagement and collaboration when working virtually. First, having an infrastructure to communicate and collaborate among the learning community is key. This can be accomplished by using a learning management system, such as Canvas, to share information and resources. The learning management system is used more for online courses and doctoral research. The mentors use more Zoom, email, and phone calls to connect with their mentees. Having group Zoom calls can create a community with the students who are all working with the same mentor.

Other resources that can be particularly helpful include anything about the online program that can provide information and support. Mentors often use degree completion plans to share information about courses and the program overall. They also share information about the dissertation, such as the dissertation handbook, to help answer questions about students' upcoming research.

Mentors also create resources that students need for their careers as they finish the program. These resources come from organizations that are critical to supporting learning and education for the discipline in which the students are focusing their graduate learning. Other resources that pertain to their careers are shared on how to locate opportunities and assess the best job fit.

Mentors provide information on how to utilize career resources, job search tools, social media, and resources such as the O*Net to gain career information that helps their preparation while in school. Mentors had specific topics that they introduced in the group sessions that would either be program specific or career specific. Some examples of group meeting topics included the use of LinkedIn, how to use O*Net, SIOP as an organizational resource, dissertation information and planning, and the SJP informational meeting.

Mentors are key to communicating what resources are needed for the online campus by working directly with their mentees. An indirect benefit that originated was that this information was shared with online program directors and used to make program and process improvements. It also helped to provide clarification to internal process or program requirements and to identify new necessary resources. Identifying any areas of confusion was key.

FINDINGS AND DISCUSSIONS

This research produced many findings. Interestingly, many areas impacted from the implementation of the mentor model. Also, the findings supported what previous research had identified as well in terms of virtual mentoring, communication, technology, and how this impacts engagement. Besides, the findings provided some crucial lessons learned as well as future recommendations both for research and practice. Challenges were experienced, but these challenges provided opportunities for learning and to make improvements to the model. There was also an understanding of needed and uncovered key resources.

The lessons learned from this research fall into a number of different categories. These categories include how the model impacted the student, the adjunct faculty, and faculty mentor, and the process impact as well. The author will review these lessons learned by category next, along with how these lessons learned are connected to current research.

Student Impacts

The first area to be reviewed is focused on how the mentor model impacted students. It was found that the mentor model made a significant impact on students. The impact was essential to providing more specific degree plan advising, more support throughout the doctoral program, specific guidance before starting the dissertation journey, and mentoring in regards to career options and the field of I/O psychology. Also, the mentor model positively affected engagement with the online program. Feedback obtained indicated that the faculty mentors lessened the anxieties around the dissertation as well as provided coaching on how to identify a topic way in advance of starting the dissertation. Feedback also indicated that students valued the coaching around using each doctoral mentor as a way to identify dissertation topics and prepare for their future careers in the field.

Students were impacted by the COVID-19 pandemic, which brought increased feelings of isolation, stress, and anxiety. The mentorship model through the use of communication and technology helped to bring engagement, which was essential during this critical time of stress and uncertainty. Sutton (2022) stated: "The COVID-19 pandemic presented new challenges and changes for everyone in higher education, but impacted adult learners differently than other traditional students. But, even within adult learners, not all groups face the same challenges" (p. 1). Sutton (2022) discussed how adult students could have a feeling of being alone in virtual classes and how the pandemic amplified this experience.

Mentor Impact

Another key theme that emerged was there were impacts on the mentor as well. These included the mentor feeling more engaged in the work they were doing. They also felt more empowered to support online students. The information obtained in the training assisted the mentors to provide more detailed feedback as it related to the students' research and career choices. They also were more informed about the program, course offerings, dissertation process, and SJP. This translated into more detailed mentoring that they felt improved the mentoring experience.

Faculty Impact

Mentoring serves to play an important role in building community among online faculty and doctoral students. One area that was not the focus of the research, but was a key takeaway, is how beneficial mentoring is for faculty engagement when working online. Heitschmidt et al. (2021) found how mentorship also enhanced faculty professional growth, development and leadership as well as improved faculty scholarship activities. Faculty mentors found that they benefited from the connections made with the students through both individual and group sessions. Similarities in terms of coaching were identified in how this also improved the engagement of the faculty mentors along with the students.

Process Impact

The mentor model was key to also identifying opportunities to improve processes within the organization. This relates to Heitschmidt et al.'s findings (2021). They implemented a mentor model which improved processes and partnerships. They found quality improvements, new collaborations, and more engagement in research projects took place. In addition, this research facilitated specific guidelines for mentoring. Guidelines are very important to improve mentoring for both the mentors and the students.

CHALLENGES

The researchers experienced some challenges when implementing this doctoral mentor model. Some of the challenges were: Determining the best methods of communication, having differences in needs among cohorts, and mentors lacking in some program knowledge that would optimize their usefulness in the mentor role. As to the first challenge, namely, determining the best methods of communication, this can vary among cohorts of students. It is determined that having a process with guidelines identifying specific methods of communication would be very useful. AS to the second challenge, namely, having differences in needs among cohorts, some students required communication often from mentors and had many questions, while other students did not. Not knowing students' communication needs in advance made it challenging to manage the workload of cohort mentors. Finally, the last challenge, namely, mentors' lack of program knowledge, impacted the level of detailed feedback with which mentors were able to provide students. Training of mentors was implemented to share more information about the program so mentors could respond more directly to student questions.

RECOMMENDATIONS

This research generated many key recommendations. There are lessons learned in terms of the mentor impacts. These lessons learned are focused on key tools and empowerment. First, more tools are needed to provide more individualized and specific advising for each student in the doctoral program. It was valuable to discuss elective options that tie to individual student goals and career plans. Students felt more empowered by the increased knowledge and tools available to them. Student feedback indicated the work with faculty mentors was critical to their career planning. They felt that this impacted the elective options they selected in the program as they aligned these electives to their career goals. Feedback indicated that working with mentors helped to provide the virtual career coaching early in their doctoral journeys, which was essential for program success, engagement in the program and career planning.

There were also key lessons learned in terms of faculty impacts. Some of the impacts identified were focused on their course curriculum and future careers. Students felt that they were able to apply the knowledge to their courses in the doctoral program. They also identified where relevant courses and work with faculty were more aligned to their career interests.

There were also lessons learned in terms of process impacts. Some of the areas identified were focused on process improvements and doctoral degree plans. One process impact involved improved overall doctoral student advising. Another key impact involved providing more specific responses to questions around student programs and degree plans. Lastly, another key process impact involved individualized doctoral plans being more aligned to individual student goals and interests.

Also, in terms of process impacts, it is identified that the specificity in terms of tasks and roles of mentors needed to be defined, so minimum expectations of mentors were set. The process guidelines identified communication responsibilities and what resources were needed to facilitate the required tasks and steps. Many tasks and guidelines were identified for mentors in terms of communication frequency and methods of communication. The process impact identified how check-ins should occur. Check-ins with assigned cohorts should occur via email on a biweekly basis. The emails should be sent to the cohort as a whole. Also, any timely research articles should be shared at this time, and writing resources should be provided among the cohort as well as networking suggestions.Any questions that students ask should be responded. Any issues to which the cohort mentor does not have the answershould be followed up via email with the program director.

Another key process impact that is identified was in the area of the utilization of virtual live sessions. It is found that mentors should facilitate virtual live sessions for assigned cohorts three times per year. The session content should match the phase that the cohort is in in terms of their overall graduate program. Utilizing a doodle poll can help to assess student availability and set up a virtual live session that works for most of the students. Two options in terms of virtual live sessions are offered to students each session. Lastly, with regard to a final process impact, was the importance of mentors conducting individual meetings with students from assigned cohorts three times a year. Mentors will set up a process for scheduling individual times with students.

Mentor will also enter notes for each meeting to be stored in a centralized location. The Program Director then reviews the notes three times per year to evaluate opportunities for change.

FUTURE DIRECTIONS

Many lessons learned helped to identify future directions for both research and application. Future directions will include the following focus areas:

- **Timing of Mentor Introduction:** It was learned that the timing of when mentors are introduced to students was a key part of the successful implementation. This will be explored further in future research. It was found that students benefitted most when students were new to the program, but were already enrolled and taking their first classes. This allowed the students to focus more on the student/mentor experience and ask questions of the mentor once the initial introductory information and orientation to the school was completed.
- **Mentor Collaboration with Faculty:** The mentor collaborating with faculty was another key learning that will be explored further. It is found that the knowledge mentors gained from working with students would also benefit the faculty teaching the online graduate courses. The knowledge that could be shared between mentors and faculty teaching graduate courses would benefit the student experience and help to resolve any issues/concerns that students were experiencing.
- **Mentor Collaboration with Program Directors:** Another key focus area will be on finding more ways for mentors to collaborate with program directors. Currently, mentors meet individually and jointly as a team with program directors to provide feedback and share their experiences with the program directors. Mentors also shared updates via email with program directors. Future opportunities for collaboration will be explored and research that will be most beneficial for the students.
- **Mentor Collaboration with Dissertation Chairs:** Mentors work with students before they work with their dissertation chairs in the online graduate programs. Opportunities that will help transition these working relationships from the mentor to the chair will be explored.
- **How the Use of Mentorship with the Goal of Student Engagement Can Be Realized by Utilizing a Co-Teaching Model:** Finally, a key takeaway is how mentors can impact student engagement through the use of a co-teaching model. This was illuminated through some of the work of the mentors working collaboratively with program directors, chairs, and adjunct faculty. Future applications will focus on co-teaching and assessing this model of engagement.

Walker and Ardell (2020) illustrated how the mentor-teacher-candidate relationship had an impact on student engagement. They also embraced how learning virtually can provide unique and new possibilities for mentorship. A co-teaching model is another way of embracing both mentorship and student engagement.

The importance of remote engagement was evident in the research conducted as well as from research from other practitioners. The use of mentors has been found to be a critical aspect of encouraging engagement with individuals working both virtually and studying online. Future research will continue to examine this important topic and identify ways and tools that mentors can work with students to build both engagement and motivation in the learning process.

Practices around co-teaching will be explored further to identify how this practice can support both engagement and motivation in learning. Creating a professional learning community where technological resources are utilized can improve the process and effectiveness of this mentoring model (Bates et al. 2016).

The assessment of the doctoral cohort mentor model evidenced it was successful at accomplishing the goals the author set forth in this study. It helped to build community within individual cohorts by creating opportunities for regular interaction between students and faculty mentors. This community continued to grow through communication and use of virtual technological tools. The model helped to cultivate relationships with students by engaging in regular individual meetings. This created engagement both with the faculty mentor and among the student cohorts. Besides, the doctoral cohort mentor model facilitated engagement within individual cohorts by hosting regular synchronous meetings. These sessions were focused on sharing specific program knowledge, information about the doctoral research process, as well as career opportunities and resources within the field. The doctoral cohort mentor model developed students by sharing resources and networking opportunities on a regular basis, which continued to cultivate a learning community. This was essential as students experienced multiple stresses and challenges during the COVID-19 pandemic. The doctoral cohort mentor model reduced the potential for student isolation by providing regular communication through diverse means, which will continue to be utilized. Finally, the doctoral cohort mentor model improved the student experience by providing meaningful opportunities for connection and shared experiences.

SOCIALLY RESPONSIBILITIES FOR ACADEMIC AFFAIRS

This mentor model and its usefulness in the online doctoral program will continue to be utilized and researched. This also sets the focus on investigating the meaning of social responsibilities for academic affairs. How can this model and focus be applied in a larger sense that can impact the academic affairs of all students? Bartusevičiene and Rupšenė (2010) appreciated how higher education is undergoing many changes in terms of paradigm shifts. The impact of COVID-19 also impacted higher education and the use of technology. The constructive learning theory emphasizes how students should assume responsibility for their own learning. Future application and research could embrace the mentor model and how it impacts student involvement, motivation, and achievement.

REFERENCES

Akojie, P., Entrekin, F., Bacon, D., & Kanai, T. (2019). Qualitative meta-data analysis: Perceptions and experiences of online doctoral students. *American Journal of Qualitative Research*, *3*(1), 117–135. doi:10.29333/ajqr/5814

American Psychological Association. (2016). Strategic mentoring. *Credit Union Management*, *39*(11), 50.

Baran, E. (2016). Investigating faculty technology mentoring as a university-wide professional development model. *Journal of Computing in Higher Education*, *28*(1), 45–71. doi:10.100712528-015-9104-7

Bartusevičiene, I., & Rupšenė, L. (2010). Periodic assessment of students' achievements as a prerequisite of social activity education under conditions of instructional/learning paradigm's shift (the case of Social Pedagogy Study Programmes). *Social Education / Socialinis Ugdymas, 12*(23), 96-105.

Bates, C. C., Huber, R., & McClure, E. (2016). Stay connected: Using technology to enhance professional learning communities. *The Reading Teacher, 70*(1), 99–102. doi:10.1002/trtr.1469

Bierema, L. L., & Merriam, S. B. (2002). E-mentoring: Using computer mediated communication to enhance the mentoring process. *Innovative Higher Education, 26*(3), 211–227. doi:10.1023/A:1017921023103

de Janasz, S. C., Ensher, E. A., & Heun, C. (2008). Virtual relationships and real benefits: Using e-mentoring to connect business students with practicing managers. *Mentoring & Tutoring, 16*(4), 394–411. doi:10.1080/13611260802433775

Ely, J. A. (2021). Modern mentorship: Developing medical writing mentor-mentee relationships in the virtual world. *Medical Writing, 30*(2), 36–39.

Gao, G., & Sai, L. (2020). Towards a "virtual" world: Social isolation and struggles during the COVID-19 pandemic as single women living alone. *Gender, Work and Organization, 27*(5), 754–762. doi:10.1111/gwao.12468 PMID:32837008

Garg, N., Murphy, W., & Singh, P. (2021). Reverse mentoring, job crafting and work-outcomes: The mediating role of work engagement. *Career Development International, 26*(2), 290–308. doi:10.1108/CDI-09-2020-0233

Ghosh, R., Shuck, B., Cumberland, D., & D'Mello, J. (2019). Building psychological capital and employee engagement: Is formal mentoring a useful strategic human resource development intervention? *Performance Improvement Quarterly, 32*(1), 37–54. doi:10.1002/piq.21285

Gibson, A., Fields, N. L., Wladkowski, S. P., Kusmaul, N., Greenfield, J. C., & Mauldin, R. L. (2019). What can an evaluation of the AGE*SW* Predissertation Fellows Program tell us about the mentoring needs of doctoral students? *Journal of Gerontological Social Work, 62*(8), 852–866. doi:10.1080/01634372.2019.1685052 PMID:31650910

Halbert, J., Dennis, M., & DiMatteo-Gibson, D. (2022). Support for distance students: Using a mentor model. *Proceedings of the Virtual Conference Transforming the Teaching and Learning Environment.*

Hamilton, B. A., & Scandura, T. A. (2003). E-mentoring: Implications for organizational learning and development in a wired world. *Organizational Dynamics, 31*(4), 388–402. doi:10.1016/S0090-2616(02)00128-6

Hart, R. K. (2016). Informal virtual mentoring for team leaders and members: Emergence, content, and impact. *Advances in Developing Human Resources, 18*(3), 352–368. doi:10.1177/1523422316645886

Heitschmidt, M., Staffileno, B. A., & Kleinpell, R. (2021). Implementing a faculty mentoring process to improve academic-clinical partnerships for nurse led evidence based practice and research projects. *Journal of Professional Nursing, 37*(2), 399–403. doi:10.1016/j.profnurs.2020.04.015 PMID:33867097

Jameson, C., & Torres, K. (2019). Fostering motivation when virtually mentoring online doctoral students. *Journal of Educational Research and Practice, 9*(1), 331–339. doi:10.5590/JERAP.2019.09.1.23

Jameson, C. M., Torres, K., & Mohammed, S. (2021). Virtual faculty strategies for supporting motivation of online doctoral students. *Journal of Educational Research and Practice*, *11*(1), 295–305. doi:10.5590/JERAP.2021.11.1.21

Kumar, S., & Coe, C. (2017). Mentoring and student support in online doctoral programs. *American Journal of Distance Education*, *31*(2), 128–142. doi:10.1080/08923647.2017.1300464

Kumar, S., & Johnson, M. (2017). Mentoring doctoral student online: Mentor strategies and challenges. *Mentoring & Tutoring*, *25*(2), 202–222. doi:10.1080/13611267.2017.1326693

Kumar, S., & Johnson, M. (2019). Online mentoring of dissertations: The role of structure and support. *Studies in Higher Education*, *44*(1), 59–71. doi:10.1080/03075079.2017.1337736

Kumar, S., Johnson, M. L., Dogan, N., & Coe, C. (2019). A framework for e-mentoring in doctoral education. In K. Sim (Ed.), *Enhancing the role of ICT in doctoral research processes* (pp. 183–208). IGI Global. doi:10.4018/978-1-5225-7065-3.ch009

Kumar, S., Kumar, V., & Taylor, S. (2020). *A guide to online supervision*. UK Council for Graduate Education. https://ukcge.ac.uk/assets/resources/A-Guide-to-Online-Supervision-Kumar-Kumar-Taylor-UK-Council-for-Graduate-Education.pdf

Lasater, K., Young, P. K., Mitchell, C. G., Delahoyde, T. M., Nick, J. M., & Siktberg, L. (2014). Connecting in distance mentoring: Communication practices that work. *Nurse Education Today*, *34*(4), 501–506. doi:10.1016/j.nedt.2013.07.009 PMID:23978777

Neely, A. R., Cotton, J. L., & Neely, A. D. (2017). E-mentoring: A model and review of the literature. *AIS Transactions on Human-Computer Interaction*, *9*(3), 220–241. doi:10.17705/1thci.00096

Pfund, C., Branchaw, J. L., McDaniels, M., Byars-Winston, A., Lee, S. P., & Birren, B. (2021). Reassess-realign-reimagine: A guide for mentors pivoting to remote research mentoring. *CBE Life Sciences Education*, *20*(1), es2. doi:10.1187/cbe.20-07-0147 PMID:33635126

Qua, K., Haider, R., Junk, D. J., & Berger, N. A. (2021). Sustaining student engagement: Successes and challenges of a virtual STEM program for high school students. *Journal of STEM Outreach*, *4*(3). Advance online publication. doi:10.15695/jstem/v4i3.09 PMID:34853829

Rohatinsky, N., Cave, J., & Krauter, C. (2020). Establishing a mentorship program in rural workplaces: Connection, communication, and support required. *Rural and Remote Health*, *20*(1), 5640. doi:10.22605/RRH5640 PMID:31928037

Sullivan, P., & Moore, K. (2013). Time talk: On small changes that enact infrastructural mentoring for undergraduate women in technical fields. *Journal of Technical Writing and Communication*, *43*(3), 333–354. doi:10.2190/TW.43.3.f

Sutton, H. (2022). Marginalized adult learners faced unique challenges during COVID-19. *Dean and Provost*, *23*(9), 7–8. doi:10.1002/dap.31032

Walker, N. T., & Ardell, A. (2020). Reconsidering preservice-mentor relationships in complex times: New possibilities for collaboration and contribution. *Issues in Teacher Education*, *29*(1/2), 132–141.

Zhang, S., Carroll, M., Li, C., & Lin, E. (2021). Supporting part-time students in doctoral programs: A technology-based situated learning model. *Studies in Graduate and Postdoctoral Education*, *12*(2), 190–205. doi:10.1108/SGPE-11-2019-0082

Zhang, S., Li, C., Carroll, M., & Schrader, P. G. (2020). Doctoral program design based on technology-based situated learning and mentoring: A comparison of part-time and full-time doctoral students. *International Journal of Doctoral Studies*, *15*(1), 393–414. doi:10.28945/4598

Chapter 14
A Case Study in Meeting the Pandemic Moment Through NGO-University Collaboration:
Creating a Global Health Virtual Practicum

Paul Shetler Fast
Mennonite Central Committee, USA

Brianne F. Brenneman
Goshen College, USA

Wade George Snowdon
Mennonite Central Committee, USA

ABSTRACT

During the COVID-19 pandemic, international exchange and service programs scrambled to adapt while students were newly engaged on topics of global health. Meeting the moment, Goshen College, a liberal arts college in northern Indiana renowned for its study abroad and community-engaged learning programs, and Mennonite Central Committee, a faith-based international relief, development, peace non-profit organization working around the world, came together to pilot a new partnership approach to web-based community-engaged learning. This chapter explores the learnings from this creative and successful pilot of a global public health virtual practicum, demonstrating the potential for mutually beneficial collaboration between higher education community-engaged learning programs and humanitarian organizations to meet the unique needs of diverse students and community-based organizations, while lowering entry costs and reducing barriers of engagement for students.

SETTING THE STAGE

In 2020, study abroad and international exchange programs ground to a halt as the COVID-19 pan-

DOI: 10.4018/978-1-6684-5190-8.ch014

demic spread around the world and governments, schools, and humanitarian organizations responded to adapt and reduce risk. Universities and students scrambled to change study plans and find alternative strategies for cross-cultural engagement and community engaged learning. Students who may never have considered global public health as a field of study were suddenly engaged in the topic. According to the Association of Schools and Program of Public Health, applications to schools of public health surged by a record 23% between 2019 and 2020 as the profession suddenly felt relevant and important to students' lives (Warnick, 2021). At the same time, international non-profit humanitarian organizations implementing global health work were struggling to resource projects in underserved communities around the world, with travel of staff difficult, increased health risks, travel restrictions unpredictable, and funding uncertainties.

Responding to this confluence of factors, Goshen College, a liberal arts college in northern Indiana renowned for its study abroad and community-engaged learning programs, and Mennonite Central Committee (MCC), a faith-based international relief, development, peace non-profit organization working in 47 countries, came together to create a new partnership and pilot a new approach to web-based global community-engaged learning. Together, Goshen College and MCC piloted a nine-month, eight-credit global public health virtual practicum course during the COVID-19 pandemic (October 2020 – June 2021). The course included a diverse cohort of 15 students from seven countries around the world. The cohort included both undergraduate students from Goshen College as well as young adults from around the world whose international exchange and service plans were delayed or canceled due to COVID-19.

Instructors came from both institutions, with MCC's global health coordinator serving as the lead instructor. A second instructor whose role with MCC focuses on international service opportunities for young adults was primarily responsible to foster team building amongst the cohort, critical self-reflection, and cross-cultural competency.

The cohort met online weekly, gaining knowledge and experience in the foundational competencies of global public health theory and practice, the complexities of implementing real-life health projects, and learning from diverse experiences of each other, their cultures, and their contexts. The course culminated with a hands-on opportunity for the students to test their skills of project design and program evaluation and cross-cultural competency.

Intersectional in its approach, the course included 4 primary modules. The first focused on building a foundational level of global public health knowledge to provide a shared language around topics and to bring students up to the same level of understanding. A second module focused on the tangible skills of public health practice, particularly skills related to participatory project evaluation. A third module connected students directly to local staff implementing health projects around the world to assist with project evaluations. Lastly, the final module tied it all together by applying these skills and learnings to assist with composing a real grant application to fund a public health program related to COVID-19 in Zimbabwe.

This successful pilot effort provides learnings for others considering creative strategies for web-based global community-engagement. The course demonstrated the mutually beneficial opportunity for collaboration between higher education community-engaged learning programs and international relief and development organizations whose values align, using a low-cost virtual format. It provided useful assistance to the humanitarian relief and development efforts of MCC community-based partner organizations, while reducing the level of burden on project staff. Additionally, the course design allowed for local staff around the world to share their knowledge to help students understand the complexities and context-specific considerations needed for successful project design and implementation. By including individuals from around the world interested in taking the course for professional development, it fostered meaningful cross-cultural exposure and experiences for a diverse set of students who otherwise may not have been able to interact or travel internationally. The course met higher-education requirements for rigor, while maintaining high student engagement, academic performance, and satisfaction.

Building on the success and learnings of the pilot, the course was replicated in the 2021-2022 academic year in a shorter modified three-credit, one-semester format, and quickly was oversubscribed by interested students. The details of this more practical and refined one-semester course are what is provided in this chapter. Details of the nine-month version are available upon request from the authors.

This chapter shares the experience of Goshen College and MCC's global public health virtual practicum that was first rolled out in the 2020-2021 academic year and then adapted and replicated for the spring 2022 semester. Details provided on the structure, implementation, challenges, and lessons learned through the implementation of the course may be helpful to others considering this type of cooperative web-based service and learning model.

COVID-19 significantly upended higher education and global public health practice, but it also provided a unique impetus to innovate and creatively try new approaches. This is a case study of one such model of collaboration between a higher education institution and an international non profit organization that can potentially help bridge the gap between their missions to meet the unique needs of diverse students and community-based organizations, while lowering entry costs and reducing barriers of engagement for students.

ORGANIZATION BACKGROUND

Effective collaboration in community engaged learning requires mutually beneficial partnerships between institutions of higher education and community organizations doing work on the ground. Each partnership is uniquely negotiated as it finds the overlap in mission, approach, and the values of the organizations involved. To understand how this new model of global web-based service and learning in the form of a Global Public Health Virtual Practicum came to be and how it succeeded in spite of, and perhaps because of, the stresses of COVID-19 requires first understanding the two organizations involved. A deeper understanding of Goshen College and MCC and their approaches to community engaged learning and mission allows readers to understand the context of this case study and how its lessons may, or may not, apply to their own settings.

Goshen College Approach to Community Engaged Learning

In 1968, Goshen College faculty affirmed the motto "Culture for Service" and the institutional core value of "global citizenship" by unanimously voting to require a service-focused international semester in the core curriculum that all students are required to complete. Goshen College's Study-Service Term (SST), "offers students the opportunity to approach life, leadership and career as global citizens, able to collaborate for the common good and respect human dignity across cultural differences" (Goshen College, 2022). In this immersive three-month program, students spend their first six weeks in "study" attending language classes and studying the country's culture, history, and a relevant global issue, all while living with a local host family. In the following six weeks, termed "service", students are scattered across the country to live with new host families and spend their days in community-engaged learning assignments, working as volunteers for a local organization. Goshen College alumni frequently reflect that the SST term was the most transformative aspect of their educational experience which highly impacted their identity as global citizens (Goshen College, 2021a).

At its height in 1983, 83% of Goshen College students participated in this immersive program. However, the SST saw a steady decline over the years and reached an all-time low participation rate of 46% by 2019 (Goshen College, 2020). Not only has SST had differential enrollment over the years, but students of color, low-income students, and first-generation students have consistently been more likely to enroll in the on-campus global studies curriculum. Through a set of meetings between administration, faculty, students, and alumni, barriers to studying abroad were identified. These barriers were named as creating a culture in which less than 20% of students of color, compared to 57% of white students, chose to study abroad (Goshen College, 2020). Equipped with institutional and national data, the global education office began to design a new way to think about SST at Goshen College. The goal was clear–continue to educate students about relevant issues concerning public health, democracy and human rights, energy and natural resources, economic development, and entrepreneurship while creating a more consistent program that gives students an immersive experience whether on-campus, in our local community, in the US, or in an international context (Goshen College, 2021a). No matter what students are learning, Goshen College is committed to requiring 12 credit hours for every student to equip students with tangible skills to navigate today's globally connected world (Goshen College, 2020). Then, the COVID-19 pandemic spread across our communities and halted all international SST programs.

The COVID-19 pandemic, while devastating, also created opportunities to rethink global education at Goshen College. Without the option for international travel, more students than ever were seeking local immersive courses focused on global citizenship. This led to additional creativity and opportunity for partnership with both local and global organizations. Around the same time, Goshen College was also rethinking the usage of the word "service" and shifting to "community-engaged learning" which denotes the values of reciprocity and relationship (Goshen College, 2021b). These values have been instilled into the partnership that Goshen College cultivated with MCC to design and pilot the Global Public Health Virtual Practicum.

Goshen College is guided by five core values in community-engaged learning partnerships (Goshen College, 2021c). The first value is alignment with mission-driven goals for all partners. Goshen College prioritizes partnering with other value-driven organizations. For this reason and many others, MCC was a clear potential partner. The second is attention to power dynamics and equity in all community interactions. Building awareness of privilege among our students, faculty, and community partners is important as we create new courses that rely on collaboration with other organizations. From the stages of planning, implementation, and evaluation, this is integrated into all stages of course development. The third is open communication and transparency facilitated by deep listening. We prioritize open communication with organizations and leaders on expectations and responsibilities for all partners. For this course, this occurred through frequent meetings between Goshen College and MCC, where we jointly drafted understandings and agreements throughout the stages of program development. The fourth value is trust developed through long-term relationships of respect and ongoing conversation. The fifth and final core value is mutually beneficial collaborations enhanced through strategic planning and evaluation. This is evidenced through the evaluation and adaptation of the pilot to shift from being a nine-month program, to a semester-long course that is cross-listed between global education and public health. This decision was made possible because of the trust and reciprocal relationship intentionally built between Goshen College and MCC.

The origins of the Global Public Health Virtual Practicum came around the same time a new public health program was being developed and launched at Goshen College. This partnership with the public health program led to the course being offered as a single semester during the 2021-2022 school year and was cross-listed between both the global education and public health programs. The course fits into a larger public health thread in the global education curriculum providing a "community-engaged learning" opportunity that is both experiential and immersive. The goal of this mode of learning is to help students develop a personal sense of social responsibility in a cross-cultural setting through engagement in an action-research project. Built into all community-engaged learning courses at Goshen College is a requirement for critical reflection by students. The assignment or assignments allow students to reflect on their work with the community partner(s), including the dynamics of cultural and power differences and the students own commitment to civic responsibility (Goshen College, 2021d). This project is made possible because MCC's approach to service learning is so similar to Goshen College's approach to community-engaged learning.

Mennonite Central Committee Approach to Service-Learning

Often referred to as an "umbrella" agency where the anabaptist church constituency comes together in service ministry, MCC is a registered faith-based non-governmental organization that accompanies local partner organizations seeking to address the critical needs of their respective communities.

From its humble beginnings in 1920 to assist those affected by war and famine in southern Russia (present-day Ukraine), MCC has since expanded to work in forty-seven countries around the world, partnering with three hundred and eighty-five local organizations to carry out humanitarian relief, development, and peace projects in their respective communities (MCC, 2021).

MCC implements its work through a network of offices located around the world, staffed by a combination of local salaried staff and international volunteers. However, apart from a couple of countries, MCC has moved away from implementing its own projects and embraced a partnership approach to fulfill its mission. Joining the good work already being done locally, MCC seeks to accompany local partner organizations and the christian church in a "process of mutual transformation, accountability and capacity building" in sustainable ways identified by the local communities themselves (MCC, 2022a). These partnerships are not purely grant based but seek to engage in a process of healthy relationship building where knowledge, skills, and expertise are mutually shared.

Commitment to Young Adult Engagement

Providing service-learning opportunities for young adults in high school, college, and university settings is not new for MCC, however, in 2020, MCC governing bodies adopted a strategic commitment that spurred renewed focus and accountability to intentionally "engage the next generation" (MCC, 2020).

Multiple entry points exist for MCC involvement of students including annual advocacy-oriented student seminars, peace essay and speech competitions, young adult learning tours, connecting study abroad programs with international MCC offices and partner organizations, and facilitating opportunities for students to fulfill internship or practicum requirements though MCC service.

Each year, opportunities are offered for young adults from around the world to serve alongside MCC partner organizations for eleven months through the International Volunteer Exchange Program (IVEP), Serving and Learning Together (SALT) and the Young Anabaptist Exchange (YAMEN) programs. The shared anticipated outcomes of these service-learning programs are: 1) promote a theology of service that encourages growth as global citizens active in social justice and peacemaking; 2) provide opportunities for learning and mutual transformation through the development of intercultural skills; 3) foster opportunities for spiritual growth, appreciation of Anabaptist values and involvement in the local and global church, and; 4) Explore and develop skills while cultivating an increased sense of vocation by serving alongside a partner organization (MCC, 2022b).

The COVID-19 pandemic placed significant barriers and abruptly halted many of the entry points for young adult engagement that are traditionally offered by MCC. Tapping into their experience of being flexible and adapting when unexpected circumstances arise, young adult program coordinators considered what alternatives could be offered to students. Given shared values and MCC's commitment to resource and engage students attending anabaptist institutions of higher learning, partnering with Goshen college to try something new was a natural fit.

Operating Principles

Along with adhering to internationally recognized core humanitarian standards, MCC's work is guided by organizational operating principles that guide the way MCC approaches its programming, administration, and governance (MCC, 2018). These operating principles informed the design of the Global Public Health Virtual Practicum.

A particularly relevant MCC operating principle is to "connect people" by building, "…bridges across cultural, political, religious, and economic divides" (MCC, 2018, pp. 3). Their young adult programming therefore intentionally seeks to bring people together who might otherwise not have crossed paths to engage in critical reflection of their own culture and worldviews, to reject ethnocentrism, acknowledge weakness and mistakes, and to engage in mutual transformation through the sharing of knowledge and resources.

MCC acknowledges that it has a lot to gain from students as they offer an additional set of eyes to provide insights, and recommendations that may not have otherwise been uncovered. As many young adults have become suspicious of the intentions and practices of international non-profits, offering intentional opportunities for young adults to have direct interaction with MCC programming models transparency and accountability to help build trust and long-term commitment to the organization.

Another MCC operating principle particularly relevant to the design of this course is dismantling oppression. MCC defines this as working to address, "barriers of racial, economic, and gender-based oppression so that all may participate in our program design, decision-making and implementation" (MCC, 2018, pp. 3). This includes not glossing over historical practices that created and continues to reinforce systemic inequalities both domestically and internationally. MCC seeks to listen to those who are marginalized, and to support their efforts to strengthen their communities. An important part of dismantling oppression includes the continual development of intercultural competency and engaging in anti-colonial practices.

CASE DESCRIPTION

Spurred by the exigencies of the COVID-19 pandemic and related travel and in-person meeting restrictions, Goshen College and MCC agreed to work together to develop and pilot a new approach to web-based community-engaged learning. The original pilot version of this program consisted of an eight-credit Global Public Health Virtual Practicum course lasting from October 2020 to June 2021. The original cohort of fifteen students were from seven countries and included both undergraduate Goshen College students seeking to participate in a SST, as well as young adults from around the world who had anticipated participating in an international service and learning assignment with MCC around the world. Due to the COVID-19 pandemic, their international exchange and service plans were delayed or canceled. The approach was then refined and shortened into a more practical and replicable one-semester, three-credit format for a larger twenty-five student cohort in the Spring 2022 semester. The details of this more practical and refined one-semester format will be explored below as it is more applicable to other institutions of higher education.

Course Structure and Approach

The course was designed to fulfill several distinct needs: 1) satisfy the academic requirements for Goshen College's global health course in line with Council for Education on Public Health (CEPH) competencies; 2) satisfy Goshen College's requirements as a intercultural service learning course, and; 3) introduce students to MCC, its approach to global public health service, and to concretely assist international partners working on health projects. Weaving these requirements together into a cohesive intersectional course for a diverse group of students, without course prerequisites or additional financial resources was a challenge.

Doing this well required building the course from scratch using an outcome-based design process, rather than building from the framework of Goshen College's existing global health course. The primary practical outcome of the course was the applied public health project evaluation that students would complete for local MCC partner organizations around the world. In order to justify MCC partners and staff around the world taking the time and energy to support the students in these evaluations the output had to be useful, rather than merely an academic exercise. By the end of the course the students, coming from diverse academic backgrounds and many without any applied public health experience, would need to be able to critically assess the academic literature supporting specific public health interventions, work effectively and respectfully across cultural, contextual, and language differences with frontline staff in these projects, be able to understand and asses complex global public health project designs and reporting, conduct basic qualitative research with project from different cultures and professional backgrounds, and to produce actionable evaluation findings and recommendations.

This practical course goal spanned all three sets of interwoven course requirements (public health competencies, intercultural service learning objectives, and practical assistance to global health partners) and provided a unifying purpose around which to design the course. With this goal in mind, the instructors worked backwards to define a detailed list of course learning objects, provided below.

Course Learning Outcomes

1. Describe key global health concepts including epidemiological transitions, measures of health status, and the burden of disease.
2. Identify key organizations and institutions in terms of their roles in global health and the manner in which they can collaborate to address key global health issues.
3. Interpret key indicators that are used to evaluate globally health priorities.

4. Describe how social, economic, environmental and political factors impact morbidity and mortality and how they present opportunities or barriers to prevention and control of disease.
5. Discuss strategies to promote maternal health, and health among all groups of the age spectrum ranging from neonates to the aged.
6. Identify health conditions with a major impact on global morbidity and mortality, including infectious diseases, cancer, cardiovascular disease, diabetes, mental health issues and injury.
7. Identify key challenges that are likely to arise in the next decade in addressing global health, especially among the poor in low- and middle-income countries, demonstrating an understanding of the history of global health, health inequalities, and links to historic and ongoing systems of oppression.
8. Use analytical tools and concepts to determine how critical health issues might be addressed in cost-effective, efficient, culturally contextualized, and sustainable ways.
9. Explore how faith, moral, and cultural values are incorporated into health systems and approaches to public health. Students demonstrate an ability to learn from and work respectfully with diverse classmates, partners, and project stakeholders, adjusting one's own attitudes and beliefs because of working within and learning from the diversity of communities and cultures.
10. Demonstrate understanding of the student's own culture(s) alongside the culture of others. This will include a description of how one's own culture(s) has influenced personal values, beliefs, and assumptions and how to approach other cultures and contexts in public health from a position of empathy and cultural humility.
11. Demonstrate an awareness of social systems and institutions from the perspective of people in other culture(s) and how those systems intersect with the structures that shape their own lives and health. Students will examine the interdependent dynamic forces shaping any society (religious, political, economic, social, artistic, geographic, and historical), and how these intersect with health and health systems.
12. Develop a plan for integrating term learnings into career and educational paths.

With these course objectives defined, the instructors then ensured compliance with public health accreditation requirements by mapping the course to the Council for Education on Public Health (CEPH) competencies (CEPH, 2021). The two CEPH competencies mapped to the course and its assessment mechanisms were:

1. Public Health Communication: Communicates public health information, in both oral and written forms and through a variety of media, to diverse audiences. Assessments: Class discussions, student news reports, global health project evaluation report and presentation.
2. Public Health Information Literacy: Locate, use, evaluate and synthesize public health information. Assessments: Reading reflections, class discussions, student news reports, global health project evaluation report and presentation.

After defining the course objectives and public health competencies, a guiding course logic was developed around three sequential modules. The first module established a foundation of shared knowledge and language relevant to global public health theory, understanding MCC as an organization and its approach to global public health practice and ethics, and a framework for understanding how public health projects are designed, funded, implemented, and evaluated. The second module sought to help students

Figure 1. Global Public Health Virtual Practicum course module logic

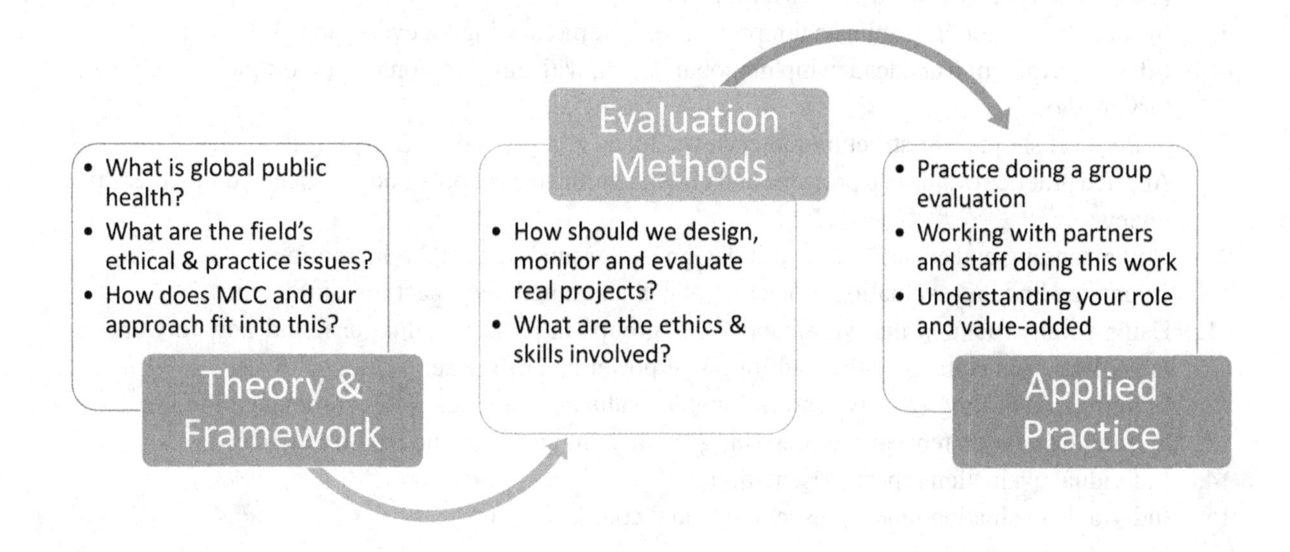

develop practical public health skills needed for project design, monitoring, and evaluation including an emphasis on the use of participatory qualitative research methods. Cross-cutting themes included the ethics of evaluation and developing intercultural competency. An understanding of implicit power dynamics and practices to help to mitigate them was also mainstreamed into this module. The final module focused on applying the knowledge and skills learned in the course to an applied cross-cultural setting. In the nine-month pilot of this course, the applied practice module included two activities: 1) individual and group water sanitation and hygiene (WASH) project evaluations; and 2) background research and writing to support a grant application for a COVID-19 project in Zimbabwe. In the subsequent, more refined one-semester version of the course, this final module consisted of one applied practice project: individual project evaluations of maternal, child, and newborn health (MNCH) projects in Africa that used the Care Group model of intervention.

Once the course logic and modules were established, a detailed week-by-week course plan was developed to lead students through this learning journey. This was a one-semester course, with weekly three-hour web-based live video sessions for fifteen weeks. Sessions were synchronously during a weekday evening to be sensitive and accommodate the existing personal and academic responsibilities of students.

Week-by-Week Course Plan

1. Course orientation, getting to know our cohort, class logistics, and orientation to Mennonite Central Committee as a global public health and humanitarian actor.
2. Introduction to global public health as a field of study and practice, its colonial history and understanding and working against systemic inequalities.
3. Understanding social and environmental determinants of health in diverse global contexts with a case study on adverse childhood events and working in situations of conflict and displacement.

4. Applying learnings to two case studies in global public health: water, sanitation, and hygiene (WASH) and mental health across cultures.
5. Introduction to global public health project design, monitoring, & evaluation skills and practice.
6. Ethics of evaluation and leadership in global health, with emphasis on the importance of participatory methods.
7. Skills workshop: semi-structured interviews, focus groups, and barrier analysis.
8. Applied practice evaluation projects, and understanding the purpose and practice of doing literature reviews.
9. Understanding and evaluating project design and grounding it in the literature
10. Understanding and evaluating project reporting and outcome/impact measurement.
11. Using semi-structured interviews and focus groups in project evaluations, and how to work effectively across cultures with mindfulness of power and privilege.
12. Framing and writing effective and actionable evaluation findings and recommendations.
13. Writing evaluation reports and presenting findings in global public health practice.
14. Individual evaluation report presentations.
15. Individual evaluation report presentations and course wrap up.

Readings for the course included videos, news articles, keeping up with global public health news (using the Johns Hopkins University Global Health Now newsletter), and review of actual project proposals, reporting, and communication pieces of MCC. The course also brought in interviews with MCC.staff and partner organizations from around the world to provide more exposure to the realities of this type of work in practice. The exposure to global staff and partners also provided students the opportunity to hear directly from the people most closely involved in the projects and gain experience in cross-cultural communication and translation processes.

Grading for the course was weighted heavily towards applied practice, with the remainder for active participation and thoughtful self reflection. Grading was done collaboratively by the interdisciplinary teaching team.

Grading Structure

Participation & weekly public health news sharing: five points each for fifteen weeks; 25% total grade

a. Each week students briefly shared one global public health story or issue from the news (generally drawn from Global Health Now) at the beginning of class. This gives practice in keeping up with global public health news, synthesizing and making sense of complex stories, and summarizing and presenting on diverse public health issues to peers.
b. Attendance and active class participation.

Reading reflection journals: fifteen points each for six weeks; 30% of total grade

a. For weeks two through seven, students submitted a concise (250 words or less) reflection on a key theme or topic raised in one of the readings and preparation materials assigned for that week. Due at the beginning of each class, these reflections were intended to hold students accountable for keeping up with course material and to foster a critical understanding of issues raised by connecting it to each student's unique experience and perspective.

Evaluation report: one hundred points; 33% of total grade

b. The evaluation report was the capstone project of the course. Each student was assigned one outcome theme and one real MCC project to evaluate during the second half of the course. Each subsequent week, students completed and submitted a draft of each evaluation element (literature review, project design summary, project reporting summary, interview notes, findings, and recommendations). Class time was provided for raising questions on each evaluation element as it was completed, and feedback was provided on submitted drafts. A complete edited and revised evaluation draft was due before the last week of class.

Evaluation Presentation: thirty-five points; 12% of total grade

c. In the final two weeks of class, each student provided a brief presentation of their evaluation including a summary of the assigned project, the project's context, and approach; connection to the literature and theory of change; project outcomes and impact; and key findings and recommendations.

RESULTS AND LEARNINGS

Both years of Global Public Health Virtual Practicum (2020-2021 and 2022) have been a success academically, in terms of student engagement, satisfaction, in the practical assistance provided to MCC's community-based health partners around the world, and in helping inspire students to further work and study in the field of public health.

The pilot cohort included students from seven countries, half of whom were taking the course for professional development and not receiving formal academic credit, met online weekly for three to four hours on Saturday mornings for nine months with an attendance rate of 83%. For the second cohort, the attendance rate for the more condensed semester-long format increased to 94%. In both these pandemic years, this attendance rate was higher than for the average Goshen College course despite including many global students not receiving formal college credit.

The quality of students' applied work was high and practically useful for MCCs community-based partners. Students completed mixed-methods project evaluations that helped inform project development and adaptation and completed a major portion of a project proposal submitted to the Canadian government funding of COVID projects in Zimbabwe. Student engagement was high, leading to oversubscription of the course in both years and a desire to continue the course in the future.

Student Feedback

During the last class, students engaged in interactive debriefing exercises to share key learnings they developed and completed an online survey to test whether the objectives of the course were achieved.

Examples of student feedback include:

- "I have learned a lot about global health. I appreciate the different perspectives shared every week. [The teachers] are really great at engaging my mind to think about the world and gain many new perspectives" (MCC, 2021, p. 3).
- "Coming into this class, I had been unsure how I should feel about MCC and other Christian-based groups who do service around the world. Part of the reason for this is because I do not like the 'white savior complex' that often accompanies those organizations. I also do not think that we, in developed nations, have everything figured out and we don't always know what is best. However, I really do appreciate that MCC is attempting to mitigate this complex by working with the leaders and others in the communities and listening to their ideas on how to best handle the problems. I think the community development piece of what MCC strives to do is important so that once MCC members leave a community, the community can continue to thrive themselves, without needing outside help. I believe that this should be one of the main goals of service for many other organizations, to help communities maintain their identity and independence" (MCC, 2021, p. 3).
- "I love this class and am so thankful to be a part of it…I did not know how to navigate serving abroad which had been placed on my heart, and how to do it in an ethical way. Working with local organizations allows members of MCC to support programs from within the community for the community. The result is impactful… I deeply believe the more we are able to meet and live among people different than us, the easier it becomes to love people from all walks of life. Therefore, I've been eager to share MCC's various young adult programs with anyone who is curious. I have found that living and serving abroad is something many people are interested in but are often too overwhelmed with the process of choosing where to go or who to work with" (MCC, 2021, p. 3-4).
- "This has been so enriching, thank you so much! What I've heard reminds me of what is inspiring about community efforts. It is so interesting. This idea of radical acceptance, seeing people as people and nothing less, and in that very fact of humanizing others, that is so healing…So many groups just look for the easiest way to intervene, and the numbers, but not the people behind them" (MCC, 2021, p. 4).
- "I just want to thank you guys so much for getting our minds stimulated during this difficult time and teaching us and being open and kind" (MCC, 2021, p. 4).
- "I have been surprised to hear these many connections around the world to my experience. For example, with COVID around Bangladesh people were saying it was just a rich person's disease and it couldn't hurt them...But then to hear it was the same in Zimbabwe was a big surprise. I have been very surprised at all the connections" (MCC, 2021, p. 4).

Learnings from the Course

Meaningful intercultural learning and exchange is possible in a virtual format. This is facilitated by having practical hands-on work to do together in diverse cohorts of students. The longer the cohort has together and the smaller the cohort, the closer these relationships can become and the more meaningful this exchange can be. However, these gains have to be balanced against the logistical needs for having larger classes and semesters that line up with the standard academic calendar.

Virtual service work in global public health is possible. Students were able to complete meaningful project evaluation and grantwriting work for community-based partners around the world after roughly two months of foundational teaching. However, the logistical demands of facilitating this service work should not be underestimated. Preexisting relationships with community partners, a shared understanding of concrete achievable work products, and the logistical capacity to set up calls and facilitate translation is required.

Group projects, with discrete individual contributions provided the best balance of benefit to burden for the global partners while being fair and manageable for students. In the first cohort,each student had their own global partner and project to evaluate. This was a rich experience for the students and gave them a strong sense of responsibility and ownership over the work product. However, the burden on each of those partners and the logistical coordination for the instrors was high for the relative value of work back. In the second cohort, students were assigned to global partners in groups, with each student doing a discrete evaluation of one element of the partner's work. This dramatically reduced the logistical burden of providing students access to project documentation, setting up interviews with partner staff, etc. It also provided each of those partners a significant body of work at the end of the semester, with 4-5 detailed evaluations of different aspects of their work.

Successful partnership between higher education and non-profit humanitarian organizations requires a clear understanding of each organization's mission, values, comparative advantages, requirements, and value proposition. Having this clear shared understanding opens up the possibility of win-win scenarios where each organization is able to benefit from the partnership. This upfront work allows for more durable and successful partnerships than the one-way transactional relationships that are more common in the field. Neither Mennonite Central Committee or Goshen College were going into this partnership to help the other, despite some shared history and overlap in mission and constituency. Transparency in the process by both sides allowed for a negotiated win-win solution that has proven durable and successful for both parties.

New partnerships and new models of community-engaged learning and cross-cultural exchange are possible, even on tight timelines in times of crisis. Virtual modalities lower barriers to piloting new approaches and making initiatives more inclusive.

SOLUTIONS AND RECOMMENDATIONS

Partnership between higher education and non-governmental organizations can be highly beneficial for student learning. Web-based models of learning are advantageous because they provide students with opportunities to learn from others outside of their physical location, and community-engaged learning gives students the opportunity to study in an immersive environment. Together, these models complement each other and create a learning environment for students that can be cross-cultural, immersive, and facilitate development of knowledge and skills.

The basis for the success of this type of learning is through a reciprocal relationship between the organizations collaborating to create the course. This reciprocity should be built through an extensive planning and relationship building process prior to piloting the course. A key first step is an assessment of the alignment between both organizations' goals and objectives for the course. This sets the stage for developing mutually agreed upon student learning objectives which should inform all other aspects of the course.

Another important aspect of the planning process is the development of evaluation processes that should be utilized to continually assess the partnership between institutions as well as evaluate the student learning experience. Evaluation should take place through both formative assessments throughout the course and a summative assessment at the end of the course, prior to the next offering of the course. Enough time should elapse between course offerings so that assessment findings can be integrated into the next iteration of the course.

Web-based learning is best implemented when students are placed in an active role within the classroom in contrast to the traditional classroom where students take a passive role in the learning process. This is exceedingly important in web-based models of learning because of the risk of students passively engaging through a screen for the entirety of the course, therefore retaining little information and building few skills. This is another reason why community-engaged learning and web-based models function strongly together. When students are given the opportunity to learn through real-world case-studies and projects, they are placed in an immersive, cross-cultural learning environment which enables them to build their skills in self-management, critical thinking, and social learning.

In community-engaged, web-based learning there should be high importance placed on scaffolding a larger assignment throughout the course. This provides an opportunity for students to build their skills in self-management, while being guided on how to complete each part of the larger project. It also helps weave the learning from class-to-class into a larger project where students apply the learnings from each class session into a cumulative deliverable. This application of the course learning objectives gives students the opportunity to use their own abilities and gifts to create something new – a method that is shown to have lasting impact on student knowledge and skill retention.

While higher levels of learning can occur in web-based models, there are limitations to this learning mode. As we learned through our pilot program, a shorter-term learning experience i.e., one semester, fifteen weeks, provides a higher level of engagement from students compared to a longer term course, i.e., two semesters, thirty weeks. Length of the course is less important, however, compared to the structure of the time spent in class. Web-based models provide flexibility in how students experience class, as instructors can use a class to lecture, show videos, invite in guest speakers from around the world, facilitate large and small group discussions, and provide time for students to work in small groups on case-studies or other project-based learning activities.

Another limitation of web-based models of education is that for students who learn best with hands-on learning, this mode of learning can be more challenging. This is where a course that focuses on case-studies and scaffolds a larger project can provide an engaging learning experience for students. This provides students with a balance of didactic and experiential learning opportunities, similar to what they would be offered in a traditional classroom. However, this mode of learning offers an additional layer of opportunity for students because they are able to engage with other students, practitioners, and projects from around the world, providing more opportunities for cross-cultural interaction and learning.

While there are limitations and challenges to web-based learning, the benefits outweigh these disadvantages. Immersive, community-engaged learning projects typically cost the higher education institution and students significant financial resources. This modality of learning is considerably less expensive and therefore increases access for students who may have less financial resources. This type of learning also increases access for students who need to work to support themselves and/or their families, and those who cannot leave the country for a variety of reasons. Web-based learning increases equitable access to higher education and to immersive, community-engaged learning opportunities.

CONCLUSION

Learning in the post pandemic, now endemic, world of COVID-19 has experienced many shifts. While cross-cultural and immersive learning remains a goal of many higher education institutions, the reality of pursuing in-person learning as the only form of community-engaged learning is no longer the reality. Web-based learning provides an opportunity for students to continue to learn alongside people from other cultures around the world while physically staying in their home environment.

While there are challenges and limitations to this type of learning, there are also many benefits to web-based learning. Not only can students have an immersive learning experience, but students can build knowledge and skills that would otherwise require much higher levels of financial resources from either the higher education institution or the participants themselves–increasing equitable access to this type of learning for students who may not otherwise be able to participate.

Collaboration between higher education and non-governmental organizations is a way for students to be taught by leaders in the field who are working on the ground in different countries and cultures – allowing students to engage in community and service-based learning while continuing to fulfill other responsibilities such as taking other courses or working. The new age of learning is here and it looks like collaborative partnerships, community-engaged, web-based education.

REFERENCES

Council on Education for Public Health. (2021). *Accreditation criteria: Schools of public health and public health programs*. https://media.ceph.org/documents/2021.Criteria.pdf

Goshen College. (2020). *2019-20 Annual Report.* Unpublished internal company document.

Goshen College. (2021a). *Building Capacity for Overcoming Barriers to Inclusive Study Abroad.* Unpublished internal company document.

Goshen College. (2021b). *Center for Community Engagement: Statement of Intent.* Unpublished internal company document.

Goshen College. (2021c). *Statement of Values.* Unpublished internal company document.

Goshen College. (2021d). *Core Refresh Curricular Proposal.* Unpublished internal company document.

Goshen College. (2022). *Study-service term (SST): global education at Goshen College*. https://www.goshen.edu/sst/

Mennonite Central Committee. (2018). *Operating principles framework.* Internal unpublished company document.

Mennonite Central Committee. (2020). *MCC shared program strategic plan, FYE 2021-FYE 2025.* Internal unpublished company document.

Mennonite Central Committee. (2021). *Learning and Reflection from Semester One of the Global Public Health Virtual Practicum*. Unpublished internal report.

Mennonite Central Committee. (2021). *Annual impact report 2020/2021.* https://mccanada.ca/learn/about/annual-reports/annual-impact-report-20202021

Mennonite Central Committee. (2022a). *MCC's principles and practices.* https://mcc.org/media/resources/1382

Mennonite Central Committee. (2022b). *Serving and Learning Together (SALT).* https://mcc.org/get-involved/serve/volunteer/salt

Warnick, A. (2021). Interest in public health degrees jumps in wake of pandemic: Applications rise. *The Nation's Health, 51*(6), 1–12.

Chapter 15
Belonging, Performance, and Engagement:
What Tomorrow's Leaders Need to Know

Andrea D. Carter
https://orcid.org/0000-0002-5712-6611
Adler University, USA

ABSTRACT

The effects of leadership on performance and engagement have long been evaluated and measured. Communities, organizations, and educational institutions seek to empower and support behaviors that lead to results, but the context of how is fundamentally different post-pandemic. Looking forward, organizational and institutional success will depend on the leader's ability to create a belonging environment. With followership no longer driven by rational rewards, leaders must learn how to connect with members' emotional needs to motivate and inspire action. This chapter examines the importance of adopting new leadership behaviors, the cost of membership loneliness and exclusion, and the impact of trust, accountability, and empathy. The chapter concludes with indicators and tactics to create belonging environments that drive engagement and high performance. In the absence of consideration of the concepts discussed in this chapter, mediocrity will rule, and a twenty percent engagement rate has the potential to become the global standard impacting performance worldwide (Gallup, 2022). These executed concepts and tactics will support community members to behave differently, empowering each other to rebound and thrive using tools that meet emotional needs and empower positive change.

INTRODUCTION

In 2020 the global economy shrank by 3.5 percent, the most significant contraction in 60 years and a direct result of the devastation of the global pandemic (Gallup, 2022). Not only did people struggle worldwide with the highest levels of stress, sadness, anger, and worry, but it became undeniably evident that people at the same workplaces and in the same communities are having radically different experiences based on the intersections of their identities (Woods et al., 2021). Additionally, the stakes and impact of holding

DOI: 10.4018/978-1-6684-5190-8.ch015

tight to a status quo that permeates exclusivity, division, polarization, and unethical behavior are now recognized as a recipe for toxic environments, decreased engagement, and low-performing organizations (Rasool et al., 2021). Post-pandemic corporate reputation is intricately tied to employee engagement and customer dedication. This shift underscores the need for leadership to focus on being accountable, trustworthy, and inclusive should they want highly engaged and high-performing organizations and institutions (Overton-de Klerk & Muir, C, 2022).

Demand is high for this shift in attention, with a 2022 Gallup survey finding that 74 percent of Americans were dissatisfied with the size and influence of major US corporations, up from 48 percent in 2001 (Gallup, 2022). Additionally, 70 percent of people worldwide believe corruption is widespread in business and impacts overall performance and life experience (Gallup, 2022). The data reinforces why leaders are looking to reinvent and step up in a new and different way. Even pre-pandemic progressive leaders were calling for changes. In a New York Times opinion piece, Salesforce CEO Marc Benioff stated that, globally, the 26 wealthiest people account for as much wealth as the 3.8 billion poorest. In the United States, income inequality has reached its highest level in approximately 50 years. The rise of equity, fairness, and transparency is now becoming an expectation to rectify injustice in the workplace. Benioff (2019) states, “It’s time for a new capitalism - a fairer, equitable and sustainable capitalism that actually works for everyone” (Benioff, 2019).

Furthermore, while this data is compelling, the trends show that the emotional connection to equity and fairness is now at the heart of engagement and performance (Kuknor & Bhattacharya, 2021; Overton-de Klerk & Muir, C, 2022). In best-run companies, 73 percent of workers are emotionally engaged and are thriving at work. Unfortunately, the global average is a mere 20 percent (Gallup, 2022).

As a society, we have moved beyond stating that we need to change how we work and interact; now, change is a matter of survival, and those who fail to adapt will lose the followership needed to succeed. Additionally, everything is more complex in the virtual world; how we communicate, how we make decisions, how we get work done, how we are accountable to others, how we create community, how we create value, and how we generate engagement - all these interactions differ in the virtual or hybrid space. These differences amplify that business, in any industry, is no longer simply transactional because for human performance to flourish, the environment must also support emotional needs. This shift demands leaders to move out of perceptually complicated strategic actions and into satisfying basic rational and emotional human needs. When organizations, institutions, and communities can fulfill emotional needs, people thrive, engage, and remain dedicated to successful outcomes (Roberson & Perry, 2022). In the workplace, rational needs are having a safe working environment, good pay, and benefits. Workplace emotional needs include autonomy, flexibility, and belonging. The gap in most workplaces is that most leaders ensure rational needs are met, but rarely do they ensure emotional needs are met (Ho & Chan, 2022).

Therefore, this chapter is a starting place for leaders who want to reinvent, inspire and motivate. A spark, if you will, for creating credibility that comes from the mind so that trust can come from the heart. I hope to provide the multi-dimensions of community inclusion and belonging, emphasizing the role of meeting followers’ emotional needs by providing leaders with awareness and methodology to activate and advocate change on systemic levels. The understanding is that, without accountable leaders who value, understand, communicate, behave and take action to include the emotional needs of their followers, leadership effectiveness will remain mediocre at best - a potential that our recent data indicates will lead to extinction for many communities, institutions, and businesses.

LEADERSHIP - THE NEW ROLE

In 1968, months before Martin Luther King, Jr. was assassinated, he said, "One of the great liabilities of history is that all too many people fail to remain awake through great periods of social change. Every society has its protectors of status quo and its fraternities of the indifferent who are notorious for sleeping through revolutions. Today, our very survival depends on our ability to stay awake, to adjust to new ideas, to remain vigilant and to face the challenge of change" (King, Jr., 1972, p.8). This quote is more prevalent in recent times since corporate reputation is now closely related to employee and customer satisfaction, a factor the media and online social outlets quickly point out (Overton-de Klerk & Muir, 2022). Leadership has fundamentally changed, and while there are still protectors of the status quo, credibility and reputation for unethical behavior are no longer tolerated by employees, shareholders, or customers. When executives and directors do not pay enough attention, or take complaints or accusations seriously enough, press releases indicating an investigation are no longer enough to maintain trust or loyalty (Overton-de Klerk & Muir, 2022; Gallup, 2022). Members perceive betrayal when trust and loyalty break (Reina & Reina, 2016).

Betrayal is often described in relation to six experiences by leadership, (1) when leaders do not do what they say they are going to do, (2) when backbiting or gossip is normalized within the organization and leaders participate in the behavior, (3) when a decision previously agreed upon is reneged, (4) when someone's agenda is hidden so they can work behind the scenes and leadership allows them the power to do so, (5) when someone spins the truth rather than speaks the truth, and (6) when leaders fail to apologize for actions, citing their intention was not meant to harm (Reina & Reina, 2016; Mahlangu, 2020).

Interestingly, many protectors of the status quo also rely on avoiding conflict by remaining silent, not realizing that their silence upholds this sense of betrayal by demonstrating complicity and perpetuating societal injustices and inequities (Reina & Reina, 2016; Mahlangu, 2020; Spiller et al., 2021). As heightening mistrust and the destruction of accountability increases, the sense of betrayal inevitably leads to damaged reputations (Gallup, 2022). All too often, change comes from ordinary citizens raising their voices in unison to foster change, a driving principle in why social media grows in its power. Leaders must cultivate camaraderie, collaboration, and a sense of unison in today's climate. Leadership increasingly depends on followership and motivation across remote and virtual spaces, regardless of the environment, context, or intention.

Inclusive Leadership

In recent years, the new role of inclusive leadership has gained heightened credibility within volatile and ever-changing markets. The reason is that inclusive leadership meets individual followers' needs, accepts differences, and is accountable for creating an environment where everyone belongs. In these climates, people become emotionally invested and remain engaged, innovative, and creative - qualities that enable crisis aversion through better decision-making (Khumalo et al., 2022; Ulmer, 2012; Zhang et al., 2012; Spiller et al., 2021). In addition, Zhang et al. (2012) contend that what leads to successful transformational changes within a community or organization, where aversion to a crisis is managed effectively, is primarily determined by the level of emotional intelligence (EI), emotional control (EC) and the quality of the leader's communication skills (LCS) because these factors speak to emotional needs. The reasoning makes sense when looking at the data. Emotionally engaged customers spend approximately 23 percent more than satisfied and dissatisfied customers (Gallup, 2022). Additionally, companies with emotionally engaged customers see (1) an average of 66 percent higher sales growth; (2) a 10 percent growth in net profit; and (3) a 25 percent increase in customer loyalty (Gallup, 2022). With these numbers driving the survival of many companies, the data support the need for leaders to take on the new role of inclusive leadership.

Inclusive leadership is "the degree to which an employee perceives that they are an esteemed member of the work group through experiencing treatment that satisfies their needs for belongingness and uniqueness" (Randel et al., 2018, p. 192). This definition is an important one to recognize as it moves the responsibility of belonging solely from the individual into a shared responsibility, a distinction that is gaining momentum (Nishii & Mayer, 2009; Shore et al., 2011; Mor Barak & Daya, 2014; Winters, 2014; Carter, 2022; Roberson & Perry, 2022). Individuals perceive inclusion when leaders facilitate and model trust, accountability, and responsibility for behaving with empathy. When leaders facilitate and model the value of belonging and uniqueness, they place effort and value in overcoming status differences, often highlighted by unique identity intersections (Randel et al., 2018; Carter, 2022; Roberson & Perry, 2022). Leaders who take action to speak to the emotional needs of their followers, including employees and customers, enable a ripple effect in the community, allowing an emotional engagement to take place (Spiller et al., 2021; Roberson & Perry, 2022).

Other researchers contend that inclusive leadership practices mediate trust in a leader through communication abilities that uphold and respect the needs of status differences (Gotsis & Grimani, 2016; Cao & Le, 2022). For example, leaders who were accountable to the Black Lives Matter movement recognized the emotional needs of their membership and released both internal and external communications speaking to the heart of the injustice. The most effective communications spoke to the emotional needs by acknowledging the pain and trauma of 400 years of oppression, racial discrimination, and the impact of white supremacy (Kiang & Tsai, 2022; Taylor et al., 2022). Leaders who acknowledged membership differences and the impact of intersecting identities on basic emotional needs more easily supported their members and recovered as an inclusive engaged entity with greater engagement and performance metrics (Powell, 2022).

The Importance of Inclusive Leadership Behaviors

Essential emotional support motivates collaboration and positive actions, a switch in conceptual thinking because leaders were long groomed to be intellectual and to reject emotions and feelings. However, the application of meeting members' emotional needs enables swift action, engagement, and positive performance (Panteli & Tucker, 2009).

Consider the move from in-person environments to virtual environments at the height of the pandemic. Leaders who leveraged inclusive leadership behaviors with emotional intelligence and repetitive communications mediated job satisfaction and engagement (Alwali & Alwali, 2022). Therefore it is no surprise that as we move into an endemic with in-person, virtual, and hybrid environments, flexibility and individual emotional needs remain as imperative as when the pandemic first hit. As constant change and transition continue in day-to-day operations, leaders must consider diverse human needs and emotional motivations as significant levers for successful outcomes. Leaders who understand unique cultural and individual needs will motivate and inspire trust. When leaders are trusted, followers take action. When followers take action, they are engaged and motivated to perform. The subsequent research could not be more transparent (Nishii & Mayer, 2009; Shore et al., 2011; Mor Barak & Daya, 2014; Winters, 2014; Alwali & Alwali, 2022; Carter, 2022; Roberson & Perry, 2022); specifically in diverse workgroups where inclusive leadership behavior is adopted to influence followers' participation and engagement so that employee turnover is reduced (Cox, 2001; Podsiadlowksi et al., 2013; Kuknor & Bhattacharya, 2021).

Inclusion, value for individuality, and belonging capabilities remain at the forefront of leadership skills because the impact of leadership communication, accountability, and trust increase with leadership promotion. As leaders move up from management to higher levels of leadership, their ability to encourage, motivate and create consistency of inclusive messaging heightens, a factor that becomes more important in virtual and hybrid environments (Kuknor & Bhattacharya, 2021). Positive high-performing cultures are supported by open communication and dialogue, learning environments, flexibility in policy, and belief and conviction about inclusive behaviors (Banks et al., 2022). The new lens of leadership goes beyond cultural competency or diversity policies. Instead, it fosters an environment where individuals are empowered to be seen, known, valued, accepted, and cared for, regardless of the intersections of their identities (Carter, 2022). While the sheer volume of research detailing inclusive leadership indicates its impact, so does the variability of Equity, Diversity, and Inclusion (EDI) initiative effectiveness (Mor Barak et al., 2022; Nishii & Leroy, 2020). The risk of not having inclusive and belonging leadership behaviors is homogeneous representation because research recognizes that homogeneous groups, teams, and communities are less rigorous in their performance and make more mistakes than diverse ones (Banks et al., 2022; Nishii & Leroy, 2020). Without an inclusionary social influence process, where followers' emotional needs are recognized, and motivation met, engagement and performance will remain elusive because diverse communities will continue to struggle to work together towards one common goal (Banks et al., 2022).

Leadership Risk - The Cost of Exclusion & Loneliness

The risk of silos in today's world is more significant than before because, as the pandemic taught us, no one thrives on an island. Isolation, ostracization, and exclusion are all social actions that cause loneliness. Loneliness starts with social exclusion, and once loneliness sets in, it is hard to overcome and return to one's pre-lonely state (Kerr, 2021). The pandemic made it cumbersome for adults (Tull et al., 2020) and adolescents (Loades et al., 2020) to interact meaningfully in ways that produce unfeigned fulfillment. Consequently, the pandemic conceived "an epidemic of loneliness" experienced with radical differences depending on the context of individualism and history. The loneliness factor weighs in on the cost of exclusion and speaks to an element of belonging that leaders now need to understand and be accountable for (Cigna, 2020; Ducharme, 2020).

In the post-pandemic environment, leaders must adapt to the realities that silos, hybrid models, and virtual workspaces provide sequestered opportunities for isolation, exclusion, and inequity. Loneliness becomes a heightened risk because loneliness breeds aggression, which emerges as an emotional response to uncertainty (Twenge et al., 2001). On top of that, loneliness independently predicts inefficiency at work amongst non-disabled, working adults in the prime of their careers (Mokros et al., 2022). As we know, market volatility and constant change will remain part of the global experience for years. With this, efficiency and collaboration will be drivers of engagement and performance for organizations, institutions, and communities to emerge from surviving into thriving actions. Therefore, the implications for leaders who need help understanding the risk of loneliness is that collaboration, innovation, and efficiency may continue to suffer unless interventions are accounted for. What is more, is that with the radical differences in the pandemic experience, where further separation of membership occurred, leaders who are unable to understand the risk and emotional needs of their members have the potential to enhance individualism further and drive aggressive behaviors and conflict (Mokros et al., 2022; Magalhães et al., 2022).

The heightened risk of leadership conflict is an emerging trend and a risk that will have a direct impact on performance. Ofei-Dodoo et al. (2021) found a significant increase in the probability of loneliness where emotional exhaustion, depersonalization, and perceived fatigue were present. Magalhães et al. (2022) analyzed the variables in leadership loneliness as they apply to the intersections of identity. In the study, gender, age, academic qualifications, function/position, years of seniority, working hours per week, and consequences on the decision-making process were examined. Findings demonstrate that a significant relationship exists between loneliness and demotivation, which means that the interaction among the leadership team is also mediated by the intersections of identity and a driver for loneliness within the workplace (Magalhães et al., 2022). The risk lies in the inefficiency of leadership behavior and awareness of loneliness impact because, at all levels, loneliness will increase aggression (whether macro or micro), reduce collaborative engagements, and drive inefficiencies. Exclusionary practices are, therefore, a direct threat to mediating emotional needs and thriving post-pandemic (Hales et al., 2021).

LONELY FOLLOWERSHIP IN THE AGE OF AGILITY

Recognizing that a threatened sense of belonging leads to binary and aggressive actions, the role of the leader is ever more critical for thriving in post-pandemic times (Hales et al., 2021). Zhou et al. (2022) argue that the greatest priority of our current reparations state should be finding the means to alleviate loneliness for others and support their ability to experience a belonging culture. With research indicating a downward trend in loneliness scale scores with increased group work and one-to-one coaching (Yildiz & Duyan, 2022), supporting inclusion and belonging behaviors and modeling their practice should be regarded as a priority practice in both the private and public sectors. Likewise, Mokros et al. (2022) found significant loneliness reductions by improving social skills, enhancing social support, increasing opportunities for social contact, and addressing maladaptive social cognition. In real workplace scenarios, many leaders accept the need for inclusive leadership and belonging behaviors in response to loneliness interventions. The struggle currently lies in addressing group work processes in the age of agility, remote and hybrid environments. While the emotional needs will be covered later in this chapter, it is crucial to understand the rational needs and structures that otherwise prevent the strategic application of these actions.

First, Gallup's (2018) analytics reveal that 84 percent of American employees work in multiple teams daily, reporting to either the same or different managers. While these structures have been popularized to enable organizations and institutions to become more agile, engaging, and collaborative, poorly managed teams become ripe with conflict quickly, especially when cultures fail to manage their people's radically different emotional needs (Gallup, 2018). Three main factors are crucial for leaders to manage group work successfully and in agile communities; work environments need to be; (1) relationship-focused, (2) fair, equitable, and transparent, and (3) collaborative with individuals perceiving a belonging community (Gallup, 2018). However, for these factors to become achievable, leaders must become accountable for the expectations and behaviors normalized within the environment because organizational culture truly matters (Gallup, 2018).

Trust, empathy, and accountability must be modeled, valued, and repetitively communicated for optimal group work. When these factors are upheld, regardless of the matrix structure, the number of teams one is on, the technology that is accessible, or the complexity of the goal, members hold the values of inclusion as foundational workplace practices, and engagement and performance flourishes (Gallup, 2018; Gallup, 2022).

However, let us be clear that performative values do not produce authentic actions. To effectively deploy a fair and equitable workplace where everyone is treated as an insider, leaders must normalize trust, accountability, and empathy by modeling and embodying belonging tactics. Only within these contexts can members unleash their full potential, responding to challenges across boundaries while improving workplace experiences (Korkmaz et al., 2022).

Building Trust

Leslie et al. (2018) reveal that 85 percent of global teams work and meet virtually. In 2018, 40 percent of those teams had never met face-to-face. Post-pandemic, approximately 56 percent of full-time employees in the U.S. are remote, with 70 million indicating that their jobs can be done from home (Gallup, 2022). Furthermore, the turnover risk is massive without the flexibility to select either hybrid or exclusively remote workplaces. Gallup (2022) asked, "If your employer decides not to offer opportunities for you to work remotely some or all of the time long term, how likely would you be to look for opportunities for employment with other organizations?" Six in 10 remote employees say they are extremely likely to search for employment elsewhere if they are not provided with remote flexibility - up from 3.7 in 10 in June 2021. This data shows that remote work is in demand and a requirement for engaged and high-performing communities. With that in mind, it is vital to understand the role of leadership in creating virtual environments where all members belong. Only 42 percent of remote workers say they have thriving relationships (Spataro, 2022). Trust will be questionable when less than half of remote workers thrive interpersonally with their peers. As such, we need to look at how virtual and hybrid environments move from initial hopes for trust into affect-based trust.

Swift Trust

Swift trust is a concept that is frequently required at the onset of forming virtual teams or communities (Blomqvist & Cook, 2018). Digital collaborative environments and temporary systems are often assembled as needed to complete everyday tasks, a concept that increases engagement and performance (Gallup, 2018). In these scenarios, tasks and teams are composed of members who must collaborate by providing fast, high-quality responses to problems under high levels of uncertainty. Post-pandemic, organizations are heavily leveraging virtual cross-functional teams making swift trust a vital concept to understand.

Swift trust provides initial confidence for a temporary team or community to interact as if the trust were present. The problem with swift trust is that it requires verification that the team can manage vulnerabilities and expectations as they progress forward (Blomqvist & Cook, 2018). Swift trust originally consisted of two components, one being more important than the other. The first component, a cognitive goal focusing, which, when outlined with clear expectations, streamlines members towards a common objective, each acting on their role to achieve it (Meyerson et al., 1996). The second component of swift trust is the emotional actions that reinforce social proof and enhance member interactions (Blomqvist & Cook, 2018). As one might predict, while the first component was initially deemed the more important focal point (Meyerson et al., 1996), researchers have recently shown that in our new hybrid climate and remote environments, emotional actions are the foundations that allow swift trust to transform into affective-based trust for longevity and commitment (Blomqvist & Cook, 2018). Unfortunately, this is where most struggle because when familiarity or prior interactions with members are unavailable, individuals rely on category-based processing that reduces the line between the individual and the group, promoting in-group beliefs and behaviors. Herein lies the importance of EDIB because leaders need to actively promote, normalize, and create environments that uphold belonging and uniqueness, virtual or hybrid environments will be able to effectively share ideas, insights, or knowledge in meaningful ways. Any bias, microaggressions, or normalized exclusivity will manifest as a breakdown of trust (Blomqvist & Cook, 2018). Therefore, the importance the leader places on increasing commonality will help facilitate the member's experience of trust and enable members to perceive an initial sense of belonging and the value of uniqueness.

Swift trust is fragile. When uncertainty is high, norms promote adaptive behavior by providing guidelines for acceptable and unacceptable behavior. For norms to exist, they must be shared, requiring leaders to ensure that members from all different social systems are seen as valuable. In these situations, valuing and modeling the importance of cross-cultural differences and individual uniqueness is imperative to sustaining swift trust. Additionally, trust relies heavily on the fairness and integrity of the actors involved, "fairness heuristics are used as early cues to the potential trustworthiness of others, especially those in authority relations" (Blomqvist & Cook, 2018, p. 18).

Affective-Based Trust

Affective-based trust is the product of swift trust being managed well. Over time, virtual and hybrid teams repetitively see the same organizational members. With this understanding in mind, it is essential for virtual, hybrid, and cross-functional teams to have leaders that demonstrate the transition from swift trust into affective-based trust. Affective-based trust is the product of accountability and normalized emotional and rational behaviors. Essentially, affective-based trust is the confidence in another generated by the level of care and concern demonstrated by that person; it is more emotional than rational (Washington, 2013). With affective-based trust, people trust because of their positive feelings for the person in question. Crisp and Jarvenpaa (2013) argue that for this type of trust to form, normative actions in virtual groups, where social influence emerges, must communicate the importance of belonging and uniqueness. Norms are the source of influence; actions are the mechanisms that explain the nature of the normative influence (Crisp & Jarvenpaa, 2013). Normative actions mediate the impact of swift trust and are seen as the adherence to emerging standards necessary for the team's success. In geographically dispersed teams, where members depend on computer-mediated cues to observe behavior, normative actions distinguish groups from one another and reinforce trusting beliefs. According to Crisp and Jarvenpaa's (2013) study, trust is the glue that holds virtual teams together (Crisp & Jarvenpaa, 2013). However, trust does more than mediate performance. Affective-based trust is critical for the emergence and exertion of normative actions, like belonging and the value of uniqueness. Swift trusting beliefs work through normative actions, not in place of them (Washington, 2013). The differentiator is whether the leader and culture monitor and maintain the beliefs that are accountable to sustain trust in virtual, hybrid and cross-functional teams. In global virtual teams, firm, trusting beliefs give members the confidence to engage in normative actions. These normative actions become a sustained basis of trusting beliefs and subsequent performance (Kroeger et al., 2021).

What Affective-Based Trust Looks Like

Kroeger, Racko & Burchell (2021) indicate clear indicators of affective-based trust. Moreover, this research intersects with Carter's (2022) Belonging in the Workplace tactics, which support members experiencing emotional comfort to perform optimally, which will be covered later in this chapter. In this section, the following are essential for leaders to work into processes and procedures as virtual, hybrid, and cross-functional teams are formed. In its most simplistic processes, leaders and groups need to, at the commencement of the formation of a group, determine the following:

1. Clarity on who is responsible for each task.
2. Clarity on who members of a team report to.
3. Clarity on how to communicate important information.
4. Clarity on how members are meant to be open and upfront without being penalized for differences in opinions.
5. Accountability for promises members make.
6. Alignment of actions and emotions being consistent with words.
7. Accountability of members being committed to the project and goal.
8. Trust that a member can ask for help and support without penalty when difficulties arise (Kroeger et al., 2021; Carter, 2022).

When these factors are accounted for early in forming groups, early trusting beliefs positively impact late trusting beliefs, and early trusting beliefs positively impact team performance (Kroeger et al., 2021). Leary & Gabriel (2021) explain that when people believe their relationship is valued by a group and members uphold processes agreed upon, they regard members as necessary, valuable, or close. These emotional needs, in turn, substantiate and improve interpersonal relationships through affective-based trust and result in higher-performing groups (Kroeger et al., 2021; Leary & Gabriel, 2021). Now that trust has been formally addressed, accountability is our next focus.

Accountability as A Broad Concept

Leaders with high trustworthiness have a more effortless ability to motivate and persuade followers to shift and change course (London, 2022). One factor that directly correlates to leaders with high trustworthiness is their ability to be accountable to the team, their members, and the outcome (Kroeger et al., 2021). Those who are trusted have fewer negative consequences for their adaptive decisions and actions, so there is more room for innovation, but they are also accountable for their actions. London (2022) notes that organizational performance is relative to responsiveness to change and adaptation. As a leader, adaptability to change becomes an important consideration in light of the turbulence of the economy, technology, global change, and organizational volatility. At the heart of leadership, adaptability means having followers who can solve problems quickly, deal with uncertainty, learn new work methods, and handle differing communication processes and stress in response to unexpected events (Ho & Chan, 2022). Still, for members to shift into this type of performance, they need leaders they can trust and who remain accountable to environments that are needs-driven. This is why accountability is so critical for the new role of leadership that demands fairness, equity, and transparency because, without leaders who model accountability, followers are often inclined to keep a status quo that is no longer working (Awali & Awali, 2022; Cao & Le, 2020; Ho & Chan, 2022).

Accountability for Power Dynamics

In 2009, Panteli and Tucker began exploring power dynamics in virtual and hybrid teams. According to their initial findings, virtual team dynamics and congruence have advanced considerably, but understanding how power dynamics are dealt with remains challenging (Bojic, 2022). These findings are logical considering the low success rates with EDIB initiatives and the inequitable use of power at its core. Interestingly, power differs in geographically distributed teams because most team dynamics are mediated by text-based and asynchronous communications (Panteli & Tucker, 2009). The challenge with these forms of communication is that most leaders need to establish processes and accountability to handle power differentials within the virtual environment. Bojic (2022) and Timperley & Schick (2022) argue that power is the ability of one member to exert influence over another.

Moreover, it is often a function of both dependence and the use of that dependence as leverage. Within virtual teams, when a leader is not accountable for managing transparency, fairness, equity, and belonging, power impacts trust development and overall team performance (Gotsis & Grimani, 2016; Panteli & Tucker, 2009). The Panteli and Tucker (2009) study examined how power is exercised in virtual and hybrid teams. The study looked at eighteen globally distributed teams within a global IT organization. Members were culturally diverse, geographically dispersed, and technology-enabled. Participants of the study were encouraged to recall experiences from working in a global virtual team where they worked well and did not work well. Open-ended questions aimed to explore performance levels, levels of trust, power distribution amongst team members, and how trust changed over time. Team dynamics were impacted by power within global virtual team interactions and were acknowledged by all teams, even where teams were considered to be successful (Panteli & Tucker, 2009). The difference with teams that worked well was that shared goals were leveraged to create guided actions. When individual members saw something more important than their individual political power needs, they refocused their energy on being more collectively productive.

Additionally, the most powerful members minimized their use of coercive power, with several members describing power originating from the knowledge and the member with the most relevant information. Where power dynamics began to transform and members could focus on the larger vision was when facilitators minimized destructive power differentials and upheld accountability for inclusion and shared understanding among members. Bringing members together and creating a shared understanding that fostered an atmosphere of belonging and trust-building within the virtual team environment was the most impactful behavior of the facilitator (Panteli & Tucker, 2009). This study speaks to the level of responsibility leaders must undertake to ensure they are accountable for creating environments where all members belong. Leaders can rebuild damaged internal relationships to develop the best possible outcomes by focusing on the main goal and reducing destructive power differentials.

Leadership Responsibility for Accountability

Leadership-followership support factors rely on trust and accountability, enabling the leader to manage change appropriately. In today's climate, followership demands more than a relationship between external events and rapid adaptive responses. Followership now expects leaders to be accountable for how they respond, which can be career-altering for certain leaders when done well or curtailing when executed poorly. So the question becomes, what are leaders responsible for and how do they demonstrate their accountability to their followers for continued followership? Many will argue that leaders attain power through achievement and dominance (London, 2022); however, those who prove their competence and others' esteem do three specific actions differently.

First, today's leaders are responsible for creating environments where all members belong. This means acting responsibly to support members towards one common goal rather than leveraging their knowledge for individual political gain and advancement. When leaders are accountable for curbing the internal motivations of power and control, leaders can motivate engagement and performance through commonality, enabling members to act more swiftly and produce positive outcomes. Simply put, instead of members using their energy to advance their individual agendas, they put that towards achieving the main objective.

Secondly, proactive leaders accountable to their members, customers, and stakeholders are more likely to course correct when they sense failure. Those who are less proactive tend to stay the course in failing strategies in the hopes of turning them around (London, 2022). Accountability allows leaders to be adaptable when dispersing power among their team members. As a result of this responsibility, a leader must explain and justify their decisions, hold themselves accountable for their decisions, and provide support to their members in fast-moving, volatile, and uncertain environments (London, 2022).

Lastly, the new leadership role must assume responsibility for being accountable to the emotional needs of their members. Specifically, belonging encompasses psychological safety and wellbeing because emotional and psychological stress is intimately related to human health and wellbeing (Carter, 2022). If leaders consider that psychological stress increases the susceptibility to the common cold, elevates the risk for major diseases, and is a strong predictor of morbidity and mortality due to its connection to the inflammatory process (Moss, 2019; World Health Organization, 2019), there is a case for leaders being accountable for empathetic environments. When accountability for empathy and belonging are effectively deployed, employees can tap into their full creative potential and respond to the challenges and changes the global environment faces daily. To do so, leaders must ensure that their members' emotional and psychological needs are met.

Empathy

Korkmaz et al. (2022) argue that inclusion is vital for sustained competitive advantage, the organization's health, and its members' wellbeing. They iterate that inclusion is experienced when the focus is placed on satisfying members' needs for uniqueness and belonging. For this to occur, social wellbeing needs to be understood, acted on, and upheld by both the culture and its individual members. Having relationships with others, being accepted, and belonging are all part of social wellbeing (Keyes, 2002, 2005; Dobusch, 2020; Slavich et al., 2022). Social evaluation and rejection are amplified with social stressors at their all-time peak. This means that to protect the health of the organization and its members, leaders need to look at developing processes that enable empathy and social wellbeing to exist as part of the cultural norms. Slavich et al. (2022) note that social stressors are mediated by perceived social threat information and the fear of rejection, increasing inflammatory activities and activating pain centers in the brain. This neuropsychological process decreases an individual's ability to perform because the fear of rejection causes them to focus more on the social environment and less on the tasks at hand. Moreover, when the fear of social rejection is high, the tendency is to either become aggressive to defend one's position or retreat for self-preservation - neither conducive to team effectiveness nor efficiency (Silverman, 2018). Members of a social organization must be responsible and accountable for empathizing with the emotional needs of their fellow members if social wellbeing is to be sustained.

There has been a rapid increase in emotional intelligence (EQ) in recent years. It refers to the ability to be aware of one's feelings, be aware of others' feelings, and differentiate between them to guide one's thinking and behavior (Polychroniou, 2009). Developing empathy and social skills is as much about one's emotional regulation as it is about emotional peer regulation because together, these converge to achieve common goals. The motivation, therefore, is to help oneself remain focused and help one's members when operating under high-pressure compounding scenarios. Work teams are more successful in achieving organizational goals if their members are empowered to do their jobs and are accountable to each other's emotional and social wellbeing (Polychroniou, 2009). Member commitment and engagement may be reduced if empathy and social wellbeing are restricted, resulting in a lack of enthusiasm for improving quality and productivity.

As a leader, empathy means recognizing the need, taking an active interest in the individual members, responding to changes in their emotional states, and working together towards goals. As a result of empathetic leadership, follower social skills become associated with encouraging peers to engage in desirable behaviors. This enhances respect while increasing emotional identity with a leader who is considerate and willing to help team members effectively handle conflict and improve job performance. Leadership empathy increases team effectiveness, motivation, social skills, and cohesion (Polychroniou, 2009).

Organizational Empathy

While leadership empathy ignites organizational empathy, organizational empathy understands the needs of its environment. It includes its challenges, accountabilities, diverse culture of stakeholders, and interdisciplinary actions of members within the community, institution, or organization. In challenging situations, what most followers react to is the empathy with which their leaders meet their needs. In conjunction, leadership approval is often defined by how well a leader has aligned the values and vision of the organization with their actions, modeling behavior for their followers. Post-pandemic, this sentiment becomes an essential mechanism to curb loneliness and isolation repercussions derived from pandemic restrictions and how virtual and hybrid teams are enacted.

Alternatively, team empathy and dedication are associated with information-sharing values and positively correlate with organizational culture. In communities where empathy is seen as a positive behavior for achieving outcomes, interpersonal competencies build inspiration and motivation for others, both externally in the environment and neurally within the individual brain circuitry (Silverman, 2018). These behaviors, therefore, reinforce individual neurocircuitry, social intelligence, and social wellbeing through empathy, setting the mood, culture, engagement, and retention for anyone participating within the community (Silverman, 2018). These behaviors improve cross-cultural competency and employee satisfaction, even in virtual and hybrid spaces.

For example, in complex organizations with cross-functional virtual and hybrid teams, many cite decreased interpersonal relationships and conflict associated with an increased perception of adversaries and competitive behaviors (Gallup, 2022). However, Silverman's (2018) research revealed that when empathy was leveraged, the perception of adversaries and grievances was reduced. A building block of organizational empathy was developing empathy skills and awareness, allowing its members to see the organization through the diverse eyes of colleagues from various disciplines and cultures. Likewise, empathy for the collective member's stress, goals, and professional values becomes an essential strategy for performance and success (Silverman, 2018).

BELONGING: HOW LEADERS CREATE ACTIONS AROUND EMOTIONAL NEEDS

Until this point, this chapter has deliberately focused on highlighting the new factors leaders must be aware of, the subsequent risks for maintaining the status quo, and the importance of trust, accountability, and empathy. Most of which is considered theoretical knowledge rather than pragmatic or practical application. Moreover, with an increased demand for leaders to adopt different behaviors to satisfy the emotional needs of their followers, many are seeking practical tools and action steps that can be transformed into sustainable practices. Consequently, we will examine belonging methodology, indicator systems, and workplace tactics in the following sections.

For belonging to be perceived, five indicators must be present. Members must be seen for their individuality, known for their strengths, skills, and knowledge, valued for their diverse intersections of identity, accepted in their current emotional and rational state, and authentically cared for (Carter, 2022). When belonging indicators are present, belonging can reduce the negative impacts of inequality through multi-dimensions of the community. This occurs through the correlation between meeting one's emotional needs and acknowledging a member's uniqueness. Belonging behaviors are needed for leaders and followers to trust, act with empathy, and be accountable to each other's social wellbeing and performance capabilities (Cao & Le, 2022).

Further, researchers have found that leadership can only be effective if leaders are proficient in equity, diversity, inclusion & belonging (Cao & Le, 2022; Gotsis & Grimani, 2016; Korkmaz, 2022). Today in order for organizations, institutions, and member-based communities to survive, they need to acknowledge, understand and be able to empower diverse teams to ensure retention, engagement, and performance. Central to this ability is where the concept of belonging emerges. However, the questions that continue to surface in the face of this need are: "How do we create a belonging culture for high performance and engagement - that can sustain spaces without walls and virtual accommodations? How do we account for flexible classrooms and workspaces where creativity, innovation, and high-level problem-solving are at the core? What would leaders need to value, think and act on so that followership consistently perceives a community where everyone belongs?"

Moreover, "What are the proficiencies in belonging, including metrics and methodology, that must be understood so that individuals within a community or culture are empowered to speak up?" By developing proficiency in belonging indicators and belonging in the workplace tactics, leaders can more easily measure and use action tools to support a belonging culture. Let us dive into those indicators and tactics next.

Belonging Workplace Tactics Leaders Need to Improve Engagement & Performance

In 2020, I set out to understand how Belonging is perceived and experienced within the workplace and explore tactics to create sustainable change. In a mixed-methods research study, I quantitatively validated a tool to measure workplace belonging and produce six scores that provide leaders with areas of focus.

Belonging indicators were validated to include; comfort, connection, contribution, psychological safety, and wellbeing. The survey provided a measurable mechanism to measure belonging in an environment and provide a score for each indicator and an overall belonging in the workplace score. Scores provide areas that are best in class (measuring above 75%), strength scores (measuring between 65-74%), and stretch scores (measuring under 64%). Based on the scores of each belonging indicator, organizations and leaders can leverage workplace tactics to improve the scores and monitor their progress over time. In conjunction with the scores, workplace tactics were developed through qualitative research from the focus groups to understand further the characterizations and workplace tactics that enable community members to experience belonging—Figure 1. Belonging-First Theory- Workplace Tactics demonstrates the initial findings reported within the research (Carter, 2022).

Each indicator produced workplace tactics based on participants' emotional needs and desire for workplace actions to achieve a belonging environment whereby all members could thrive. While each indicator produced characterizations of what that indicator should look like and how leaders and members would articulate its existence, the tactics to achieve the characterizations are of critical importance.

Comfort, for example, produced the need to be seen as one's authentic self. If comfort is present, members will feel seen for who they are and their unique skills and abilities. However, for comfort to be leveraged within the workplace, participants indicated they needed their leaders and managers to ensure they had a clear vision and mission, job description, policies and procedures, and someone to model role expectations. These elements of understanding are necessary for comfort to be achieved (Carter, 2022). Tactical elements are easily put into practical actions for leaders and managers, noting that once the comfort factors are achieved, fear of rejection for doing something wrong is minimized. These tactics take care of normalizing what is expected of members. Therefore, at the commencement of a team, comfort elements enable trust, accountability, and empathy to be established at the formation stage. In so doing, valuing diverse perspectives and cultural differences is outlined for optimal behavioral actions and focus.

Connection, which characterizes the need to be known and therefore trusted, requires the workplace tactics of aligning communications, values, goals, and objectives while being solution-focused and empathetically generous (Carter, 2022). The underlying descriptions indicated by participants spoke to the concept that when members have an interpersonal connection with each other, their social wellbeing and unique experiences based on their intersections of identity and cultural experiences are known and accepted. While other members do not need to ascribe to their differences, the differences need to be known and valued so that despite the diverse perspectives and across boundaries of space, members focus on solving problems over politicking for status. Therefore, the connection is between the members of the interaction and the organization itself (Carter, 2022).

Contribution characterizes valuing the unique contributions of each member. Interestingly, 95 percent of participants with ethnocultural diversity indicated that for members to experience the perception of belonging, they had to be valued for their unique contributions and endorsed for their skill, knowledge, and ability (Carter, 2022). The tactics to achieve recognition for one's unique contributions are often where power imbalances are realized. As such, leaders must provide equal and fair access to key people for growth and development. Moreover, members need to be asked and sought out for their input and perspectives when facing challenges and change. When these elements are formally put in place, members are fairly celebrated for their unique contributions while being highlighted for the group contributions and team efforts (Carter, 2022).

Psychological Safety included the need for empathy. However, its emphasis was distributed throughout maintaining social wellbeing by ensuring that peers and leaders checked in on workload and pressure,

Figure 1. Belonging-First Theory – Workplace Tactics
Note: This model shows that Belonging in the workplace is experienced with the necessity of having five key indicators present. Moreover, how each indicator is characterized differs yet corresponds with critical workplace tactics to increase the perception of Belonging in the workplace. Depending on the associated scores of belonging acquired through the Belonging-First survey, tactics are actioned in order of importance and priority. Reproduced from "Belonging within the Workplace: Mixed Methods Constructivist Grounded Theory Study For Instrument Validation and Behavioural Indicators For Performance & Governance," by A.D.Carter (2022).[Master's thesis, Adler University]. ProQuest Dissertations and Theses Global. p.137. https://www.proquest.com/docview/2716586005/FECBB62CCA204025PQ/1. Copyright 2022. Reprinted with permission.

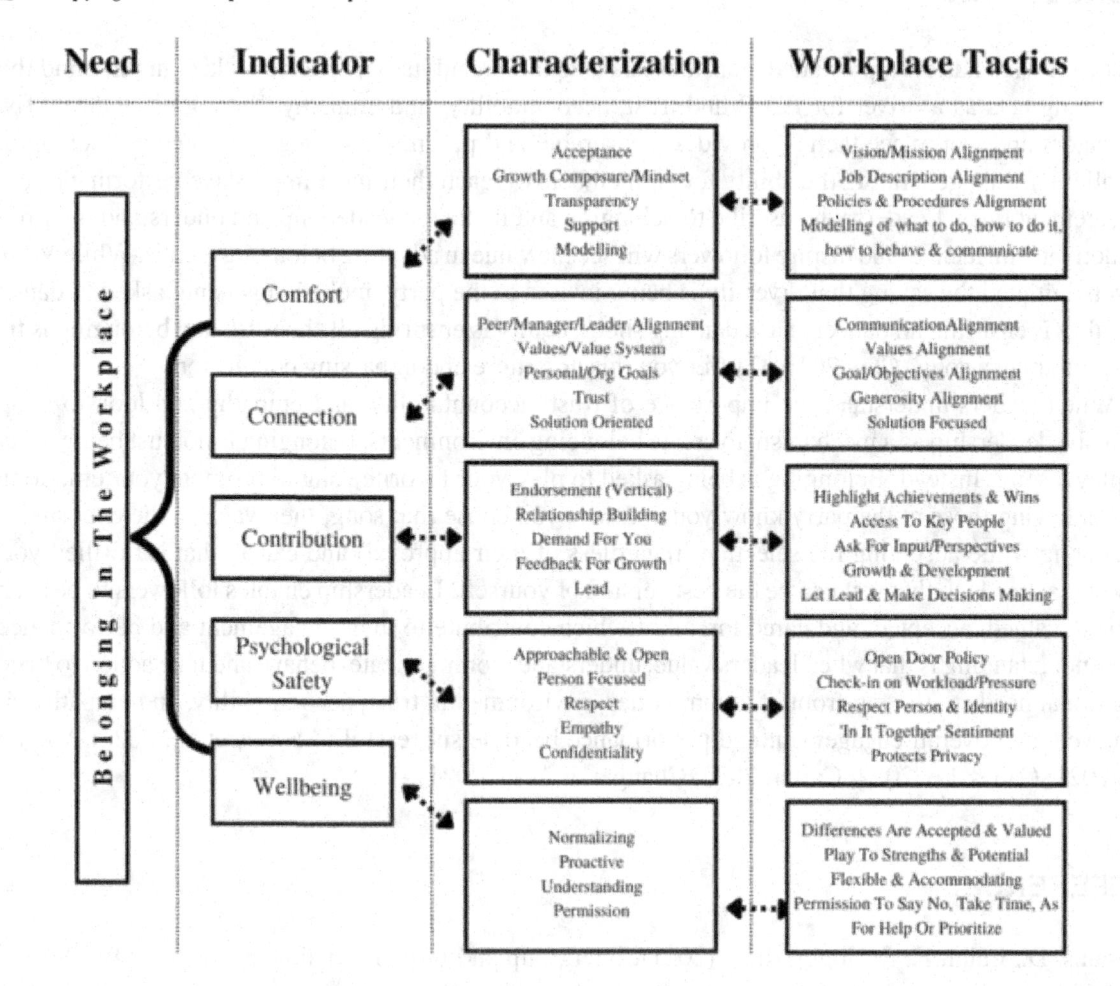

having open door policies where respect and identity were valued, and perhaps most importantly, where an 'in-it-together' sentiment was not only upheld but also actively promoted. These elements speak to the tactics of understanding how to communicate and accept people for where they are and what they are facing. Acceptance includes being cared for and having one's emotional needs addressed and respected (Carter, 2022).

Wellbeing, the last of the indicators and workplace tactics, ensures that all members are cared for. In so doing, workplace tactics emphasized creating communication and processes around playing to members' strengths and potential, empowering members with the permission to decline more work, ask for a more extended deadline, or help to get the work done. Additionally, wellbeing included the ability for members to ask leaders for help prioritizing items to action based on vision and objective alignment (Carter, 2022).

Based on these indicators, workplace tactics, and overall scores, organizations can construct achievable objectives to fulfill the tactics and produce a belonging environment.

CONCLUSION

Based on what has been presented, consider this adaptation and its importance by keeping in mind that belonging acts as a driver for EDIB and trust, accountability, and empathy. Belonging, with its five indicators and workplace tactics, provides measurable and pragmatic strategies to increase belonging for all members, generating the ability for all members to reach their most impressive performance and engagement rates. Leaders who ascribe to belonging and inclusion leadership and understand their role to motivate, influence, and inspire followers will see the value in adapting belonging tactics. While Verna Myers initiated the saying that diversity is being invited to the party, inclusion is being asked to dance. Equality is ensuring all dancers have dancing shoes, equity is ensuring all shoes fit, and belonging is the ability to play a song (Cho, 2016); I offer you this as a more encompassing conclusion.

When leaders understand the importance of trust, accountability, and empathy and look to adopt inclusive leadership as a mechanism to create belonging environments, belonging is not just being asked to play a song. Instead, Belonging is being asked to play your favorite song to promote your emotional needs, having those at the party know you and why you chose that song, then valuing the exposure to your song while accepting the selection (regardless of their approval) and caring that it fulfilled your emotional needs so that you can be the best version of yourself. Leadership enables followers to be seen, known, valued, accepted, and cared for, all of which contribute to their engagement and performance. The understanding is that when leaders value, understand, communicate, behave and take action to bring emotional needs to the forefront of communities, environmental trust, accountability, and empathy are achieved, and overall engagement and performance become successful (Abrams et al., 2021; Allen et al., 2021; Cao & Le, 2022; Carter, 2022; Changar & Atan, 2021).

REFERENCES

Abrams, D., Lalot, F., & Hogg, M. A. (2021). Intergroup and intragroup dimensions of COVID-19: A social identity perspective on social fragmentation and unity. *Group Processes & Intergroup Relations, 24*(2), 201–209. doi:10.1177/1368430220983440

Allen, K., Kern, M. L., Rozek, C. S., McInerney, D. M., & Slavich, G. M. (2021). Belonging: A review of conceptual issues, an integrative framework, and directions for future research. *Australian Journal of Psychology, 73*(1), 87–102. doi:10.1080/00049530.2021.1883409 PMID:33958811

Alwali, J., & Alwali, W. (2022). The relationship between emotional intelligence, transformational leadership, and performance: A test of the mediating role of job satisfaction. *Leadership and Organization Development Journal, 43*(6), 928–952. doi:10.1108/LODJ-10-2021-0486

Banks, G. C., Dionne, S. D., Schmid Mast, M., & Sayama, H. (2022). Leadership in the digital era: A review of who, what, when, where, and why. *The Leadership Quarterly*, *33*(5), 101643–101639. doi:10.1016/j.leaqua.2022.101634

Benioff, M. (2019, Oct 14). Benioff: We Need A New Capitalism, The current system has led to profound inequity. To fix it, we need businesses and executives to value purpose alongside profit. *The New York Times*. Retrieved on October 23, 2022 from https://www.nytimes.com/2019/10/14/opinion/benioff-salesforce-capitalism.html

Blomqvist, K., & Cook, K. S. (2018). Swift Trust - State-of-the-Art and Future Research Directions. In R. H. A. N. Searle & S. Sitkin (Eds.), *Routledge Companion to trust* (pp. 29–49). Routledge. doi:10.4324/9781315745572-4

Bojic, L. (2022). Metaverse through the prism of power and addiction: What will happen when the virtual world becomes more attractive than reality? *European Journal of Futures Research*, *10*(1), 1–24. doi:10.118640309-022-00208-4

Cao, T.T. & Le, P.B., (2022), Impacts of transformational leadership on organizational change capability: a two-path mediating role of trust in leadership, *European Journal of Management and Business Economics, Vol. ahead-of-print No. ahead-of-print*. doi:10.1108/EJMBE-06-2021-0180

Carter, A. (2022). *Belonging within the workplace* (Publication No. 29393403) [Adler University]. ProQuest. https://www.proquest.com/openview/df3bede74f97b31ba91c846f316515e6/1.pdf?pq-origsite=gscholar&cbl=18750&diss=y

Changar, M., & Atan, T. (2021). The role of transformational and transactional leadership approaches on environmental and ethical aspects of CSR. *Sustainability*, *13*(3), 1411. doi:10.3390u13031411

Chen, C., Arakawa, Y., Watanabe, K., & Ishimaru, S. (2022). Quantitative Evaluation System for Online Meetings Based on Multimodal Microbehavior Analysis. *Sensors and Materials*, *34*(8), 3017–3027. doi:10.18494/SAM3959

Cho, J. (2016, May 25). *"Diversity is being invited to the party; inclusion is being asked to dance," Verna Myers tells Cleveland Bar*. Retrieved on July 22, 2021 from https://www.cleveland.com/business/2016/05/diversity_is_being_invited_to.html#incart_more_business

Cigna. (2020). *Loneliness and the Workplace*. 2020 U.S Report. Retrieved April 9 from https://www.cigna.com/static/www-cigna-com/docs/about-us/newsroom/studies-and-reports/combatting-loneliness/cigna-2020-1
oneliness-report.pdf

Cox, T. Jr. (2001). *Creating the multicultural organization: A strategy for capturing the power of diversity* (Vol. 6). John Wiley & Sons.

Crisp, C., & Jarvenpaa, S. (2013). Swift Trust in Global Virtual Teams: Trusting Beliefs and Normative Actions. *Journal of Personnel Psychology*, *12*(1), 45–56. doi:10.1027/1866-5888/a000075

Dobusch, L. (2020). The inclusivity of inclusion approaches: A relational perspective on inclusion and exclusion in organizations. *Gender, Work and Organization, 28*(1), 379–396. doi:10.1111/gwao.12574

Ducharme. (2020, March 11). World Health Organization Declares COVID-19 a 'Pandemic.' Here's What That Means. *Time.* Retrieved on March 30, 2020 from https://time.com/5791661/who-coronavirus-pandemic-declaration/

Gallup. (2022). *State of the global workplace: 2022 report.* Gallup. https://www.gallup.com/workplace/349484/state-of-the-global-workplace-2022-report.aspx

Gallup. (2018). *2018 Clifton Strengths® Meta-Analysis Report.* Gallup. https://www.gallup.com/cliftonstrengths/en/270350/2018-cliftonstrengths-meta-analysis-report.aspx

Gotsis, G., & Grimani, K. (2016). Diversity as an aspect of effective leadership: Integrating and moving forward. *Leadership and Organization Development Journal, 37*(2), 241–264. doi:10.1108/LODJ-06-2014-0107

Hales, A. H., McIntyre, M. M., Rudert, S. C., Williams, K. D., & Thomas, H. (2021). Ostracized and observed: The presence of an audience affects the experience of being excluded. *Self and Identity, 20*(1), 94–115. doi:10.1080/15298868.2020.1807403

Ho, H. C., & Chan, Y. C. (2022). Flourishing in the workplace: A one-year prospective study on the effects of perceived organizational support and psychological capital. *International Journal of Environmental Research and Public Health, 19*(2), 922. doi:10.3390/ijerph19020922 PMID:35055747

Kerr, N. A., & Stanley, T. B. (2021). Revisiting the social stigma of loneliness. *Personality and Individual Differences, 171*, 110482. doi:10.1016/j.paid.2020.110482

Keyes, C. L. M. (2002). The mental health continuum: From languishing to flourishing in life. *Journal of Health and Social Behavior, 43*(2), 207–222. doi:10.2307/3090197 PMID:12096700

Keyes, C. L. (2005). Mental illness and/or mental health? Investigating axioms of the complete state model of health. *Journal of Consulting and Clinical Psychology, 73*(3), 539–548. doi:10.1037/0022-006X.73.3.539 PMID:15982151

Khumalo, N., Dumont, K. B., & Waldzus, S. (2022). Leaders' influence on collective action: An identity leadership perspective. *The Leadership Quarterly, 33*(4), 101609–101621. doi:10.1016/j.leaqua.2022.101609

Kiang, M. V., & Tsai, A. C. (2020). Statements issued by academic medical institutions after George Floyd's killing by police and subsequent unrest in the United States: Cross-sectional study. MedRxiv. doi:10.1101/2020.06.22.20137844

King, M. L. Jr. (1972). A New Sense of Direction. *Worldview, 15*(4), 5–12. doi:10.1017/S0084255900014893

Klerk, N. O.-d., & Muir, C. (2022). Corporate Brand Reputation and Ethic, Sustainability and Inclusion. The Shift in Post Pandemic Corporate Narrative: From Corporate Brand Reputation to Corporate Sustainability. In The Emerald Handbook of Multi-Stakeholder Communication. Emerald Publishing Limited. doi:10.1108/978-1-80071-897-520221031

Korkmaz, A. V., van Engen, M. L., Knappert, L., & Schalk, R. (2022). About and beyond leading uniqueness and belongingness: A systematic review of inclusive leadership research. *Human Resource Management Review*, *100894*(4). Advance online publication. doi:10.1016/j.hrmr.2022.100894

Kroeger, F., Racko, G., & Burchell, B. (2021). How to create trust quickly: A comparative empirical investigation of the bases of swift trust. *Cambridge Journal of Economics*, *45*(1), 129–150. doi:10.1093/cje/beaa041

Kuknor, S., & Bhattacharya, S. (2021). Organizational inclusion and leadership in times of global crisis. *Australasian Accounting. Business and Finance Journal*, *15*(1), 93–112. doi:10.14453/aabfj.v15i1.7

Leary, M. R., & Gabriel, S. (2021). The relentless pursuit of acceptance and belonging. *Elsevier*.

Leslie, J. B., Luciano, M. M., Mathieu, J. E., & Hoole, E. (2018). Challenge accepted: Managing polarities to enhance virtual team effectiveness. *People & Strategy*, *41*(2), 22–28.

Loades, M. E., Chatburn, E., Higson-Sweeney, N., Reynolds, S., Shafran, R., Brigden, A., Linney, C., McManus, M. N., Borwick, C., & Crawley, E. (2020). Rapid systematic review: The impact of social isolation and loneliness on the mental health of children and adolescents in the context of COVID-19. *Journal of the American Academy of Child and Adolescent Psychiatry*, *59*(11), 1218–1239. doi:10.1016/j.jaac.2020.05.009 PMID:32504808

London, M. (2022). Causes and consequences of adaptive leadership: A model of leaders' rapid responses to unexpected events. *Psychology of Leaders and Leadership*. Advance online publication. doi:10.1037/mgr0000136

Magalhães, C. M., Machado, C. F., & Nunes, C. P. (2022). Loneliness in Leadership: A Study Applied to the Portuguese Banking Sector. *Administrative Sciences*, *12*(4), 130. doi:10.3390/admsci12040130

Mahlangu, V. P. (2020). *Understanding Toxic Leadership in Higher Education Work Places through Betrayal Trauma Theory*. Bulgarian Comparative Education Society.

Meyerson, D., Weick, K. E., & Kramer, R. M. (1996). Swift trust and tempo- rary groups. In R. M. Kramer & T. R. Tyler (Eds.), *Trust in organizations: Frontiers of theory and research* (pp. 166–195). Sage. doi:10.4135/9781452243610.n9

Mokros, Ł., Świtaj, P., Bieńkowski, P., Święcicki, Ł., & Sienkiewicz-Jarosz, H. (2022, October). Depression and loneliness may predict work inefficiency among professionally active adults. *International Archives of Occupational and Environmental Health*, *95*(8), 1775–1783. doi:10.100700420-022-01869-1 PMID:35503113

Mor Barak, M. E., & Daya, P. (2014). Fostering inclusion from the inside out to create an inclusive workplace: Corporate and organizational efforts in the community and the global society. In B. M. Ferdman & B. R. Deane (Eds.), *Diversity at work: The practice of inclusion* (pp. 391–412). Jossey-Bass.

Mor Barak, M. E., Luria, G., & Brimhall, K. C. (2022). What leaders say versus what they do: Inclusive leadership, policy-practice decoupling, and the anomaly of climate for inclusion. *Group & Organization Management*, *47*(4), 840–871. doi:10.1177/10596011211005916

Moss, J. (2019). *Burnout Is About Your Workplace, Not Your People.* Harvard Business School Publishing Corporation. Retrieved on November 1, 2022 from https://egn.com/dk/wp-content/uploads/sites/3/2020/08/Burnout-is-about-your-workplace-not-your-people-1.pdf

Nishii, L. H., & Leroy, H. L. (2020). Inclusive leadership: Leaders as architects of inclusive workgroup climates. In *Inclusive Leadership* (pp. 162–178). Routledge. doi:10.4324/9780429449673-12

Nishii, L. H., & Mayer, D. M. (2009). Do inclusive leaders help to reduce turnover in diverse groups? The moderating role of leader–member exchange in the diversity to turnover relationship. *The Journal of Applied Psychology*, *94*(6), 1412–1426. doi:10.1037/a0017190 PMID:19916652

Ofei-Dodoo, S., Mullen, R., Pasternak, A., Hester, C. M., Callen, E., Bujold, E. J., Carroll, J. K., & Kimminau, K. S. (2021). Loneliness, burnout, and other types of emotional distress among family medicine physicians: Results from a national survey. *Journal of the American Board of Family Medicine*, *34*(3), 531–541. doi:10.3122/jabfm.2021.03.200566 PMID:34088813

Overton-de Klerk, N., & Muir, C. (2022). Corporate Brand Reputation and Ethic, Sustainability and Inclusion. The Shift in Post Pandemic Corporate Narrative: From Corporate Brand Reputation to Corporate Sustainability. In *The Emerald Handbook of Multi-Stakeholder Communication* (pp. 365–391). Emerald Publishing Limited. doi:10.1108/978-1-80071-897-520221031

Panteli, N., & Tucker, R. (2009). Power and trust in global virtual teams. *Communications of the ACM*, *52*(12), 113–115. doi:10.1145/1610252.1610282

Podsiadlowski, A., Gröschke, D., Kogler, M., Springer, C., & Van Der Zee, K. (2013). Managing a culturally diverse workforce: Diversity perspectives in organizations. *International Journal of Intercultural Relations*, *37*(2), 159–175. doi:10.1016/j.ijintrel.2012.09.001

Polychroniou, V. P. (2009). Relationship between emotional intelligence and transformational leadership of supervisors The impact on team effectiveness. *Team Performance Management*, *15*(7/8), 343–356. doi:10.1108/13527590911002122

Powell, A. (2022). Two-step flow and protesters: Understanding what influenced participation in a George Floyd protests. *Communication Quarterly*, *70*(4), 1–22. doi:10.1080/01463373.2022.2077122

Randel, A. R., Galvin, B. M., Shore, L. M., Holcombe Erhart, K., Chung, B. G., Dean, M. A., & Kedharnath, U. (2018). Inclusive leadership: Realizing positive outcomes through belongingness and being valued for uniqueness. *Human Resource Management Review*, *28*(2), 190–203. doi:10.1016/j.hrmr.2017.07.002

Rasool, S. F., Wang, M., Tang, M., Saeed, A., & Iqbal, J. (2021). How Toxic Workplace Environment Effects the Employee Engagement: The Mediating Role of Organizational Support and Employee Wellbeing. *International Journal of Environmental Research and Public Health*, *18*(5), 2294. doi:10.3390/ijerph18052294 PMID:33652564

Reina, D. S., & Reina, M. L. (2016). *Trust and betrayal in the workplace: Building effective relationships in your organization*. Recorded Books.

Roberson, Q., & Perry, J. L. (2022). Inclusive leadership in thought and action: A thematic analysis. *Group & Organization Management*, *47*(4), 755–778. doi:10.1177/10596011211013161

Shore, L. M., Randel, A. E., Chung, B. G., Dean, M. A., Ehrhart, K. H., & Singh, G. (2011). Inclusion and diversity in work groups: A review and model for future research. *Journal of Management*, *37*(4), 1262–1289. doi:10.1177/0149206310385943

Silverman, E. (2018). Moving beyond collaboration: A Model for enhancing social work's organizational empathy. *Social Work*, *63*(4), 297–304. doi:10.1093wwy034 PMID:30113662

Slavich, G. M., Roos, L. G., & Zaki, J. (2022). Social belonging, compassion, and kindness: Key ingredients for fostering resilience, recovery, and growth from the COVID-19 pandemic. *Anxiety, Stress, and Coping*, *35*(1), 1–8. doi:10.1080/10615806.2021.1950695 PMID:34369221

Spataro, J. (2022). 5 Key Trends Leaders Need to Understand to Get Hybrid Right. *Harvard Business Review*.

Spiller, C., Evans, M., Schuyler, K. G., & Watson, L. W. (2021). What silence can teach us about race and leadership. *Leadership*, *17*(1), 81–98. doi:10.1177/1742715020976003

Taylor, Z. W., Pereira, M., Rainey, L., Gururaj, S., Gibbs, B., Wiser, J., Benson, C., Childs, J., & Somers, P. (2022). Saying His Name: How Faith-Based Higher Education Leaders Addressed the George Floyd Murder. *Religious Education (Chicago, Ill.)*, 1–31. doi:10.1080/15507394.2022.2156262

Timperley, C., & Schick, K. (2022). Hiding in Plain Sight: Pedagogy and Power. *International Studies Perspectives*, *23*(2), 113–128. doi:10.1093/isp/ekab002

Tull, M. T., Edmonds, K. A., Scamaldo, K. M., Richmond, J. R., Rose, J. P., & Gratz, K. L. (2020). Psychological outcomes associated with stay-at-home orders and the perceived impact of COVID-19 on daily life. *Psychiatry Research*, *289*, 113098. doi:10.1016/j.psychres.2020.113098 PMID:32434092

Twenge, J. M., Baumeister, R. F., Tice, D. M., & Stucke, T. S. (2001). If you can't join them, beat them: Effects of social exclusion on aggressive behavior. *Journal of Personality and Social Psychology*, *81*(6), 1058–1069. doi:10.1037/0022-3514.81.6.1058 PMID:11761307

Ulmer, R. R. (2012). Increasing the impact of thought leadership in crisis communication. *Management Communication Quarterly*, *26*(4), 523–542. doi:10.1177/0893318912461907

Washington, M. G. (2013). Trust and project performance: The effects of cognitive and affective-based trust on client-project manager engagements. *University of Pennsylvania Scholarly Commons*. Retrieved on October 29, 2022 from https://repository.upenn.edu/od_theses_msod/67

Winters, M. (2014). From diversity to inclusion: An inclusion equation. In B. M. Ferdman & B. R. Deane (Eds.), *Diversity at work: The practice of inclusion* (pp. 205–228). Jossey-Bass.

Woods, D. R., Benschop, Y., & van den Brink, M. (2021). What is intersectional equality? A definition and goal of equality for organizations. *Gender, Work & Organization*, 1–18. doi:10.1111/gwao.12760

World Health Organization. (2019). Burn-out an "occupational phenomenon": International Classification of Diseases. *World Health Organization Departmental News*. Retrieved on October 22, 2022 from https://www.who.int/news/item/28-05-2019-burn-out-an-occupational-phenomenon-international-classification-of-diseases

Yildiz, H., & Duyan, V. (2022). Effect of group work on coping with loneliness. *Social Work with Groups*, *45*(2), 132–144. doi:10.1080/01609513.2021.1990192

Zhang, Z., Wang, M. O., & Shi, J. (2012). Leader-follower congruence in proactive personality and work outcomes: The mediating role of leader-member exchange. *Academy of Management Journal*, *55*(1), 111–130. doi:10.5465/amj.2009.0865

Zhou, X., Sedikides, C., Mo, T., Li, W., Hong, E. K., & Wildschut, T. (2022). The restorative power of nostalgia: Thwarting loneliness by raising happiness during the COVID-19 pandemic. *Social Psychological & Personality Science*, *13*(4), 803–815. doi:10.1177/19485506211041830

Compilation of References

Abedini, A., Abedin, B., & Zowghi, D. (2021). Adult learning in online communities of practice: A systematic review. *British Journal of Educational Technology*, *52*(4), 1663–1694. doi:10.1111/bjet.13120

Aboagye, E. (2021). Transitioning from face-to-face to online instruction in the COVID-19 era: Challenges of tutors at colleges of education in Ghana. *Social Education Research*, *2*(1), 9–19. doi:10.37256er.212021545

Abrams, D., & Hogg, M. A. (1988). Comments on the motivational status of self-esteem in social identity and intergroup discrimination. *European Journal of Social Psychology*, *18*(4), 317–334. doi:10.1002/ejsp.2420180403

Abrams, D., Lalot, F., & Hogg, M. A. (2021). Intergroup and intragroup dimensions of COVID-19: A social identity perspective on social fragmentation and unity. *Group Processes & Intergroup Relations*, *24*(2), 201–209. doi:10.1177/1368430220983440

Ackerman, G., & Peterson, H. (2020). Terrorism and COVID-19. *Perspectives on Terrorism*, *14*(3), 59–73. doi:10.2307/26918300

Adebisi, Y. A., Rabe, A., & Lucero-Prisno Iii, D. E. (2021). Risk communication and community engagement strategies for COVID-19 in 13 African countries. *Health Promotion Perspectives*, *11*(2), 137–147. doi:10.34172/hpp.2021.18 PMID:34195037

Adhikari, B., Pell, C., & Cheah, P. Y. (2020). Community engagement and ethical global health research. *Global Bioethics*, *31*(1), 1–12. doi:10.1080/11287462.2019.1703504 PMID:32002019

Adler, A. (1931). The meaning of life. *Lancet*, *217*(5605), 225–228. Advance online publication. doi:10.1016/S0140-6736(00)87829-0

Afuape, T. (2011). *Power, Resistance and Liberation in Therapy with Survivors of Trauma: To Have Our Hearts Broken*. Routledge.

Akojie, P., Entrekin, F., Bacon, D., & Kanai, T. (2019). Qualitative meta-data analysis: Perceptions and experiences of online doctoral students. *American Journal of Qualitative Research*, *3*(1), 117–135. doi:10.29333/ajqr/5814

Aktı Aslan, S., & Turgut, Y. E. (2021). Effectiveness of community of inquiry based online course: Cognitive, social and teaching presence. *Journal of Pedagogical Research*, *5*(3), 187–197. doi:10.33902/JPR.2021371365

Albuquerque, T. P. D. (2018). *Strategic marketing on museum audience attraction–a comparative study between Portugal and the UK* [Doctoral dissertation].

Alcañiz, E. B., García, I. S., & Blas, S. S. (2009). The functional-psychological continuum in the cognitive image of a destination: A confirmatory analysis. *Tourism Management*, *30*(5), 715–723. doi:10.1016/j.tourman.2008.10.020

Alemán-Saravia, A., & Deroncele-Acosta, A. (2021). Technology, Pedagogy and Content (TPACK framework): Systematic Literature Review. *XVI Latin American Conference on Learning Technologies* (pp. 104-111). IEEE. 10.1109/LACLO54177.2021.00069

Alexandra, Y., & Fahmi Choirisa, S. (2021). Understanding college 'students' e-loyalty to online practicum courses in hospitality programs during COVID-19. *Journal Of Learning Development in Higher Education*, *21*(21). Advance online publication. doi:10.47408/jldhe.vi21.627

Alghamdi, J. (2022). Equipping Student Teachers with Remote Teaching Competencies Through an Online Practicum: A Case Study. In M. S. Khine (Ed.), *Handbook of research on teacher education* (pp. 187–206). Springer. doi:10.1007/978-981-19-2400-2_12

Allen, I. E., Seaman, J., Poulin, R., & Straut, T. T. (2016). *Online report card: Tracking online education in the United States*. Babson Survey Research Group and Quahog Research Group, LLC.

Allen, I., & Seaman, J. (2014). *Grade change: Tracking online education in the United States*. Babson Survey Research Group and Quahog Research Group, LLC.

Allen, K., Kern, M. L., Rozek, C. S., McInerney, D. M., & Slavich, G. M. (2021). Belonging: A review of conceptual issues, an integrative framework, and directions for future research. *Australian Journal of Psychology*, *73*(1), 87–102. doi:10.1080/00049530.2021.1883409 PMID:33958811

Almahasees, Z., Mohsen, K., & Amin, M. O. (2021, May). Faculty's and students' perceptions of online learning during COVID-19. *Frontiers in Education.*, *6*(3), 638470. Advance online publication. doi:10.3389/feduc.2021.638470

Aloni, M., & Harrington, C. (2018). Research based practices for improving the effectiveness of asynchronous online discussion boards. *Scholarship of Teaching and Learning in Psychology*, *4*(4), 271–289. doi:10.1037tl0000121

Alsofyani, M., bin Aris, B., & Eynon, R. (2013). A Preliminary Evaluation of a Short Online Training Workshop for TPACK Development. *International Journal on Teaching and Learning in Higher Education*, *25*(1), 118–128.

Alwali, J., & Alwali, W. (2022). The relationship between emotional intelligence, transformational leadership, and performance: A test of the mediating role of job satisfaction. *Leadership and Organization Development Journal*, *43*(6), 928–952. doi:10.1108/LODJ-10-2021-0486

American Psychological Association. (2016). Strategic mentoring. *Credit Union Management*, *39*(11), 50.

Andersen, J. C., Lampley, J. H., & Good, D. W. (2013). Learner satisfaction in online learning: An analysis of the perceived impact of learner-social media and learner-instructor interaction. *Review of Higher Education and Self-Learning*, *6*(21), 81–96.

Anderson, A., Barham, N., & Northcote, M. (2013). Using the TPACK framework to unite disciplines in online learning. *Australasian Journal of Educational Technology*, *29*(4), 549–565. doi:10.14742/ajet.24

Andreasen, A. R. (Ed.). (2006). *Social marketing in the 21st century*. Sage.

Ansbacher, H. L. (1968). The concept of social interest. *Journal of Individual Psychology*, *24*(2), 131. PMID:5724576

Archambault, L., & Crippen, K. (2009). Examining TPACK among K-12 online distance educators in the United States. *Contemporary Issues in Technology & Teacher Education*, *9*(1), 71–88.

Archer-Kuhn, B., Judge-Stasiak, A., Letkemann, L., Hewson, J., & Ayala, J. (2022). The self-directed practicum: An innovative response to COVID-19 and a crisis in field education. In R. Baikady, S. M. Sajid, V. Nadesan, & M. R. Islam (Eds.), *The Routledge handbook of field work education in social work* (pp. 531–540). doi:10.4324/9781032164946-40

Armstrong, M. (2020). Moving classes online is hard—Especially in prisons: Higher education in prison programs get creative to keep classes going. *Slate.* https://slate.com/technology/2020/05/remote-learning-prisons.amp

Ashikali, T., Groeneveld, S., & Kuipers, B. (2021). The role of inclusive leadership in supporting an inclusive climate in diverse public sector teams. *Review of Public Personnel Administration, 41*(3), 497–519. doi:10.1177/0734371X19899722

Aslan, I., Çınar, O., & Kumpikaitė, V. (2012). Creating strategies from tows matrix for strategic sustainable development of Kipaş Group. *Journal of Business Economics and Management, 13*(1), 95–110. doi:10.3846/16111699.2011.620134

Astin, A. W. (1984). Student involvement: A developmental theory for higher education. *Journal of College Student Personnel, 25*(4), 297–308.

Au, C. H., Ho, K. K. W., & Chiu, D. K. W. (2021). The role of online misinformation and fake news in ideological polarization: Barriers, catalysts, and implication. *Information Systems Frontiers.* Advance online publication. doi:10.100710796-021-10133-9

Australian Universities Community Engagement Alliance. (2008). *Position Paper 2008-2010: Universities and Community Engagement.* Retrieved from http://admin.sun.ac.za/ ci/resources/AUCEA_universities_CE.pdf

Awang-Hashim, R., & Valdez, N. (2019). Strategizing inclusivity in teaching diverse learners in higher education. *Malaysian Journal of Learning and Instruction, 16*(1), 105–128. doi:10.32890/mjli2019.16.1.5

Ayu, M. (2020). Online learning: Leading e-learning at higher education. *The Journal of English Literacy and Education, 7*(1), 47–54. doi:10.36706/jele.v7i1.11515

Baah, F. O., Teitelman, A. M., & Riegel, B. (2019). Marginalization: Conceptualizing patient vulnerabilities in the framework of social determinants of health— An integrative review. *Nursing Inquiry, 26*(1), e12268. doi:10.1111/nin.12268 PMID:30488635

Babapour Chafi, M., Hultberg, A., & Bozic Yams, N. (2022). Post-pandemic office work: Perceived challenges and opportunities for a sustainable work environment. *Sustainability, 14*(1), 294. doi:10.3390u14010294

Baciu, E. L., & Trancă, L. M. (2021). Re-framing challenges as opportunities: Moving a social work practicum program in an online format and making it work. *Social Work Research, 1*, 179–191.

Bandura, A. (1989). Human agency in social cognitive theory. *The American Psychologist, 44*(9), 1175–1184. doi:10.1037/0003-066X.44.9.1175 PMID:2782727

Banks, G. C., Dionne, S. D., Schmid Mast, M., & Sayama, H. (2022). Leadership in the digital era: A review of who, what, when, where, and why. *The Leadership Quarterly, 33*(5), 101643–101639. doi:10.1016/j.leaqua.2022.101634

Baran, E. (2016). Investigating faculty technology mentoring as a university-wide professional development model. *Journal of Computing in Higher Education, 28*(1), 45–71. doi:10.100712528-015-9104-7

Baran, E., & Correia, A. P. (2009). Student-led facilitation strategies in online discussions. *Distance Education, 30*(3), 339–361. doi:10.1080/01587910903236510

Barge, K., & Pearce, W. B. (2004). A reconnaissance of CMM research. *Human Systems, 1*, 13-32. https://cmminstitute.org/wp-content/uploads/2020/12/Human-Systems-2004-A-reconnaissance-of-CMM-research-Barge_W.-Pearce.pdf

Barnes, M., & Schmitz, P. (2016). Community engagement matters (now more than ever). *Stanford Social Innovation Review, 14*(2), 32–39.

Barry, T. E. (1987). The development of the hierarchy of effects: An historical perspective. *Current Issues and Research in Advertising, 10*(1-2), 251-295.

Bart, M. (2012). *Online student engagement tools and strategies*. Magna Publications.

Bartusevičiene, I., & Rupšenė, L. (2010). Periodic assessment of students' achievements as a prerequisite of social activity education under conditions of instructional/learning paradigm's shift (the case of Social Pedagogy Study Programmes). *Social Education / Socialinis Ugdymas, 12*(23), 96-105.

Bates, C. C., Huber, R., & McClure, E. (2016). Stay connected: Using technology to enhance professional learning communities. *The Reading Teacher, 70*(1), 99–102. doi:10.1002/trtr.1469

Baumeister, R. F., & Leary, M. R. (1995). The need to belong: Desire for interpersonal attachments as a fundamental human motivation. *Psychological Bulletin, 117*(3), 497–529. doi:10.1037/0033-2909.117.3.497 PMID:7777651

Baur, T., Kniffin, L. E., & Priest, K. L. (2015). The future of service-learning and community engagement: Asset-based approaches and student learning in first-year courses. *Michigan Journal of Community Service Learning*, 89–92.

Beckett, K.S. (2019). Dewey Online: A Critical Examination of the Communities of Inquiry Approach to Online Discussions. *Philosophical Studies in Education, 50*, 46-58.

Bednar, J. (2021). Best of both worlds: Hybrid work schedules are fast becoming the new norm. *BusinessWest, 38*(6), 25–28.

Beecher, W. (1990). Beyond Success and Failure. Ways to Self-Reliance and Maturity. Richardson, TX: Willard & Marguerite Beecher Foundation.

Beere, C., Votruba, J. C., & Wells, G. W. (2011). *Becoming an engaged campus: A practical guide for institutionalizing public engagement*. John Wiley & Sons, Inc.

Belanger, P. (2015). *Self-construction and social transformation: lifelong, lifewide, and life-deep learning*. Hamburg, Germany: UNESCO Institute for Lifelong Learning.

Benioff, M. (2019, Oct 14). Benioff: We Need A New Capitalism, The current system has led to profound inequity. To fix it, we need businesses and executives to value purpose alongside profit. *The New York Times*. Retrieved on October 23, 2022 from https://www.nytimes.com/2019/10/14/opinion/benioff-salesforce-capitalism.html

Benitz, M. A., & Yang, L. (2020). Adapting a community engagement project in engineering and education to remote learning in the era of COVID-19. *Advances in Engineering Education, 8*(4), 1–8.

Benson, S., & Ward, C. (2013). Teaching with technology: Using TPACK to understand teaching expertise in online higher education. *Journal of Educational Computing Research, 48*(2), 153–172. doi:10.2190/EC.48.2.c

Berger, P., & Luckmann, T. (1967). *The social construction of reality: A treatise in the sociology of knowledge*. Anchor Books.

Berge, Z. L. (1995). Facilitating computer conferencing: Recommendations from the field. *Educational Technology, 35*(1), 22–30.

Berge, Z. L. (2008). Changing instructor's roles in virtual worlds. *Quarterly Review of Distance Education, 9*(4), 408–414.

Berry, S. (2017). *Exploring community in an online doctoral program: A digital case study* (Publication No. 10257431) [Doctoral dissertation, University of Southern California]. ProQuest Dissertations & Theses Global.

Besse, R., Whitaker, W. K., & Brannon, L. A. (2021). Loneliness among college students: The influence of targeted messages on befriending. *Psychological Reports*, 1–24. doi:10.1177/0033294121993067 PMID:33593152

Bickhoff, N., Hollensen, S., & Opresnik, M. (2014). *The quintessence of marketing: What you really need to know to manage your marketing activities*. Springer. doi:10.1007/978-3-642-45444-8

Bidandi, F., Anthony, A. N., & Mukong, C. (2022). Collaboration and partnerships between South African higher education institutions and stakeholders: A case study of a post-apartheid university. *Discover Education, 1*(1), 1–14. doi:10.100744217-022-00001-2 PMID:35795019

Bierema, L. L., & Merriam, S. B. (2002). E-mentoring: Using computer mediated communication to enhance the mentoring process. *Innovative Higher Education, 26*(3), 211–227. doi:10.1023/A:1017921023103

Black, G. F., Davies, A., Iskander, D., & Chambers, M. (2018). Reflections on the ethics of participatory visual methods to engage communities in global health research. *Global Bioethics, 29*(1), 22–38. doi:10.1080/11287462.2017.1415722 PMID:29434532

Blery, E. K., Katseli, E., & Tsara, N. (2010). Marketing for a nonprofit organization. *International Review on Public and Nonprofit Marketing, 7*(1), 57–68. doi:10.100712208-010-0049-2

Blomqvist, K., & Cook, K. S. (2018). Swift Trust - State-of-the-Art and Future Research Directions. In R. H. A. N. Searle & S. Sitkin (Eds.), *Routledge Companion to trust* (pp. 29–49). Routledge. doi:10.4324/9781315745572-4

Blondel, J., & Inoguchi, T. (2006). *Political Cultures in Asia and Europe: Citizens, States and Societal Values* (1st ed.). Routledge. doi:10.4324/9780203966907

Bloomberg, L. D. (2022). Designing and delivering effective online instruction, how to engage the adult learner. *Adult Learning, 34*(1), 55–56. doi:10.1177/10451595211069079

Bojic, L. (2022). Metaverse through the prism of power and addiction: What will happen when the virtual world becomes more attractive than reality? *European Journal of Futures Research, 10*(1), 1–24. doi:10.118640309-022-00208-4

Bolton, A., Goosen, L., & Kritzinger, E. (2021). An Empirical Study into the Impact on Innovation and Productivity Towards the Post-COVID-19 Era: Digital Transformation of an Automotive Enterprise. In *Handbook of Research on Entrepreneurship, Innovation, Sustainability, and ICTs in the Post-COVID-19 Era* (pp. 133–159). IGI Global. doi:10.4018/978-1-7998-6776-0.ch007

Bond, M. H., Leung, K., Au, A., Tong, K.-K., de Carrasquel, S. R., Murakami, F., Yamaguchi, S., Bierbrauer, G., Singelis, T. M., Broer, M., Boen, F., Lambert, S. M., Ferreira, M. C., Noels, K. A., van Bavel, J., Safdar, S., Zhang, J., Chen, L., Solcova, I., ... Lewis, J. R. (2004). Culture-Level Dimensions of Social Axioms and Their Correlates across 41 Cultures. *Journal of Cross-Cultural Psychology, 35*(5), 548–570. doi:10.1177/0022022104268388

Bond, R., & Smith, P. B. (1980). Culture and conformity: A meta-analysis of studies using Asch's (1952b, 1956) line judgment task. *Psychological Bulletin, 119*(1), 111–137. doi:10.1037/0033-2909.119.1.111

Bonsaksen, T., Ruffolo, M., Leung, J., Price, D., Thygesen, H., Schoultz, M., & Geridal, A. O. (2021). Loneliness and its association with social media use during the COVID-19 outbreak. *Social Media + Society, 7*(3), 1–10. doi:10.1177/20563051211033821

Books, J. P. (2014, Jan 9). *There's no such thing as a kid who hates reading. There are kids who love reading, and kids who are reading the wrong books* [Twitter post]. Retrieved 9 June 2022 from https://twitter.com/jp_books/status/421344811224817664?lang=en-GB

Bortree, D. S., & Seltzer, T. (2009). Dialogic strategies and outcomes: An analysis of environmental advocacy groups' Facebook profiles. *Public Relations Review, 35*(3), 317–319. doi:10.1016/j.pubrev.2009.05.002

Bozkurt, A., & Sharma, R. C. (2020). Emergency remote teaching in a time of global crisis due to Coronavirus pandemic. *Asian Journal of Distance Education, 15*(1), 1–4.

Braithwaite, R. L., Akintobi, T. H., Blumenthal, D. S., & Langley, W. M. (2020). *The Morehouse Model: How one school of medicine revolutionized community engagement and health equity.* JHU Press. doi:10.1353/book.75006

Brantley-Dias, L., & Ertmer, P. (2013). Goldilocks and TPACK: Is the construct 'just right'? *Journal of Research on Technology in Education, 46*(2), 103–128. doi:10.1080/15391523.2013.10782615

Brindley, J. E., Blaschke, L. M., & Walti, C. (2009). Creating effective collaborative learning groups in an online environment. *International Review of Research in Open and Distributed Learning, 10*(3). Advance online publication. doi:10.19173/irrodl.v10i3.675

Bringle, R. G., & Hatcher, J. A. (2011). Student engagement trends over time. In H. E. Fitzgerald, C. Burack, & S. D. Seifer (Eds.), Handbook of engaged scholarships: Contemporary landscapes, future directions: Vol. 2. Community-campus partnerships (pp. 411-430). Michigan State University Press.

Bringle, R. G., & Hatcher, J. A. (2000). Campus-community partnerships: The terms of engagement. *The Journal of Higher Education, 71*(3), 273–290. doi:10.2307/2649291

Brinkley-Etzkorn, K. (2018). Learning to teach online: Measuring the influence of faculty development training on teaching effectiveness through a TPACK lens. *The Internet and Higher Education, 38*, 28–35. doi:10.1016/j.iheduc.2018.04.004

Brooks, C. D., & Jeong, A. (2006). Effects of pre-structuring discussion threads on group interaction and group performance in computer-supported collaborative argumentation. *Distance Education, 27*(3), 371–390. doi:10.1080/01587910600940448

Brosens, D., Croux, F., & De Donder, L. (2019). Barriers to prisoner participation in educational courses: Insights from a remand prison in Belgium. *International Review of Education, 65*(5), 735–754. doi:10.100711159-018-9727-9

Brower, T. (2021, February 7). *How to sustain company culture in a hybrid work model.* Forbes. https://www.forbes.com/sites/tracybrower/2021/02/07/how-to-sustain-company-culture-in-a-hybrid-work-model/?sh=565f883610
09

Bruhn, M., Schoenmueller, V., & Schäfer, D. B. (2012). Are social media replacing traditional media in terms of brand equity creation? *Management Research Review, 35*(9), 770–790. doi:10.1108/01409171211255948

Buechner, B., Van Middendorp, S., & Spann, R. (2018). Moral Injury on the Front Lines of Truth: Encounters with Liminal Experience and the Transformation of Meaning. *Journal of Schutzian Research, 10*, 51–84. doi:10.5840chutz2018104

Bueddefeld, J., Murphy, M., Ostrem, J., & Halpenny, E. (2021). Methodological bricolage and COVID-19: An illustration from innovative, novel, and adaptive environmental behavior change research. *Journal of Mixed Methods Research, 15*(3), 437–461. doi:10.1177/15586898211019496

Burton, C., & Winter, M. A. (2021). Benefits of service-learning for students during the COVID-19 crisis: Two case studies. *Scholarship of Teaching and Learning in Psychology.* Advance online publication. doi:10.1037tl0000292

Butler, P. (2000). By popular demand: Marketing the arts. *Journal of Marketing Management, 16*(4), 343–364. doi:10.1362/026725700784772871

Byrne, J. V. (2016). Outreach, engagement, and the changing culture of the university. *Journal of Higher Education Outreach & Engagement, 20*(1), 53–58.

Cacioppo, J. T. (2018). Loneliness in the Modern Age: An Evolutionary Theory of Loneliness (ETL). Advances in Experimental Social Psychology, 58C, 127.

Cacioppo, J. T., & Hawkley, L. C. (2009). Perceived social isolation and cognition. *Trends in Cognitive Sciences, 13*(10), 447–454. doi:10.1016/j.tics.2009.06.005 PMID:19726219

Cacioppo, J. T., & Patrick, W. (2008). *Loneliness: Human nature and the need for social connection.* WW Norton & Company.

Cacioppo, S., Bangee, M., Balogh, S., Cardenas-Iniguez, C., Qualter, P., & Cacioppo, J. T. (2016). Loneliness and implicit attention to social threat: A high-performance electrical neuroimaging study. *Cognitive Neuroscience, 7*(1-4), 1–4, 138–159. doi:10.1080/17588928.2015.1070136 PMID:26274315

Cacioppo, S., Capitanio, J. P., & Cacioppo, J. T. (2014). Toward a neurology of loneliness. *Psychological Bulletin, 140*(6), 1464–1504. doi:10.1037/a0037618 PMID:25222636

Cao, T.T. & Le, P.B., (2022), Impacts of transformational leadership on organizational change capability: a two-path mediating role of trust in leadership, *European Journal of Management and Business Economics, Vol. ahead-of-print No. ahead-of-print.* doi:10.1108/EJMBE-06-2021-0180

Cao, J., Yang, T., Lai, I. K. W., & Wu, J. (2021). Is online education more welcomed during COVID-19? An empirical study of social impact theory on online tutoring platforms. *International Journal of Electrical Engineering Education.* Advance online publication. doi:10.1177/0020720920984001

Carter, A. (2022). *Belonging within the workplace* (Publication No. 29393403) [Adler University]. ProQuest. https://www.proquest.com/openview/df3bede74f97b31ba91c846f316515e6/1.pdf?pq-origsite=gscholar&cbl=18750&diss=y

Cash, C., Cox, T., & Hahs-Vaughn, D. (2021). Distance educators' attitudes and actions towards inclusive teaching practices. *The Journal of Scholarship of Teaching and Learning, 21*(2). Advance online publication. doi:10.14434/josotl.v21i2.27949

Cattapan, A., Acker-Verney, J. M., Dobrowolsky, A., Findlay, T., & Mandrona, A. (2020, October 1). Community engagement in a time of confinement. *Canadian Public Policy, 46*(S3), S287–S299. doi:10.3138/cpp.2020-064

Çelik, S. (2013). Unspoken social dynamics in an online discussion group: The disconnect between attitudes and overt behavior of English language teaching graduate students. *Educational Technology Research and Development, 61*(4), 665–683. doi:10.100711423-013-9288-3

Chabbott, C., & Sinclair, M. (2020). SDG 4 and the COVID-19 emergency: Textbooks, tutoring, and teachers. *Prospects, 49*(1), 51–57. doi:10.100711125-020-09485-y PMID:32836424

Chaffey, D. (2022). *Global social media statistics research summary 2022.* https://www.smartinsights.com/social-media-marketing/social-media-strategy/new-global-social-media-research/

Chambers, D. (2020). An interdisciplinary approach to language learning through community engagement. *13th International Conference Innovation in Language Learning Virtual Edition*. Retrieved 30 May 2022 from https://conference.pixel-online.net/ICT4LL/files/ict4ll/ed0013/FP/6968-EMO4902-FP-ICT4LL13.pdf

Changar, M., & Atan, T. (2021). The role of transformational and transactional leadership approaches on environmental and ethical aspects of CSR. *Sustainability*, *13*(3), 1411. doi:10.3390u13031411

Chan, M. M. W., & Chiu, D. K. W. (2022). Alert Driven Customer Relationship Management in Online Travel Agencies: Event-Condition-Actions rules and Key Performance Indicators. In A. Naim & S. Kautish (Eds.), *Building a Brand Image Through Electronic Customer Relationship Management* (pp. 286–303). IGI Global. doi:10.4018/978-1-6684-5386-5.ch012

Chansanam, W., Tuamsuk, K., Poonpon, K., & Ngootip, T. (2021). Development of online learning platform for Thai university students. *International Journal of Information and Education Technology (IJIET)*, *11*(8), 348–355. http://www.ijiet.org/vol11/1534-IJIET-1860.pdf. doi:10.18178/ijiet.2021.11.8.1534

Chan, T. T. W., Lam, A. H. C., & Chiu, D. K. W. (2020). From Facebook to Instagram: Exploring user engagement in an academic library. *Journal of Academic Librarianship*, *46*(6), 102229. doi:10.1016/j.acalib.2020.102229 PMID:34173399

Chan, V. H. Y., Ho, K. K. W., & Chiu, D. K. W. (2022). Mediating effects on the relationship between perceived service quality and public library app loyalty during the COVID-19 era. *Journal of Retailing and Consumer Services*, *67*, 102960. doi:10.1016/j.jretconser.2022.102960

Chauke, M. T. (2015). The role of women in traditional leadership with special reference to the Valoyi tribe. *Studies of Tribes and Tribals*, *13*(1), 34–39. doi:10.1080/0972639X.2015.11886709

Chen, C., Arakawa, Y., Watanabe, K., & Ishimaru, S. (2022). Quantitative Evaluation System for Online Meetings Based on Multimodal Microbehavior Analysis. *Sensors and Materials*, *34*(8), 3017–3027. doi:10.18494/SAM3959

Cheng, C., Paré, D., Collimore, L., & Joordens, S. (2011). Assessing the effectiveness of a voluntary online discussion forum on improving students' course performance. *Computers & Education*, *56*(1), 253–261. doi:10.1016/j.compedu.2010.07.024

Cheng, W. W. H., Lam, E. T. H., & Chiu, D. K. W. (2020). Social media as a platform in academic library marketing: A comparative study. *Journal of Academic Librarianship*, *46*(5), 102188. doi:10.1016/j.acalib.2020.102188

Chen, J. C., & Bogachenko, T. (2022). Online community building in distance education: The case of social presence in the blackboard discussion board versus multimodal VoiceThread interaction. *Journal of Educational Technology & Society*, *25*(2), 62–75. https://www.researchgate.net/publication/359278705_Online_Community_Building_in_Distance_Education_The_Case_of_Social_Presence_in_the_Blackboard_Discussion_Board_versus_Multimodal_VoiceThread_Interaction

Chen, T., Peng, L., Jing, B., Wu, C., Yang, J., & Cong, G. (2020). The Impact of the COVID-19 Pandemic on User Experience with Online Education Platforms in China. *MDPI Sustainability*, *12*(18), 4–17. doi:10.3390u12187329

Cheung, T. Y., Ye, Z., & Chiu, D. K. W. (2021). Value chain analysis of information services for the visually impaired: A case study of contemporary technological solutions. *Library Hi Tech*, *39*(2), 625–642. doi:10.1108/LHT-08-2020-0185

Cheung, V. S. Y., Lo, J. C. Y., Chiu, D. K. W., & Ho, K. K. W. (2022). Predicting Facebook's influence on travel products marketing based on the AIDA model. *Information Discovery and Delivery*. Advance online publication. doi:10.1108/IDD-10-2021-0117

Cho, J. (2016, May 25). *"Diversity is being invited to the party; inclusion is being asked to dance," Verna Myers tells Cleveland Bar*. Retrieved on July 22, 2021 from https://www.cleveland.com/business/2016/05/diversity_is_being_invited_to.html#incart_more_business

Chung, C., Chiu, D. K. W., Ho, K. K. W., & Au, C. H. (2020). Applying social media to environmental education: Is it more impactful than traditional media? *Information Discovery and Delivery*, *48*(4), 255–266. doi:10.1108/IDD-04-2020-0047

Cialdini, R. B. (2009). *Influence: Science and practice* (Vol. 4). Pearson Education.

Cigna. (2020). *Loneliness and the Workplace*. 2020 U.S Report. Retrieved April 9 from https://www.cigna.com/static/www-cigna-com/docs/about-us/newsroom/studies-and-reports/combatting-loneliness/cigna-2020-loneliness-report.pdf

Clark, J. L., Algoe, S. B., & Green, M. C. (2018). Social network sites and wellbeing: The role of social connection. *Association for Psychological Science*, *27*(1), 32–37. doi:10.1177/096372147730833 PMID:30407887

Clark, M. S., Oullette, R., Powell, M. C., & Milberg, S. (1987). Recipient's mood, relationship type, and helping. *Journal of Personality and Social Psychology*, *53*(1), 94–103. doi:10.1037/0022-3514.53.1.94 PMID:3612495

Cohen, A. (2021, December 30). *What really happens when workers are given a flexible hybrid schedule?* Bloomberg. https://www.bloomberg.com/news/articles/2021-12-30/the-flexible-hybrid-work-schedule-that-employees-actually-want

Collins, R. (2004). *Interaction ritual chains*. Princeton University Press. doi:10.1515/9781400851744

Commission on Public Purpose in Higher Education. (2022). public-purpose.org

Coronado, G., Chio-Lauri, J., Dela Cruz, R., & Roman, Y. M. (2021). Health disparities of cardiometabolic disorders among Filipino Americans: Implications for health equity and community-based genetic research. *Journal of Racial and Ethnic Health Disparities*, 1–8. PMID:34837163

Corrigan, T. (2020). *Africa's ICT Infrastructure: Its present and prospects*. Academic Press.

Council on Education for Public Health. (2021). *Accreditation criteria: Schools of public health and public health programs*. https://media.ceph.org/documents/2021.Criteria.pdf

Cowie, H., & Myers, C. A. (2021). The impact of the COVID-19 pandemic on the mental health and wellbeing of children and young people. *Children & Society*, *35*(1), 62–74. doi:10.1111/chso.12430 PMID:33362362

Cox, B., & Cox, B. (2008). Developing interpersonal and group dynamics through asynchronous threaded discussions: The use of discussion board in collaborative learning. *Education*, *128*(4), 553–565.

Cox, M. (2001). Faculty learning communities: Change agents for transforming institutions into learning organizations. In D. Lieberman & C. Wehlburg (Eds.), *To Improve the Academy* (Vol. 19). Anker.

Cox, T. Jr. (2001). *Creating the multicultural organization: A strategy for capturing the power of diversity* (Vol. 6). John Wiley & Sons.

Cranton, P. (1994). *Understanding and promoting transformative learning: A guide for educators of adults*. Jossey-Bass.

Creswell, J. (1998). *Qualitative inquiry and research design: Choosing among five traditions*. Sage.

Crisp, C., & Jarvenpaa, S. (2013). Swift Trust in Global Virtual Teams: Trusting Beliefs and Normative Actions. *Journal of Personnel Psychology*, *12*(1), 45–56. doi:10.1027/1866-5888/a000075

Cristofoletti, E. C., & Pinheiro, R. (2022). Taking stock: The impacts of the COVID-19 pandemic on university-community engagement. *Industry and Higher Education*. doi:10.1177/09504222221119927

Cronen, V. (2001). Practical Theory, practical art, and the pragmatic-systemic account of inquiry. *Communication Theory, 11*(1), 14–35. doi:10.1111/j.1468-2885.2001.tb00231.x

Cunningham, H. R., & Smith, P. C. (2020). Community engagement plans: A tool for institutionalizing community engagement. *Journal of Higher Education Outreach & Engagement, 24*(2), 53–68.

Dam, R. F. (2021). *Five stages in the design thinking process*. The Interaction Design Foundation. Retrieved June 2, 2022, from https://www.interaction-design.org/literature/article/5-stages-in-the-design-thinking-process

Darabi, A., & Jin, L. (2013). Improving the quality of online discussion: The effects of strategies designed based on cognitive load theory principles. *Distance Education, 34*(1), 21–36. doi:10.1080/01587919.2013.770429

Davies, J., & Graff, M. (2005). Performance in e-learning: Online participation and student grades. *British Journal of Educational Technology, 36*(4), 657–663. doi:10.1111/j.1467-8535.2005.00542.x

Daymont, T., Blau, G., & Campbell, D. (2011). Deciding between traditional and online formats: Exploring the role of learning advantages, flexibility, and compensatory adaptation. *Journal of Behavioral and Applied Management, 12*(2), 156.

de Janasz, S. C., Ensher, E. A., & Heun, C. (2008). Virtual relationships and real benefits: Using e-mentoring to connect business students with practicing managers. *Mentoring & Tutoring, 16*(4), 394–411. doi:10.1080/13611260802433775

De Miranda, M. A. (2004). The grounding of a discipline: Cognition and instruction in technology education. *International Journal of Technology and Design Education, 14*(1), 61–77. doi:10.1023/B:ITDE.0000007363.44114.3b

De Weger, E., Baan, C., Bos, C., Luijkx, K., & Drewes, H. (2022). 'They need to ask me first'. Community engagement with low-income citizens. A realist qualitative case-study. *Health Expectations, 25*(2), 684–696. Advance online publication. doi:10.1111/hex.13415 PMID:35032414

Delahunty, J., Verenikina, I., & Jones, P. (2014). Socio-emotional connections: Identity, belonging and learning in online interactions. A literature review. *Technology, Pedagogy and Education, 23*(2), 243–265. doi:10.1080/1475939X.2013.813405

DeLoach, S. B., & Greenlaw, S. A. (2007). Effectively moderating electronic discussions. *The Journal of Economic Education, 38*(4), 419–434. doi:10.3200/JECE.38.4.419-434

Dempsey, S. (2010). Critiquing community engagement. *Management Communication Quarterly, 24*(3), 359–390. doi:10.1177/0893318909352247

Deng, S., & Chiu, D. K. W. (2022). Analyzing Hong Kong Philharmonic Orchestra's Facebook Community Engagement with the Honeycomb Model. In M. Dennis & J. Halbert (Eds.), *Community Engagement in the Online Space*. IGI Global.

Dennis, M. (2021a). Best Practices for Emergency Remote Teaching. In Handbook of Research on Emerging Pedagogies for the Future of Education: Trauma-Informed, Care, and Pandemic Pedagogy (pp. 82-100). IGI Global. doi:10.4018/978-1-7998-7275-7.ch005

Dennis, M. (2021b). Supporting Faculty and Students During Pandemic Conditions: An Online Department Chair's Perspective. In Handbook of Research on Developing a Post-Pandemic Paradigm for Virtual Technologies in Higher Education (pp. 329-346). IGI Global.

Dennis, M., Halbert, J., & Fornero, S. (2022). Structured Development and Support for Online Adjunct Faculty: Case Studies and Best Practices. In Quality in Online Programs (pp. 211-228). Brill.

Dennis, M., DiMatteo-Gibson, D., Halbert, J., Gonzalez, M., & Byrd, I. (2020). Building Faculty Community: Implementation of a Research Colloquium Series. *Journal of Higher Education Theory & Practice*, *20*(6), 19–30.

Dennis, M., Gordon, E., DiMatteo-Gibson, D., & Halbert, J. D. (2022). Social justice practicum in non-clinical online programs: Engagement strategies and lessons learned. *Journal of Leadership, Accountability and Ethics*, *19*(3), 147–156.

Dennis, M., & Halbert, J. D. (2022). Effective online course delivery in correctional settings: A pilot. *Journal of Higher Education Theory and Practice*, *22*(8), 91–99.

Dennis, M., Halbert, J., DiMatteo-Gibson, D., Agada, C., & Fornero, S. (2020). Implementation of a faculty evaluation model. *Journal of Leadership, Accountability and Ethics*, *17*(5), 30–35.

deNoyelles, A., Zydney, J. M., & Chen, B. (2014). Strategies for creating a community of inquiry through online asynchronous discussions. *Journal of Online Learning and Teaching*, *10*(1), 153–165.

Denzin, N. K., & Lincoln, Y. S. (2018). *The SAGE handbook of qualitative research* (5th ed.). SAGE Publications.

Deslauriers, L., McCarty, L. S., Miller, K., Callaghan, K., & Kestin, G. (2019). Measuring actual learning versus feeling of learning in response to being actively engaged in the classroom. *Proceedings of the National Academy of Sciences of the United States of America*, *116*(39), 19251–19257. doi:10.1073/pnas.1821936116 PMID:31484770

Dhir, A., & Torsheim, T. (2016). Age and gender differences in photo tagging gratifications. *Computers in Human Behavior*, *63*(October), 630–638. doi:10.1016/j.chb.2016.05.044

Dick, W., Carey, L., & Carey, J. O. (2015). *The systematic design of instruction* (8th ed.). Pearson.

Dierberger, J., Everett, O., Kehrberg, R. S., & Greene, J. (2019). University, school district, and service-learning community partnerships that work. In J. A. Allen & R. Reiter-Palmon (Eds.), *The Cambridge handbook of organizational community engagement and outreach* (pp. 244–260). Cambridge University Press. doi:10.1017/9781108277693.014

Dobusch, L. (2020). The inclusivity of inclusion approaches: A relational perspective on inclusion and exclusion in organizations. *Gender, Work and Organization*, *28*(1), 379–396. doi:10.1111/gwao.12574

Dodd, A. (2017). Finding the community in sustainable online community engagement: Not-for-profit organization websites, service-learning and research. *Gateways: International Journal of Community Research & Engagement*, *10*, 185–203. doi:10.5130/ijcre.v10i0.5278

Dolan, J., Kain, K., Reilly, J., & Bansal, G. (2017). How do you build community and foster engagement in online courses? *New Directions for Teaching and Learning*, *151*(151), 45–60. doi:10.1002/tl.20248

Dolan, V. (2011). The isolation of online adjunct faculty and its impact on their performance. *International Review of Research in Open and Distance Learning*, *12*(2), 62–77. doi:10.19173/irrodl.v12i2.793

Dong, G., Chiu, D. K. W., Huang, P.-S., Lung, M. M., Ho, K. K. W., & Geng, Y. (2021). Relationships between Research Supervisors and Students from Coursework-based Master's Degrees: Information Usage under Social Media. *Information Discovery and Delivery*, *49*(4), 319–327. doi:10.1108/IDD-08-2020-0100

Dostilio, L. D., & Perry, L. G. (2017). An explanation of community engagement professionals as professionals and leaders. In L. D. Dostilio (Ed.), *The community engagement professional in higher education: A competency model for an emerging field*. Campus Compact.

Dostilio, L. D., & Welch, M. (2019). *The community engagement professional's guidebook: A companion to the community engagement professional in higher education*. Campus Compact.

Doucette, B., Sanabria, A., Sheplak, A., & Aydin, H. (2021). The perceptions of culturally diverse graduate students on multicultural education: Implication for inclusion and Diversity Awareness in higher education. *European Journal of Educational Research, 10*(3), 1259–1273. doi:10.12973/eu-jer.10.3.1259

Douglas, T., James, A., Earwaker, L., Mather, C., & Murray, S. (2020). Online discussion boards: Improving practice and student engagement by Harnessing facilitator perceptions. *Journal of University Teaching & Learning Practice, 17*(3), 86–100. doi:10.53761/1.17.3.7

Dreikurs, R. (2000). *Social Equality: The Challenge of Today*. Adler School of Professional Psychology.

Driscoll, A. (2008). Carnegie's community-engagement classification: Intentions and insights. *Change. The Magazine of Higher Learning, 40*(1), 38–41. doi:10.3200/CHNG.40.1.38-41

Driscoll, A. (2009). Carnegie new community engagement classification: Affirming higher education's role in community. In L. R. Sandmann, C. H. Thornton, & A. J. Jaeger (Eds.), *Institutionalizing community engagement in higher education: The first wave of Carnegie classified institutions* (pp. 5–12). Wiley Periodicals, Inc.

Driscoll, A. (2014). Analysis of the Carnegie classification of community engagement: Patterns and impact on institutions. In D. G. Terkla & L. S. O'Leary (Eds.), *Assessing civic engagement* (pp. 3–16). Wiley Periodicals, Inc. doi:10.1002/ir.20072

Ducharme. (2020, March 11). World Health Organization Declares COVID-19 a 'Pandemic.' Here's What That Means. *Time.* Retrieved on March 30, 2020 from https://time.com/5791661/who-coronavirus-pandemic-declaration/

Duffy, T. M., & Cunningham, D. J. (1996). Constructivism: implications for the design and delivery of instruction. In D. Jonassen (Ed.), *Handbook of research on educational communications and technology* (1st ed., pp. 1–31). Routledge/ Taylor & Francis Group.

Du, J., Yu, C., & Olinzock, A. (2011). Enhancing collaborative learning: Impact of "question Prompts" design for online discussion. *Delta Pi Epsilon Journal, 53*(1), 28–41.

Dukesmith, F. H. (1904). Three natural fields of salesmanship. *Salesmanship, 2*(1), 14.

Dziuban, C., Moskal, P., Kramer, L., & Thompson, J. (2013). Student satisfaction with online learning in the presence of ambivalence: Looking for the will-o'-the-wisp. *Internet and Higher Education, 17*, 1–8. doi:10.1016/j.iheduc.2012.08.001

Eagan, M. K., & Jaeger, A. J. (2008). Closing the gate: Part-time faculty instruction in gatekeeper courses and first-year persistence. *New Directions for Teaching and Learning, 2008*(115), 39–53. doi:10.1002/tl.324

Earl, W. R. (1988). Intrusive advising of freshmen in academic difficulty. *NACADA Journal, 8*(2), 27–33. doi:10.12930/0271-9517-8.2.27

Economist Intelligence Unit (EIU). (2007a). *Beyond loyalty: Meeting the challenge of customer engagement, part 1*. EIU.

Economist Intelligence Unit (EIU). (2007b). *Beyond loyalty: meeting the challenge of customer engagement, part 2*. EIU.

Economist Intelligence Unit (EIU). (2007c). *The engaged constituent: meeting the challenge of engagement in the public sector, part 1*. EIU.

Economist Intelligence Unit (EIU). (2007d). *The engaged constituent: meeting the challenge of engagement in the public sector, part 2*. EIU.

Edmondson, J. (1999). *America Reads: A critical policy analysis*. The Pennsylvania State University. Dissertation. Retrieved 4 June 2022 from https://www.proquest.com/docview/304521148?pq-origsite=gscholar&fromopenview=true

Edwards, S. J., Silaigwana, B., Asogun, D., Mugwagwa, J., Ntoumi, F., Ansumana, R., Bardosh, K., & Ambe, J. (2022). An ethics of anthropology-informed community engagement with COVID-19 clinical trials in Africa. *Developing World Bioethics*, dewb.12367. doi:10.1111/dewb.12367 PMID:35944158

Ehrenberg, R. G., & Zhang, L. (2005). Do tenured and tenure-track faculty matter? *The Journal of Human Resources*, *40*(3), 647–659. doi:10.3368/jhr.XL.3.647

Eisenberg, N. I. (2012). Broken hearts and broken bones: A neural perspective on the similarities between social and physical pain. *Current Directions in Psychological Science*, *21*(1), 42–47. doi:10.1177/0963721411429455

Ely, J. A. (2021). Modern mentorship: Developing medical writing mentor-mentee relationships in the virtual world. *Medical Writing*, *30*(2), 36–39.

Engage, C. (2022). *Community engaged teaching and learning*. https://www.campusengage.ie/our-work/making-an-impact/community-engaged-teaching-and-learning

Entringer, T. M., & Gosling, S. D. (2021). Loneliness during a nationwide lockdown and the moderating effect of extroversion. *Social Psychological & Personality Science*, •••, 1–12.

Ertmer, P. A., Richardson, J. C., Belland, B., Camin, D., Connolly, P., Coulthard, G., Lei, K., & Mong, C. (2007). Using peer feedback to enhance the quality of student online postings: An exploratory study. *Journal of Computer-Mediated Communication*, *12*(2), 412–433. doi:10.1111/j.1083-6101.2007.00331.x

Evans, E. (2022). Cracking the hybrid work culture conundrum: How to create a strong culture across a workforce you may never even see. *Strategic HR Review*, *21*(2), 46–49. doi:10.1108/SHR-12-2021-0065

Ewell, P. T. (2010). The U.S. national survey of student engagement (NSSE). In *Public policy for academic quality* (pp. 83–97). Springer. doi:10.1007/978-90-481-3754-1_5

Faculty Forward Academy. (2022). *Fellowship*. https://facultyforward.jhu.edu/fellowship

Fairclough, N. (1989). *Language and Power*. Longman.

Fairclough, N. (2001). Critical discourse analysis as a method in social scientific research. In R. Wodak & M. Meyer (Eds.), *Introducing qualitative methods: Methods of critical discourse analysis* (pp. 121–138). SAGE Publications Ltd.

Falk, A., & Olwell, R. (2019). Institutionalizing community engagement in higher education: The community engagement institute. In J. W. Moravec (Ed.), *Emerging education futures: Experiences and visions from the field* (pp. 59–82). Education Futures.

Farner, K. (2019). Institutionalizing community engagement in higher education: A case study of processes toward engagement. *Journal of Higher Education Outreach & Engagement*, *23*(2), 147–152.

Fatoni, N. A., Nurkhayati, E., Nurdiawati, E., Fidziah, G. P., Adha, S., Irawan, A. P., . . . Azizi, E. (2020). University Students Online Learning System During Covid-19 Pandemic: Advantages, Constraints and Solutions. *Systematic Reviews in Pharmacy*, *11*(7), 570-576. Retrieved from: https://www.sysrevpharm.org/articles/university-students-online-learning-system-during- covid19-pandemic-advantages-constraints-and-solutions.pdf

Fayard, A. L., Weeks, J., Khan, M., & Hines, F. (2021). Designing the hybrid office: From workplace to "culture space." *Harvard Business Review*, *99*(2), 114–123.

Ferguson. (1989). Adler's Motivational Theory: An Historical Perspective on Belonging and the Fundamental Human Striving. *Individual Psychology*, *45*(3).

Ferguson, E. D. (2006). A Long-Awaited Book on Adlerian Psychotherapy for the Modern Reader. *PsycCRITIQUES*, *51*(8).

Ferguson, E. D. (2010). Adler's Innovative Contributions Regarding the Need to Belong. *Journal of Individual Psychology*, *66*(1). https://web.p.ebscohost.com/ehost/pdfviewer/pdfviewer?vid=1&sid=2d4830a9-1a93-4502-917b-ee9a8991535a%40redis

Figueroa, I. (2014). The value of connectedness in inclusive teaching. *New Directions for Teaching and Learning*, *2014*(140), 45–49. doi:10.1002/tl.20112

Finegold, A., & Cooke, L. (2006). Exploring the attitudes, experiences and dynamics of interaction in online groups. *The Internet and Higher Education*, *9*(3), 201–215. doi:10.1016/j.iheduc.2006.06.003

Fitzgerald, H. E., Bruns, K., Sonka, S. T., Furco, A., & Swanson, L. (2012). The centrality of engagement in higher education. *Journal of Higher Education Outreach & Engagement*, *16*(3), 7–27.

Flynn, S., & Noonan, G. (2020). Mind the gap: Academic staff experiences of remote teaching during the Covid 19 emergency. *All Ireland Journal of Higher Education*, *12*(3).

Fong, K. C. H., Au, C. H., Lam, E. T. H., & Chiu, D. K. W. (2020). Social network services for academic libraries: A study based on social capital and social proof. *Journal of Academic Librarianship*, *46*(1), 102091. doi:10.1016/j.acalib.2019.102091

Foshay, W., Silber, K., & Westgaard, O. (1986). *Instructional Design Competencies: The Standards*. International Board of Standards for Training, Performance and Instruction.

France-Harris, A., Burton, C., & Mooney, M. (2019). Putting theory into practice: Incorporating a community engagement model into online pre-professional courses in legal studies and human resources management. *Online Learning*, *23*(2), 21–39. doi:10.24059/olj.v23i2.1448

Froiland, J. M., & Worrell, F. C. (2017). Parental autonomy support, community feeling and student expectations as contributors to later achievement among adolescents. *Educational Psychology*, *37*(3), 261–271. doi:10.1080/01443410.2016.1214687

Gaborov, M., & Ivetić, D. (2022). The importance of integrating Thinking Design, User Experience and Agile methodologies to increase profitability. *jATES. Journal of Applied Technical and Educational Sciences*, *12*(1), 1–17. doi:10.24368/jates286

Gagné, R. M., Wager, W. W., Golas, K. C., Keller, J. M., & Russell, J. D. (2004). *Principles of instructional design* (5th ed.). Wadsworth Publishing Company.

Gallup. (2018). *2018 Clifton Strengths® Meta-Analysis Report*. Gallup. https://www.gallup.com/cliftonstrengths/en/270350/2018-cliftonstrengths-meta-analysis-report.aspx

Gallup. (2022). *State of the global workplace: 2022 report*. Gallup. https://www.gallup.com/workplace/349484/state-of-the-global-workplace-2022-report.aspx

Gao, G., & Sai, L. (2020). Towards a "virtual" world: Social isolation and struggles during the COVID-19 pandemic as single women living alone. *Gender, Work and Organization*, *27*(5), 754–762. doi:10.1111/gwao.12468 PMID:32837008

Gao, W., Lam, K. M., Chiu, D. K. W., & Ho, K. K. W. (2020). A Big data Analysis of the Factors Influencing Movie Box Office in China. In Z. Sun (Ed.), *Handbook of Research on Intelligent Analytics with Multi-Industry Applications* (pp. 232–249). IGI Global.

Garcia-O'Neill, E. (2021, May 12). *Social presence in online learning: 7 things instructional designers can do to improve it*. eLearning Industry. Retrieved November 13, 2022, from https://elearningindustry.com/social-presence-in-online-learning-7-things-instructional-designers-can-improve

Garcia-Rivera, D., Matamoros-Rojas, S., Pezoa-Fuentes, C., Veas-González, I., & Vidal-Silva, C. (2022). Engagement on Twitter, a Closer Look from the Consumer Electronics Industry. *Journal of Theoretical and Applied Electronic Commerce Research*, *17*(2), 558–570. doi:10.3390/jtaer17020029

Garg, N., Murphy, W., & Singh, P. (2021). Reverse mentoring, job crafting and work-outcomes: The mediating role of work engagement. *Career Development International*, *26*(2), 290–308. doi:10.1108/CDI-09-2020-0233

Garrett, R., Legon, R., Fredericksen, E. E., & Simunich, B. (2020). *CHLOE 5: The pivot to remote teaching in spring 2020 and its impact*. Quality Matters. https://www.qualitymatters.org/qa-resources/resource-center/articles-resources/CHLOE-5-report-2020

Garrison, D. R. (2006). Online collaboration principles. *Journal of Asynchronous Learning Networks*, *10*(1), 25–34.

Garrison, D. R., Anderson, T., & Archer, W. (2001). Critical thinking, cognitive presence, and computer conferencing in distance education. *American Journal of Distance Education*, *15*(1), 7–23. doi:10.1080/08923640109527071

Gaventa, J., & Cornwall, A. (2008). Power and knowledge. The Sage handbook of action research: Participative inquiry and practice, 2, 172-189.

Gelmon, S. B., Holland, B. A., & Spring, A. (2018). *Assessing service-learning and civic engagement: Principles and techniques*. Campus Compact.

Germain, M. (2019). *Integrating service-learning and consulting in distance education*. Emerald Publishing Limited. doi:10.1108/9781787694095

Ghosh, R., Shuck, B., Cumberland, D., & D'Mello, J. (2019). Building psychological capital and employee engagement: Is formal mentoring a useful strategic human resource development intervention? *Performance Improvement Quarterly*, *32*(1), 37–54. doi:10.1002/piq.21285

Gibson, A., Fields, N. L., Wladkowski, S. P., Kusmaul, N., Greenfield, J. C., & Mauldin, R. L. (2019). What can an evaluation of the AGE*SW* Predissertation Fellows Program tell us about the mentoring needs of doctoral students? *Journal of Gerontological Social Work*, *62*(8), 852–866. doi:10.1080/01634372.2019.1685052 PMID:31650910

Gilbert, C., & Bower, J. (2002, May). Disruptive Change: When Trying Harder Is Part of the Problem. *Harvard Business Review*. PMID:12024762

Giles, D., Stommel, W., Paulus, T., Lester, J., & Reed, D. (2015). Microanalysis of online data: The methodological development of "digital CA". *Discourse, Context & Media*, *7*, 45-51.

Giles, D. E. (2008). Understanding an emerging field of scholarship: Toward a research agenda for engaged, public scholarship. *Journal of Higher Education Outreach & Engagement*, *12*(2), 97–106.

Gilmore, B., Ndejjo, R., Tchetchia, A., De Claro, V., Mago, E., Lopes, C., & Bhattacharyya, S. (2020). Community engagement for COVID-19 prevention and control: A rapid evidence synthesis. *BMJ Global Health*, *5*(10), e003188. doi:10.1136/bmjgh-2020-003188 PMID:33051285

Goggins, J., & Hajdukiewicz, M. (2020). *Community-engaged learning: A building engineering case study.* https://sword.cit.ie/cgi/viewcontent.cgi?article=1069&context=ceri

Goldberg, A., Hannan, M. T., & Kovács, B. (2016). What does it mean to span cultural boundaries? Variety and atypicality in cultural consumption. *American Sociological Review*, *81*(2), 215–241. doi:10.1177/0003122416632787

Google Chrome. (n.d.). Available from: https://www.google.com/chrome/

Google Drive. (n.d.). *Extendable Functioning Waiting Room Project.* Available from: https://drive.google.com/drive/folders/1hdmd8E5VC8TbID0ASD4U_Qj7Oql9iVwx?usp=sharing

Google Jamboard. (n.d.). Available from: https://jamboard.google.com/d/1- pTwY1vbAVNgsuVyrtplqZ6zgPfiJ-AiRmx8X9-yKUM/edit?usp=sharing

Google Meet Help. (2021). *Raise your hand in Google Meet.* Available from: https://support.google.com/meet/answer/10159750?hl=en&co=GENIE.Platform%3DDesktop

Google Meet. (n.d.). Available from: https://meet.google.com/

Goosen, L. (2015). Educational Technologies for Growing Innovative e-Schools in the 21st Century: A Community Engagement Project. *Proceedings of the South Africa International Conference on Educational Technologies* (pp. 49-61). African Academic Research Forum (AARF).

Goosen, L. (2016, February 18). *"We don't need no education"? Yes, they DO want e-learning in Basic and Higher Education!* Retrieved from https://uir.unisa.ac.za/handle/10500/20999

Goosen, L. (2018b). Trans-Disciplinary Approaches to Action Research for e-Schools, Community Engagement, and ICT4D. In Cross-Disciplinary Approaches to Action Research and Action Learning (pp. 97-110). IGI Global.

Goosen, L. (2018d). Ethical Information and Communication Technologies for Development Solutions: Research Integrity for Massive Open Online Courses. In Ensuring Research Integrity and the Ethical Management of Data (pp. 155-173). IGI Global.

Goosen, L., & Mukasa-Lwanga, T. (2017). Educational Technologies in Distance Education: Beyond the Horizon with Qualitative Perspectives. In U. I. Ogbonnaya, & S. Simelane-Mnisi (Ed.), *Proceedings of the South Africa International Conference on Educational Technologies* (pp. 41 - 54). African Academic Research Forum.

Goosen, L., & Van der Merwe, R. (2015). e-Learners, Teachers and Managers at e-Schools in South Africa. In *Proceedings of the 10th International Conference on e-Learning* (pp. 127-134). Academic Conferences and Publishing International.

Goosen, L. (2004). *Criteria and Guidelines for the Selection and Implementation of a First Programming Language in High Schools. 178.* Potchefstroom Campus: North West University.

Goosen, L. (2018a). Sustainable and Inclusive Quality Education Through Research Informed Practice on Information and Communication Technologies in Education. In *Proceedings of the 26th Conference of the Southern African Association for Research in Mathematics, Science and Technology Education (SAARMSTE)* (pp. 215-228). University of Botswana.

Goosen, L. (2018c). Ethical Data Management and Research Integrity in the Context of E-Schools and Community Engagement. In *Ensuring Research Integrity and the Ethical Management of Data* (pp. 14–45). IGI Global. doi:10.4018/978-1-5225-2730-5.ch002

Goosen, L., & Molotsi, A. (2019). Student Support Towards Rethinking Teaching and Learning in the 21st Century: A Collaborative Approach Involving e-Tutors. In *Proceedings of the South Africa International Conference on Education* (pp. 43-55). AARF.

Goosen, L., & Naidoo, L. (2014). Computer Lecturers Using Their Institutional LMS for ICT Education in the Cyber World. In C. Burger, & K. Naudé (Ed.), *Proceedings of the 43rd Conference of the Southern African Computer Lecturers' Association (SACLA)* (pp. 99-108). Nelson Mandela Metropolitan University.

Gorelick, P. (2022). Community Engagement: Lessons Learned From the AAASPS and SDBA. *Stroke, 53*(3), 654–662. Advance online publication. doi:10.1161/STROKEAHA.121.034554 PMID:35078349

Goshen College. (2020). *2019-20 Annual Report.* Unpublished internal company document.

Goshen College. (2021a). *Building Capacity for Overcoming Barriers to Inclusive Study Abroad.* Unpublished internal company document.

Goshen College. (2021b). *Center for Community Engagement: Statement of Intent.* Unpublished internal company document.

Goshen College. (2021c). *Statement of Values.* Unpublished internal company document.

Goshen College. (2021d). *Core Refresh Curricular Proposal.* Unpublished internal company document.

Goshen College. (2022). *Study-service term (SST): global education at Goshen College.* https://www.goshen.edu/sst/

Gotsis, G., & Grimani, K. (2016). Diversity as an aspect of effective leadership: Integrating and moving forward. *Leadership and Organization Development Journal, 37*(2), 241–264. doi:10.1108/LODJ-06-2014-0107

Graham, R., Burgoyne, N., Cantrell, P., Smith, L., St Clair, L., & Harris, R. (2009). Measuring the TPACK confidence of inservice science teachers. *TechTrends, 53*(5), 70–79. doi:10.100711528-009-0328-0

Grant, M., & Judy, C. (2017, February 15). *Viewpoint: 3 steps to cultivating a customized culture.* Society for Human Resource Management. https://www.shrm.org/resourcesandtools/hr-topics/employee-relations/pages/viewpoint-3-steps-to-cultivating-a-customized-culture.aspx

Grant, K., & Lee, V. (2014). Teacher educators wrestling with issues of diversity in online courses. *Qualitative Report.* Advance online publication. doi:10.46743/2160-3715/2014.1275

Gratton, L. (2021). Four principles to ensure hybrid work is productive work. *MIT Sloan Management Review, 62*(2), 11A–16A.

Gray, B. (2005). Informal learning in an online community of practice. *International Journal of E-Learning & Distance Education, 19*(1), 20–35.

Greenberg, J., & MacAulay, M. (2009). NPO 2.0? Exploring the web presence of environmental nonprofit organizations in Canada. Global Media Journal: Canadian Edition, 2(1), 63-88.

Gross Domestic Product and its major components. (n.d.). *Hong Kong Economy.* https://www.hkeconomy.gov.hk/en/situation/development/index.htm

Gur, H. (2015). A short review of TPACK for teacher education. *Educational Research Review*, *10*(7), 777–789. doi:10.5897/ERR2014.1982

Hadiyati, E. (2016). Study of marketing mix and AIDA model to purchasing on line product in Indonesia. *British Journal of Marketing Studies*, *4*(7), 49–62.

Halbert, J., Dennis, M., & DiMatteo-Gibson, D. (2022). Support for distance students: Using a mentor model. *Proceedings of the Virtual Conference Transforming the Teaching and Learning Environment.*

Hales, A. H., McIntyre, M. M., Rudert, S. C., Williams, K. D., & Thomas, H. (2021). Ostracized and observed: The presence of an audience affects the experience of being excluded. *Self and Identity*, *20*(1), 94–115. doi:10.1080/15298868.2020.1807403

Hales, A. H., Wood, N. R., & Williams, K. D. (2021). Navigating COVID-19: Insights from research on social ostracism. *Group Processes & Intergroup Relations*, *24*(2), 306–310. doi:10.1177/1368430220981408

Halupa, C. (2019). Differentiation of roles: Instructional designers and faculty in the creation of online courses. *International Journal of Higher Education*, *8*(1), 55–68. doi:10.5430/ijhe.v8n1p55

Hamilton, B. A., & Scandura, T. A. (2003). E-mentoring: Implications for organizational learning and development in a wired world. *Organizational Dynamics*, *31*(4), 388–402. doi:10.1016/S0090-2616(02)00128-6

Hamza, C. A., Ewing, L., Heath, N. L., & Goldstein, A. L. (2021). When social isolation is nothing new: A longitudinal study on psychological distress during COVID-19 among university students with and without preexisting mental health concerns. *Canadian Psychology*, *62*(1), 20–30. doi:10.1037/cap0000255

Han, J., Jiang, Y., Mentzer, N., & Kelley, T. (2022). The role of sense of community and motivation in the collaborative learning: An examination of the first-year design course. *International Journal of Technology and Design Education*, *32*(3), 1837–1852. doi:10.100710798-021-09658-6

Hannay, M., & Newvine, T. (2006). Perceptions of distance learning: A comparison of online and traditional learning. *Journal of Online Learning and Teaching*, *2*(1), 1–11.

Han, R., & Xu, J. (2020). A Comparative Study of the Role of Interpersonal Communication, Traditional Media and Social Media in Pro-Environmental Behavior: A China-Based Study. *International Journal of Environmental Research and Public Health*, *17*(6), 1883. doi:10.3390/ijerph17061883 PMID:32183217

Harasim, L. (2000). Shift happens: Online education as a new paradigm in learning. *The Internet and Higher Education*, *3*(1-2), 41–61. doi:10.1016/S1096-7516(00)00032-4

Hart, R. K. (2016). Informal virtual mentoring for team leaders and members: Emergence, content, and impact. *Advances in Developing Human Resources*, *18*(3), 352–368. doi:10.1177/1523422316645886

Harvey, D. (1989). *The condition of postmodernity: An enquiry into the origins of cultural change.* Academic Press.

Hatcher, J. A., & Bringle, R. G. (2010). Reflection: Bridging the gap between service and learning. *College Teaching*, *45*(4), 153–158. doi:10.1080/87567559709596221

Hauser, L., & Darrow, R. (2013). Cultivating a Doctoral Community of Inquiry and Practice: Designing and Facilitating Discussion Board Online Learning Communities. *Education Leadership Review*, *14*(3).

Hausmann, A. (2012). Creating 'buzz': Opportunities and limitations of social media for arts institutions and their viral marketing. *International Journal of Nonprofit and Voluntary Sector Marketing*, *17*(3), 173–182. doi:10.1002/nvsm.1420

Haverila, K., Haverila, M., & McLaughlin, C. (2022). Development of a brand community engagement model: A service-dominant logic perspective. *Journal of Consumer Marketing*, *39*(2), 166–179. Advance online publication. doi:10.1108/JCM-01-2021-4390

Head, B. (2007). Community engagement: Participation on whose terms? *Australian Journal of Political Science*, *42*(3), 441–454. doi:10.1080/10361140701513570

Heisserer, D. L., & Parette, P. (2002, March). Advising at-risk students in college and university settings. *College Student Journal*, *36*(1), 69–84.

Heitschmidt, M., Staffileno, B. A., & Kleinpell, R. (2021). Implementing a faculty mentoring process to improve academic-clinical partnerships for nurse led evidence based practice and research projects. *Journal of Professional Nursing*, *37*(2), 399–403. doi:10.1016/j.profnurs.2020.04.015 PMID:33867097

Heng, K., & Sol, K. (2021). Online learning during COVID-19: Key challenges and suggestions to enhance effectiveness. *Cambodian Journal of Educational Research*, *1*(1), 3-16. Retrieved from https://www.academia.edu/51722743/The_Roles_of_Parental_to_Promote_Inclusive_Education_During_COVID_19?bulkDownload=thisPaper-topRelated-sameAuthor-citingThis-citedByThis-secondOrderCitations&from=cover_page

Hernandez, K., & Pasquesi, K. (2017). Critical perspectives and commitments deserving attention from community engagement professionals. In L. D. Dostilio (Ed.), *The community engagement professional in higher education: A competency model for an emerging field* (pp. 56–78). Compact Campus.

Herrera-Franco, G., Carrión-Mero, P., Alvarado, N., Morante-Carballo, F., Maldonado, A., Caldevilla, P., Briones-Bitar, J., & Berrezueta, E. (2020). Geosites and Georesources to Foster Geotourism in Communities: Case Study of the Santa Elena Peninsula Geopark Project in Ecuador. *Sustainability (Basel, Switzerland)*, *12*(11), 4484. doi:10.3390u12114484

Herring, S. (1999). Interactional coherence in CMC. *Journal of Computer-Mediated Communication*, *4*(4), 0. Advance online publication. doi:10.1111/j.1083-6101.1999.tb00106.x

He, Z., Chiu, D. K. W., & Ho, K. K. W. (2022). Weibo Analysis on Chinese Cultural Knowledge for Gaming. In Z. Sun (Ed.), *Handbook of Research on Foundations and Applications of Intelligent Business Analytics* (pp. 320–349). doi:10.4018/978-1-7998-9016-4.ch015

Hobart & William Smith Colleges. (2014, May 29). America Reads 25th Anniversary [video]. *YouTube*. Retrieved 4 June 2022 from https://www.youtube.com/watch?v=fgvRVDMzilo&t=4s

Hobart & William Smith Colleges. (2019, May 18). The legacy of America Reads at HWS. *HWS Update*. Hobart & William Smith Colleges. Retrieved 4 June 2022 from https://www2.hws.edu/the-legacy-of-america-reads-at-HWS-2/

Hobart & William Smith Colleges. (2020a, April 17). HWS virtual tutoring for Geneva students. *HWS Update*. Hobart & William Smith Colleges. Retrieved 5 June 2022 from https://www2.hws.edu/hws-virtual-tutoring-for-Geneva-students/

Hobart & William Smith Colleges. (2020b, February 7). *HWS earns Carnegie Community Engagement Designation*. Retrieved 10 June 2022 from https://www2.HWS.edu/HWS-earns-carnegie-community-engagement-designation/

Hobart & William Smith Colleges. (2021a, January 15). Service-learning and social science in action. *HWS Update*. Hobart & William Smith Colleges. Retrieved 6 June 2022 from https://www2.HWS.edu/service-learning-and-social-science-in-action/

Hobart & William Smith Colleges. (2021b, September 9). HWS ranked no. 3 in the nation for service. *HWS Update*. Hobart & William Smith Colleges. Retrieved 10 June 2022 from https://www2.HWS.edu/HWS-ranked-no-3-in-nation-for-service/

Hobart & William Smith Colleges. (2022, May 9). The grassroots initiative that grew into America Reads. *HWS Update*. Hobart & William Smith Colleges. Retrieved 4 June 2022 from https://www2.HWS.edu/the-grassroots-initiative-that-grew-into-america-reads/

Hofer, M., & Harris, J. (2012, March). TPACK research with inservice teachers: Where's the TCK? In *Society for Information Technology & Teacher Education International Conference* (pp. 4704-4709). Association for the Advancement of Computing in Education.

Hoffman, D., & Fodor, M. (2010). Can you measure the ROI of your social media marketing? *MIT Sloan Management Review*, *52*(1), 55–61.

Hofstede, G. (1980). Culture and organizations. *International Studies of Management & Organization*, *10*(4), 15–41. doi:10.1080/00208825.1980.11656300

Ho, H. C., & Chan, Y. C. (2022). Flourishing in the workplace: A one-year prospective study on the effects of perceived organizational support and psychological capital. *International Journal of Environmental Research and Public Health*, *19*(2), 922. doi:10.3390/ijerph19020922 PMID:35055747

Ho, K. K. W., Chan, J. Y., & Chiu, D. K. W. (2022). Fake News and Misinformation During the Pandemic: What We Know, and What We Don't Know. *IT Professional*, *24*(2), 19–24. doi:10.1109/MITP.2022.3142814

Holzer, J., Ellis, L., & Merritt, M. (2014). Why we need community engagement in medical research. *Journal of Investigative Medicine*, *62*(6), 851–855. doi:10.1097/JIM.0000000000000097 PMID:24979468

Honig, C. A., & Salmon, D. (2021). Learner presence matters: A learner-centered exploration into the community of Inquiry Framework. *Online Learning*, *25*(2). Advance online publication. doi:10.24059/olj.v25i2.2237

Hortulanus, R., Machielse, A., & Meeuwesen, L. (2009). *Social isolation in modern society*. Routledge.

Hove, P., & Grobbelaar, S. S. (2020). Innovation for inclusive development: Mapping and auditing the use of ICTs in the South African primary education system. *South African Journal of Industrial Engineering*, *31*(1), 47–64. doi:10.7166/31-1-2119

Huang, K. Y., Kwon, S. C., Cheng, S., Kamboukos, D., Shelley, D., Brotman, L. M., Kaplan, S. A., Olugbenga, O., & Hoagwood, K. (2018). Unpacking partnership, engagement, and collaboration research to inform implementation strategies development: Theoretical frameworks and emerging methodologies. *Frontiers in Public Health*, *6*, 190. doi:10.3389/fpubh.2018.00190 PMID:30050895

Huang, P. S., Paulino, Y., So, S., Chiu, D. K. W., & Ho, K. K. W. (2021). Special Issue Editorial - COVID-19 Pandemic and Health Informatics (Part 1). *Library Hi Tech*, *39*(3), 693–695. doi:10.1108/LHT-09-2021-324

Huang, P.-S., Paulino, Y. C., So, S., Chiu, D. K. W., & Ho, K. K. W. (2022). Guest editorial: COVID-19 Pandemic and Health Informatics Part 2. *Library Hi Tech*, *40*(2), 281–285. doi:10.1108/LHT-04-2022-447

Hudson, E., Hamlin, E., & Cummings, J. (2020, September 11). *How to cultivate belonging in online learning*. GOA. Retrieved May 27, 2022, from https://globalonlineacademy.org/insights/articles/how-to-cultivate-belonging-in-online-learning

Hudson, S., Wang, Y., & Gil, S. M. (2011). The influence of a film on destination image and the desire to travel: A cross-cultural comparison. *International Journal of Tourism Research*, *13*(2), 177–190.

Huertas-Abril, C., & García-Molina, M. (2022). Spanish Teacher Attitudes Towards Digital Game-Based Learning: An Exploratory Study Based on the TPACK Model. In Handbook of Research on Acquiring 21st Century Literacy Skills Through Game-Based Learning (pp. 554-578). IGI Global.

Hughes, E. M., Brooker, H., Gambrell, L. B., & Foster, V. (2011). Tutoring and mentoring: The results of an America Reads program on struggling readers' motivation and achievement. *Literacy promises: Association of Literacy Educators and Researchers Yearbook, 33*, 205-218. Retrieved 6 June 2022 from https://files.eric.ed.gov/fulltext/ED527009.pdf#page=217

Husky, M. M., Kovess-Masfety, V., & Swendsen, J. D. (2020). Stress and anxiety among university students in France during Covid-19 mandatory confinement. *Comprehensive Psychiatry, 102*, 152191. doi:10.1016/j.comppsych.2020.152191 PMID:32688023

Hutson, N., York, T., Kim, D., Fiester, H., & Workman, J. L. (n.d.). Institutionalizing community engagement: A quantitative approach to identifying patterns of engagement based on institutional characteristics. *Journal of Community Engagement and Higher Education, 11*(2), 3-15.

Ilgaz, H., & Aşkar, P. (2013). The contribution of technology acceptance and community feeling to learner satisfaction in distance education. *Procedia: Social and Behavioral Sciences, 106*, 2671–2680. doi:10.1016/j.sbspro.2013.12.308

Inglehart, R. (2000). Culture and democracy. *Culture matters: How values shape human progress*, 80-97. Retrieved from https://www.academia.edu/download/55039033/Culture_Matters_How_Values_Shape_Human_Progress_-_SAMUEL_HUNTINGTON.pdf#page=115

Inglehart, R., & Baker, W. E. (2000). Modernization, cultural change, and the persistence of traditional values. *American Sociological Review, 65*(1), 19–51. doi:10.2307/2657288

International Technology and Engineering Educators Association. (2020). *Standards for technological and engineering literacy: Defining the role of technology and engineering in STEM education*. Author.

Jacobs, S., Mishra, C. E., Doherty, E., Nelson, J., Duncan, E., Fraser, E. D., Hodgins, K., Mactaggart, W., & Gillis, D. (2021). Transdisciplinary, community-engaged pedagogy for undergraduate and graduate student engagement in challenging times. *International Journal of Higher Education, 10*(7), 84–95. doi:10.5430/ijhe.v10n7p84

Jacoby, B. (2015). *Service-learning essentials: Questions, answers, and lesson learned.* John Wiley & Sons, Inc.

Jacoby, D. (2006). Effects of part-time faculty employment on community college graduation rates. *The Journal of Higher Education, 77*(6), 1081–1103. doi:10.1353/jhe.2006.0050

Jaeger, A. J., & Eagan, M. K. (2011). Examining retention and contingent faculty use in a state system of public higher education. *Educational Policy, 25*(3), 507–537. doi:10.1177/0895904810361723

Jameson, C. M., Torres, K., & Mohammed, S. (2021). Virtual faculty strategies for supporting motivation of online doctoral students. *Journal of Educational Research and Practice, 11*(1), 295–305. doi:10.5590/JERAP.2021.11.1.21

Jameson, C., & Torres, K. (2019). Fostering motivation when virtually mentoring online doctoral students. *Journal of Educational Research and Practice, 9*(1), 331–339. doi:10.5590/JERAP.2019.09.1.23

Jang, S., & Tsai, M. (2013). Exploring the TPACK of Taiwanese secondary school science teachers using a new contextualized TPACK model. *Australasian Journal of Educational Technology, 29*(4). Advance online publication. doi:10.14742/ajet.282

Janke, E. M., & Domagal-Goldman, J. M. (2017). Institutional characteristics and students civic outcomes. In J. A. Hatcher, R. G. Bringle, & T. W. Hahn (Eds.), *Research on student civic outcomes in service learning: Conceptual frameworks and methods*. Stylus Publishing, Inc.

Jan, S. K., & Vlachopoulos, P. (2018). Influence of learning design of the formation of online communities of learning. *The International Review of Research in Open and Distributed Learning*, *19*(4). Advance online publication. doi:10.19173/irrodl.v19i4.3620

Jansen van Rensburg, S.K. (2021). Doing gender well: Women's perceptions on gender equality and career progression in the South African security industry. *SA Journal of Industrial Psychology/SA Tydskrif vir Bedryfsielkunde, 47*(0), a1815. doi:10.4102/sajip.v47i0.181

Järvelä, S., Järvenoja, H., & Veermans, M. (2008). Understanding the dynamics of motivation in socially shared learning. *International Journal of Educational Research*, *47*(2), 122–135. doi:10.1016/j.ijer.2007.11.012

Jee, Y. (2020). WHO international health regulations emergency committee for the COVID-19 outbreak. *Epidemiology and Health*, *42*, 42. doi:10.4178/epih.e2020013 PMID:32192278

Jeong, A. (2004). The combined effects of response time and message content on growth patterns of discussion threads in computer supported collaborative argumentation. *Journal of Distance Education*, *19*(1), 36–53.

Jiang, T., Lo, P., Cheuk, M. K., Chiu, D. K. W., Chu, M. Y., Zhang, X., Zhou, Q., Liu, Q., Tang, J., Zhang, X., Sun, X., Ye, Z., Yang, M., & Lam, S. K. (2019). 文化新語:兩岸四地傑出圖書館、檔案館及博物館傑出工作者訪談 [New Cultural Dialog: Interviews with Outstanding Librarians, Archivists, and Curators in Greater China]. Hong Kong: Systech publications.

Jiang, X., Chiu, D. K. W., & Chan, C. T. (2022). Application of the AIDA model in social media promotion and community engagement for small cultural organizations: A case study of the Choi Chang Sau Qin Society. In M. Dennis & J. Halbert (Eds.), *Community Engagement in the Online Space*. IGI Global.

Jinhong, J., & Gilson, T. A. (2014). Online threaded discussion: Benefits, issues, and strategies. *Kinesiology Review (Champaign, Ill.)*, *3*(4), 241–246. doi:10.1123/kr.2014-0062

Jobber, D., & Ellis-Chadwick, F. (2012). *Principles and practice of marketing*. McGraw-Hill Europe.

Johns Hopkins Engineering for Professionals. (2022a). *Applied Biomedical Engineering Master's Program Online*. Johns Hopkins Engineering Online. https://ep.jhu.edu/programs/applied-biomedical-engineering

Johns Hopkins Engineering for Professionals. (2022b). *Executive Technical Leadership Online*. Johns Hopkins Engineering Online. https://ep.jhu.edu/programs/engineering-management

Johns, C., & Mills, M. (2021). Online mathematics tutoring during the COVID-19 pandemic: Recommendations for best practices. *PRIMUS (Terre Haute, Ind.)*, *31*(1), 99–117. doi:10.1080/10511970.2020.1818336

Jones, A., & Issroff, K. (2005). Learning technologies: Affective and social issues in computer-supported collaborative learning. *Computers & Education*, *44*(4), 395–408. doi:10.1016/j.compedu.2004.04.004

Joseph, C. (2019, February 5). *A.I.D.A model in marketing communication*. Small Business - Chron.com. https://smallbusiness.chron.com/aida-model-marketing-communication-10863.html

Kahneman, D. (1973). *Attention and effort* (Vol. 1063). Prentice-Hall.

Kaleli, Y. (2021). The Effect of Individualized Online Instruction on TPACK Skills and Achievement in Piano Lessons. *International Journal of Technology in Education*, *4*(3), 399–412. doi:10.46328/ijte.143

Kang, M., Shin, D., & Gong, T. (2016). The role of personalization, engagement, and trust in online communities. *Information Technology & People*, *29*(3), 580–596. doi:10.1108/ITP-01-2015-0023

Kaplan, A. M., & Haenlein, M. (2010). Users of the world, unite! The challenges and opportunities of Social Media. *Business Horizons*, *53*(1), 59–68. doi:10.1016/j.bushor.2009.09.003

Karamunya, J., & Cheben, P. (2016). Socio-cultural factors influencing community participation in community projects among the residents Inpokot South Sub-County, Kenya. *American Based Research Journal, 5*(11).

Kassop, M. (2003). Ten ways online education matches, or surpasses, face-to-face learning. *The Technology Source, 3*.

Kay, R. H. (2006). Developing a comprehensive metric for assessing discussion board effectiveness. *British Journal of Educational Technology*, *37*(5), 761–783. doi:10.1111/j.1467-8535.2006.00560.x

Kebble, P. G. (2017). Assessing online asynchronous communication strategies designed to enhance large student cohort engagement and foster a community of learning. *Journal of Education and Training Studies*, *5*(8), 92. doi:10.11114/jets.v5i8.2539

Kehrwald, B. (2008). Understanding social presence in text-based online learning environments. *Distance Education*, *29*(1), 89–106. doi:10.1080/01587910802004860

Kemmis, S., McTaggart, R., & Nixon, R. (2014). *The action research planner: Doing critical participatory action research*. Springer Science. doi:10.1007/978-981-4560-67-2

Kemp, S. (2020). *Digital 2020: Hong Kong*. Available at: https://datareportal.com/reports/digital-2020-hong-kong

Kemp, S. (2020). *Digital 2020: July Global Statshot*. Available at: https://datareportal.com / reports /digital-2020

Kent, M. L., Taylor, M., & White, W. J. (2003). The relationship between Web site design and organizational responsiveness to stakeholders. *Public Relations Review*, *29*(1), 63–77. doi:10.1016/S0363-8111(02)00194-7

Kerr, N. A. (2021). Teaching in a lonely world: Educating students about the nature of loneliness and promoting social connection in the classroom. *Teaching of Psychology*, 1–11. doi:10.1177/00986283211043787

Kerr, N. A., & Stanley, T. B. (2021). Revisiting the social stigma of loneliness. *Personality and Individual Differences*, *171*, 110482. doi:10.1016/j.paid.2020.110482

Keshtgar, A., Hania, M., & Sharif, M. O. (2022). Consent and parental responsibility: The past, the present and the future. *British Dental Journal*, *232*(2), 115–119. doi:10.103841415-022-3877-7 PMID:35091615

Keyes, C. L. (2005). Mental illness and/or mental health? Investigating axioms of the complete state model of health. *Journal of Consulting and Clinical Psychology*, *73*(3), 539–548. doi:10.1037/0022-006X.73.3.539 PMID:15982151

Keyes, C. L. M. (2002). The mental health continuum: From languishing to flourishing in life. *Journal of Health and Social Behavior*, *43*(2), 207–222. doi:10.2307/3090197 PMID:12096700

Khumalo, N., Dumont, K. B., & Waldzus, S. (2022). Leaders' influence on collective action: An identity leadership perspective. *The Leadership Quarterly*, *33*(4), 101609–101621. doi:10.1016/j.leaqua.2022.101609

Kiang, M. V., & Tsai, A. C. (2020). Statements issued by academic medical institutions after George Floyd's killing by police and subsequent unrest in the United States: Cross-sectional study. MedRxiv. doi:10.1101/2020.06.22.20137844

Kietzmann, J. H., Hermkens, K., McCarthy, I., & Silvestre, B. (2011). Social media? Get serious! Understanding the functional building blocks of social media. *Business Horizons*, *54*(3), 241–251. doi:10.1016/j.bushor.2011.01.005

Kilis, S., & Yıldırım, Z. (2019). Posting patterns of students' social presence, cognitive presence, and teaching presence in online learning. *Online Learning*, *23*(2). Advance online publication. doi:10.24059/olj.v23i2.1460

Kim, T. L., Wah, W. K., & Lee, C, T. A. (2007). Asynchronous electronic discussion group: Analysis of postings and perception of inservice teachers. *Turkish Online Journal of Distance Education*, *8*(1), 33–41.

Kim, U., Triandis, H. C., Kâğitçibaşi, Ç., Choi, S.-C., & Yoon, G. (Eds.). (1994). *Individualism and collectivism: Theory, method, and applications*. Sage Publications, Inc.

King, M. L. Jr. (1972). A New Sense of Direction. *Worldview*, *15*(4), 5–12. doi:10.1017/S0084255900014893

King, R. A., & Shelley, C. A. (2008). Community feeling and social interest: Adlerian parallels, synergy and differences with the field of community psychology. *Journal of Community & Applied Social Psychology*, *18*(2), 96–107. doi:10.1002/casp.962

Klerk, N. O.-d., & Muir, C. (2022). Corporate Brand Reputation and Ethic, Sustainability and Inclusion. The Shift in Post Pandemic Corporate Narrative: From Corporate Brand Reputation to Corporate Sustainability. In The Emerald Handbook of Multi-Stakeholder Communication. Emerald Publishing Limited. doi:10.1108/978-1-80071-897-520221031

Knowles, M. S. (1984). *The adult learner: A neglected species* (3rd ed.). Gulf Publishing Co.

Knudson, D. (2020). A tale of two instructional experiences: Student engagement in active learning and emergency remote learning of biomechanics. *Sports Biomechanics*, 1–11. doi:10.1080/14763141.2020.1810306 PMID:32924795

Koehler, M., & Mishra, P. (2008). Introducing TPCK. In A. C. Technology (Ed.), *Handbook of technological pedagogical content knowledge (TPCK) for educators* (pp. 3–29). Routledge.

Koehler, M., & Mishra, P. (2009). What is technological pedagogical content knowledge (TPACK)? *Contemporary Issues in Technology & Teacher Education*, *9*(1), 60–70.

Koehler, M., Mishra, P., & Cain, W. (2013). What is technological pedagogical content knowledge (TPACK)? *Journal of Education*, *193*(3), 13–19. doi:10.1177/002205741319300303

Koehler, M., Shin, T., & Mishra, P. (2012). How do we measure TPACK? Let me count the ways. In *Educational technology, teacher knowledge, and classroom impact: A research handbook on frameworks and approaches* (pp. 16–31). IGI Global. doi:10.4018/978-1-60960-750-0.ch002

Koh, J., & Chai, C. (2014). Teacher clusters and their perceptions of technological pedagogical content knowledge (TPACK) development through ICT lesson design. *Computers & Education*, *70*, 222–232. doi:10.1016/j.compedu.2013.08.017

Koh, J., Chai, C., & Tay, L. (2014). TPACK-in-Action: Unpacking the contextual influences of teachers' construction of technological pedagogical content knowledge (TPACK). *Computers & Education*, *78*, 20–29. doi:10.1016/j.compedu.2014.04.022

Koh, J., Chai, C., & Tsai, C. (2014). Demographic factors, TPACK constructs, and teachers' perceptions of constructivist-oriented TPACK. *Journal of Educational Technology & Society*, *17*(1), 185–196.

Korkmaz, A. V., van Engen, M. L., Knappert, L., & Schalk, R. (2022). About and beyond leading uniqueness and belongingness: A systematic review of inclusive leadership research. *Human Resource Management Review*, *100894*(4). Advance online publication. doi:10.1016/j.hrmr.2022.100894

Kotler, P., Rackham, N., & Krishnaswamy, S. (2006). Ending the war between sales and marketing. *Harvard Business Review*, *84*(7/8), 68. PMID:16846190

Kovacs, B., Caplan, N., Grob, S., & King, M. (2021). Social networks and loneliness during the COVID-19 pandemic. *Socius: Sociological Research for a Dynamic World*, *7*, 1–16. doi:10.1177/2378023120985254

Krathwohl, D. R. (2002). A revision of Bloom's taxonomy: An overview. *Theory into Practice*, *41*(4), 212–218. doi:10.120715430421tip4104_2

Krings, V. C., Steeden, B., Abrams, D., & Hogg, M. A. (2021). Social attitudes and behavior in the COIVD-19 pandemic: Evidence and prospects from research on group processes and intergroup relations. *Group Processes & Intergroup Relations*, *24*(4), 195–200. doi:10.1177/1368430220986673

Kroeger, F., Racko, G., & Burchell, B. (2021). How to create trust quickly: A comparative empirical investigation of the bases of swift trust. *Cambridge Journal of Economics*, *45*(1), 129–150. doi:10.1093/cje/beaa041

Kshetri, N. (2020). China's emergence as the global fintech capital and implications for southeast asia. *Asia Policy*, *27*(1), 61–81. doi:10.1353/asp.2020.0004

Kuem, J., Khansa, L., & Kim, S. S. (2020). Prominence and engagement: Different mechanisms regulating continuance and contribution in online communities. *Journal of Management Information Systems*, *37*(1), 162–190. doi:10.1080/07421222.2019.1705510

Kuh, G. D. (2003). What we're learning about student engagement from NSSE: Benchmarks for effective educational practices. *Change: The Magazine of Higher Learning*, *35*(2), 24-32.

Kuknor, S., & Bhattacharya, S. (2021). Organizational inclusion and leadership in times of global crisis. *Australasian Accounting. Business and Finance Journal*, *15*(1), 93–112. doi:10.14453/aabfj.v15i1.7

Kumar, S., Kumar, V., & Taylor, S. (2020). *A guide to online supervision*. UK Council for Graduate Education. https://ukcge.ac.uk/assets/resources/A-Guide-to-Online-Supervision-Kumar-Kumar-Taylor-UK-Council-for-Graduate-Education.pdf

Kumar, S., & Coe, C. (2017). Mentoring and student support in online doctoral programs. *American Journal of Distance Education*, *31*(2), 128–142. doi:10.1080/08923647.2017.1300464

Kumar, S., & Johnson, M. (2017). Mentoring doctoral student online: Mentor strategies and challenges. *Mentoring & Tutoring*, *25*(2), 202–222. doi:10.1080/13611267.2017.1326693

Kumar, S., & Johnson, M. (2019). Online mentoring of dissertations: The role of structure and support. *Studies in Higher Education*, *44*(1), 59–71. doi:10.1080/03075079.2017.1337736

Kumar, S., Johnson, M. L., Dogan, N., & Coe, C. (2019). A framework for e-mentoring in doctoral education. In K. Sim (Ed.), *Enhancing the role of ICT in doctoral research processes* (pp. 183–208). IGI Global. doi:10.4018/978-1-5225-7065-3.ch009

Kupczynski, L., Mundy, M.-A., & Maxwell, G. (2012). Faculty perceptions of cooperative learning and traditional discussion strategies in online courses. *Turkish Online Journal of Distance Education*, *13*(2), 84–95.

Lam, A. H. C., Chiu, D. K. W., & Ho, K. K. W. (2022). *Instagram for student learning and library promotions? A quantitative study using the 5E Instructional Model*. Aslib Journal of Information Management.

Lam, A. H. C., Chiu, D. K. W., & Ho, K. K. W. (2022). Instagram for student learning and library promotions? A quantitative study using the 5E Instructional Model. *Aslib Journal of Information Management*. Advance online publication. doi:10.1108/AJIM-12-2021-0389

Lamb, C. W., Hair, J. F., & McDaniel, C. (2016). MKTG 10 (10th ed.). Cengage Learning.

Lamb, C. W., Hair, J. F., & McDaniel, C. (2012). *Marketing* (12th ed.). Cengage Learning.

Lam, E. T. H., Au, C. H., & Chiu, D. K. W. (2019). Analyzing the use of Facebook among university libraries in Hong Kong. *Journal of Academic Librarianship*, *45*(3), 175–183. doi:10.1016/j.acalib.2019.02.007

Lasater, K., Young, P. K., Mitchell, C. G., Delahoyde, T. M., Nick, J. M., & Siktberg, L. (2014). Connecting in distance mentoring: Communication practices that work. *Nurse Education Today*, *34*(4), 501–506. doi:10.1016/j.nedt.2013.07.009 PMID:23978777

Lau, K. S. N., Lo, P., Chiu, D. K. W., Ho, K. K. W., Jiang, T., Zhou, Q., Percy, P., & Allard, B. (2020). Library, Learning, and Recreational Experiences Turned Mobile: A Comparative Study between LIS and non-LIS students. *Journal of Academic Librarianship*, *46*(2), 102103. doi:10.1016/j.acalib.2019.102103

Lave, J. (1991). Situating learning in communities of practice. In L. B. Resnick, J. M. Levine, & S. D. Teasley (Eds.), *Perspectives on socially shared cognition* (pp. 63–82). American Psychological Association. doi:10.1037/10096-003

Lavidas, K., Katsidima, M., Theodoratou, S., Komis, V., & Nikolopoulou, K. (2021). Preschool teachers' perceptions about TPACK in Greek educational context. *Journal of Computers in Education*, *8*(3), 395–410. doi:10.100740692-021-00184-x

Lean Enterprise Institute. (n.d.). *What is lean?* https://www.lean.org/explore-lean/what-is-lean/

Leander, M. (2014). *What Is a Good Facebook Engagement Rate on a Facebook? Here Is a Benchmark for You.* Michael Leander Company. Available at https://www.michaelleander.me/blog/facebook-engagement-rate-benchmark

Leary, M. R., & Gabriel, S. (2021). The relentless pursuit of acceptance and belonging. *Elsevier.*

Lee, K. (2007). Online collaborative case study learning. *Journal of College Reading and Learning*, *37*(2), 82–100. doi:10.1080/10790195.2007.10850199

Lee, S. H., & Hoffman, K. D. (2015). Learning the ShamWow: Creating infomercials to teach the AIDA model. *Marketing Education Review*, *25*(1), 9–14. doi:10.1080/10528008.2015.999586

Leijerholt, U., Biedenbach, G., & Hultén, P. (2019). Branding in the public sector: A systematic literature review and directions for future research. *Journal of Brand Management*, *26*(2), 126–140. doi:10.105741262-018-0116-2

Lei, S. Y., Chiu, D. K. W., Lung, M. M., & Chan, C. T. (2021). Exploring the Aids of Social Media for Musical Instrument Education. *International Journal of Music Education*, *39*(2), 187–201. doi:10.1177/0255761420986217

Leslie, J. B., Luciano, M. M., Mathieu, J. E., & Hoole, E. (2018). Challenge accepted: Managing polarities to enhance virtual team effectiveness. *People & Strategy*, *41*(2), 22–28.

Letven, E., Ostheimer, J., & Statham, A. (2001). *Institutionalizing university-community engagement.* Retrieved February 22, 2022, from https://journals.iupui.edu/index.php/muj/article/download/19907/19601/27754

Leung, T. N., Hui, Y. M., Luk, C. K. L., Chiu, D. K. W., & Kevin, K. K. W. (2022). An Empirical Study on the Aid of Facebook for Japanese Learning. *Library Hi Tech.*

Leung, T. N., Luk, C. K. L., Chiu, D. K. W., & Kevin, K. K. W. (2022). User perceptions, academic library usage, and social capital: A correlation analysis under COVID-19 after library renovation. *Library Hi Tech*, *40*(2), 304–322. doi:10.1108/LHT-04-2021-0122

Lewis, E. S. E. (1909). The Duty and Privilege of Advertisng a Bank. *Bankers' Magazine, 78*(4), 710-711.

Li, J., & Yu, H. (2013). An Innovative Marketing Model Based on AIDA:-A Case from E-bank Campus-marketing by China Construction Bank. *I-Business*, *5*(03, 3B), 47–51. doi:10.4236/ib.2013.53B010

Lin, C.-H., Chiu, D. K. W., & Lam, K. T. (2022). Hong Kong Academic Librarians'. *Attitudes Towards Robotic Process Automation*. Advance online publication. doi:10.1108/LHT-03-2022-0141

Linders, B. (2018, May 11). *Psychological safety in training games*. InfoQ. https://infoq.com/articles/psychological-safety-games

Lin, Y. S., & Huang, J. Y. (2006). Internet blogs as a tourism marketing medium: A case study. *Journal of Business Research*, *59*(10-11), 1201–1205. doi:10.1016/j.jbusres.2005.11.005 PMID:32287521

Lioubimtseva, E., & Cunha, C. (2022). Community engagement and equity in climate adaptation planning: experience of small-and mid-sized cities in the United States and in France. In *Justice in climate action planning* (pp. 257–276). Springer. doi:10.1007/978-3-030-73939-3_13

Litchman, M. L., Edelman, L. S., & Donaldson, G. W. (2018). Effect of diabetes online community engagement on health indicators: Cross-sectional study. *JMIR Diabetes*, *3*(2), e8603. doi:10.2196/diabetes.8603 PMID:30291079

Liu, H., Zhang, M., Yang, Q., & Yu, B. (2020). Gender differences in the influence of social isolation and loneliness on depressive symptoms in college students: A longitudinal study. *Social Psychiatry and Psychiatric Epidemiology*, *55*(2), 251–257. doi:10.100700127-019-01726-6 PMID:31115597

Liu, Z.-Y., Lomovtseva, N., & Korobeynikova, E. (2020). Online Learning Platforms: Reconstructing Modern Higher Education. *International Journal of Emerging Technologies in Learning*, *15*(13), 4–21. doi:10.3991/ijet.v15i13.14645

Lo, P., Cheuk, M. K., Ng, C. H., Lam, S. K., & Chiu, D. K. W. (2021). 文武之道【下冊】:以拳入哲 [The Tao of Arts and Warriorship: The Philosophy of Fists]. Systech Publications.

Loades, M. E., Chatburn, E., Higson-Sweeney, N., Reynolds, S., Shafran, R., Brigden, A., Linney, C., McManus, M. N., Borwick, C., & Crawley, E. (2020). Rapid systematic review: The impact of social isolation and loneliness on the mental health of children and adolescents in the context of COVID-19. *Journal of the American Academy of Child and Adolescent Psychiatry*, *59*(11), 1218–1239. doi:10.1016/j.jaac.2020.05.009 PMID:32504808

Lobo, I., & Vélez, M. (2022). From strong leadership to active community engagement: Effective resistance to illegal coca crops in Afro-Colombian collective territories. *The International Journal on Drug Policy*, *102*, 103579. Advance online publication. doi:10.1016/j.drugpo.2022.103579 PMID:35121354

London, M. (2022). Causes and consequences of adaptive leadership: A model of leaders' rapid responses to unexpected events. *Psychology of Leaders and Leadership*. Advance online publication. doi:10.1037/mgr0000136

Lo, P., Chan, H. H. Y., Tang, A. W. M., Chiu, D. K. W., Cho, A., Ho, K. K. W., See-To, E., He, J., Kenderdine, S., & Shaw, J. (2019). Visualising and Revitalising Traditional Chinese Martial Arts – Visitors' Engagement and Learning Experience at the 300 Years of Hakka KungFu. *Library Hi Tech*, *37*(2), 273–292. doi:10.1108/LHT-05-2018-0071

Lo, P., Hsu, W.-E., Wu, S. H. S., Travis, J., & Chiu, D. K. W. (2021). *Creating a Global Cultural City via Public Participation in the Arts: Conversations with Hong Kong's Leading Arts and Cultural Administrators*. Nova Science Publishers.

Lorenzo, G. (2001). Learning Anytime, Anywhere... for Everybody? Making Online Learning Accessible. *Distance Education Report*, *5*(11), 1–3.

Lovejoy, K., & Saxton, G. D. (2012). Information, community, and action: How nonprofit organizations use social media. *Journal of Computer-Mediated Communication*, *17*(3), 337–353. doi:10.1111/j.1083-6101.2012.01576.x

Luchetti, M., Hyun Lee, J., Aschwanden, D., Esker, A., Strickhouser, J. E., Terracciano, A., & Sutin, A. R. (2020). *The trajectory of loneliness in response to COVID-19*. American Psychological Association. https://www.apa.org/pubs/journals/releases/amp-amp0000690.pdf

Lyublinskaya, I., & Kaplon-Schilis, A. (2022). Analysis of Differences in the Levels of TPACK: Unpacking Performance Indicators in the TPACK Levels Rubric. *Education Sciences*, *12*(2), 79. doi:10.3390/educsci12020079

Machimana, E. G., Sefotho, M. M., Ebersöhn, L., & Shultz, L. (2021). Higher education uses community engagement-partnership as a research space to build knowledge. *Educational Research for Policy and Practice*, *20*(1), 45–62. doi:10.100710671-020-09266-6

Macias, W., Hilyard, K., & Freimuth, V. (2009). Blog functions as risk and crisis communication during Hurricane Katrina. *Journal of Computer-Mediated Communication*, *15*(1), 1–31. doi:10.1111/j.1083-6101.2009.01490.x

Magalhães, C. M., Machado, C. F., & Nunes, C. P. (2022). Loneliness in Leadership: A Study Applied to the Portuguese Banking Sector. *Administrative Sciences*, *12*(4), 130. doi:10.3390/admsci12040130

Mager, R. F. (1997). *Preparing Instructional Objectives*. Center for Effective Performance.

Magnuson, C. (2005). Experiential learning and the discussion board: A strategy, a rubric, and management techniques. *Distance Learning*, *2*(2), 15–20.

Mahlangu, V. P. (2020). *Understanding Toxic Leadership in Higher Education Work Places through Betrayal Trauma Theory*. Bulgarian Comparative Education Society.

Mahlomaholo, S. (2009). Critical emancipatory research and academic identity. *Africa Education Review*, *6*(2), 224–237. doi:10.1080/18146620903274555

Mahlomaholo, S., & Nkoane, M. (2002). The case for emancipatory qualitative research on assessment of quality. *Education as Change*, *6*(1), 69–84.

Maipita, I., Dongoran, F., Syah, D., & Hafiz, G. (2022). TPACK Knowledge Mastery of Pre-Service Teacher Students in the Faculty of Economics Universitas Negeri Medan. *The Age (Melbourne, Vic.)*, *18*(27), 6–43. doi:10.2991/aebmr.k.220104.018

Mak, M. Y. C., Poon, A. Y. M., & Chiu, D. K. W. (2022). Using Social Media as Learning Aids and Preservation: Chinese Martial Arts in Hong Kong. In S. Papadakis & A. Kapaniaris (Eds.), *The Digital Folklore of Cyberculture and Digital Humanities* (pp. 171–185). IGI Global. doi:10.4018/978-1-6684-4461-0.ch010

Mali, D., & Lim, H. (2021). How do students perceive face-to-face/blended learning as a result of the Covid-19 pandemic? *International Journal of Management Education*, *19*(3), 100552. https://doi.org/10.1016/j.ijme.2021.100552

Mangold, W. G., & Faulds, D. J. (2009). Social media: The new hybrid element of the promotion mix. *Business Horizons*, *52*(4), 357–365. doi:10.1016/j.bushor.2009.03.002

Manok, G. (2018). Key lessons and guiding questions. In J. Saltmarsh & M. B. Johnson (Eds.), *The elective Carnegie community engagement classification: Constructing a successful application for first-time and re-classification applicants* (pp. 50–54). Campus Compact.

Maor, D. (2017). Using TPACK to develop digital pedagogues: A higher education experience. *Journal of Computers in Education, 4*(1), 71–86. doi:10.100740692-016-0055-4

Marinoni, G., van't Land, H., & Jensen, T. (2020). *The impact of Covid-19 on higher education around the world.* International Association of Universities. https://www.iau-aiu.net/IMG/pdf/iau_covid19_and_he_survey_report_final_may_2020.pdf

Markus, H. R., & Kitayama, S. (1991). Culture and the self: Implications for cognition, emotion, and motivation. *Psychological Review, 98*(2), 224–253. doi:10.1037/0033-295X.98.2.224

Markus, H. R., Mullally, P. R., & Kitayama, S. (1997). *Selfways: Diversity in modes of cultural participation.* The Conceptual Self. In *Context: Culture, Experience, Self-Understanding.* University Cambridge Press.

Marzano, R. J. (Ed.). (2009). *On excellence in teaching.* Solution Tree Press.

Maslow, A. H. (1968). *Toward a psychology of being.* Van Nostrand.

Masters, K., & Oberprieler, G. (2004). Encouraging equitable online participation through curriculum articulation. *Computers & Education, 42*(4), 319–332. doi:10.1016/j.compedu.2003.09.001

Matters, Q. (2011). *The Quality Matters higher education rubric.* https://www.qualitymatters.org

Matters, Q. (2013). *Rubric and standards.* https://www.qualitymatters.org/rubric

McBrien, J. L., Cheng, R., & Jones, P. (2009). Virtual spaces: Employing a synchronous online classroom to facilitate student engagement in online learning. *International Review of Research in Open and Distributed Learning, 10*(3).

McDougall, J. (2015). The quest for authenticity: A study of an online discussion forum and the needs of adult learners. *Australian Journal of Adult Learning, 55*(1), 94–113.

McElhone, R. (2020, January 25). *Using the AIDA model to get 'buy in'.* B Online Learning. https://bonlinelearning.com/using-the-aida-model-to-get-buy-in-elearning/

McGinty, A. S., Justice, L., & Rimm-Kaufman, S. E. (2008). Sense of school community for preschool teachers serving at-risk children. *Early Education and Development, 19*(2), 361–384. doi:10.1080/10409280801964036

McGinty, E. E., Presskreischer, R., Anderson, K. E., Han, H., & Barry, C. L. (2020). Psychological distress and COVID-19–related stressors reported in a longitudinal cohort of U.S. adults in April and July 2020. *Journal of the American Medical Association, 324*(24), 2555–2557. doi:10.1001/jama.2020.21231 PMID:33226420

McInroy, L. B., McCloskey, R. J., Craig, S. L., & Eaton, A. D. (2019). LGBTQ+ youths' community engagement and resource seeking online versus offline. *Journal of Technology in Human Services, 37*(4), 315–333. doi:10.1080/15228835.2019.1617823

McIntosh, S., Braul, B., & Chao, T. (2003). A case study in asynchronous voice conferencing for language instruction. *Educational Media International, 40*(1-2), 63–74. doi:10.1080/0952398032000092125

McIsaac, M. S., Blocher, J. M., Mahes, V., & Vrasidas, C. (1999). Student and teacher perceptions of interaction in online computer-mediated communication. *Educational Media International, 36*(2), 121–131. doi:10.1080/0952398990360206

Mcluhan, M. (1995). *Essential McLuhan*. Anansi.

McMillan, D. W., & Chavis, D. M. (1986). Sense of community: A definition and theory. *Journal of Community Psychology*, *14*(1), 6–23. doi:10.1002/1520-6629(198601)14:1<6::AID-JCOP2290140103>3.0.CO;2-I

McNeil, R. C. (2011). A Program Evaluation Model: Using Bloom's Taxonomy to Identify Outcome Indicators in Outcomes-Based Program Evaluations. *Journal of Adult Education*, *40*(2), 24–29.

McRae, H. (2015). Situation Engagement in Canadian Higher Education. In O. Delano-Oriaran, M. W. Penick-Parks, & S. Fondie (Eds.), *The SAGE sourcebook of service-learning and civic engagement* (pp. 401–406). SAGE Publications, Inc. doi:10.4135/9781483346625.n68

Mee, G. (2019). *What is a good engagement rate on Twitter?* https://www.scrunch.com/blog/what-is-a-good-engagement-rate-on-twitter

Mehta, S. S., & Ahmed, I. (2018). Planning academic community engagement courses. In H. K. Evans (Ed.), *Community engagement best practices across the disciplines: Applying course content to community needs* (pp. 1–16). Rowman& Littlefield.

Meng, Y., Chu, M.Y., & Chiu, D.K.W. (2022). The impact of COVID-19 on museums in the digital era: Practices and Challenges in Hong Kong. *Library Hi Tech*. doi:10.1108/LHT-05-2022-0273

Men, L. R., & Muralidharan, S. (2017). Understanding social media peer communication and organization–public relationships: Evidence from China and the United States. *Journalism & Mass Communication Quarterly*, *94*(1), 81–101. doi:10.1177/1077699016674187

Men, L. R., & Tsai, W. H. S. (2013). Toward an integrated model of public engagement on corporate social networking sites: Antecedents, the process, and relational outcomes. *International Journal of Strategic Communication*, *7*(4), 257–273. doi:10.1080/1553118X.2013.822373

Men, L. R., & Tsai, W. H. S. (2014). Perceptual, attitudinal, and behavioral outcomes of organization–public engagement on corporate social networking sites. *Journal of Public Relations Research*, *26*(5), 417–435. doi:10.1080/1062726X.2014.951047

Mennonite Central Committee. (2018). *Operating principles framework*. Internal unpublished company document.

Mennonite Central Committee. (2020). *MCC shared program strategic plan, FYE 2021-FYE 2025*. Internal unpublished company document.

Mennonite Central Committee. (2021). *Annual impact report 2020/2021*. https://mccanada.ca/learn/about/annual-reports/annual-impac
t-report-20202021

Mennonite Central Committee. (2021). *Learning and Reflection from Semester One of the Global Public Health Virtual Practicum*. Unpublished internal report.

Mennonite Central Committee. (2022a). *MCC's principles and practices*. https://mcc.org/media/resources/1382

Mennonite Central Committee. (2022b). *Serving and Learning Together (SALT)*. https://mcc.org/get-involved/serve/volunteer/salt

Mensah, B., Poku, A., & Quashigah, A. (2022). Technology Integration into the Teaching and Learning of Geography in Senior High Schools in Ghana: A TPACK Assessment. *Social Education Research*, *3*(1), 80–90.

Mentz, E., & Goosen, L. (2007). Are groups working in the Information Technology class? *South African Journal of Education*, *27*(2), 329–343.

Mentzer, N. (2014). Team Based Engineering Design Thinking. *Journal of Technology Education*, *25*(2), 52–72. doi:10.21061/jte.v25i2.a.4

Meshi, D., Tamir, D. I., & Heekeren, H. R. (2015). The emerging neuroscience of social media. *Trends in Cognitive Sciences*, *19*(12), 771–782. doi:10.1016/j.tics.2015.09.004 PMID:26578288

Meyer, K. A. (2014). Student engagement in online learning: What works and why. *ASHE Higher Education Report*, *40*(6), 1–114. doi:10.1002/aehe.20018

Meyerson, D., Weick, K. E., & Kramer, R. M. (1996). Swift trust and tempo- rary groups. In R. M. Kramer & T. R. Tyler (Eds.), *Trust in organizations: Frontiers of theory and research* (pp. 166–195). Sage. doi:10.4135/9781452243610.n9

Microsoft ASP.NET. (n.d.). Available from: https://dotnet.microsoft.com/en-us/apps/aspnet

Microsoft Visual Studio Community 2019 v.16.11.7. (2021). Available from: https://docs.microsoft.com/tr-tr/visualstudio/releases/2019/release-notes#16.11.7

Microsoft Visual Studio. (2022). Available from: https://visualstudio.microsoft.com/

Milliman, J., Czaplewski, A. J., & Ferguson, J. (2003). Workplace spirituality and employee work attitudes: An exploratory empirical assessment. *Journal of Organizational Change Management*, *16*(4), 426–447. doi:10.1108/09534810310484172

Mishra, P. (2019). Considering contextual knowledge: The TPACK diagram gets an upgrade. *Journal of Digital Learning in Teacher Education*, *35*(2), 76–78. doi:10.1080/21532974.2019.1588611

Mishra, P., & Koehler, M. J. (2006). Technology pedagogical content knowledge: A framework for teacher knowledge. *Teachers College Record*, *108*(6), 1017–1054. doi:10.1111/j.1467-9620.2006.00684.x

Mohezar, S., Moghavvemi, S., & Zailani, S. (2017). Malaysian Islamic medical tourism market: A SWOT analysis. *Journal of Islamic Marketing*, *8*(3), 444–460. doi:10.1108/JIMA-04-2015-0027

Mohmmed, A. O., Khidhir, B. A., Nazeer, A., & Vijayan, V. J. (2020). Emergency remote teaching during Coronavirus pandemic: The current trend and future directive at Middle East College Oman. *Innovative Infrastructure Solutions*, *5*(3), 1–11. doi:10.100741062-020-00326-7

Mokros, Ł., Świtaj, P., Bieńkowski, P., Święcicki, Ł., & Sienkiewicz-Jarosz, H. (2022, October). Depression and loneliness may predict work inefficiency among professionally active adults. *International Archives of Occupational and Environmental Health*, *95*(8), 1775–1783. doi:10.100700420-022-01869-1 PMID:35503113

Molotsi, A., & Goosen, L. (2022). Teachers Using Disruptive Methodologies in Teaching and Learning to Foster Learner Skills: Technological, Pedagogical, and Content Knowledge. In Handbook of Research on Using Disruptive Methodologies and Game-Based Learning to Foster Transversal Skills (pp. 1-24). IGI Global.

Molotsi, A., & Goosen, L. (2019). e-Tutors' Perspectives on the Collaborative Learning Approach as a Means to Support Students of Computing Matters of Course! In *Proceedings of the 48th Annual Conference of the Southern African Computer Lecturers' Association* (pp. 37-54). University of South Africa.

Moore, M. (1989). Editorial: Three types of interaction. *American Journal of Distance Education*, *3*(2), 1–7. doi:10.1080/08923648909526659

Moore, T. L., & Mendez, J. P. (2014). Civic engagement and organizational learning strategies for student success. *New Directions for Higher Education*, *165*(165), 31–40. doi:10.1002/he.20081

Mor Barak, M. E., & Daya, P. (2014). Fostering inclusion from the inside out to create an inclusive workplace: Corporate and organizational efforts in the community and the global society. In B. M. Ferdman & B. R. Deane (Eds.), *Diversity at work: The practice of inclusion* (pp. 391–412). Jossey-Bass.

Mor Barak, M. E., Luria, G., & Brimhall, K. C. (2022). What leaders say versus what they do: Inclusive leadership, policy-practice decoupling, and the anomaly of climate for inclusion. *Group & Organization Management*, *47*(4), 840–871. doi:10.1177/10596011211005916

Morris, T. A. (2010). Anytime/anywhere online learning: does it remove barriers for adult learners? In *Online education and adult learning: New frontiers for teaching practices* (pp. 115–123). IGI Global. doi:10.4018/978-1-60566-830-7.ch009

Morrow, J., & Ackermann, M. (2012). Intention to persist and retention of first-year students: The importance of motivation and sense of belonging. *College Student Journal*, *46*(3), 483–491.

Morton, P., & Rosenfeld, D. (2021). Reconceptualizing service-learning during the COVID-19 pandemic: reflections and recommendations. In C. Tosone (Ed.), Shared trauma, shared resilience during a pandemic (pp. 331–339). Springer. https://doi.org/10.1007/978-3-030-61442-3_34.

Moss, J. (2019). *Burnout Is About Your Workplace, Not Your People.* Harvard Business School Publishing Corporation. Retrieved on November 1, 2022 from https://egn.com/dk/wp-content/uploads/sites/3/2020/08/Burnout-is-about-your-workplace-not-your-people-1.pdf

Mpungose, C. B. (2020). Emergent transition from face-to-face to online learning in a South African university in the context of the coronavirus pandemic. *Humanities and Social Sciences Communications*, *7*(1), 1–9. doi:10.105741599-020-00603-x

Mtawa, N. N., Fongwa, S. N., & Wangenge-Ouma, G. (2016). The scholarship of university-community engagement: Interrogating Boyer's model. *International Journal of Educational Development*, *49*, 126–133. doi:10.1016/j.ijedudev.2016.01.007

Mudau, T. S. (2019). *Enhancing self-regulation among teenage mothers: A university-community engagement approach* [Unpublished PhD thesis]. University of the Free State.

Mughal, R., Thomson, L. J., Daykin, N., & Chatterjee, H. J. (2022). Rapid evidence review of community engagement and resources in the UK during the COVID-19 Pandemic: How can community assets redress health inequities? *International Journal of Environmental Research and Public Health*, *19*(7), 4086. https://doi.org/10.3390/ijerph19074086

Muilenburg, L. Y., & Berge, Z. L. (2002). Designing discussion for the online classroom. In *Designing instruction for technology-enhanced learning* (pp. 100–113). IGI Global. doi:10.4018/978-1-930708-28-0.ch006

Müller, A. M., Goh, C., Lim, L. Z., & Gao, X. (2021). Covid-19 emergency elearning and beyond: Experiences and perspectives of university educators. *Education Sciences*, *11*(1), 19. doi:10.3390/educsci11010019

Mullet, D. R. (2018). A general critical discourse analysis framework for educational research. *Journal of Advanced Academics*, *29*(2), 116–142. doi:10.1177/1932202X18758260

Mund, M., Freuding, M. M., Möbius, K., Horn, N., & Neyer, F. J. (2020). The Stability and Change of Loneliness Across the Life Span: A Meta-Analysis of Longitudinal Studies. *Personality and Social Psychology Review*, *24*(1), 24–52. doi:10.1177/1088868319850738 PMID:31179872

Murphy, K. L., & Cifuentes, L. (2001). Using Web tools, collaborating, and learning online. *Distance Education, 22*(2), 285–305. doi:10.1080/0158791010220207

Musesengwa, R., & Chimbari, M. J. (2017). Experiences of community members and researchers on community engagement in an Ecohealth project in South Africa and Zimbabwe. *BMC Medical Ethics, 18*(1), 1–15. doi:10.118612910-017-0236-3 PMID:29237440

Muthuprasad, T., Aiswarya, S., Aditya, K. S., & Jha, G. K. (2021). Students' perception and preference for online education in India during COVID -19 pandemic. *Social Sciences & Humanities Open, 3*(1), 100101. doi:10.1016/j.ssaho.2020.100101 PMID:34173507

Nambiar, D. (2020). The impact of online learning during COVID-19: Students' and educators' perspective. *International Journal of Indian Psychology, 8*(2), 783–793. https://ijip.in/?s=10.25215%2F0802.094

Nasir, M. K. (2020). The influence of social presence on students' satisfaction toward online course. *Open Praxis, 12*(4), 485. doi:10.5944/openpraxis.12.4.1141

Neely, A. R., Cotton, J. L., & Neely, A. D. (2017). E-mentoring: A model and review of the literature. *AIS Transactions on Human-Computer Interaction, 9*(3), 220–241. doi:10.17705/1thci.00096

Nerds Chalk. (2021). *Google Meet Hand Raise Not Available? Here's Why and What to do.* Available from: https://nerdschalk.com/google-meet-hand-raise-not-available-heres-why-and-what-to-do/

Newman, P. A. (2006). Towards a science of community engagement. *Lancet, 367*(9507), 302. doi:10.1016/S0140-6736(06)68067-7 PMID:16443036

Ngugi, J., & Goosen, L. (2021). Innovation, Entrepreneurship, and Sustainability for ICT Students Towards the Post-COVID-19 Era. In Handbook of Research on Entrepreneurship, Innovation, Sustainability, and ICTs in the Post-COVID-19 Era (pp. 110-131). IGI Global. doi:http://doi:10.4018/978-1-7998-6776-0.ch006

Niess, M. L. (2011). Investigating TPACK: Knowledge growth in teaching with technology. *Journal of Educational Computing Research, 44*(3), 299–317. doi:10.2190/EC.44.3.c

Niess, M., van Zee, E., & Gillow-Wiles, H. (2010). Knowledge growth in teaching mathematics/science with spreadsheets: Moving PCK to TPACK through online professional development. *Journal of Digital Learning in Teacher Education, 27*(2), 42–52. doi:10.1080/21532974.2010.10784657

Ni, J., Chiu, D. K. W., & Ho, K. K. W. (2022). Exploring Information Search Behavior among Self-Drive Tourists. *Information Discovery and Delivery, 50*(3), 285–296. doi:10.1108/IDD-05-2020-0054

Nishii, L. H., & Leroy, H. L. (2020). Inclusive leadership: Leaders as architects of inclusive workgroup climates. In *Inclusive Leadership* (pp. 162–178). Routledge. doi:10.4324/9780429449673-12

Nishii, L. H., & Mayer, D. M. (2009). Do inclusive leaders help to reduce turnover in diverse groups? The moderating role of leader–member exchange in the diversity to turnover relationship. *The Journal of Applied Psychology, 94*(6), 1412–1426. doi:10.1037/a0017190 PMID:19916652

Nkoana, E. M., & Dichaba, M. M. (2016). Are we heading in the right direction? Towards excellence in educational practices. *South Africa International Conference on Education 2016, 15*(7), 213-227.

Noel, J., & Earwicker, D. P. (2015). Documenting community engagement practices and outcomes: Insights from recipients of the 2010 Carnegie community engagement classification. *Journal of Higher Education Outreach & Engagement, 19*(3), 33–62.

O'Driscoll, D. (2018). *Transformation of marginalised through inclusion.* University of Manchester.

Oakes, S., Dennis, N., & Oakes, H. (2013). Web-based forums and metaphysical branding. *Journal of Marketing Management, 29*(5-6), 607–624. doi:10.1080/0267257X.2013.774289

Ofei-Dodoo, S., Mullen, R., Pasternak, A., Hester, C. M., Callen, E., Bujold, E. J., Carroll, J. K., & Kimminau, K. S. (2021). Loneliness, burnout, and other types of emotional distress among family medicine physicians: Results from a national survey. *Journal of the American Board of Family Medicine, 34*(3), 531–541. doi:10.3122/jabfm.2021.03.200566 PMID:34088813

Office of Postsecondary Education. (2021). *Higher education emergency fund III: Frequently asked questions.* United States Department of Education. https://www2.ed.gov/about/offices/list/ope/arpfaq.pdf

Ohmer, M., Carrie, F., Lina, D., Aliya, D., & Alicia, M. (2021). University-community engagement during a pandemic: Moving beyond "helping" to public problem solving. *Metropolitan Universities, 32*(3), 81-91. doi:10.18060/25329

Okoro, C. S., Takawira, O., & Baur, P. (2021). An assessment of tutoring performance, challenges and support during COVID-19: A qualitative study in a South African university. *Journal of University Teaching & Learning Practice, 18*(8). https://doi.org/10.53761/1.18.8.4

Omodan, B. I., & Ige, O. A. (2021). Re-constructing the tutors-tutees relationships for better academic performance in universities amidst COVID-19 new normal. *Mediterranean Journal of Social Sciences, 12*(2), 30–40. https://doi.org/10.36941/mjss-2021-0010

Online Learning Consortium. (2014). *Quality scorecard for administration of online programs.* https://onlinelearningconsortium.org/consult/quality-scorecard

Online Learning Consortium. (2022). *Quality framework.* https://onlinelearningconsortium.org/about/quality-framework-five-pillars

Online Learning Consortium. (n.d.). *OLC quality scorecard: Online student support.* https://onlinelearningconsortium.org/consult/olc-quality-scorecard-student-support/

Orcutt, J. M., & Dringus, L. P. (2017). Beyond being there: Practices that establish presence, engage students and influence intellectual curiosity in a structured online learning environment. *Online Learning, 21*(3). Advance online publication. doi:10.24059/olj.v21i3.1231

Oregon State University. (2022). *Ecampus essentials.* https://ecampus.oregonstate.edu/faculty/courses/Best_Practices_Online_Course_Design.pdf

Ouzts, K. (2006). Sense of community in online courses. *Quarterly Review of Distance Education, 7*(3).

Over, H. (2016). The origins of belonging: social motivation in infants and young children. *Philosophical Transactions of the Royal Society B: Biological Sciences, 371*(1686).

Ovesleová, H. (2015). E-Learning Platforms and Lacking Motivation in Students: Concept of Adaptable UI for Online Courses. In A. Marcus (Ed.), Lecture Notes in Computer Science: Vol. 9188. *Design, User Experience, and Usability: Interactive Experience Design. DUXU 2015.* Springer. doi:10.1007/978-3-319-20889-3_21

Oyserman, D., & Lee, S. W. (2008). Does culture influence what and how we think? Effects of priming individualism and collectivism. *Psychological Bulletin, 134*(2), 311–342. doi:10.1037/0033-2909.134.2.311 PMID:18298274

Palloff, R. M., & Pratt, K. (2010). *Collaborating online: Learning together in community* (Vol. 32). John Wiley & Sons.

Panteli, N., & Tucker, R. (2009). Power and trust in global virtual teams. *Communications of the ACM*, *52*(12), 113–115. doi:10.1145/1610252.1610282

Paquin, J. L. (2006). *How service-learning can enhance the pedagogy and culture of engineering programs at institutions of higher education: A review of the literature*. University of Nebraska Omaha. https://digitalcommons.unomaha.edu/slcedt/19/

Pashootanizadeh, M., & Khalilian, S. (2018). Application of the AIDA model: Measuring the effectiveness of television programs in encouraging teenagers to use public libraries. *Information and Learning Science*, *119*(11), 635–651. doi:10.1108/ILS-04-2018-0028

Patrut, M., & Patrut, B. (2013). *Social media in higher education: teaching in Web 2.0*. Information Science Reference. doi:10.4018/978-1-4666-2970-7

Patsiaouras, G., Veneti, A., & Green, W. (2018). Marketing, art and voices of dissent: Promotional methods of protest art by the 2014 Hong Kong's Umbrella Movement. *Marketing Theory*, *18*(1), 75–100. doi:10.1177/1470593117724609

Payne, A. L. (2021). A resource for E-moderators on fostering participatory engagement within discussion boards for online students in Higher Education. *Student Success*, *12*(1), 93–101. doi:10.5204sj.1865

Peacock, S., & Cowan, J. (2019). Promoting sense of belonging in online learning communities of inquiry in accredited courses. *Online Learning*, *23*(2), 67–81. doi:10.24059/olj.v23i2.1488

Peacock, S., Cowan, J., Irvine, L., & Williams, J. (2020). An exploration into the importance of a sense of belonging for online learners. *International Review of Research in Open and Distributed Learning*, *21*(2), 18–35. doi:10.19173/irrodl.v20i5.4539

Pearce, W. B. (2007). *Making social worlds: A communication perspective*. Blackwell.

Peat, J. A., & Helland, K. R. (2004). *The Competitive Advantage of Online versus Traditional Education*. Online Submission.

Penman, R., & Jensen. (2019). *Making better social worlds: Inspirations from the Theory of the Coordinated Management of Meaning*. CMMI Press.

Penman, R. (2021). *A cosmopolitan sensibility: Compelling stories from a communication perspective*. CMM Institute Press.

Peplau, L. A., & Perlman, D. (1982). Perspectives on loneliness. In L. A. Peplau & D. Perlman (Eds.), *Loneliness* (pp. 1–18). Wiley.

Pérez-Jorge, D., Rodríguez-Jiménez, M. D. C., Ariño-Mateo, E., & Barragán-Medero, F. (2020). The effect of COVID-19 in university tutoring models. *Sustainability*, *12*(20), 8631. https://doi.org/10.3390/su12208631

Perrotta, D. (2021). Universities and COVID-19 in Argentina: From community engagement to regulation. *Studies in Higher Education*, *46*(1), 30–43. https://doi.org/10.1080/03075079.2020.1859679

Perrotta, K., & Bohan, C. H. (2020). A Reflective Study of Online Faculty Teaching Experiences in Higher Education. *Journal of Effective Teaching in Higher Education.*, *3*(1), 50–66. doi:10.36021/jethe.v3i1.9

Petko, D. (2020, June). Quo vadis TPACK? Scouting the road ahead. In *EdMedia+ innovate learning* (pp. 1349-1358). Association for the Advancement of Computing in Education.

Pfund, C., Branchaw, J. L., McDaniels, M., Byars-Winston, A., Lee, S. P., & Birren, B. (2021). Reassess-realign-reimagine: A guide for mentors pivoting to remote research mentoring. *CBE Life Sciences Education*, *20*(1), es2. doi:10.1187/cbe.20-07-0147 PMID:33635126

Pierce, S., Blanton, J., & Gould, D. (2018). An online program for high school student-athlete leadership development: Community engagement, collaboration, and course creation. *Case Studies in Sport and Exercise Psychology, 2*(1), 23-29.

Pigza, J., & Troppe, M. (2003). Developing an infrastructure for service-learning and community engagement. In *B. Jacoby & Associates, Building partnerships for service- learning* (pp. 106–130). Jossey-Bass.

Pillay, N., & Luckan, Y. (2019). The rural school as a place for sustainable community development. *Sustainable Urbanisation through Research, Innovation and Partnerships*, 342-353.

Podsiadlowski, A., Gröschke, D., Kogler, M., Springer, C., & Van Der Zee, K. (2013). Managing a culturally diverse workforce: Diversity perspectives in organizations. *International Journal of Intercultural Relations*, *37*(2), 159–175. doi:10.1016/j.ijintrel.2012.09.001

Polk, X. L. (2018). Marketing: The key to successful teaching and learning. *Journal of Marketing Development and Competitiveness*, *12*(2), 49–57.

Polychroniou, V. P. (2009). Relationship between emotional intelligence and transformational leadership of supervisors The impact on team effectiveness. *Team Performance Management*, *15*(7/8), 343–356. doi:10.1108/13527590911002122

Powell, A. (2022). Two-step flow and protesters: Understanding what influenced participation in a George Floyd protests. *Communication Quarterly*, *70*(4), 1–22. doi:10.1080/01463373.2022.2077122

Pratt, B., & de Vries, J. (2018). Community engagement in global health research that advances health equity. *Bioethics*, *32*(7), 454–463. doi:10.1111/bioe.12465 PMID:30035349

Pressman, S. D., Jenkins, B., & Moskowitz, J. (2019). Positive Affect and Health: What do we know and where next should we go? *Annual Review of Psychology*, *70*(1), 627–650. doi:10.1146/annurev-psych-010418-102955 PMID:30260746

Project Management Institute. (2021). *A guide to the project management body of knowledge* (7th ed.).

Public Health Reports. (2021). COVID-19 pandemic underscores the need to address social isolation and loneliness. *Association of Schools and Programs of Public Health, 136*(6), 653-655.

Puzyreva, K., Henning, Z., Schelwald, R., Rassman, H., Borgnino, E., de Beus, P., Casartelli, S., & Leon, D. (2022). Professionalization of community engagement in flood risk management: Insights from four European countries. *International Journal of Disaster Risk Reduction*, *71*, 102811. Advance online publication. doi:10.1016/j.ijdrr.2022.102811

Qua, K., Haider, R., Junk, D. J., & Berger, N. A. (2021). Sustaining student engagement: Successes and challenges of a virtual STEM program for high school students. *Journal of STEM Outreach*, *4*(3). Advance online publication. doi:10.15695/jstem/v4i3.09 PMID:34853829

Quality Matters. (n.d.). *QM program review annotated criterion*. https://www.qualitymatters.org/sites/default/files/program-review-docs-pdfs/Annotated-Program-Criteria.pdf

Quimí, J., & Alexandra, J. (2022). *Face-to-face vs online learning advantages and disadvantages* [Master's thesis]. Universidad Estatal Península de Santa Elena. Retrieved from https://repositorio.upse.edu.ec/handle/46000/6928

Quinn, E., Cotter, K., Kurin, K., & Brown, K. (2022). Conducting a Community Engagement Studio to Adapt Enhanced Milieu Teaching. *American Journal of Speech-Language Pathology*, *31*(3), 1–19. doi:10.1044/2021_AJSLP-21-00100 PMID:35007426

Rajendran, L., & Thesinghraja, P. (2014). The impact of new media on traditional media. *Middle East Journal of Scientific Research, 22*(4), 609–616.

Ramusetheli, M. D. (2019). *The relevance of nyambedzano as an effective process for promoting morality among the youth* [Unpublished PhD thesis]. University of Venda.

Randel, A. R., Galvin, B. M., Shore, L. M., Holcombe Erhart, K., Chung, B. G., Dean, M. A., & Kedharnath, U. (2018). Inclusive leadership: Realizing positive outcomes through belongingness and being valued for uniqueness. *Human Resource Management Review, 28*(2), 190–203. doi:10.1016/j.hrmr.2017.07.002

Rasool, S. F., Wang, M., Tang, M., Saeed, A., & Iqbal, J. (2021). How Toxic Workplace Environment Effects the Employee Engagement: The Mediating Role of Organizational Support and Employee Wellbeing. *International Journal of Environmental Research and Public Health, 18*(5), 2294. doi:10.3390/ijerph18052294 PMID:33652564

Rathor, D. (2022). Students' and Educators' Perspectives on the Impact of Online Education during Covid- 19. *Social Science Journal for Advanced Research, 2*(1), 1-5. Retrieved from https://www.ssjar.org/index.php/ojs/article/view/29

Ray, S., Kim, S. S., & Morris, J. G. (2014). The central role of engagement in online communities. *Information Systems Research, 25*(3), 528–546. doi:10.1287/isre.2014.0525

Redmon, R. J., & Burger, M. (2004). WEB CT discussion forums: Asynchronous group reflection of the student teaching experience. *Curriculum and Teaching Dialogue, 6*(2), 157–166.

Reeves, T., & Gomm, P. (2015). Community and contribution: Factors motivating students to participate in an extra-curricular online activity and implications for learning. *E-Learning and Digital Media, 12*(3–4), 391–409. doi:10.1177/2042753015571828

Reilly, J. R., Vandenhouten, C., Gallagher-Lepak, S., & Ralston-Berg, P. (2012). Faculty development for e-learning: A multi-campus community of practice (COP) approach. *Journal of Asynchronous Learning Networks, 16*(2), 99–110. doi:10.24059/olj.v16i2.249

Reilly, S., & Langley-Turnbaugh, S. (2021). *The intersection of high-impact practices: What's next for higher education?* Lexington Books.

Reina, D. S., & Reina, M. L. (2016). *Trust and betrayal in the workplace: Building effective relationships in your organization.* Recorded Books.

Republic of South Africa. (2021). *Department of Statistics South Africa. Marginalized Groups Indicator Report, 2019.* Statistics South Africa.

Rienties, B., Brouwer, N., Bohle Carbonell, K., Townsend, D., Rozendal, A., van der Loo, J., Dekker, P., & Lygo-Baker, S. (2013). Online training of TPACK skills of higher education scholars: A cross-institutional impact study. *European Journal of Teacher Education, 36*(4), 480–495. doi:10.1080/02619768.2013.801073

Rienties, B., Lewis, T., O'Dowd, R., Rets, I., & Rogaten, J. (2020). The impact of virtual exchange on TPACK and foreign language competence: Reviewing a large-scale implementation across 23 virtual exchanges. *Computer Assisted Language Learning, 35*(3), 577–603. doi:10.1080/09588221.2020.1737546

Rigby, A. (2021, August 12). *A manager's guide for creating a hybrid work schedule.* Trello. https://blog.trello.com/creating-a-hybrid-work-schedule

Right Attitudes. (2008). Available from: https://www.rightattitudes.com/2008/10/04/7-38-55-rule- pers onal-communication/

Rizzolatti, G., & Fogassi, L. (2014). The mirror mechanism: Recent findings and perspectives. *Philosophical Transactions of the Royal Society of London. Series B, Biological Sciences*, *369*(1644). doi:10.1098/rstb.2013.0420 PMID:24778385

Roach, M., & Fritz, J. (2022). Breaking barriers and building bridges: Increasing community engagement in program evaluation. *Evaluation and Program Planning*, *90*, 101997. Advance online publication. doi:10.1016/j.evalprogplan.2021.101997 PMID:34503853

Roberson, Q., & Perry, J. L. (2022). Inclusive leadership in thought and action: A thematic analysis. *Group & Organization Management*, *47*(4), 755–778. doi:10.1177/10596011211013161

Robinson, C. C., & Hullinger, H. (2008). New benchmarks in higher education: Student engagement in online learning. *Journal of Education for Business*, *84*(2), 101–109. doi:10.3200/JOEB.84.2.101-109

Rodríguez Moreno, J., Agreda Montoro, M., & Ortiz Colón, A. M. (2019). Changes in teacher training within the TPACK model framework: A systematic review. *Sustainability*, *11*(7), 1870. Advance online publication. doi:10.3390u11071870

Rohatinsky, N., Cave, J., & Krauter, C. (2020). Establishing a mentorship program in rural workplaces: Connection, communication, and support required. *Rural and Remote Health*, *20*(1), 5640. doi:10.22605/RRH5640 PMID:31928037

Rosenberg, J. M., & Koehler, M. J. (2015). Context and technological pedagogical content knowledge (TPACK): A systematic review. *Journal of Research on Technology in Education*, *47*(3), 186–210. doi:10.1080/15391523.2015.1052663

Rourke, L., Anderson, T., Garrison, D. R., & Archer, W. (1999). Assessing social presence in asynchronous text-based computer conferencing. *Journal of Distance Education*, *14*(2), 50–71.

Rovai, A. P. (2001). Building classroom community at a distance: A case study. *Educational Technology Research and Development*, *49*(4), 33–48. doi:10.1007/BF02504946

Rovai, A. P. (2002). Building sense of community at a distance. *International Review of Research in Open and Distance Learning*, *3*(1), 1–16. doi:10.19173/irrodl.v3i1.79

Rowley, J. (2002). Information marketing in a digital world. *Library Hi Tech*, *20*(3), 352–358. doi:10.1108/07378830210444540

Royal, M. A., & Rossi, R. J. (1996). Individual-level correlations of sense of community: Findings from workplace and school. *Journal of Community Psychology*, *24*(4), 395–416. doi:10.1002/(SICI)1520-6629(199610)24:4<395::AID-JCOP8>3.0.CO;2-T

Ruane, R., & Lee, V. (2016). Analysis of Discussion Board Interaction in an online peer-mentoring site. *Online Learning*, *20*(4). Advance online publication. doi:10.24059/olj.v20i4.1052

Sabbott. (2013, August 29). *Synchronous learning definition*. The Glossary of Education Reform. Retrieved July 29, 2022, from https://www.edglossary.org/synchronous-learning/

Sackolick, I. (2020, February 25). *What is agile methodology? Modern software development explained*. InfoWorld. https://www.infoworld.com/article/3237508/what-is-agile-methodology-modern-software-development-explained.html

Saha, S. M., Pranty, S. A., Rana, M. J., Islam, M. J., & Hossain, M. E. (2022). Teaching during a pandemic: Do university educators prefer online teaching? *Heliyon*, *8*(1), e08663. doi:10.1016/j.heliyon.2021.e08663 PMID:35028450

Sahrir, M., Hamid, M., Zaini, A., Hamat, Z., & Ismail, T. (2022). Investigating the technological pedagogical content knowledge (TPACK) skill among Arabic school trainee teachers in online assessment during COVID-19 pandemic. *Journal of Language and Linguistic Studies*, *18*(2), 1111–1126.

Saltmarsh, J. (2017). A collaborative turn: Trends and directions in community engagement. In J. Sachs & L. Clarke (Eds.), *Learning through community engagement: Vision and practice in higher education* (pp. 3–16). Springer. doi:10.1007/978-981-10-0999-0_1

Saltmarsh, J., & Johnson, M. (2020). Campus classification, identity, and change: The elective Carnegie classification for community engagement. *Journal of Higher Education Outreach & Engagement, 24*(3), 105–114.

Saltmarsh, J., & Johnson, M. B. (2018). An introduction to the elective Carnegie community engagement classification. In J. Saltmarsh & M. B. Johnson (Eds.), *The elective Carnegie community engagement classification* (pp. 1–18). Campus Compact.

Samson, P. L. (2015). Fostering student engagement: Creative problem-solving in small group facilitations. *Collected Essays on Learning and Teaching, 8*, 153-164

San Tham, S. Y. (2020). *Exploring the Self-Reflection of America Reads Tutors* [Doctoral dissertation]. University of Kansas. Retrieved 6 June 2022 from https://www.proquest.com/docview/2426575455

Sandmann, L. R., Furco, A., & Adams, K. R. (2019). Building the field of higher education engagement. In L. R. Sandmann & D. O. Jones (Eds.), *Building the field of higher education engagement: Foundational ideas and future directions*. Stylus Publishing.

Sandmann, L. R., & Weerts, D. J. (2008). Reshaping institutional boundaries to accommodate an engagement agenda. *Innovative Higher Education, 33*(3), 181–196. doi:10.100710755-008-9077-9

Sashi, C. M. (2012). Customer engagement, buyer-seller relationships, and social media. *Management Decision, 50*(2), 253–272. doi:10.1108/00251741211203551

Saubern, R., Henderson, M., Heinrich, E., & Redmond, P. (2020). TPACK–time to reboot? *Australasian Journal of Educational Technology, 36*(3), 1–9. doi:10.14742/ajet.6378

Saxton, G. D., Guo, S. C., & Brown, W. A. (2007). New dimensions of nonprofit responsiveness: The application and promise of Internet-based technologies. *Public Performance & Management Review, 31*(2), 144–173. doi:10.2753/PMR1530-9576310201

Schein, E. (2004). *Organizational culture and leadership.* John Wiley & Sons.

Schmid, M., Brianza, E., & Petko, D. (2020). Developing a short assessment instrument for Technological Pedagogical Content Knowledge (TPACK. xs) and comparing the factor structure of an integrative and a transformative model. *Computers & Education, 157*, 103967. Advance online publication. doi:10.1016/j.compedu.2020.103967

Schmid, M., Brianza, E., & Petko, D. (2021). Self-reported technological pedagogical content knowledge (TPACK) of pre-service teachers in relation to digital technology use in lesson plans. *Computers in Human Behavior, 115*, 106586. Advance online publication. doi:10.1016/j.chb.2020.106586

Schreiber, R., & Tomm-Bonde, L. (2015). Ubuntu and constructivist grounded theory: An African methodology package. *Journal of Research in Nursing, 20*(8), 655–664. doi:10.1177/1744987115619207

Schweder, R. A., & Sullivan, M. A. (1993). Cultural psychology: Who needs it? *Annual Review of Psychology, 44*(1), 497–523. doi:10.1146/annurev.ps.44.020193.002433

Seaman, J. E., Allen, I. E., & Seaman, J. (2018). *Grade increase: Tracking distance education in the United States.* Babson Survey Research Group. https://onlinelearningsurvey.com/reports/gradeincrease.pdf

Seifert, T. (2016). Involvement, collaboration and engagement: Social networks through a pedagogical lens. *Journal of Learning Design*, *9*(2), 31–45. doi:10.5204/jld.v9i2.272

Selingo, J., Clark, C., Noone, D., & Wittmayer, A. (2021) *The hybrid campus: Three major shifts for the post COVID university*. The Deloitte Center for Higher Education Excellence. https://www2.deloitte.com/content/dam/insights/articles/6756_CGI-Higher-ed-COVID/DI_CGI-Higher-ed-COVID.pdf

Selvaraj, A., Vishnu, R., Ka, N., Benson, N., & Mathew, A. J. (2021). Effect of pandemic based online education on teaching and learning system. *International Journal of Educational Development*, *85*, 102444. doi:10.1016/j.ijedudev.2021.102444 PMID:34518732

Serrano-Solano, B., Föll, M. C., Gallardo-Alba, C., Erxleben, A., Rasche, H., Hiltermann, S., ... Grüning, B. A. (2021). Fostering accessible online education using Galaxy as an e-learning platform. *PLoS Computational Biology*, *17*(5), e1008923. doi:10.1371/journal.pcbi.1008923 PMID:33983944

Seru, E. J. (2021). Critical, interdisciplinary, and collaborative approaches to virtual community-engaged learning during the COVID-19 pandemic and social unrest in the Twin Cities. *Journal of Higher Education Outreach & Engagement*, *25*(3), 79–90. https://openjournals.libs.uga.edu/jheoe/article/download/2560/2689

Shea, P., Li, C. S., & Pickett, A. (2006). A study of teaching presence and student sense of learning community in fully online and web-enhanced college courses. *The Internet and Higher Education*, *9*(3), 175–190. doi:10.1016/j.iheduc.2006.06.005

Sherer, P. D., Shea, T. P., & Kristensen, E. (2003). Online communities of practice: A catalyst for faculty development. *Innovative Higher Education*, *27*(3), 183–194. doi:10.1023/A:1022355226924

Shevlin, M., McBride, O., Murphy, J., Miller, J. G., Hartman, T. K., Levita, L., Mason, L., Martinez, A. P., McKay, R., Stocks, T., Bennett, K. M., Hyland, P., Karatzias, T., & Bentall, R. P. (2020). Anxiety, depression, traumatic stress and COVID-19-related anxiety in the U.K. general population during the COVID-19 pandemic. *BJPsych Open*, *6*(6), e125. doi:10.1192/bjo.2020.109 PMID:33070797

Shim, T. E., & Lee, S. Y. (2020). College students' experience of emergency remote teaching due to COVID-19. *Children and Youth Services Review*, *119*, 105578. doi:10.1016/j.childyouth.2020.105578 PMID:33071405

Shore, L. M., Randel, A. E., Chung, B. G., Dean, M. A., Ehrhart, K. H., & Singh, G. (2011). Inclusion and diversity in work groups: A review and model for future research. *Journal of Management*, *37*(4), 1262–1289. doi:10.1177/0149206310385943

Shostack, L. (1984, January). Designing services that deliver. *Harvard Business Review Magazine*. https://hbr.org/1984/01/designing-services-that-deliver

Shuib, M., & Azizan, S. N. (2017). University-community engagement via m-learning. In M. Shuib & K. Y. Lie (Eds.), *The role of the university with a focus on university-community engagement*. EPUB.

Shulman, L. (1986). Those who understand: Knowledge growth in teaching. *Educational Researcher*, *15*(2), 4–14. doi:10.3102/0013189X015002004

Shutterstock. (n.d.a). *Gülümseyen kendine güvenen Latin genç kız okul öğrencisi*. Available from: https://www.shutterstock.com/tr/image-photo/smiling-confident-latin-teen-girl-school-1740966191

Shutterstock. (n.d.b). *Young happy Hispanic Indian Latin student*. Available from: https://www.shutterstock.com/tr/image-photo/young-happy-hispanic-indian-latin-student-2007080243

Silva, L., Shuttlesworth, M., & Ice, P. (2021). Moderating relationships: Non-designer instructor's teaching presence and distance learners' cognitive presence. *Online Learning*, *25*(2). Advance online publication. doi:10.24059/olj.v25i2.2222

Silverman, E. (2018). Moving beyond collaboration: A Model for enhancing social work's organizational empathy. *Social Work*, *63*(4), 297–304. doi:10.1093wwy034 PMID:30113662

Simkovich, J. (2022, March 1). *Collaborative learning strategies for professors in 2022*. Ment.io. Retrieved July 30, 2022, from https://www.ment.io/collaborative-learning-strategies-for-professors/

Simunich, B. (2017). Service learning in online courses. In C. Crosby & F. Brockmeier (Eds.), *Community engagement program implementation and teacher preparation for 21st century education*. IGI Global. doi:10.4018/978-1-5225-0871-7.ch009

Singh, J., Evans, E., Reed, A., Karch, L., Qualey, K., Singh, L., & Wiersma, H. (2021). Online, Hybrid, and Face-to-Face Learning Through the Eyes of Faculty, Students, Administrators, and Instructional Designers: Lessons Learned and Directions for the Post-Vaccine and Post-Pandemic/COVID-19 World. *Journal of Educational Technology Systems*.

Slavich, G. M., Roos, L. G., & Zaki, J. (2022). Social belonging, compassion, and kindness: Key ingredients for fostering resilience, recovery, and growth from the COVID-19 pandemic. *Anxiety, Stress, and Coping*, *35*(1), 1–8. doi:10.1080/10615806.2021.1950695 PMID:34369221

Smith, D. A. (2011). Strategic Marketing of Library Resources and Services. *College & Undergraduate Libraries*, *18*(4), 333–349. doi:10.1080/10691316.2011.624937

Soler-Costa, R., Moreno-Guerrero, A. J., López-Belmonte, J., & Marín-Marín, J. A. (2021). Co-word analysis and academic performance of the term TPACK in web of science. *Sustainability*, *13*(3), 1481. Advance online publication. doi:10.3390u13031481

Spataro, J. (2022). 5 Key Trends Leaders Need to Understand to Get Hybrid Right. *Harvard Business Review*.

Spiller, C., Evans, M., Schuyler, K. G., & Watson, L. W. (2021). What silence can teach us about race and leadership. *Leadership*, *17*(1), 81–98. doi:10.1177/1742715020976003

Statista. (2019, December). *Statista TrendCompass 2020*. https://www.statista.com/study/69166/statista-trendcompass/

Statista. (2020). *Social media usage in the United Kingdom (UK)*. https://www.statista.com/study/21322/social-media-usage-in-the-united-kingdom-statista-dossier/

Statistics South Africa. (2019). *Quarterly Labour Force Survey*. Retrieved from https:// www.statssa.gov.za/publications/P0211/P02114thQuarter2019.pdf

Steen, S., Mackenzie, L., & Buechner, B. (2018). Incorporating Cosmopolitan Communication into Diverse Teaching and Training Contexts: Considerations from Our Work with Military Students and Veterans. *The Routledge Handbook of Communication Training*. https://www.routledge.com/The-Handbook-of-Communication-Training-A-Best-Practices-Framework-for/Wallace-Becker/p/book/97
81138736528

Stein, H. T. (2002). A Psychology for Democracy*: Presentation at the IAIP Congress*. Alfred Adler Institutes of San Fransisco and Northwestern Washington.

Stepaniuk, K. (2017). Blog content management in shaping pro recreational attitudes. *Journal of Business Economics and Management*, *18*(1), 146–162. doi:10.3846/16111699.2017.1280693

Stojan, J., Haas, M., Thammasitboon, S., Lander, L., Evans, S., Pawlik, C., ... Daniel, M. (2022). Online learning developments in undergraduate medical education in response to the COVID-19 pandemic: A BEME systematic review: BEME Guide No. 69. *Medical Teacher*, *44*(2), 109–129.

Strong, E. K. (1925). *The psychology of selling and advertising*. McGraw-Hill book Company, Incorporated.

Sullivan, P., & Moore, K. (2013). Time talk: On small changes that enact infrastructural mentoring for undergraduate women in technical fields. *Journal of Technical Writing and Communication*, *43*(3), 333–354. doi:10.2190/TW.43.3.f

Sun, A., & Chen, X. (2016). Online education and its effective practice: A research review. *Journal of Information Technology Education*, *15*, 15. doi:10.28945/3502

Sun, L., Tang, Y., & Zuo, W. (2020). Coronavirus pushes education online. *Nature Materials*, *19*(6), 687–687. doi:10.103841563-020-0678-8 PMID:32341513

Sun, X., Chiu, D. K. W., & Chan, C. T. (2022). Recent Digitalization Development of Buddhist Libraries: A Comparative Case Study. In S. Papadakis & A. Kapaniaris (Eds.), *The Digital Folklore of Cyberculture and Digital Humanities* (pp. 251–266). IGI Global. doi:10.4018/978-1-6684-4461-0.ch014

Sutia, C., Wulan, A. R., & Solihat, R. (2019, February). Students' response to project learning with online guidance through google classroom on biology projects. *Journal of Physics: Conference Series*, *1157*(2), 022084. doi:10.1088/1742-6596/1157/2/022084

Sutton, H. (2022). Marginalized adult learners faced unique challenges during COVID-19. *Dean and Provost*, *23*(9), 7–8. doi:10.1002/dap.31032

Swallow, M., & Olofson, M. (2017). Contextual understandings in the TPACK framework. *Journal of Research on Technology in Education*, *49*(3-4), 228–244. doi:10.1080/15391523.2017.1347537

Swani, K., Milne, G. R., Brown, B. P., Assaf, A. G., & Donthu, N. (2017). What messages to post? Evaluating the popularity of social media communications in business versus consumer markets. *Industrial Marketing Management*, *62*(April), 77–87. doi:10.1016/j.indmarman.2016.07.006

Swan, K. (2002). Building learning communities in online courses: The importance of interaction. *Education Communication and Information*, *2*(1), 23–49. doi:10.1080/1463631022000005016

Swan, K., & Shih, L. (2005). On the nature and development of social presence in online course discussions. *Journal of Asynchronous Learning Networks*, *9*(3), 115–136. https://olj.onlinelearningconsortium.org/index.php/olj/article/view/1788

Sybing, R. (2022). Dialogic validation: a discourse analysis for conceptual development within dialogic classroom interaction. *Classroom Discourse*, 1-17.

Talmage, C. A., Baker, A. L., Guest, M. A., & Knopf, R. C. (2020). Responding to social isolation among older adults through lifelong learning: Lessons and questions during COVID-19. *Local Development & Society*, *1*(1), 26-33. doi:10.1080/26883597.2020.1794757

Talmage, C. A., Annear, C., Equinozzi, K., Flowers, K., Hammett, G., Jackson, A., ... Turino, C. (2021). Rapid community innovation: A small urban liberal arts community response to COVID-19. *International Journal of Community Well-Being*, *4*(3), 323–337. https://doi.org/10.1007/s42413-020-00074-7

Tanner, O. C. (2021). *Hybrid workplace: The future of work is a combination of workplaces*. https://www.octanner.com/global-culture-report/2022/hybrid-workplace.html

Taylor, Z. W., Pereira, M., Rainey, L., Gururaj, S., Gibbs, B., Wiser, J., Benson, C., Childs, J., & Somers, P. (2022). Saying His Name: How Faith-Based Higher Education Leaders Addressed the George Floyd Murder. *Religious Education (Chicago, Ill.)*, 1–31. doi:10.1080/15507394.2022.2156262

Telles, A. B. (2019). Community engagement vs. racial equity: Can community engagement work be racially equitable? *Metropolitan Universities*, *30*(2), 95–108. doi:10.18060/22787

Temporal, P. (2015). *Branding for the public sector: Brand Communications Strategy*. John Wiley & Sons.

Tenorio, E. H. (2011). Critical discourse analysis: An overview. *Nordic Journal of English Studies*, *10*(1), 183–210. doi:10.35360/njes.247

Tess, P. A. (2013). The role of social media in higher education classes (real and virtual) – A literature review. *Computers in Human Behavior*, *29*(5), A60–A68. doi:10.1016/j.chb.2012.12.032

Thayer, B. (2021). *Planning for higher ed's digital-first, hybrid future: A call to action for college and university cabinet leaders*. Education Advisory Board. https://eab.com/research/strategy/whitepaper/plan-digital-first-hybrid-future-higher-ed/

Thomas, L., Herbert, J., & Teras, M. (2014). *A sense of belonging to enhance participation, success and retention in online programs*. https://ro.uow.edu.au/asdpapers/475/

Thompson, H. M., Clement, A. M., Ortiz, R., Preston, T. M., Wells Quantrell, A. L., Enfield, M., King, A. J., Klosinski, L., Reback, K. J., Hamilton, A., & Milburn, N. (2022). Community engagement to improve access to healthcare: A comparative case study to advance implementation science for transgender health equity. *International Journal for Equity in Health*, *21*(1), 1–14. doi:10.118612939-022-01702-8 PMID:35907962

Tiller, W. (n.d.). *The theory of positive disintegration by Kazimierz Dąbrowski*. Retrieved from http://www.positivedisintegration.com/

Timperley, C., & Schick, K. (2022). Hiding in Plain Sight: Pedagogy and Power. *International Studies Perspectives*, *23*(2), 113–128. doi:10.1093/isp/ekab002

Tømte, C., Enochsson, A., Buskqvist, U., & Kårstein, A. (2015). Educating online student teachers to master professional digital competence: The TPACK-framework goes online. *Computers & Education*, *84*, 26–35. doi:10.1016/j.compedu.2015.01.005

Tragodara, K. S. C. (2021, March). *Virtual tutoring from the comprehensive training model to Engineering students during the COVID-19 pandemic*. Conference paper at the 2021 IEEE World Conference on Engineering Education (EDUNINE). https://ieeexplore.ieee.org/abstract/document/9429115

Triandis, H. C. (1995). A theoretical framework for the study of diversity. In M. M. Chemers, S. Oskamp, & M. A. Costanzo (Eds.), *Diversity in organizations: New perspectives for a changing workplace* (pp. 11–13). Sage Publications Inc. doi:10.4135/9781452243405.n2

Tsang, A. L. Y., & Chiu, D. K. W. (2022). Effectiveness of Virtual Reference Services in Academic Libraries: A Qualitative Study based on the 5E Learning Model. *Journal of Academic Librarianship, 48*(4), 102533. doi:10.1016/j.acalib.2022.102533

Tse, H. L., Chiu, D. K., & Lam, A. H. (2022). From Reading Promotion to Digital Literacy: An Analysis of Digitalizing Mobile Library Services With the 5E Instructional Model. In A. Almeida & S. Esteves (Eds.), *Modern Reading Practices and Collaboration Between Schools, Family, and Community* (pp. 239–256). IGI Global. doi:10.4018/978-1-7998-9750-7.ch011

Tshishonga, N. S. (2020). Forging University social responsibility through community engagement in higher education. In S. Chhabra & M. Kumar (Eds.), *Civic engagement frameworks and strategic leadership practices for organization development* (pp. 96–115). IGI Global. doi:10.4018/978-1-7998-2372-8.ch005

Tuck, N. L., Adams, K. S., Pressman, S. D., & Consedine, N. S. (2017). Greater ability to express positive emotion is associated with lower projected cardiovascular disease risk. *Journal of Behavioral Medicine, 40*(6), 855–864. doi:10.100710865-017-9852-0 PMID:28455831

Tull, M. T., Edmonds, K. A., Scamaldo, K. M., Richmond, J. R., Rose, J. P., & Gratz, K. L. (2020). Psychological outcomes associated with stay-at-home orders and the perceived impact of COVID-19 on daily life. *Psychiatry Research, 289*, 113098. doi:10.1016/j.psychres.2020.113098 PMID:32434092

Twenge, J. M., Baumeister, R. F., Tice, D. M., & Stucke, T. S. (2001). If you can't join them, beat them: Effects of social exclusion on aggressive behavior. *Journal of Personality and Social Psychology, 81*(6), 1058–1069. doi:10.1037/0022-3514.81.6.1058 PMID:11761307

Twenge, J. M., Spitzberg, B. H., & Campbell, W. K. (2019). Less in-person social interaction with peers among U.S. adolescents in the 21st century and links to loneliness. *Journal of Social and Personal Relationships, 36*(6), 1892–1913. doi:10.1177/0265407519836170

Ulmer, R. R. (2012). Increasing the impact of thought leadership in crisis communication. *Management Communication Quarterly, 26*(4), 523–542. doi:10.1177/0893318912461907

United States Department of Education. (n.d.). *Programs: Federal Work-Study (FWS) Program*. Retrieved 10 June 2022 from https://www2.ed.gov/programs/fws/index.html#:~:text=Institutions%20must%20use%20at%20least%207%20percent%20of,family%20literacy%20activities%3B%20or%20emergency%20preparedness%20and%20response

Upcraft, M. L., & Kramer, G. (1995). Intrusive advising as discussed in the first-year academic advising: Patterns in the present, pathways to the future. Academic Advising; Barton College.

Utomo, E. (2022). The synthesis of qualitative evidence-based learning by design model to improve TPACK of prospective mathematics teacher. In *The 5th International Conference on Combinatorics, Graph Theory, and Network Topology 2021*. IOP Publishing. 10.1088/1742-6596/2157/1/012044

Vakratsas, D., & Ambler, T. (1999). How advertising works: What do we really know? *Journal of Marketing, 63*(1), 26–43. doi:10.1177/002224299906300103

Valentine, A., Gemin, B., Vashaw, L., Watson, J., Harrington, C., & LeBlanc, E. (2021). Digital learning in rural K–12 settings: A survey of challenges and progress in the United States. *Research Anthology on Developing Effective Online Learning Courses*, 1987-2019.

Valtonen, T., Leppänen, U., Hyypiä, M., Sointu, E., Smits, A., & Tondeur, J. (2020). Fresh perspectives on TPACK: Pre-service teachers' own appraisal of their challenging and confident TPACK areas. *Education and Information Technologies*, *25*(4), 2823–2842. doi:10.100710639-019-10092-4

van Dyk, T. A. (2008). *Discourse and context.* Cambridge University Press.

Van Heerden, D., & Goosen, L. (2021). Students' Perceptions of e-Assessment in the Context of Covid-19: The Case of UNISA. In *Proceedings of the 29th Conference of SAARMSTE* (pp. 291-305). SAARMSTE.

Van Maaren, J., Jensen, M., & Foster, A. (2022). Tutoring in higher education during COVID-19: Lessons from a private university's transition to remote learning. *Journal of College Reading and Learning*, *52*(1), 3–22. https://doi.org/10.1080/10790195.2021.2007175

Vasiliki, B., & Psoni, P. (2021). Online teaching practicum during COVID-19: The case of a teacher education program in Greece. *Journal of Applied Research in Higher Education*. Advance online publication. doi:10.1108/JARHE-07-2020-0223

Verma, R. B. S., & Singh, A. P. (2015). The abstract book of 3rd Indian social work congress. In *Community engagement, social responsibility and social work profession.* Rapid Book Service.

Voogt, J., Fisser, P., Tondeur, J., & van Braak, J. (2016). Using theoretical perspectives in developing an understanding of TPACK. In Handbook of technological pedagogical content knowledge (TPACK) for educators (p. 33). Academic Press.

Vuong, T., Hoyt, L., Newcomb-Rowe, A., & Carrier, C. (2017). Faculty perspectives rewards and incentives for community-engaged work. *International Journal of Community Research and Engagement*, *10*, 249–264.

W3schools.com Tryit Editor v2.2-Show razor. (2021). Available from: https://www.w3schools.com/asp/showfile_c.asp?filename=try_webpages_cs_010

W3schools.com. (2021). Available from: https://www.w3schools.com/asp/webpages_forms.asp

Walker, C. H. (2016). *The correlation between types of instructor-student communication in online graduate courses and student satisfaction levels in the private university setting* [Doctoral dissertation]. Carson-Newman University. https://classic.cn.edu/libraries/tiny_mce/tiny_mce/plugins/filemanager/files/Dissertations/Christy_Walker.pdf

Walker, N. T., & Ardell, A. (2020). Reconsidering preservice-mentor relationships in complex times: New possibilities for collaboration and contribution. *Issues in Teacher Education*, *29*(1/2), 132–141.

Wang, J., Deng, S., Chiu, D. K. W., & Chan, C. T. (2022) Social Network Customer Relationship Management for Orchestras: A Case Study on Hong Kong Philharmonic Orchestra. In Social Customer Relationship Management (Social-CRM) in the Era of Web 4.0. IGI Global. doi:10.4018/978-1-7998-9553-4.ch012

Wang, M., Sierra, C., & Folger, T. (2003). Building a dynamic online learning community among adult learners. *Educational Media International*, *40*(1-2), 49–62. doi:10.1080/0952398032000092116

Wang, P., Chiu, D. K. W., Ho, K. K., & Lo, P. (2016). Why read it on your mobile device? Change in reading habit of electronic magazines for university students. *Journal of Academic Librarianship*, *42*(6), 664–669. doi:10.1016/j.acalib.2016.08.007

Wang, W., Lam, E. T. H., Chiu, D. K. W., Lung, M. M., & Ho, K. K. W. (2021). Supporting Higher Education with Social Networks: Trust and Privacy vs. Perceived Effectiveness. *Online Information Review*, *45*(1), 207–219. doi:10.1108/OIR-02-2020-0042

Wang, W., Schmidt-Crawford, D., & Jin, Y. (2018). Preservice teachers' TPACK development: A review of literature. *Journal of Digital Learning in Teacher Education, 34*(4), 234–258. doi:10.1080/21532974.2018.1498039

Ward, C., & Benson, S. (2010). Developing new schemas for online teaching and learning: TPACK. *Journal of Online Learning and Teaching, 6*(2), 482–490.

Ward, E., Buglione, S., Giles, D. E., & Saltmarsh, J. (2013). The Carnegie classification for community engagement. In P. Benneworth (Ed.), *University engagement with socially excluded communities* (pp. 285–308). Springer. doi:10.1007/978-94-007-4875-0_15

Warnick, A. (2021). Interest in public health degrees jumps in wake of pandemic: Applications rise. *The Nation's Health, 51*(6), 1–12.

Warren, C., Barsky, A., & McGraw, A. P. (2021). What makes things funny? An integrative review of the antecedents of laughter and amusement. *Personality and Social Psychology Review, 25*(1), 41–65. doi:10.1177/1088868320961909 PMID:33342368

Washington, M. G. (2013). Trust and project performance: The effects of cognitive and affective-based trust on client-project manager engagements. *University of Pennsylvania Scholarly Commons.* Retrieved on October 29, 2022 from https://repository.upenn.edu/od_theses_msod/67

Wasserman, I. (2014). Strengthening interpersonal awareness and fostering relational eloquence. In B. Ferdman & B. Deane (Eds.), *Diversity at work: the practice of inclusion* (pp. 128–154). Jossey-Bass. https://cmminstitute.org/wp-content/uploads/2019/07/4-Wasserman-Strengthening-Interpersonal-Awareness-and-Relational-Eloquence-in-Ferdman-and-Deane-Diversity-at-Work-The-Practice-of-Inclusion.pdf

Watson, D. (2007). *Managing civic and community engagement.* McGraw-Hill Education.

Watson, S., & Waterton, E. (2010). Heritage and community engagement. *International Journal of Heritage Studies, 16*(1-2), 1–3. doi:10.1080/13527250903441655

Watts, R., Williamson, D., & Williamson, J. (2004). *Adlerian Psychology: A Relational Constructivist Approach. Adlerian Yearbook. 2004.* Adlerian Society and Institute for Individual Psychology.

Ways to create belonging in a virtual classroom (n.d.). UVA School of Nursing. Retrieved December 3, 2023, from https://www.nursing.virginia.edu/diversity/inclusive-teaching/belonging/

Weisman, M. (2021). Remote community engagement in the time of COVID-19, a surging racial Justice movement, wildfires, and an election year. *Higher Learning Research Communications, 11,* 6.

Weisman, M. (2021). Remote community engagement in the time of COVID-19, a surging racial justice movement, wildfires, and an election year. *Higher Learning Research Communications, 11,* 88-95. https://scholarworks.waldenu.edu/cgi/viewcontent.cgi?article=1225&context=hlrc

Welch, M., & Saltmarsh, J. (2013). Best practices and infrastructures for campus centers of community engagement. In A. Hoy & M. Johnson (Eds.), *Deepening community engagement in higher education: Forging new pathways.* Palgrave MacMillian. doi:10.1057/9781137315984_14

Weltzer-Ward, L., Baltes, B., & Lynn, L. K. (2009). Assessing quality of critical thought in online discussion. *Campus-Wide Information Systems*, *26*(3), 168–177. doi:10.1108/10650740910967357

Wenger, E. C. (2011). *Communities of practice: A brief introduction*. University of Oregon. http://hdl.handle.net/1794/11736

Wenger, E. C. (1998). *Communities of practice: Learning, meaning, and identity*. Cambridge University Press. doi:10.1017/CBO9780511803932

Wenger, E. C., McDermott, R., & Snyder, W. (2002). *Cultivating communities of practice: A guide to managing knowledge*. Harvard Business School Press.

Wenger, E. C., & Snyder, W. M. (2000). Communities of practice: The organizational frontier. *Harvard Business Review*, *78*(1), 139–145. https://hbr.org/2000/01/communities-of-practice-the-organizational-frontier

Westerhof, G. J., Dittmann-Kohli, F., & Katzko, M. W. (2000). Individualism and Collectivism in the Personal Meaning System of Elderly Adults: The United States and Congo/Zaire as an Example. *Journal of Cross-Cultural Psychology*, *31*(6), 649–676. doi:10.1177/0022022100031006001

Wexler, M. N. (2005). *Leadership in context: The four faces of capitalism*. Edward Elgar Publishing.

What is asynchronous learning? (n.d.). Coursera. Retrieved July 29, 2022, from https://www.coursera.org/articles/what-is-asynchronous-learning

Wiliam, D. (2013). *Principled curriculum design*. SSAT (The Schools Network) Limited.

Wilkialis, L., Rodrigues, N. B., Cha, D. S., Siegel, A., Majeed, A., Lui, L. M. W., Tamura, J. K., Gill, B., Teopiz, K., & McIntyre, R. S. (2021). Social isolation, loneliness and generalized anxiety: Implications and associations during the COVID-19 quarantine. *Brain Sciences*, *11*(12), 1620. doi:10.3390/brainsci11121620 PMID:34942920

Williams, K. D. (2009). Ostracism: Effects of being excluded and ignored. In M. P. Zanna (Ed.), Advances in experimental social psychology (Vol. 41, pp. 275–314). Academic Press.

Williams, S. A. S., Hanssen, D. V., Rinke, C. R., & Kinlaw, C. R. (2019). Promoting race pedagogy in Higher Education: Creating an inclusive community. *Journal of Educational & Psychological Consultation*, *30*(3), 369–393. doi:10.1080/10474412.2019.1669451

Williams, S., Jaramillo, A., & Pesko, J. (2015). Improving depth of thinking in online discussion boards. *Quarterly Review of Distance Education*, *16*(3), 45–66.

Wilson, D., Heaslip, V., & Jackson, D. (2018). Improving equity and cultural responsiveness with marginalized communities: Understanding competing worldviews. *Journal of Clinical Nursing*, *27*(19-20), 3810–3819. doi:10.1111/jocn.14546 PMID:29869819

Winters, M. (2014). From diversity to inclusion: An inclusion equation. In B. M. Ferdman & B. R. Deane (Eds.), *Diversity at work: The practice of inclusion* (pp. 205–228). Jossey-Bass.

Wirtz, B. W., Schilke, O., & Ullrich, S. (2010). Strategic development of business models: Implications of the Web 2.0 for creating value on the internet. *Long Range Planning*, *43*(2-3), 272–290. doi:10.1016/j.lrp.2010.01.005

Wodak, R. (2001). The discourse-historical approach. In R. Wodak & M. Meyer (Eds.), *Methods of critical discourse analysis* (pp. 63–94). SAGE Publications.

Women, U. N. (2020). Commission on the status of women. *Fiftieth Session, 27.*

Wong, J., Chiu, D. K. W., Leung, T. N., & Kevin, K. K. W. (in press). Exploring the Associations of Addiction as a Motive for Using Facebook with Social Capital Perceptions. *Online Information Review*.

Wood, L., & Zuber-Skerrit, O. (2013). PALAR as a methodology for community engagement by faculties of education. *South African Journal of Education*, *33*(4), 1–15. doi:10.15700/201412171322

Woods, D. R., Benschop, Y., & van den Brink, M. (2021). What is intersectional equality? A definition and goal of equality for organizations. *Gender, Work & Organization*, 1–18. doi:10.1111/gwao.12760

World Health Organization. (2019). Burn-out an "occupational phenomenon": International Classification of Diseases. *World Health Organization Departmental News*. Retrieved on October 22, 2022 from https://www.who.int/news/item/28-05-2019-burn-out-an-occupational-phenomenon-international-classification-of-diseases

Yang, Q., Gu, J., & Hong, J. C. (2021). Parental social comparison related to tutoring anxiety, and guided approaches to assisting their children's home online learning during the COVID-19 lockdown. *Frontiers in Psychology*. doi:10.3389/fpsyg.2021.708221

Yang, Z., Zhou, Q., Chiu, D. K. W., & Wang, Y. (2022). Exploring the factors influencing continuance intention to use academic social network sites. *Online Information Review*, *46*(7), 1225–1241. Advance online publication. doi:10.1108/OIR-01-2021-0015

Yao, L., Lei, J., Chiu, D. K. W., & Xie, Z. (2022). Adult Learners' Perception of Online Language English Learning Platforms in China. In A. Garcés-Manzanera & M. E. C. García (Eds.), *New Approaches to the Investigation of Language Teaching and Literature*. IGI Global.

Yarbrough, J. R. (2018). Adapting adult learning theory to support innovative, advanced, online learning—WVMD model. *Research in Higher Education*, *35*, 1–15.

Yeh, Y., Chan, K., & Hsu, Y. (2021). Toward a framework that connects individual TPACK and collective TPACK: A systematic review of TPACK studies investigating teacher collaborative discourse in the learning by design process. *Computers & Education*, *171*, 104238. Advance online publication. doi:10.1016/j.compedu.2021.104238

Yeung, M. W., & Yau, A. H. (2022). A thematic analysis of higher education students' perceptions of online learning in Hong Kong under COVID-19: Challenges, strategies and support. *Education and Information Technologies*, *27*(1), 181–208. doi:10.100710639-021-10656-3 PMID:34421326

Yildiz, H., & Duyan, V. (2022). Effect of group work on coping with loneliness. *Social Work with Groups*, *45*(2), 132–144. doi:10.1080/01609513.2021.1990192

Yip, K. H. T., Chiu, D. K. W., Ho, K. K. W., & Lo, P. (2021). Adoption of Mobile Library Apps as Learning Tools in Higher Education: A Tale between Hong Kong and Japan. *Online Information Review*, *45*(2), 389–405. doi:10.1108/OIR-07-2020-0287

Yu, H. H. K., Chiu, D. K. W., & Chan, C. T. (2022). Resilience of symphony orchestras to challenges in the COVID-19 era: Analyzing the Hong Kong Philharmonic Orchestra with Porter's five force model. In W. Aloulou (Ed.), *Handbook of Research on Entrepreneurship and Organizational Resilience During Unprecedented Times*. IGI Global. doi:10.4018/978-1-6684-4605-8.ch026

Yu, H. Y., Tsoi, Y. Y., Rhim, A. H. R., Chiu, D. K., & Lung, M. M. W. (2021). (in press). Changes in habits of electronic news usage on mobile devices in university students: A comparative survey. *Library Hi Tech*. Advance online publication. doi:10.1108/LHT-03-2021-0085

Yu, P. Y., Lam, E. T. H., & Chiu, D. K. W. (2022). (in press). Operation management of academic libraries in Hong Kong under COVID-19. *Library Hi Tech*. Advance online publication. doi:10.1108/LHT-10-2021-0342

Yusnilita, N. (2020). The impact of online learning: Student's views. *ETERNAL*, *11*(1). Advance online publication. doi:10.26877/eternal.v11i1.6069

Yusof, N., Hashim, N. L., & Hussain, A. (2022). A Conceptual User Experience Evaluation Model on Online Systems. *International Journal of Advanced Computer Science and Applications*, *13*(1). Advance online publication. doi:10.14569/IJACSA.2022.0130153

Yusuf, N. (2021). The effect of online tutoring applications on student learning outcomes during the COVID-19 Pandemic. *ITALIENISCH*, *11*(2), 81-88. doi:10.1115/italienisch.v11i2.100

Zhang, M., & Chen, S. (2022). Modeling dichotomous technology use among university EFL teachers in China: The roles of TPACK, affective and evaluative attitudes towards technology. *Cogent Education*, *9*(1), 2013396. Advance online publication. doi:10.1080/2331186X.2021.2013396

Zhang, Q., Huang, B., & Chiu, D.K.W., & Ho, K. W. (2015). Learning Japanese through social network sites: A case study of Chinese learners' perceptions. *Micronesian Educators*, *21*, 55–71.

Zhang, S., Carroll, M., Li, C., & Lin, E. (2021). Supporting part-time students in doctoral programs: A technology-based situated learning model. *Studies in Graduate and Postdoctoral Education*, *12*(2), 190–205. doi:10.1108/SGPE-11-2019-0082

Zhang, S., Li, C., Carroll, M., & Schrader, P. G. (2020). Doctoral program design based on technology-based situated learning and mentoring: A comparison of part-time and full-time doctoral students. *International Journal of Doctoral Studies*, *15*(1), 393–414. doi:10.28945/4598

Zhang, Y., Lo, P., So, S., & Chiu, D. K. W. (2020). Relating Library User Education to Business Students' Information Needs and Learning Practices: A Comparative Study. *RSR. Reference Services Review*, *48*(4), 537–558. doi:10.1108/RSR-12-2019-0084

Zhang, Z., Wang, M. O., & Shi, J. (2012). Leader-follower congruence in proactive personality and work outcomes: The mediating role of leader-member exchange. *Academy of Management Journal*, *55*(1), 111–130. doi:10.5465/amj.2009.0865

Zhou, A., Sedikides, C., Mo, T., Li, W., Hong, E. K., & Wildschut, T. (2021). The restorative power of nostalgia: Thwarting loneliness by raising happiness. *Social Psychological & Personality Science*, 1–13.

Zhou, X., Sedikides, C., Mo, T., Li, W., Hong, E. K., & Wildschut, T. (2022). The restorative power of nostalgia: Thwarting loneliness by raising happiness during the COVID-19 pandemic. *Social Psychological & Personality Science*, *13*(4), 803–815. doi:10.1177/19485506211041830

Zoltán, A. (2011). Alfred Adler's Individual Psychology-towards an Integrative psychosocial foundation of the education in the 21st Century. *Reconnect-Electronic Journal of Social, Environmental and Cultural Studies*, *3*(1), 4–19.

Zoom Video Communications, Inc. (2021). Available from: https://zoom.us/

Zuiches, J. J., Cowling, E., Clark, J., Clayton, P., Helm, K., Henry, B., ... Warren, A. (2008). Attaining Carnegie's community engagement classification. *Change*, *40*(1), 42–45. doi:10.3200/CHNG.40.1.42-45

About the Contributors

Michelle Dennis serves as Department Chair of the Online Campus of Adler University, where she recently held a 22-month interim appointment as Executive Dean and Chief Academic Officer. She previously served this campus as Director of the MA and Ph.D. in Industrial and Organizational Psychology programs. She has held various positions in the field of higher education over the past 22 years and she previously worked as a statistician. Dr. Dennis provides service as a Peer Reviewer for the Higher Learning Commission and as a Reviewer for the International Journal of Smart Education and Urban Society. She previously served as Treasurer of the Board of Directors for Anafiel House, a nonprofit serving survivors of intimate partner violence. Dr. Dennis also previously served the Online Administration Network of the Association for Professional, Continuing and Online Education as Vice Chair of Diversity and Inclusion and she is a member of the Illinois Coalition for Higher Education in Prisons. She earned her Ph.D. in Clinical Psychology from Marquette University, her MA in Training and Development from Roosevelt University, and her M.Ed. in Higher Education Administration from the American College of Education. Dr. Dennis has published in the areas of online faculty management, virtual student and faculty engagement, online faculty training and evaluation and best practices for online instruction.

James D. Halbert is the Program Director for the Masters of Arts in Industrial-Organizational Psychology at Adler University in Chicago, IL, and the board chairman for Thrive Academy in Augusta, GA. He has a strong background in clinical research, in which most of his focus has been on cardiovascular disease and mental stress. His current research and interest has been to investigate and improve Online Education as it is becoming the more popular choice to higher education. Dr. Halbert has an MS in General Psychology and a Ph.D. in Industrial Organizational Psychology. He completed his post-doc in research at the Medical College of Georgia / Augusta University.

* * *

Bettyjo Bouchey is associate professor and vice provost of digital strategy and operations at National Louis University, where she is responsible for standards of quality and service for online programming across the institution. She holds a B.A. in psychology from the University at Albany, an M.B.A. in entrepreneurship from Rensselaer Polytechnic Institute, and a Doctor of Education from Northeastern University. Her scholarship is concerned with the evolving nature of online education as it pertains to organizational structures, pedagogy, and its prominence in higher education.

Brianne F. Brenneman is an Assistant Professor of Public Health and directs the undergraduate public health program at Goshen College. In this role, she specializes in the development and evaluation of community-engaged learning courses. Prior to working in higher education, Brenneman worked in community and economic development. She currently sits on the board of Center for Healing and Hope, a local clinic serving those without health insurance. She is also a commission member on the City of Goshen's Redevelopment Commission.

Peter Budmen has extensive experience in higher-ed student affairs from student life, admissions, orientation, community living, and service learning. Peter worked at Brandeis University then worked in elementary/special education for a number of years before returning to Community Engagement & Service Learning work. Peter's academic focus and interest is on equity within the classroom and early literacy strategies.

Barton Buechner is the Interim Director of the MA in Psychology with emphasis in Military Psychology (MAMP) program at Adler University. He earned his doctorate from Fielding Graduate University's school of Human and Organizational Development in 2014. Buechner is also a 1978 graduate of the US Naval Academy, and earned a Masters in Organization Development and Assessment degree from Case Western Reserve University in 1993. He served in the US Navy for 30 years in a mix of active duty and reserve capacities, including the Office of the Secretary of Defense for Reserve Affairs, Office of the Inspector General of the Department of Defense, and the Armed Forces Information Service. He retired at the rank of Captain in 2008. Dr. Buechner is the Co-Chair of both the Military and Moral Injury Special Interest Groups of the International Society for Traumatic Stress Studies (ISTSS), and also serves on the Boards of Directors of YourNexStage (transition support for women veterans); the National Veterans Foundation (NVF); and the Coordinated Management of Meaning (CMM) Institute for Personal and Social Evolution. He is the co-editor of a 2016 monograph publication of veteran-related scholarship from Fielding Graduate University titled "Veteran and family reintegration: Identity, healing and reconciliation (2016), and is co-editor of a special edition of the Journal of Community Engagement and Scholarship (JCES) featuring research by military-connected graduate students (2021).

Andrea Carter is a Neuroscience-based Equity, Diversity, Inclusion & Belonging Senior Consultant & Strategist. Andrea holds her Master of Arts Degree in Industrial & Organizational Psychology with a specialization in Human Resource Management. Her thesis research mapped belonging in the workplace by examining the organizational structure and integrating employee market changes affecting human capital post-pandemic performance. Over the past two years, she has created the only validated organizational belongingness metric tool to be used for Equity, Diversity, Inclusion and Belonging strategy that impacts corporate culture and governance. This tool allows organizations to score their belongingness rates and integrate sustainable and measurable metrics into their business. Andrea is now finalizing her research that delivers practical training for each level of the organization, ensuring belonging-first behaviours that are measurable and accountable. She enables a sustainable, reliable, and valid methodology for enhancing culture and growth within organizations. Andrea has worked within Global and National organizations in the healthcare, transportation and warehousing, manufacturing, mining, finance and insurance, and global spirits industries. Andrea brings over 18 years of research and practical application to her training, public speaking, and proven methods. She supports leaders and their teams to create a culture of belonging to produce good work despite high-pressure situations and imperfect conditions.

Cheuk Ting Chan received the degree of Bachelor of Arts in Music at The Chinese University of Hong Kong in 1995, and Master of Arts in Music Education at the Hong Kong Baptist University in 2007. She was the concertmaster of the Hong Kong Youth Chinese Orchestra 1991-1995, and was a freelance musician of the Hong Kong Chinese Orchestra 1994-1997. She has also been a music teacher for over 20 years. Her research interest is in music education.

Dickson K. W. Chiu received the B.Sc. (Hons.) degree in Computer Studies from the University of Hong Kong in 1987. He received the M.Sc. (1994) and the Ph.D. (2000) degrees in Computer Science from the Hong Kong University of Science and Technology (HKUST). He started his own computer consultant company while studying part-time. He has also taught at several universities in Hong Kong. His teaching and research interest is in Library & Information Management, Service Computing, and E-learning with a cross-disciplinary approach, involving library and information management, e-learning, e-business, service sciences, and databases. The results have been widely published in around 300 international publications (most of them have been indexed by SCI/-E, SSCI, and EI, such as top journals MIS Quarterly, Computer & Education, Government Information Quarterly, Decision Support Systems, Information Sciences, Knowledge-Based Systems, Expert Systems with Application, Information Systems Frontiers, IEEE Transactions, including many taught master and undergraduate project results and around 20 edited books. He received a best paper award in the 37th Hawaii International Conference on System Sciences in 2004. He is an Editor (-in-chief) of Library Hi Tech, a prestigious journal indexed by SSCI (impact factor 2.357). He is the Editor-in-chief Emeritus of the International Journal on Systems and Service-Oriented Engineering (founding) and International Journal of Organizational and Collective Intelligence, and serves in the editorial boards of several international journals. He co-founded several international workshops and co-edited several journal special issues. He also served as a program committee member for around 300 international conferences and workshops. Dr. Chiu is a Senior Member of both the ACM and the IEEE, and a life member of the Hong Kong Computer Society. According to Google Scholar, he has over 5,000 citations, h-index 38, i-10 index 109, ranked worldwide 1st in "LIS," "m-learning," and "e-services".

Alex Cottrell (HWS '20) joined The Center for Community Engagement and Service Learning (CCESL) as a Program Assistant in December 2021. Alex graduated from HWS in 2020 with a double major in ancient Greek and economics.

Suming Deng received the B.Sc. (Hons.) in Data Science from the Beijing Normal University Hong Kong Baptist University United International College (UIC). He has done some projects related to machine learning, data mining, text mining, recommendation system, etc. He is currently a candidate of MSc in Library and Information Management at the University of Hong Kong.

Paul Shetler Fast is Global Health Coordinator at Mennonite Central Committee (MCC). In his work with MCC he oversees and supports the organization's health and public health programming in the 52 countries it works in and serves as the public health technical lead in health emergencies, including COVID-19. Shetler Fast is also an affiliate faculty of public health at Goshen College in Goshen Indiana. He has masters degrees in public health and international development from the University of Pittsburgh. He was born and raised in Tanzania, worked in Haiti from 2015-2020 and is now based in Goshen, Indiana.

Kathleen "Katie" Flowers completed her undergraduate degree in Communications from Stonehill College and her EdM in Higher Education Administration from the University at Buffalo. She joined Hobart and William Smith Colleges (HWS) in 2004 and is responsible for leading the Center for Community Engagement and Service-Learning (CCESL). She oversees HWS's service-learning, community-engaged teaching, volunteer, and community-based research programs. Her purview also includes federal work-study tutoring programs, Days of Service, Alternative Spring Break trips, and postgraduate service opportunities, and local collective impact initiatives. She is an AmeriCorps VISTA alumna and serves on the board for multiple local nonprofit agencies.

Jonathan Garcia is senior at Hobart & William Smith Colleges majoring in Mathematics and minoring in Computer Science. Jonathan worked as a tutor for the program discussed in this chapter.

Leila Goosen is a full professor in the Department of Science and Technology Education of the University of South Africa. Prof. Goosen was an Associate Professor in the School of Computing, and the module leader and head designer of the fully online signature module for the College for Science, Engineering and Technology, rolled out to over 92,000 registered students since the first semester of 2013. She also supervises ten Masters and Doctoral students, and has successfully completed supervision of 43 students at postgraduate level. Previously, she was a Deputy Director at the South African national Department of Education. In this capacity, she was required to develop ICT strategies for implementation. She also promoted, coordinated, managed, monitored and evaluated ICT policies and strategies, and drove the research agenda in this area. Before that, she had been a lecturer of Information Technology (IT) in the Department for Science, Mathematics and Technology Education in the Faculty of Education of the University of Pretoria. Her research interests have included cooperative work in IT, effective teaching and learning of programming and teacher professional development.

Michael Graham is vice president of operations and technology at National Louis University (NLU), where he is responsible for enhancing operational efficiency, supporting academic and administrative technology, and leading instructional design and online programming. Mike holds a B.A. and an M.A. in history from Youngstown State University and a Doctorate in Education from National Louis University. His research interests span the rise and change of online program management organizations, as well as leadership and operational efficiency, as the disciplines of information technology and academic affairs continue to collide in higher education.

Nathan Graham is currently the assistant dean of media and technology at The Johns Hopkins Whiting School of Engineering. He received an M.F.A. in Poetry at New Mexico State University and is a Ph.D. candidate in Information Science at Rutgers University. His teaching and research interests include instructional technology, history of electronic publishing, and cross-media adaptation.

Paul Huckett is the assistant dean of learning design and innovation and lecturer with The Johns Hopkins Whiting School of Engineering. As assistant dean, Huckett provides strategic vision in teaching, learning, and technology across the school and leads curriculum development for new and existing programs. He also oversees the Center for Learning Design & Technology (CLDT), which provides learning design and teaching support to faculty. As a lecturer, Huckett teaches the course entitled "Strategic Communications in Technical Organizations" in the Engineering Management program within the

Whiting School of Engineering. He additionally developed and taught the Coursera MOOC Inclusive Online Teaching ().

Xinyu Jiang received the Master of Science in Library and Information Management from the University of Hong Kong in 2021. Her research interest is in social media marketing.

Kaan Kırlı was graduated from Yeditepe University in the department of Information Systems and Technologies in Turkey.

Abueng Rachael Molotsi is a senior lecturer at the University of South Africa (UNISA) in the Department of Science and Technology Education, under the College of Education. She is the College of Education (CEDU) ICT chairperson and also coordinating technology enhanced teaching. She is currently a Research Project Leader for a project titled ''The use of Technological, Pedagogical and Content Knowledge (TPACK) model to improve teachers' delivery of lessons. A focus is on developing teachers to use the four components of TPACK to deliver lessons efficiently and successfully. Dr Molotsi uses her specialisation to moderate and externally evaluate other Higher Education Institutions' students' dissertation and theses. Her research focus is on Open Distance e- Learning and transforming teacher education in the context of 4IR through the internet of things and social presence. She presented papers at international and nationally conferences. Her publications include book chapters, conference proceedings and articles.

Tshimangadzo Selina Mudau holds PhD in Higher Education, a Master's in health studies, Bachelor of Nursing, and a Diploma in Primary Health Care. She is a lecturer at the University of KwaZulu-Natal, School of Nursing and Public Health. In 2021 she received the VC's Distinguished Community Engagement Scholar Award. Her research interests are on teenage mothers, youth, and women, higher education, bricolage, community engagement, and action research.

Cagla Ozen has taken her PhD from Computer Science Department of University of York, UK. She has more than 40 publications and one of her publications awarded as Outstanding Paper Award Winner at the Literati Network Awards for Excellence 2011. Her research interests are user experience, user interface, technology adoption and IoT.

La Toya L. Patterson was born and raised In Chicago, IL. Dr. Patterson received her bachelor's in Psychology and Master of Arts in Clinical Professional Psychology from Roosevelt University. She went on to receive her second Master of Arts in Forensic Psychology from Argosy University and her Ph.D. in Counselor Education & Supervision from Adler University. She is a Licensed Clinical Professor Counselor and a National Certified Counselor. Dr. Patterson 12 years of experience in the counseling field providing individual therapy. Dr. Patterson currently works as an Assistant Professor for Chicago State University.

Lilya Shienko is a Canadian doctorate student and graduate of Adler University's MA program in Industrial and Organizational Psychology. She has received a bachelor's of science in psychology and business studies from MacEwan University. She is interested in research concerning how mindfulness, mental health, burnout, trauma, and bullying affect employee performance and cardiac health.

Wade Snowdon, based in Winnipeg, Canada, is currently part of the Global Service Learning (GSL) department of Mennonite Central Committee (MCC) that provides oversight and leadership to their young adult service-learning opportunities around the world. Prior to this role, he worked with MCC in Kenya and Uganda to assist grassroot peacebuilding organizations. His research interests include intercultural competency, anti-colonial practice, reconciliation, and trauma healing.

Heather Stewart is the Academic Affairs Manager for the Whiting School of Engineering - Engineering for Professionals programs at Johns Hopkins University.

Craig Talmage holds a B.S. in Family Studies and Human Development from the University of Arizona to focus on nonprofit and community engagement. To increase his skills and knowledge, he completed an M.A. in Industrial/Organizational Psychology at Minnesota State University (MSU), Mankato conducing projects with local school districts and MSU, Mankato. He then sought to further apply industrial/organizational psychology to understand nonprofit and community spaces. He completed his Ph.D. in Community Resources and Development at Arizona State University and now teaches courses in local development, tourism development, organizational behavior/development, nonprofit management and leadership, and social entrepreneurship. Talmage has more than fifteen years of combined not-for-profit, nonprofit, small business, and public sector research and leadership experience. Before becoming a professor, he founded and operated a research consulting practice for small businesses and nonprofits in Maricopa County, Arizona. He now is the editor-in-chief of Community Development. At HWS, he serves as the faculty liaison for the Center for Community Engagement and Service Learning in addition to many other service roles. He still teaches asynchronously for ASU's School of Public Affairs on community resilience, organizational behavior, and community conflict resolution.

Jasmine Webb-Pellegrin is senior at Hobart & William Smith Colleges majoring in Biochemistry and minoring in American Studies. Jasmine worked as a tutor for the program discussed in this chapter.

Index

A

Adult Learners 53, 70, 92, 96-97, 178-179, 181-182, 230, 232, 262, 267
AIDA Model 44, 48, 50-51, 53-54, 60, 63, 65-69
America Reads 235, 237-238, 240, 247, 249-252
and Content Knowledge (TPACK) 127
ASP.NET 149-150, 154, 168, 172, 175, 177
Asynchronous Tutoring Modalities 252

B

Belonging 10, 49, 73, 76, 80, 82, 84-87, 92-93, 120, 179, 182, 184, 187-188, 203, 209-210, 213-225, 227-228, 230, 232, 246, 285-286, 288-295, 298-301, 303, 305
Belongingness 71-73, 76, 78, 81-83, 87, 210, 212-213, 218, 222-223, 225, 288, 303-304

C

C# 150, 168, 170, 177
Chinese Music 48, 57
Civic Engagement 1, 7, 14-16, 29
Classical Music 31, 40
Classification 1-7, 13-16, 96, 113, 133, 143, 306
Co-Construction 71, 73, 75, 87
Cohort Mentors 254-255, 263
Collaboration 1-3, 5-6, 11, 19, 22, 27, 46, 70, 73, 77-79, 87, 92-93, 101-102, 104, 107, 109, 111-113, 139, 176, 190, 206, 209, 215-217, 219, 222-225, 228, 231, 236-237, 260, 264, 267, 269-272, 283, 287-289, 305
Collective Identity 73, 75, 87
Community 1-31, 33-37, 40-42, 44, 46, 48-52, 56-58, 60, 63-65, 68, 70-72, 77-79, 83, 85-86, 91-94, 97-98, 100-107, 110-114, 116-125, 127-133, 137, 141-148, 168, 173, 175, 178-182, 184-193, 195-197, 199-200, 203-205, 209-210, 212-213, 215-220, 222-225, 230, 232, 234-239, 243-252, 257-258, 260, 262, 265, 270-273, 280-281, 283, 285-288, 290-291, 296-298, 303
Community Engagement 1-31, 33-34, 44, 48, 65, 91, 102, 127-133, 137, 141-148, 173, 200, 204-205, 234-238, 246-252, 283
Community of Practice 37, 40, 64, 78-79, 85-86, 91-92, 94, 102-103, 118, 122, 125, 180
Community Partnerships 14, 26
Community Well-Being 130, 234, 251
Community-Engaged Learning 101, 121, 251, 269-273, 275, 281-283
Concentration 150, 152-153, 167, 173, 177
Coordinated Management of Meaning (CMM) 71-72, 75, 77-79, 87
Cosmopolitan Communication 77, 84, 89
Course Room 178, 181-183, 185-188, 190
COVID-19 12, 17-19, 21-22, 25, 27-28, 32, 34, 42, 44-45, 47, 49, 57-59, 65-66, 68, 70, 75, 77, 83-84, 118, 123, 130-131, 147-151, 153, 165, 167, 174-177, 194-196, 198-199, 211, 215, 220, 225-232, 234-241, 244, 246-252, 256, 262, 265-267, 269-272, 274-275, 277, 283, 300, 302-303, 305-306
Critical Discourse Analysis 17-18, 20, 24, 28-29, 121
Cross-Cultural Competency 270, 296
Customer Relationship Management 43, 46-47, 65, 200

D

Digitalization 46-47, 69
Discussion Posts 102, 178-180
Distance Education 15, 91, 97, 119-126, 134, 144, 191, 226, 229-230, 267

E

Efficiency 33, 149-151, 165-167, 177, 196, 289, 296
Engagement 1-31, 33-36, 42-45, 48, 50, 52, 55, 60, 62-65, 68-69, 72, 80, 91-94, 101-103, 116, 119,

122, 124, 127-133, 137, 141-150, 165-167, 173, 177-179, 181-183, 185, 187, 189-192, 197-198, 200-201, 203-205, 209-210, 213, 215-220, 222-224, 227-231, 234-238, 240-241, 246-252, 254-267, 269-271, 273-274, 279, 282-283, 285-286, 288-289, 291, 295-297, 300, 304
Engineering 28, 74, 91-93, 97, 100-102, 104-105, 107-108, 118-119, 121-122, 124, 249, 251-252
Experience 9-10, 13, 18, 45, 51-52, 56, 59, 68, 71-74, 76-77, 79-80, 82-88, 92-94, 96-100, 103, 108, 113, 124, 132, 136, 145, 149-150, 153, 166-167, 172-184, 186-187, 189, 191, 195-199, 202, 210-216, 218, 220-223, 229, 236, 239, 241-248, 254-256, 259-260, 262, 264-265, 270-272, 274-275, 278-283, 286, 289-290, 292, 298, 302

F

Faculty Community 91, 102, 106, 111-113, 116-117, 131, 143
Faculty Mentors 254, 259, 261-263, 265
Formation 184-185, 192, 197, 199, 213, 293, 298

G

Google Meet 149-150, 153, 158, 167-168, 173, 175, 177
Goshen College 269-275, 279, 281, 283

H

Higher Education Institutions 1-13, 18, 27, 92, 130, 165, 188, 235, 283
Higher-Education 269, 271
Honeycomb Model 31-33, 36, 42, 47, 65
Hong Kong Philharmonic Orchestra (HKPhil) 32, 47
HTML 67, 150, 168-169, 177, 208, 251-252, 301
Hybrid 45, 106, 108, 114, 194, 196-201, 203-208, 223, 231, 239, 241, 285-286, 288-294, 296, 305
Hybrid and Virtual Communities 285

I

Inclusive Leadership 221-222, 226, 285, 287-290, 300, 303-305
Institutional Change 7, 194
Interaction 11, 20, 24, 29, 37, 56, 63, 72, 76, 80, 88, 91-106, 108, 111, 116, 118-121, 124-125, 150, 165-166, 178-183, 186-187, 190-192, 209, 211, 215-218, 220, 223-224, 232, 240, 248, 258, 265, 267, 274, 282, 290, 298
Isolation 10-11, 95, 120, 151, 179, 209-211, 220, 223, 227-229, 231-232, 251, 254, 259, 262, 265-266, 289, 296, 303

L

Learning 3-16, 18-19, 28-29, 33, 36-37, 41, 45-47, 55-57, 60-61, 63-64, 67-71, 73, 75, 77-79, 81, 85-88, 91-108, 110-113, 116, 118-153, 165-167, 173-176, 178-188, 191-193, 195-196, 198-199, 202, 206-207, 215-232, 236-246, 248-252, 254-258, 260-261, 264-266, 268-277, 280-284, 289
Loneliness 83, 151, 209-213, 220-221, 224, 226-227, 229-232, 254, 285, 289-290, 296, 301-304, 306

M

Marginalized Group 18, 20, 24, 30
Marginalized Groups 17-18, 23, 26, 29, 188
Mennonite Central Committee 269-270, 273, 277, 281, 283-284

O

Online 6, 9, 11, 14, 16-18, 28-29, 31-32, 34, 36, 40, 42-47, 49, 51-52, 56-58, 60-65, 67-68, 70-74, 76, 78-80, 82-87, 92-95, 97-99, 101-108, 110-111, 113-114, 116, 118-134, 137, 140-153, 165-168, 172-192, 195-196, 198-200, 202, 207, 209-210, 215-223, 225-232, 235, 238-241, 245, 248-252, 254-262, 264-267, 270, 279, 287, 301, 303
Online Community 17-18, 83, 122, 179-182, 184-185, 187, 191, 209, 215-218, 220, 222-223
Online Education Platform 149-150, 177
Online Education Platforms 149-150, 153, 165, 167-168, 174, 177
Online Learning 28-29, 68, 71, 87, 94-95, 98, 101-104, 106, 111, 118-121, 123-124, 126, 134, 140, 142, 148-153, 165-166, 173-176, 178-190, 192, 195, 199, 207, 215, 220, 222, 228-232, 251, 256
Organizational Culture 194, 208, 290, 296
Ostracization 209-211, 289

P

Pandemic 12, 17, 19, 22, 26-28, 44, 47, 49, 54, 58, 66, 75, 77-78, 83-84, 119, 128, 130-131, 142-143, 147, 149-151, 153, 165, 173-177, 181, 194-204, 206, 209-213, 227, 229, 231, 234-236, 239, 241, 243-246, 250-252, 262, 265-266, 269-270, 272, 274-275, 279, 283-285, 288-289, 296, 302-306
Parent-Tutor Interactions 248

Participation 10, 18-20, 22-24, 26, 28, 30-31, 34-35, 38, 42, 48-49, 53, 55, 57, 59, 63, 68, 80, 91-92, 94, 97-100, 102, 105, 108, 120, 123, 129, 144, 150, 152, 159, 166-167, 173, 177, 179-180, 185-186, 188-189, 196, 201, 216, 229, 232, 235, 239, 272, 278, 289, 304
Partnership 1-4, 6-7, 9, 62, 101, 196, 228, 238, 243-244, 246-247, 269-273, 281
Pedagogical 11, 101, 108, 118, 127-128, 131, 133-137, 144-148, 181, 188, 191, 231
Performance 2, 10, 32, 40-41, 43, 63, 65, 69, 81, 98-99, 120-121, 123, 135-136, 145, 147, 186, 195, 205, 213-215, 218, 225, 250-251, 257, 266, 271, 285-286, 288-297, 299-300, 304-305
PEST Analysis 48
Program Development 272
Program Evaluation 124, 147, 270
Promotion 5, 27, 31, 42, 44-46, 48-50, 54, 56-57, 60-61, 63-64, 70, 205, 289

R

Remote and Virtual Tutoring 252
Remote Engagement 235, 238, 254, 264
Remote Learning 152, 165-166, 234, 239, 241, 244, 246, 251
Rural Areas 17-19, 21-22
Rural Community 20, 24, 30, 57
Rural Leader 30

S

Sense of Community 91-92, 102, 104, 118, 122, 124-125, 179, 181-191, 199, 210, 224, 230
Service 3-16, 29, 33, 41, 47, 55, 60-61, 65, 110-111, 122, 177, 199-200, 202-203, 206, 234, 236-237, 246, 250, 269-275, 280
Service-Learning 6-7, 9-16, 124, 234, 236, 243, 248-250, 273-274
Social Construction 71-72, 74, 77, 79, 81, 86, 88, 97
Social Justice 9-10, 19, 22-26, 188, 259, 274
Social Media 30-38, 42-54, 56, 58, 60-69, 211, 224, 226, 239, 261, 287
Social Media Analysis 31
STEM Education 122, 235, 252
Student 3-4, 7, 9-12, 14-16, 45, 53, 67, 91-93, 96, 98-106, 108, 110-111, 113-122, 124-125, 132, 134, 143, 147, 151-153, 156, 161, 166-167, 176, 178-180, 182-183, 185-192, 195-197, 199-204, 206-207, 211, 216-217, 219-220, 226-227, 229-231, 235-238, 240-241, 243-248, 252, 254-256, 258-261, 263-265, 267, 271-273, 275-276, 279, 281-282
Student Community 91, 102, 105-106, 110-111, 113, 116-117, 220
SWOT Analysis 48, 54-55, 69
Synchronous Tutoring Modalities 252

T

Teachers 32, 40-42, 91-92, 123, 127-129, 131-141, 144-148, 150, 222, 239-241, 244-245, 249, 258, 280
Technological 17, 21, 54, 65, 73, 122, 127-128, 131, 133-136, 140, 144-148, 152-153, 173, 194-195, 197-198, 200, 202, 237, 265
Teenage Mothers 17, 19, 23, 25, 28

U

User Engagement 42-43, 50, 65, 149-150, 167, 173, 177
User Experience (UX) 149-150, 167, 172-173, 177

V

Videoconference Platforms 149
Virtual Community Engagement 17-18, 20-21, 23-26, 30, 238
Virtual Ethics 17
Virtual Tutoring 234, 241-243, 248-249, 251-252
Virtual Work 254, 256
Visual Studio 149-150, 168, 175, 177
Vygotsky's Zone of Proximal Development 240, 252

W

Waiting Room 149-150, 168, 173-174, 177
Web-Based Learning 269, 282-283

Z

Zoom 19, 21, 79, 83-84, 87, 114, 149-150, 153, 155, 159, 165, 167-168, 173, 176-177, 182, 195-196, 199, 219, 235, 238-241, 243, 245-247, 252, 260

Printed in the United States
by Baker & Taylor Publisher Services